Pearlite in Steels

Pearlitic steel is the strongest commercially available mass-produced alloy, and pearlite is used in a myriad of applications, from the reinforcement of tyres in every single road vehicle in the world to the creation of pathways across troubled waters. *Pearlite in Steels* is the first book dedicated to this vast subject. The book uniquely covers the topic in-depth, providing comprehensive coverage of fundamentals and technologies that make pearlite vital as an engineered material. This book

- covers the 3D structure of pearlite over a wide range of length scales,
- details the relationship between imperfect structure and consequences,
- discusses all aspects of structure and complex properties,
- covers the evolution of structure during traumatic processing,
- highlights the engineering aspects: the twisting of pearlitic wires, for the safe application of the strongest of ropes in safety-critical scenarios and
- highlights difficulties and inconsistencies, thereby helping to develop research.

Aimed at materials engineers and metallurgists, this book by a leading global authority on the subject offers readers the deepest and broadest overview of pearlite in steels to date.

Harshad K. D. H. Bhadeshia is the Emeritus Tata Steel Professor of Metallurgy at the University of Cambridge and Professor of Metallurgy at Queen Mary University of London. His main interest has been in the theory of solid-state phase transformations with an emphasis on the prediction and verification of microstructural development in complex metallic alloys, particularly multicomponent steels. He has authored or co-authored some 700 research publications. His books include *Bainite in Steels* (3rd edition, 2015), *Steels* (5th edition, 2024, with Robert Honeycombe), *Geometry of Crystals, Poly-crystals and Phase Transformations* (2018), *Theory of Transformations in Steels* (2021), *Innovations in Everyday Engineering Materials* (2021, with Tarashankar DebRoy) and *Phase Transitions* (2023 with Haixue Yan). He became a Fellow of the Royal Society in 1998, of the Royal Academy of Engineering in 2002, a Foreign Fellow of the National Academy of Engineering (India) in 2004 and was awarded the Bessemer Gold Medal in 2006. Professor Bhadeshia was appointed a Knights Bachelor in the Queen's Birthday Honours for services to science and technology.

Pearlite in Steels

Harshad K. D. H. Bhadeshia

CRC Press is an imprint of the
Taylor & Francis Group, an **informa** business

Designed cover image: © Shutterstock, TAW4

First edition published 2025
by CRC Press
2385 Executive Center Drive, Suite 320, Boca Raton, FL 33431, U.S.A.

and by CRC Press
4 Park Square, Milton Park, Abingdon, Oxon, OX14 4RN

CRC Press is an imprint of Taylor & Francis Group, LLC

Library of Congress Cataloging-in-Publication Data
Names: Bhadeshia, H. K. D. H. (Harshad Kumar Dharamshi Hansraj), 1953- author.
Title: Pearlite in steels / Harshad K.D.H. Bhadeshia.
Description: First edition. | Boca Raton, FL : CRC Press, 2025. | Includes bibliographical references and indexes. | Summary: "Pearlitic steel is the strongest commercially available mass-produced alloy, and pearlite is routinely used for tyre reinforcement in every single road vehicle in the world. Pearlite in Steels is the first book dedicated to pearlitic steels and uniquely covers the topic in depth to provide readers with comprehensive coverage of fundamentals and advances in this vital and wide-reaching engineered material. Details the relationship between transformations and properties. Discusses all aspects of structure and properties. Covers the twisting of pearlitic wires, particularly important in the creation of the strongest of ropes in large-scale engineering applications. Highlights difficulties and inconsistencies, thereby helping to develop research. Aimed at materials engineers and metallurgists, this book by a leading global authority on the subject offers readers the deepest and broadest overview of pearlite in steels to date"-- Provided by publisher.
Identifiers: LCCN 2024018172 (print) | LCCN 2024018173 (ebook) | ISBN 9781032631929 (hardback) | ISBN 9781032631974 (paperback) | ISBN 9781032631981 (ebook)
Subjects: LCSH: Pearlitic steel.
Classification: LCC TN731 .B425 2025 (print) | LCC TN731 (ebook) | DDC 669/.96142--dc23/eng/20240512
LC record available at https://lccn.loc.gov/2024018172
LC ebook record available at https://lccn.loc.gov/2024018173

ISBN: 978-1-032-63192-9 (hbk)
ISBN: 978-1-032-63197-4 (pbk)
ISBN: 978-1-032-63198-1 (ebk)

DOI: 10.1201/9781032631981

Typeset in Nimbus Roman
by KnowledgeWorks Global Ltd.

Publisher's note: This book has been prepared from camera-ready copy provided by the authors.

To my late father and mother.

Contents

Preface

There exist a lovely pair of crystals that evolve hand-in-hand, with an intimacy that is alluring and iridescent. They complement each other in more ways than one. It is as if they were made for each other. One of the pair is so strong, that on its own, it does not survive if tormented. But together, they reign supreme, able to support awe-inspiring structures made by mortals, structures that survive when even the mantle beneath jerks.

This is the story of pearlite

The strongest available, mass-produced 'micro'-structure. Pearlite with a strength exceeding 3 GPa is used routinely in the manufacture of robust tyres where the adhesion of the steel to rubber is important, while facilitating a hermetic seal at the tyre-rim interface. Strength approaching seven billion pascals has been achieved with pearlite, but this is unlikely to see the light of day given the trauma that the necessarily tiny sample of steel has to experience in order to reach this lofty ability to support a load.

The term *lamellar* was used some twenty-three decades ago, to describe mica as a 'smooth substance, but not greasy to the touch, ... lamellar or scaly, and its appearance, sometimes splendidly metallic'.[1] According to the Oxford English Dictionary, the guardian of the English language, the word characterises ... thin plates or scales. Pearlite often is said to consist of 'fine laminations, of ferrite and cementite side by side'.[2] But we shall see that it in fact has a much more convoluted structure, making any assumption that the lamellae are plates, untenable. Why does this matter? Well, toughness depends on the continuity of the structure and this continuity is not defined by the interlamellar spacing.

A pearlite colony in reality is a bi-crystal. Although, there exists an *artificial* form fabricated by joining layers of cementite and ferrite together, i.e. the idealised structure described schematically in many books and teaching materials. The interlamellar spacing of this artificial composite is the coarsest ever recorded ($\approx 400\,\mu$m), by a large margin. And the cementite layers are polycrystalline as are those of ferrite! The finest interlamellar spacing generated by phase transformation alone is probably $\approx 0.05\,\mu$m where the pearlite is induced in a high-carbon alloy under the influence of a 30 T magnetic field. A typical spacing in commercial steel might be $0.2\,\mu$m. Although nodules of pearlite can grow to sizes greater than that of the parent austenite grains that typically have a size in the range 10–40 μm.

Cementite is meant to be brittle, in part because it has a large, primitive crystallographic unit-cell, making the displacement between successive equilibrium positions of a dislocation correspondingly large. Large samples of single-phase, pure polycrystalline cementite have been manufactured by Umemoto's team at Toyohashi University. When tested at ambient temperatures, it is indeed brittle. But when the cementite is cosseted within pearlite, it surprises by tolerating huge plastic strains, so much so that it ultimately loses the will to exist and enters solution within the defect-rich ferrite, which normally shuns carbon.

Much has been written about the fitting of a microstructural length parameter to strength, including the notion that all such fits can be described by a function created to represent the relaxation of an epitaxially deposited film, where the space available for dislocation sources is the determining factor. There even is the suggestion that the relationship between strength and interlamellar spacing

[1] R. J. Sulivan, *A view of nature, in letters to a traveller among the Alps*, (6 volumes), 1794, pp. 438-439.

[2] A. Morley, *Strength of Materials*, Longmans, Green & Co., 9th edition, 1940, p. 47.

is a spurious correlation, with α/θ elastic strains being responsible for the strength of pearlite. An examination of minutiae and cumulative data, however, shows that the Hall-Petch relationship rules supreme, both from mechanistic and observational points of view.

Strength is important but only if a bank of other properties, such as torsional ductility, ensure that it can be applied safely. Steel ropes based entirely on pearlite support the largest single-span bridges in the world and in earthquake zones; they support massive tuned dampers hanging within skyscrapers, that stop the buildings from swaying to the tune of winds and tectonic perturbations. The Earth is the only planet in the solar system that has colliding plates forming its crust. Rails for high-speed ($300\,\mathrm{km\,h^{-1}}$) transit systems in Japan are ... pearlitic. Structural steels, such as for bridges, buildings and ships, on the other hand, are a mixture of proeutectoid-ferrite and pearlite because design requirements are less onerous with respect to strength – toughness, fatigue performance and the ability to weld using reliable and high-productivity methods are crucial. In contrast, rail steels of eutectoid composition have to be joined using bizarre methods which would be wholly unsuitable for most applications.

We often talk about the importance of interlamellar spacing as if it is a single defining quantity. Belaiew in his study of the stereometry of the 'pearlite grain' (1922) showed that there can be a distribution of *true* interlamellar spacings rather than some unique value determined as an average of intercepts. His paper also notes, perhaps for the first time, that for pearlite to grow, the alloy must be supersaturated with respect to both of its phases. He included the first phase diagram showing the extrapolation of the $\gamma/\theta+\gamma$ and $\gamma/\alpha+\gamma$ phase boundaries below the eutectoid temperature, a construction perhaps unfairly referred to today as the Hultgren-extrapolation.

The kinetic theory for pearlite is a mess in the sense that all useful steels are alloyed with substitutional solutes, but the phases within are not inclined to respect local equilibrium with the austenite. All experimental measurements show that the compositions deviate from expectations based on either the equilibrium or para-equilibrium phase diagram. In fact, para-equilibrium should not even be regarded as an appropriate state for a reconstructive transformation.

Pearlite does behave differently when shocked, with beautiful observations of twins in its ferritic component, causing systematic shears of the interleaving cementite, almost like the deflection of scratches on an otherwise flat surface, caused by partial transformation into martensite. Mechanical twinning manifests if the strain rate is large enough, but adiabatic shear bands can also form in a manner that seems to ignore *micro*structure. High-velocity impacts can cause internal spalling of pearlite-containing steels when the speed of sound within the metal is greater than the impact.

The reader may wonder why I have bothered to include eutectoids other than pearlite in the final chapter, for example in the Ag_2HgI_4-Cu_2HgI_4, Ar-CO and many other systems. Why is a eutectoid absent in the Ar-N system in its solid state? I have a straightforward response: do we understand which systems should exhibit eutectoids? Are there two different eutectoid reactions possible in steels? And is there a eutectoid that evolves from austenite with more than two product phases?

I hope that I have covered most aspects of pearlite, only a handful of which are touched upon in this preface, the primary aim of which is to titillate the imagination – the subject, of course, requires careful study given its incredible contributions to the quality of life – hence, this book. I have, where possible, derived and provided simple relationships that can be applied on a calculator with reasonable expectations. I am grateful to many for helpful discussions and information, as acknowledged in the text. A particular applaud for Apparao Chintha and Minoru Umemoto who responded so well, to each and every one of my many requests.

Harshad K. D. H. Bhadeshia
The year 2024
London and Cambridge, United Kingdom

Author

Harshad K. D. H. Bhadeshia is the Emeritus Tata Steel Professor of Metallurgy at the University of Cambridge and Professor of Metallurgy at Queen Mary University of London. His main interest has been on the theory of solid-state phase transformations with emphasis on the prediction and verification of microstructural development in complex metallic alloys, particularly multicomponent steels. He has authored or co-authored some 700 research publications. His books include *Bainite in steels* (3rd edition 2015), *Steels* (5th edition 2024, with Robert Honeycombe), *Geometry of crystals, polycrystals and phase transformations* (2018), *Theory of transformations in steels* (2021), *Innovations in everyday engineering materials* (2021, with Tarashankar DebRoy) and *Phase transitions* (2023 with Haixue Yan). He became a Fellow of the Royal Society in 1998, of the Royal Academy of Engineering in 2002, a Foreign Fellow of the National Academy of Engineering (India) in 2004 and was awarded the Bessemer Gold Medal in 2006. Professor Bhadeshia was appointed a Knights Bachelor in the Queen's Birthday Honours for services to science and technology.

1 Introduction

The name *perlite* was in use in the early 19th century, well before the metallographic study of iron based alloys. It was invented by François-Sulpice Beudant in 1822 [1] from a translation of the German *Perlstein* where *Perl* stands for pearl and *stein* for rock. Charles Darwin referred to Beudant's 'perlite lithoide globulaire' as appropriate for the 'little brown sphærulitic globules of rocks of Ascension' [2]. Extinct Hungarian volcanos were known to be rich in the minerals silex, perlite, obsidian and pitchstone [3]. The quartz-based minerals were later listed as being in glassy or semi-vitreous states [4]. Rutley in his description of perlite adds in brackets the alternative names pearlite, pearlstone, to describe the glassy mineral that mostly is silica and may enclose other minerals [p. 192, 5].

The association of pearls and the structure of steels comes from the strikingly original work of Henry Clifton Sorby (Figure 1.1a), who founded the subject of metallography, during his progression from the transmitted-light examination of minerals to the reflected-light microscopy of steel samples that were polished flat and etched. In one of his lectures, he described a *pearly* compound that he observed in steel, which when examined at high power revealed fine straight or curved parallel lines attributed to the presence of alternating thin plates of varying hardness – his feeling at this discovery is epitomised in his own words, that reveal the pedigree of a careful scientist [6]:

> *I felt almost certain that these thin plates were iron free from carbon, and the intensely hard substance seen so well in blister steel; but the facts were so extraordinary, and so unlike anything I had ever seen or heard of in any mineral substance, that it was not until after several months devoted to the careful study of all the chief kinds of iron and steel that I felt confidence in the results.*

He explained that this structure resembled that of pearl. A natural pearl consists of onion-like layers of nacre that has a layered structure, with each layer consisting of parallel hexagonal platelets of aragonite ($CaCO_3$) about 0.5 μm thick and an order of magnitude longer [7] (Figure 1.1b). The layers are separated by an organic material. The thickness of the aragonite crystals is similar to the wavelength of light, so incident and waves reflected from a depth within the sample interfere either constructively or destructively, an effect dependent on the wavelength of light. Therefore, when viewed in white light, some components of the light will be removed by destructive interference, rendering the colours in the remaining components visible. Since this also depends on the angle of incidence, the pearl can exhibit iridescence (Figure 1.1c). A similar though not identical effect occurs with the steel-pearlite when viewed in an optical microscope because unlike aragonite nothing in steel is transparent.

In order to record his observations, Sorby followed a procedure that had previously been used in Austria in the context of the structure of meteorites, known at the time as 'nature printing', whereby the polished and etched samples were used as blocks to print on paper [11]. The magnification of such images would of course be $\times 1$. It was during the period 1863–1865 that Sorby enlisted the help of a Sheffield photographer to capture images from his microscope, probably at a magnification of $\times 9$. Probably the first photomicrograph ever published was in his 1887 paper [12]. Figure 1.2 shows this image, which is from a polished and etched sample of mild steel, and hence exhibits a mixture of ferrite and pearlite (the dark regions).

Sorby originally studied minerals – the nomenclature applied to steels: austenite, ferrite and cementite – is characterised in each case with an 'ite' ending. This originates from the naming conventions that existed for minerals [13]. The -ite ending apparently was used by Aristotle who

DOI: 10.1201/9781032631981-1

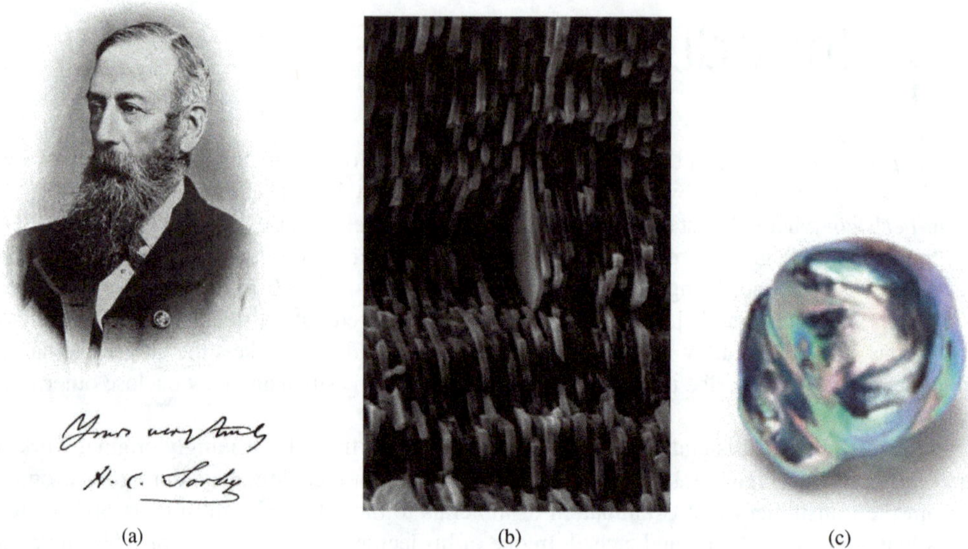

Figure 1.1 Henry Clifton Sorby, reproduced with permission from [8] 2009 © Cambridge University Press. (b) Scanning electron micrograph of mother of pearl lamellae consisting of aragonite platelets (about 0.5 μm thick) stacked in layers. Image courtesy of Fabian Heinemann [9]. (c) An illustration of the iridescence associate with pearls. Image courtesy of Masayuki Kato, reproduced under the GFDL CC-BY 3.0 licence [10].

Figure 1.2 This might be the first photomicrograph showing pearlite, in a mild steel 'in natural state', transverse section. The pearlite is represented by the dark regions, with the remaining microstructure being ferrite. After Sorby [12], reproduced with the permission of I● M3.

described hematite as a stone that appears like dried blood.[1] The Greek word *itēs* is an adjectival form of *lithos* which means rock or stone.

It was in 1888 that Henry Marion Howe first described the constituents of steels as pearlite (at that time also referred to as *pearlyte*), ferrite and cementite [16]. He felt uncomfortable with calling the pearlite reaction a eutectic, which implies a lowest melting point. In a letter to Sauveur on the 4th of June 1903, he wrote [17]:

[1]This popular attribution to Aristotle may not be accurate. Theophrastus, who was a follower and colleague of Aristotle, wrote a treatise entitled *On Stones* in which he refers to a stone called *haematitis*, because of the colour of dried blood. Theophrastus may have been referring not to hematite but to a silica stone coloured by the presence of iron oxide [14]. The regular use of *haematitis* is in connection with blood vessels [15].

Let us use the word 'eutectoid' to denote alloys of lowest transformation point. The suffix 'oid' clearly indicates that the eutectoid has the form of other important properties of the eutectic; it keeps the resemblance of the eutectoid to a eutectic before the mind, while it allows us to preserve the initial meaning for eutectic and to distinguish between these two really distinct though related entities.

Howe did refer to steel containing as much as 0.92C wt% as eutectoid in subsequent work [18], presumably because the phase diagram for the pearlite reaction was not at that time, well-defined. A part of Rosenhain's 1911 Fe-C phase diagram [13] is reproduced in Figure 1.3. It is not therefore surprising to suggest that a 0.92C wt% is a eutectoid steel. The book *Strength of Materials* by Morely [19] has sold some 52,000 copies. It mentions that the proportion of cementite is such that the carbon content of pearlite is about 0.9 percent. The hugely popular book *Metals in the Service of Man* published in 1944 by Alexander and Street shows the Fe-C diagram with the eutectoid temperature at 700 °C and eutectoid composition at 0.9 wt% [p. 107, 20].

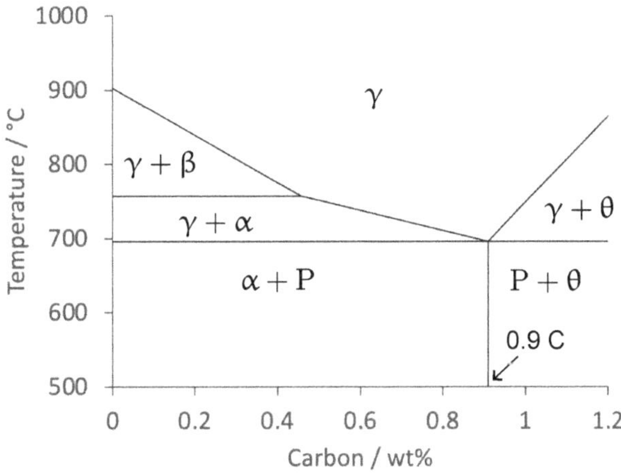

Figure 1.3 A selected part of Rosenhain's Fe-C equilibrium phase diagram as known in 1911 with annotation added. γ, α, θ and P represent austenite, ferrite, cementite and pearlite, respectively. At the time, the Curie transition was associated with a $\beta \rightarrow \alpha$ allotropic transformation with the β phase being much harder than α; quenching the steel was supposed to retain the β and hence to harden the steel. The eutectoid composition was set approximately at 0.9 wt% carbon, with the eutectoid temperature somewhat below 700 °C. The α and β single-phase fields are missing.

Colonel Belaiew in 1907–1908 had a number of steels prepared at Russian steel works 'under conditions of extreme slowness of cooling', aiming to bring 'into closer harmony the processes of crystallisation with various areas of the equilibrium diagram' [21]. His teacher was Professor Dmitry Konstantinovitch Tschernoff who at one time joined a newly founded steel plant, where he made the discovery of the critical temperatures in steel [22], in particular, the temperatures now known as Ae_3 and Ae_1 which form the foundation of the present Fe-C phase diagram. For an appropriate steel, Ae_1 is the temperature at or below which pearlite can form. The work was announced to the Imperial Russian Technical Society and became internationally renowned, translated and honoured.

Belaiew had hoped the slow cooling of the steels might lead to pearlite that is discernible to the naked eye, rather like some of the structures in meteorites, but the prospect 'still seemed very distant' [23]. But the studies led to some remarkable insights: that the pearlite proceeds with a certain constant velocity, that it only can form if the alloy is supersaturated simultaneously with regard to both components (it is common to refer to this as the Hultgren extrapolation on the Fe-C diagram, but Belaiew was much earlier). He may have been the first to deal with the 'stereometry of the pearlite grain', when he wondered why the interlamellar spacings varied across the field of observation – he worked out correctly that the plane of section matters.[2] He further considered the case where the true interlamellar spacing may not be uniform.

To finish this introduction, it is worth highlighting how pearlite has changed life as we know it. The mathematician Pappus of Alexandria in the 4th century CE wrote of the 'needs of life', for example, a crane with a compound pulley worked by a treadmill [pp.104-106, 24]. Ropes for such a device might in ancient times have been made by weaving natural fibres. Today, steel ropes based on pearlite feature in elevators, bridges, aircraft, cranes, ski lifts, brake cables for cars & bicycles … The first such ropes were in fact wrought-iron wires twisted together, invented by Wilhelm August Julius Albert for a mine in Germany, around 1834 [25]. This idea, over time, morphed into the huge industry for really strong ropes that are made of cold-drawn pearlitic steel, assembled in imaginative ways that make them as flexible and tolerant to flexing as meets the 'needs of life'.

REFERENCES

1. F. S. Beudant: Voyage Mineralogique Et Geologique, En Hongrie, vol. 1: 1822.
2. C. Darwin: Geological observations on coral reefs, volcanic islands, and on South America: Being the geology of the voyage of the Beagle, under the command of Captain Fitzroy, RN, during the years 1832 to 1836: Smith, Elder & Co., 1851.
3. C. Lyell: Principles of Geology, vol. 3: London: John Murray, 1833.
4. F. Zirkel: Microscopical petrography: Washington USA: US Government Printing Office, 1876.
5. F. Rutley: The study of rocks: New York, USA: D. Appleton and Co., 1879.
6. H. C. Sorby: 'On the application of very high powers to the study of the microscopical structure of steel', *Journal of the Iron and Steel Institute*, 1886, **1**, 140–147.
7. F. Nudelman, B. A. Gotliv, L. Addadi, and S. Weiner: 'Mollusk shell formation: mapping the distribution of organic matrix components underlying a single aragonitic tablet in nacre', *Journal of Structural Biology*, 2006, **153**, 176–187.
8. J. W. Judd: 'Henry Clifton Sorby, and the birth of microscopical petrology', *Geological Magazine*, 1908, **5**, 193–204.
9. F. Heinemann: 'Electron microscopy image of a fractured surface of nacre': https://en.wikipedia.org/wiki/Nacre, 2017.
10. M. Kato: 'Various pearls': https://commons.wikimedia.org/wiki/File:Various_pearls.jpg, 2011.
11. R. Edyvean, and C. Hammond: 'The metallurgical work of Henry Clifton Sorby and an annotated catalogue of his extant metallurgical specimens', *Historical metallurgy*, 1997, **31**, 54–85.
12. H. C. Sorby: 'On the microscopical structure of iron and steel', *Journal of the Iron and Steel Institute*, 1887, **31**, 255–288.
13. W. Rosenhain: 'Two lectures on steel: First lecture', *Proceedings of the Institution of Mechanical Engineers*, 1911, **80**, 241–275.
14. E. R. Caley, and J. F. Richards: Theophrastus on stones: introduction, Greek text, English translation, and commentary: Ohio, U. S. A.: The Ohio State University Press, 1956.
15. G. E. R. Lloyd: 'Private communication to H. K. D. H. Bhadeshia', 2023: On the origin of the term 'hematite'.
16. F. P. Peters: 'Into the alloy age', *Scientific American*, 1945, **172**, 15.
17. A. Sauveur: Metallurgical reminiscences: U.S.A.: American Institute of Mining and Metallurgical Engineers (published for the Seeley W. Mudd fund), 1937.
18. H. M. Howe, and A. G. Levy: 'Notes on pearlite', *Journal of the Iron and Steel Institute*, 1916, **94**, 210.

[2]How sad that so many contemporary publications using automated measures of areal or lineal intercepts on metallographic sections take the intercepts to represent true size.

19. A. Morely: Strength of Materials: 9 ed., London, U.K.: Longmans, 1940.
20. W. Alexander, and A. Street: Metals in the service of man: New York, USA: Penguin Books, 1944.
21. N. Belaiew: 'On the genesis of Widmanstätten structure in meteorites and in iron-nickel and iron-carbon alloys', *Mineralogical Magazine and Journal of the Mineralogical Society*, 1924, **20**, 173–185.
22. N. T. Belaiew: 'The Russian contribution, in the nineteenth century, to the metallurgy of steel', *Journal of the Royal Society of Arts*, 1921, **69**, 833–836.
23. N. T. Belaiew: 'The inner structure of the pearlite grain', *Journal of the Iron and Steel Institute*, 1922, **105**, 201–227.
24. G. E. R. Lloyd: The ideals of inquiry: An ancient history: OUP Oxford, 2014.
25. H. Dickinson: 'A condensed history of rope-making', *Transactions of the Newcomen Society*, 1942, **23**, 71–91.

2 Structure

Classic observations of pearlite based on polished and etched planar-sections are taken to reveal a lamellar mixture of ferrite and cementite. Figure 2.1 illustrates light directed at an idealised etched-surface of strictly lamellar pearlite, assuming that the cementite and ferrite etch at different rates, and that the resulting topography acts as a diffraction grating. The angle θ_1 is between the normal to the average etched-surface and the incident wave front; θ_2 represents the angle between the reflected wave front and the same normal. Rays 1 and 2 from the incident wave front have a path difference 'cd' here taken to be positive because it is ahead of the incident wave front marked 'bc'; rays 3 and 4 have a path difference 'ab' which therefore is negative (behind the wave front marked 'ad'), resulting in the diffraction condition for constructive interference as

$$S_{\mathrm{I}}(\sin\{\theta_1\} - \sin\{\theta_2\}) = n\lambda \tag{2.1}$$

where S_{I} is the interlamellar spacing, λ is the wavelength, n an integer since a phase difference of any whole number of wavelengths gives constructive interference. Colours arise when the pearlite is viewed in white light, since some wavelengths will interfere destructively, so the mixture of wavelengths that constitute white light is depleted. The colours vary as a function of the apparent spacing of the lamellae at the surface, and of the angle of incidence of the light, giving rise to iridescence.

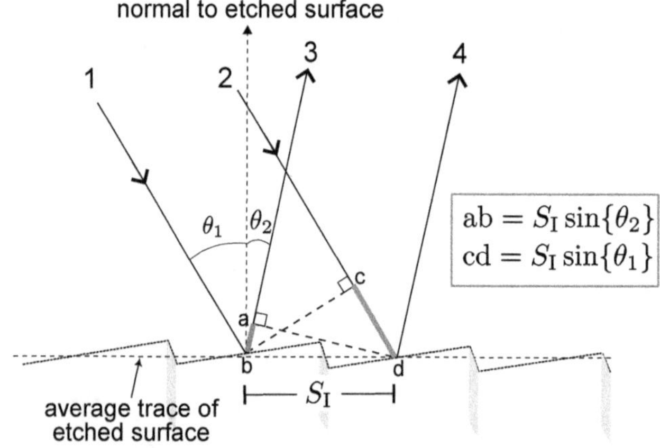

Figure 2.1 Schematic illustration of path differences arising from the reflection of an incident beam from the rough surface created by the etching of lamellar pearlite. Rays 1 and 2 impinge the surface at the same height, a distance S_{I}, which is the interlamellar spacing, apart.

We have seen that the term *pearlite* is an adaptation of the fact that natural pearls that have layered structures also exhibit iridescence as noted originally by Sorby (Chapter 1); the spelling ranged from pearlyte, perlite before settling on pearlite in the context of steel [1].[1]

A *colony* of pearlite has all its lamellae growing in approximately the same direction when observed on a two-dimensional section, whereas a *nodule* consists of several colonies growing along

[1]Other languages based on the Roman alphabet, such as Italian, French, German, Spanish, Galician, Corsican, Danish, Swedish etc. use perlite, perlita or perlit, though Swahili, Xhosa and Welsh use pearlite. Norwegian uses perlitt, Icelandic pelulit, Latvian perlīts and Finish perliitti. This list is not intended to be comprehensive, simply interesting.

DOI: 10.1201/9781032631981-2

different directions, permitting the nodule to acquire an approximately spherical shape in the absence of impingement. The generation of different colonies within a nodule must involve the branching of cementite, although the role of the ferrite in this process is not known.

Figure 2.2a is an attempt to illustrate the iridescence observed using optical microscopy, together with the form of the pearlite nodules when the steel is not fully transformed. The lamellae within each nodule are not always resolved, except when the plane of section is inclined sufficiently with respect to the normal to the lamellae within a small region. It is noticeable that although the general form of an isolated nodule approximates a sphere, the transformation front with the austenite has many protrusions. The relationship between a nodule and the colonies within is illustrated schematically in Figure 2.2b.

Figure 2.2 (a) Pearlite nodules in Fe-1.21C-2.2Mn-1.7Si-1Ni-0.4Mo-0.2Cr wt% steel, partially transformed into pearlite, with the intervening lighter-regions consisting of a mixture of bainite and martensite plates embedded in retained austenite. The pearlite iridescence ranges from bronze to blue-green and a tinge of violet, not unlike the colours associated with male peacocks [2]. Micrograph courtesy of Hala Salman Hasan. (b) Simplified illustration of the colonies within the unhindered growth of a nodule of pearlite.

Intuition might indicate that the probability a given colony of pearlite lies at some angle between ϕ and $\phi + \mathrm{d}\phi$ to a random reference-plane of observation should be independent of ϕ. This in not in fact the case [3]. There is a greater chance of the lamellae being steeply inclined to the reference plane. Consider Figure 2.3a, where $\mathbf{n}_\ell$ represents the normal to a cementite lamella. The density of $\mathbf{n}_\ell$ will be equal over the whole hemispherical surface if there is a random spatial-distribution of pearlite colonies. Therefore, dividing the area of the annular ring in Figure 2.3a by that of the hemispherical surface gives the probability p_ϕ of $\mathbf{n}_\ell$ lying in the range ϕ and $\phi + \mathrm{d}\phi$, where ϕ is the angle of $\mathbf{n}_\ell$ relative to the reference plane normal $\mathbf{n}_\mathrm{r}$. The derivation that follows is after Pellissier et al. [3]:

$$p_\phi = \frac{2\pi \sin\{\phi\}}{2\pi} \mathrm{d}\phi = \sin\{\phi\}\,\mathrm{d}\phi. \tag{2.2a}$$

Lamellae inclined at an angle between ϕ and $\phi + \mathrm{d}\phi$ will have an apparent spacing between $S_{\mathrm{I,a}}$ and $S_{\mathrm{I,a}} + \mathrm{d}S_{\mathrm{I,a}}$ on the reference surface of

$$S_{\mathrm{I,a}} = \frac{S_{\mathrm{I,t}}}{\sin\{\phi\}} \qquad \text{so that} \qquad \mathrm{d}S_{\mathrm{I,a}} = -\frac{S_{\mathrm{I,t}}\cos\{\phi\}}{\sin^2\{\phi\}}\mathrm{d}\phi \tag{2.2b}$$

where $S_{\mathrm{I,t}}$ is the true interlamellar spacing. On substituting these into Equation 2.2a, it follows that

$$p_\phi = -\sin\{\phi\}\frac{\sin^2\{\phi\}}{S_{\mathrm{I,t}}\cos\{\phi\}}\mathrm{d}S_{\mathrm{I,a}} = -\frac{\sin^3\{\phi\}}{\cos\{\phi\}}\frac{\mathrm{d}S_{\mathrm{I,a}}}{S_{\mathrm{I,t}}}, \tag{2.2c}$$

and given that $\sin\{\phi\} = S_{\mathrm{I,t}}/S_{\mathrm{I,a}}$ and $\cos\{\phi\} = \sqrt{S_{\mathrm{I,a}}^2 - S_{\mathrm{I,t}}^2}/S_{\mathrm{I,a}}$,

$$p_\phi = -\frac{S_{\mathrm{I,t}}^2}{S_{\mathrm{I,a}}^2\sqrt{S_{\mathrm{I,a}}^2 - S_{\mathrm{I,t}}^2}}\mathrm{d}S_{\mathrm{I,a}}. \tag{2.2d}$$

The conversion of variables enables the probability that the apparent spacing lies between $S_{\mathrm{I,a}}$ and $S_{\mathrm{I,a}} + \mathrm{d}S_{\mathrm{I,a}}$ i.e., p_ϕ to be estimated; assuming a random spatial distribution of colonies, Figure 2.3b show that the chances of obtaining a fine spacing on a random plane of observation are greater than of seeing large apparent interlamellar spacings.

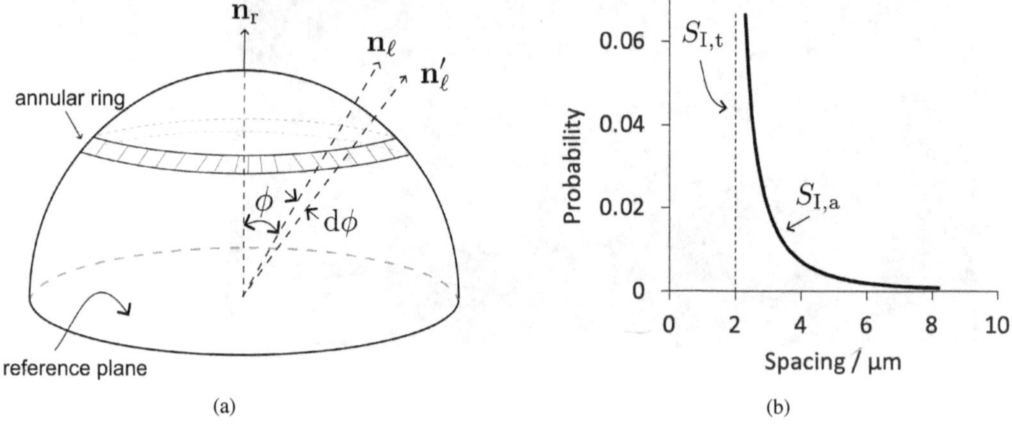

Figure 2.3 (a) Geometrical construction relating the normals to the cementite lamellae ($\mathbf{n}_\ell$) and the normal ($\mathbf{n}_\mathrm{r}$) to the reference plane. Adapted from Pellissier et al. [3]. (b) Probability p_ϕ of observing a particular apparent interlamellar spacing when the true spacing is 2 μm; the increment $\mathrm{d}S_{\mathrm{I,a}}$ was taken to be 0.1 μm.

The fractional area A_A of the reference plane that is occupied by pearlite that has an apparent spacing up to a certain value of $S'_{\mathrm{I,a}}$ is [3]:

$$A_\mathrm{A} = \int_{S_{\mathrm{I,t}}}^{S'_{\mathrm{I,a}}} -\frac{S_{\mathrm{I,t}}^2}{S_{\mathrm{I,a}}^2\sqrt{S_{\mathrm{I,a}}^2 - S_{\mathrm{I,t}}^2}}\,\mathrm{d}S_{\mathrm{I,a}} \equiv -\int_{\pi/2}^{\phi'} \sin\{\phi\}\,\mathrm{d}\phi = \cos\{\phi'\} \tag{2.2e}$$

where the equivalence can be derived using the identities in Equations 2.2b-2.2d.

Some calculations using Equation 2.2e are presented in Figure 2.4; it was a comparison of such a curve against experimental measurements that first revealed the fact that the true spacing is not constant in a given sample, but that there is a distribution about a mean value that is greater than the minimum observed spacing; the subject is explained in rather more detail on page 10. Equation 2.2 is of course derived assuming that there is a single value of the true spacing.

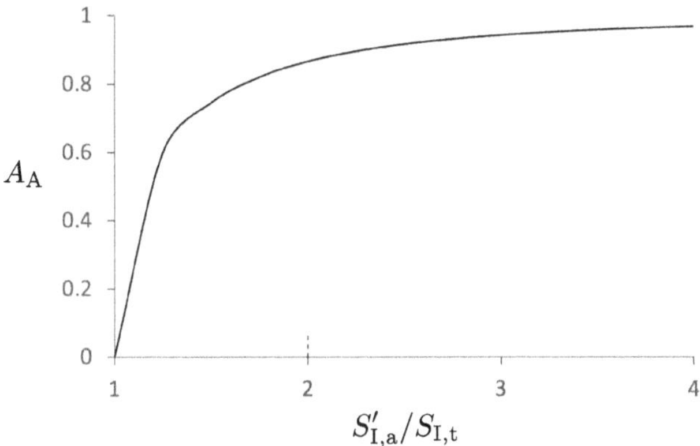

Figure 2.4 The fractional area occupied by pearlite colonies with an apparent spacing that is less than or equal to a particular value $S'_{I,a}$. The calculations have been carried out assuming that $S_{I,t} = 2\,\mu m$.

2.1 MEASUREMENT OF INTERLAMELLAR SPACING

Obviously, the apparent interlamellar-spacing as measured on a two-dimensional section will in general be larger than the true spacing $S_{I,t}$ assuming that the structure of pearlite can be described as alternating plates of ferrite and cementite. For a structure with a constant true spacing, $S_{I,t}$ is independent of the plane of section as illustrated in Figure 2.5a. For this case, Saltykov [p. 182, 4],[5] showed that the true spacing is related to the mean random spacing $S_{I,r}$ by

$$S_{I,t} = 0.5\overline{S}_{I,r} \tag{2.3a}$$

where the random measurements would be made at any orientation relative to adjacent lamellae.

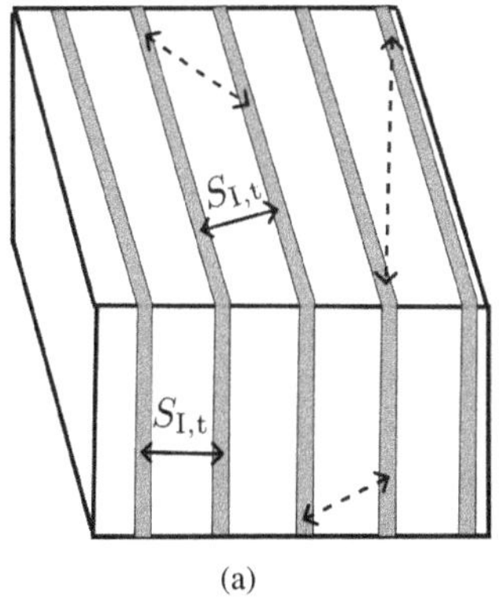

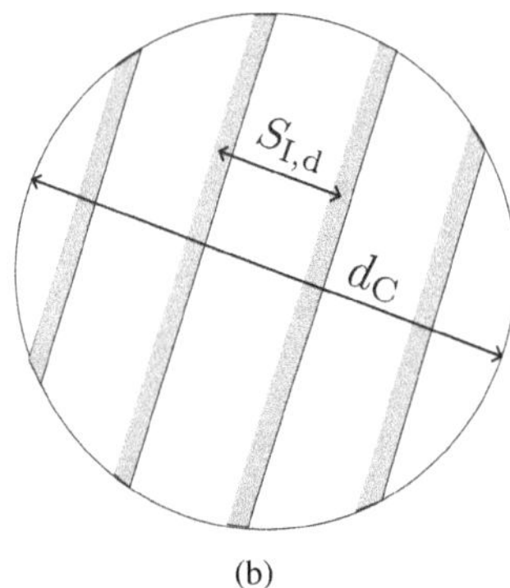

Figure 2.5 (a) $S_{I,t}$ represents the true spacing measured normal to the plates of cementite. Measurements made at random orientations on any plane section are represented by dashed lines. (b) Test circle of diameter d_C drawn on a random section of the microstructure, and directed spacing $S_{I,d}$ measured normal to the intercept of the cementite lamellae with the free surface. The test circle here intersects the lamellae on eight occasions whereas the diameter (drawn normal to the lamella traces) intersects on four occasions.

An alternative method [6] of measuring the spacing is to draw a test circle of diameter d_C and determine the number n_r of times the circumference of length πd_C intersects a cementite lamella, Figure 2.5b. Since a circumference covers all orientations in the plane, the mean random intercept is given by

$$\overline{S}_{\mathrm{I,r}} = \pi d_{\mathrm{C}}/n_{\mathrm{r}}. \tag{2.3b}$$

If $S_{\mathrm{I,d}}$ is an intercept measured normal to the trace of the cementite plates, then the mean directed spacing is given by

$$\overline{S}_{\mathrm{I,d}} = d_{\mathrm{C}}/n_{\mathrm{d}} \tag{2.3c}$$

where n_{d} is the number of cementite plates intercepted by the diameter d_{C}. From Figure 2.5b, $n_{\mathrm{r}} = 2n_{\mathrm{d}}{}^{2}$ which when substituted into Equation 2.3b gives $\overline{S}_{\mathrm{I,r}} = \pi d_{\mathrm{C}}/2n_{\mathrm{d}}$. It follows from Equation 2.3a that $S_{\mathrm{I,t}} = \pi \overline{S}_{\mathrm{I,d}}/4$.

2.1.1 NON-UNIFORM INTERLAMELLAR SPACING

It was shown some time ago [3] that S_{I} need not be constant even when the pearlite has been generated by isothermal transformation. By following the trace of the α/θ interface as a function of depth, the measured distribution of true interlamellar spacing is shown in Figure 2.6 – obviously, if the true spacing were to be constant, then a distribution of apparent spacings would fall monotonically from the value of the true spacing [3]. Transmission electron microscopy of thin foils of pearlite, which in effect gives a three-dimensional view of the lamellae so that sample tilting can be used to make the lamellae parallel to the electron bean in order to fix the true spacing, also shows a distribution of true spacings between colonies of isothermally generated pearlite [7]. The real structure of pearlite after all, does not consist of a set of flat, parallel plates of alternating α and θ plates (Section 2.2).

It is also possible that the $S_{\mathrm{I,t}}$ in isothermally generated pearlite is influenced by the recalescence that accompanies rapid transformation, so in spite of intentions, the transformation may occur over a range of temperatures. Figure 2.6b shows typical time-temperature data recorded during the transformation of an eutectoid steel – clearly, there can be significant excursions from the intended isothermal transformation temperature.

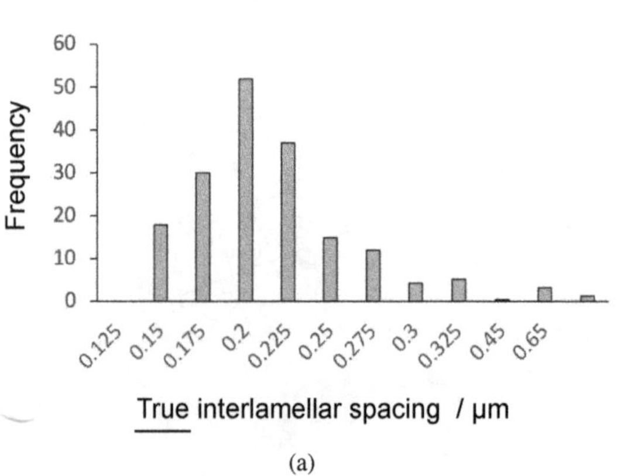

(a)

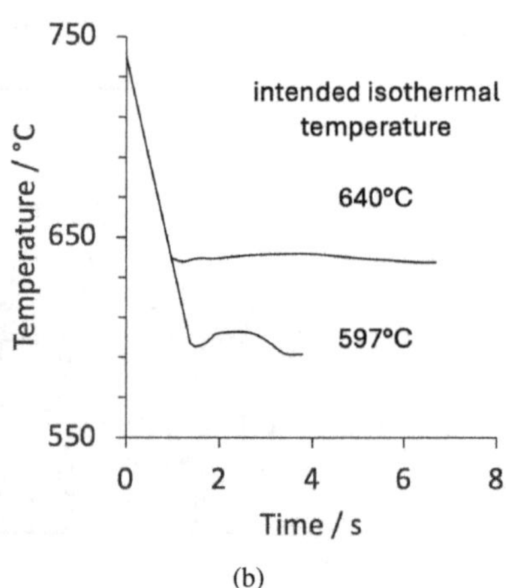

(b)

Figure 2.6 (a) The frequency of the true interlamellar spacing, measured following the transformation of Fe-0.82C-0.71Mn-0.25Si wt% steel at 700 °C. Adapted using data from Zhang and Enomoto [8]. (b) Recalescence during 'isothermal' transformation of Fe-0.8C-0.91Mn-0.49Si wt% steel. Adapted using selected data from Hawbolt et al. [9].

[2] Writing $n_{\mathrm{r}} = 2n_{\mathrm{d}}$ assumes extreme regularity in the structure of pearlite, so the relation should be regarded as an approximation.

2.1.2 DERIVATION OF TRUE SPACING DISTRIBUTION

The function A_A defined by Equation 2.2e and plotted in Figure 2.4 is calculated assuming that there is a unique value of the true spacing $S_{I,t}$ in the observed microstructure. However, there may exist a distribution of spacings (Section 2.1.1). The function $A_A\{S'_{I,a}\}$ that defines the proportion of the observed area in which the apparent interlamellar spacings are in the range $S_{I,t} \rightarrow S'_{I,a}$, given a unique true-spacing S_{I,t_o} can be written, on the basis of the integral in Equation 2.2e,[3]

$$A_A = \sqrt{1 - \left(\frac{S_{I,t_o}}{S'_{I,a}}\right)^2}. \tag{2.4a}$$

When the true interlamellar spacing is not the same in different colonies, this equation is scaled by a fraction A_i which represents the proportion of area occupied by the ith value of S_{I,t_i}

$${}^iA_A = A_i\sqrt{1 - \left(\frac{S_{I,t_i}}{S'_{I,a}}\right)^2}. \tag{2.4b}$$

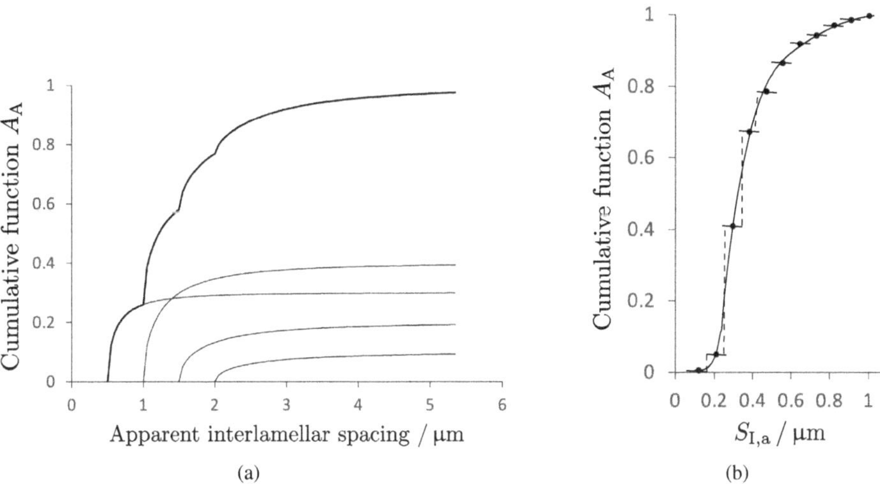

Figure 2.7 (a) Example in which there are four kinds of colonies of pearlite with true interlamellar spacings 0.5, 1.0, 1.5 and 2.0 μm, with proportions A_i equal to 0.3, 0.4, 0.2 and 0.1 respectively. The horizontal axis represents $S_{I,a}$. The light lines represent the individual functions and the heavy line is the cumulative function. (b) The points represent experimental data from Roosz and Baan [10], the horizontal bars on the points the spread of apparent interlamellar spacing within the plotting interval, and the dashed lines illustrate the fact that the function is not continuous.

The example in Figure 2.7a illustrates how the cumulative function evolves from an assumed distribution of true interlamellar spacings. The problem is more difficult to determine experimentally. An approximate cumulative function can be measured experimentally by conducting a large number of apparent interlamellar spacing measurements within individual colonies observed on metallographic sections. The data are then divided into discrete, equal intervals to estimate ${}^iA_A$ for each i, Figure 2.7b. The measured function would therefore be stepped rather than continuous. To determine the distribution of true interlamellar spacings, it has been assumed [10] that the mean of

3 $\int \frac{a^2}{x^2\sqrt{x^2-a^2}} = \sqrt{1-\frac{a^2}{s^2}} + \text{constant}$

each interval gives $S_{\mathrm{I,t}_i}$ and therefore ${}^iA_{\mathrm{A}}$ from the value on the vertical axis. The procedure does not seem justified, but the alternative would be much more difficult, involving measurements based on serial sectioning to unambiguously determine the distribution.

It is difficult to see the relevance of variations in true interlamellar spacings on properties. The distribution of true spacings is likely to be narrow when the pearlite is generated isothermally; even if it is generated during continuous cooling transformation, the temperature interval over which pearlite forms is usually rather small when considering structural steels. Furthermore, there are many other heterogeneities expected in the microstructure, such as the branching of lamellae, the fact that they are not idealised lamellae but contain holes, etc. It probably is sufficient to measure data assuming that there is a unique true spacing when that information is to be used for correlating against macroscopic mechanical-properties.

2.1.3 MEASUREMENT ERRORS IN INTERLAMELLAR SPACING

The term *noise* in the present context refers to errors caused by unknown factors, which may, for example, arise because different techniques have been used to characterise the same feature. If a large quantity of published data on S_{I} is analysed as a function of the steel composition and transformation temperature, using a neural network in a Bayesian framework, the level of noise is found to be about ±5-6% [11]. This value can be doubled for 95% confidence, giving a similar error to that reported in [12] for $S_{\mathrm{I,t}}$. So it might be reasonable to assume that most measurements are likely to be associated with a ±12% uncertainty.

There essentially are two categories of error, the first being statistical and the second related to experimental issues [e.g., 13]. The statistical error is the simplest to understand, basically a function of the number of measurements made.

During any stereological measurements, the fractional error E in the mean depends on the number of intercept measurements N done by point counting; and for 67% confidence, $E^2 = 1/N$ [5, 14]. Based on a hundred random measurements of $S_{\mathrm{I,a}}$, a mean of 0.2 μm would therefore be associated with an uncertainty ±0.02 μm, so the calculated proof strength would be reported as 517 ± 20 MPa (Equation 4.17); the errors should be doubled for 95% confidence.

This calculation assumes perfect measurements, but there are experimental issues to consider. It is often the case that the microstructure is not homogeneous, either because of solidification-induced chemical segregation in the steel, or because it is generated by continuous cooling transformation. In the former case, it is necessary to examine sufficient fields that extend significantly over the length scale of the segregation. There may be bias because of limited resolution, though the higher magnifications available today reduce this problem but create another one where the area of the field visible in the microscope becomes smaller [13]. Bias also arises when measurements are conducted only on pearlite colonies that exhibit a clear separation between the lamellae. Heavy etching that leaves a surface relief on the metallographic specimen can lead to projection errors.

In attempting to discover a size distribution of the true spacing $S_{\mathrm{I,t}}$, it is recommended that at least 500 random measurements are carried out of apparent spacings on random sections, and that the range of apparent spacing is divided into sufficient segments (say 20) to ensure accuracy in the estimation of true spacing [15].

It is as well to bear in mind, that pearlite does not consist of parallel, flat, alternating plates of ferrite and cementite, but has convoluted three-dimensional form with features such as branching, twisting and the passage of ferrite through the cementite [16]. Therefore, any spacing that is measured is an imperfect representation of the real structure; it is likely, nevertheless, that the measured spacing will correlate well with macroscopic properties, because such measurements represent the collective behaviour of a polyphase, polycrystalline sample.

2.2 LAMELLAE

The notion that pearlite is lamellar with flat and alternating plates of α and θ is an approximation. Each θ-lamella has curvature and contains holes through which ferrite permeates. A lamella can appear to terminate when observed in a two-dimensional section that intersects a hole, though as Figure 2.8 shows, it remains extended beyond the holes [17–21].

Why does the cementite contain holes? Their presence implies a temporary breakdown of the $\alpha+\theta$ cooperative growth. This is perhaps more likely during rapid growth when the ferrite may occasionally protrude further into the austenite and then grow laterally to block off the cementite, followed by a recovery of cooperation when the α slows down due to the accumulation of partitioned carbon. Unfortunately, the only quantitative characterisation of the number density of holes is by Wang et al. [20, 21] for a plain carbon Fe-0.8 wt% steel where the growth rate would be fast, but similar data for slow rates are not available.

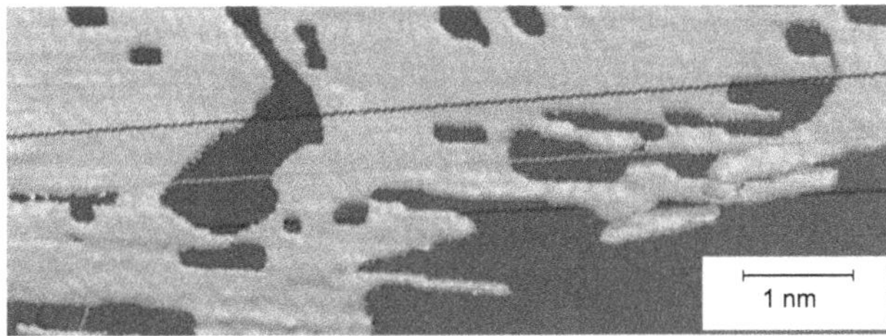

Figure 2.8 Three-dimensional, high-resolution image of a single lamella of cementite (light grey) in Fe-0.8C wt%. The lamella is not a continuous slab but contains holes. The image is generated using serial sectioning and provided courtesy of Yoshitaka Adachi. The straight lines in the image are artefacts.

Cementite lamellae twist [19] in an attempt to fill space in three dimensions. There obviously is no reproducible crystallographic *habit plane* on which the cementite shares the major part of its interface with ferrite. This is unlike displacive transformations where the fit between the parent and product lattices is essential in order to maintain a glissile interface that can glide without the need for diffusion. Also unlike displacive transformations, pearlite is not dominated by strain energy, its growth is reconstructive with the diffusion of both solvent and solute atoms. Its morphology is controlled instead by the need to maintain cooperative growth, where factors such as the entropy production rate or those that maximise the rate of transformation are influential. The θ and α are connected in three dimensions, with nodules best visualised as interpenetrating bi-crystals of the two phases [22].

The idea that a colony of pearlite is a *bi-crystal* has caveats. Observations of a hypoeutectoid steel air cooled from the austenitisation temperature have revealed otherwise. The colony shown in Figure 2.9 has a homogeneous ferrite orientation but the cementite lamellae have two sets of orientations, one of which is absent in the dark-field image, at the positions identified by the arrowheads. The relative orientation of the two sets of cementite was within $\approx 7^\circ$, which is small but significant, especially because many such cases were observed. It is possible that there were two independent nucleation events for the cementite, though any differences in crystallography did not affect the direction of cooperative growth with the shared ferrite, indicating vividly the minor role of crystal orientation in the growth process and emphasising the importance of maintaining a common growth direction.

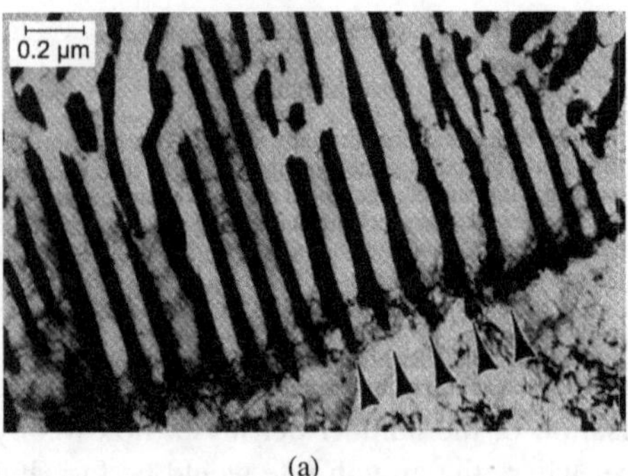

(a)

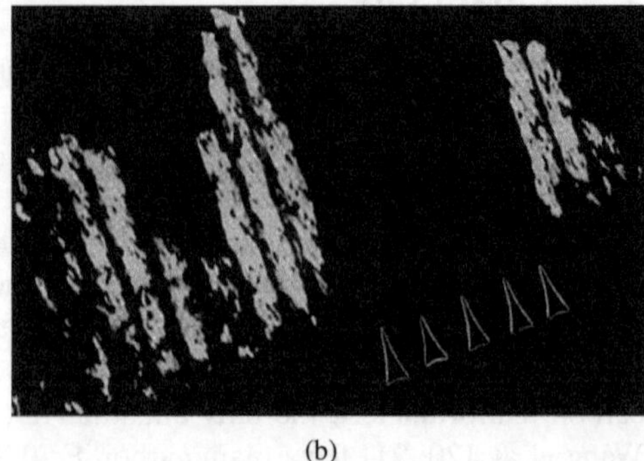

(b)

Figure 2.9 Pearlite colony generated in Fe-0.15C-1.49Mn-0.26Si wt% during air cooling from the austenitisation temperature. (a) Bright-field image showing both the proeutectoid ferrite at the bottom and the pearlite colony at the top. (b) Dark field image of cementite. After Thompson and Howell [23], with the images slightly truncated for clarity, reproduced with the permission of Elsevier.

2.3 AUSTENITE GRAIN AND PEARLITE COLONY SIZE

The austenite grain size is determined by heat treatment and the composition of the steel; pearlitic steels are not hot-worked in a manner that leaves the γ-grain structure in a deformed state prior to transformation. They also are not microalloyed so it is safe to assume that the austenite grains will be in a recrystallised, equiaxed shape before pearlite is generated.

A colony of pearlite is defined as the region within which the traces of the lamellae of cementite observed in two-dimensional sections, may fan out from an origin but are roughly parallel, Figure 2.10a. The colony size is used sometimes in correlation with mechanical properties. Its size is characterised as a mean lineal intercept $\overline{L}_{\text{colony}}$, though it may not always be straightforward to discern a colony structure in fully pearlitic steel, Figure 2.10b.

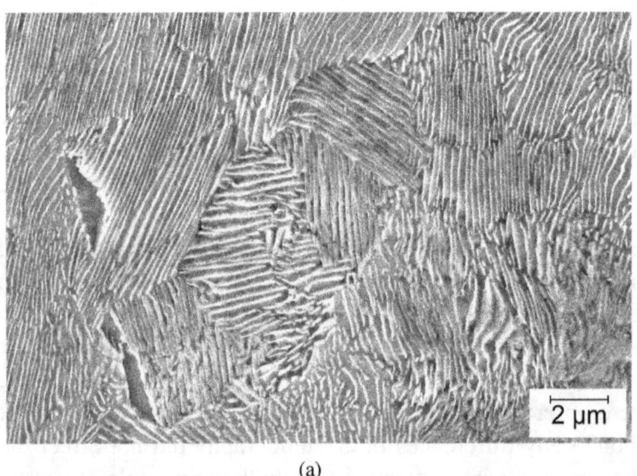

(a)

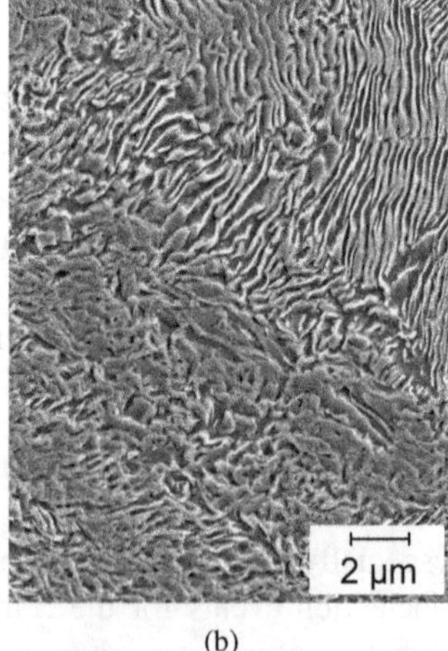

(b)

Figure 2.10 Pearlite colonies in a fully transformed eutectoid steel. Each colony is defined by a region in which the lamellae traces are approximately parallel. (a) Case where the colonies are clearly defined. (b) The idealised structure of a pearlite colony as parallel 'plates' of the constituent phases is an oversimplification of its three-dimensional form. This two-dimensional image shows that it often is difficult to identify the boundaries between colonies. Micrographs courtesy of Apparao Chintha.

The need to fill space as transformation progresses, can cause large changes in the direction of cementite growth, without altering its crystallographic orientation, nor that of the ferrite. The process does not require a new nucleation event, simply a change in growth direction within the bounds of a nodule. The α/θ lattice misfit at the transformation temperature causes the two phases to curve elastically along the length of the lamellae, by about $0.6^\circ\ \mu m^{-1}$ [24]. This curvature is much bigger than expected during cooling from the transformation temperature [25]. The sign of the elastic strain changes when the cementite alters its growth direction to generate another colony within the nodule. These effects are responsible for the small orientation gradients observed within a nodule. Figure 2.11a illustrates one such nodule, that judging from its shape, was able to develop freely before colliding with other nodules. The colour, which indicates crystallographic orientation, is not uniform within the nodule, but the deviations are small when compared with the surrounding pearlite. The nodule is able to grow in all directions while maintaining the cementite lamellae approximately normal to the local growth-front with the austenite.

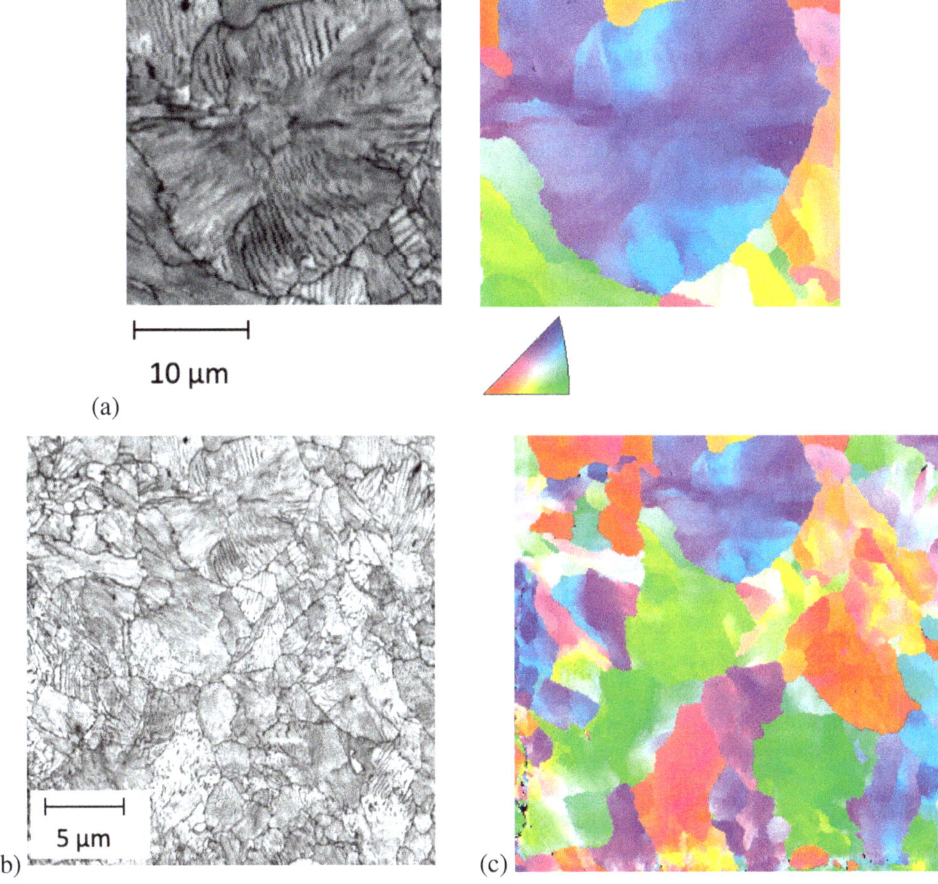

Figure 2.11 A pearlite nodule in a eutectoid steel. (a) Image showing the colonies within the nodule, and corresponding electron back-scattered diffraction image illustrating the gradients of crystallographic orientation within the ferrite. (b,c) Larger views of fully transformed structure. The nodule shown in (a) is located at top-centre. Pearlite nodules in fully transformed microstructures will not always exhibit a spheroidal shape.

The structure of a nodule is best observed in partly-transformed samples but when using optical microscopy, requires careful etching when dealing with fully-transformed specimens [26]. This involves tempering the partially or fully transformed steel at 510 °C for 16 h followed by metallographic preparation and etching in a saturated, aqueous picric-acid solution containing a wetting agent [26], Figure 2.12.

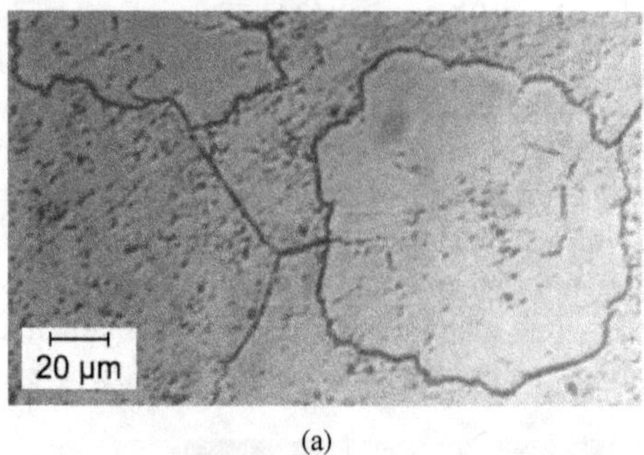

(a)

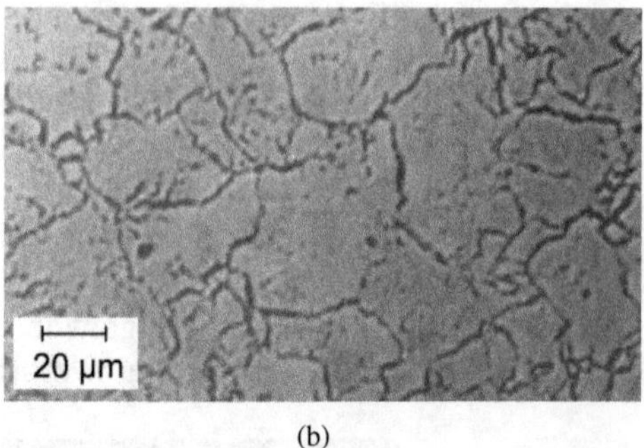

(b)

Figure 2.12 Special tempering and etching used to reveal nodule boundaries in a eutectoid steel 'AISI 1080'. (a) Partially transformed, showing the free-growth shape of the nodule; austenite grain boundaries are also revealed, showing that the nodule can grow across the γ-boundaries, as if they do not exist. (b) Fully transformed sample where space-filling dominates the nodule shapes observed. The structure resembles that of a single-phase material containing equiaxed grains. Reprinted in truncated form, from Doi and Kestenbach [26], with permission of Elsevier.

A combination of scanning electron microscopy and orientation imaging makes it possible to identify crystallographically homogeneous domains, which have a role in determining the resistance to brittle fracture, Figure 2.11c. Unfortunately, a number of such studies incorrectly identify colony size by the areal *intercept* of these domains, neglecting sectioning effects [e.g., 27, 28]. Published distributions that assume an intercept can be identified with a true size, are wrong – the automatically generated data should be averaged to a mean value that indicates true size. Obtaining distributions of colony size would require serial sectioning but approximate methods (and their limitations) that avoid a lot of work are described in [5]. The utility of measuring a distribution rather than a mean value may lie in understanding the scatter in mechanical properties during repeated testing, although such experiments have yet to be conducted.

The crystallographic grain-size, i.e., the domain over which there exist only small misorientations, can influence the mechanical properties that depend on crystal structure, for example, cleavage fracture [29, 30]. Given that it is the nodule rather than colony size that determines the crystallographic grain-size, it is not surprising that the fraction of large misorientation ($> 15°$) α_P/α_P boundaries in a fully pearlitic eutectoid steel is only ≈ 0.35 [31].[4] The remaining fraction (0.65) represents colony-colony interfaces which typically are better regarded as sub-boundaries with a misorientation of $< 10°$ [19].

Pearlite mostly nucleates at the austenite grain surfaces so the scale of the structure obtained on transformation may depend on $\overline{L}_\gamma$. While the colony size does decrease as $\overline{L}_\gamma$ is reduced, the change is surprisingly small. Even though pearlite is not limited by γ/γ boundaries, the colonies remain much smaller than $\overline{L}_\gamma$, Figure 2.14. The ratio $\overline{L}_{\text{colony}}/\overline{L}_\gamma$ never exceeds $\frac{1}{3}$ even for the smallest of austenite grain sizes, probably because of rapid nucleation during transformation. The absolute size of a pearlite colony for any of the austenite grain sizes ($\overline{L}_\gamma = 15$-$400\,\mu$m) illustrated, never exceeds $\approx 13\,\mu$m with an average of $\approx 7\,\mu$m.

[4]The emphasis of α_P/α_P boundaries at nodule junctions is because the cementite often is not indexed in electron back-scatter diffraction experiments.

The nodule size is measured as the pearlite region in which the α_P/α_P misorientation does not exceed $\approx 15°$. Figure 2.13 shows that the dependence of the nodule size on that of the austenite grains is largest at the smaller grain sizes, with the nodule size almost identical to the austenite grain size for $\overline{L}_\gamma \lessapprox 60\,\mu$m. It may then be reasonable to conclude that each austenite grain is transformed by just one nucleation event so the ferrite orientation is relatively uniform throughout the γ-grain. The nodule dimensions become insensitive to coarse austenite grain sizes, greater than about $100\,\mu$m. Nodules nucleating at opposite faces of an austenite grain would be limited in size to $\frac{1}{2}\overline{L}_\gamma$, but the observed $\overline{L}_{\text{nodule}}$ in the coarsest austenite grains is much smaller than this limit. It is possible, therefore, that there is a contribution from heterogeneous intragranular-nucleation. High-manganese steel containing concentrations of sulphur or vanadium have been found to exhibit intragranular nucleation [32]; whether this extrapolates to low-alloy steels remains to be demonstrated.

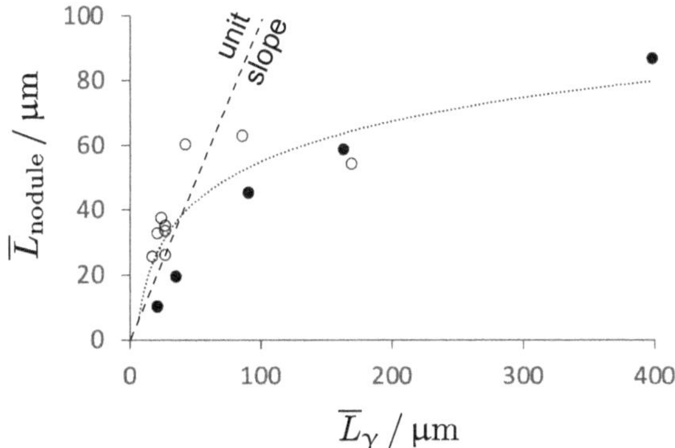

Figure 2.13 Open points represent the variation in the pearlite nodule size as determined by setting a crystallographic misorientation threshold of 12° to define the nodule boundary, as a function of the austenite grain size. Data adapted from Behera et al. [33] on a Fe-0.81C-0.73Mn-0.15Si-0.009S wt% steel with a fully pearlitic microstructure. Filled points are from Zhou et al. [27], where the threshold was 15°, in an essentially pearlitic Fe-0.55C-0.78Mn-0.89Si-0.07V wt% steel. The dashed line shows that nodules and austenite grains have similar sizes for $\overline{L}_\gamma \lessapprox 100\,\mu$m.

The size of the nodule does depend on the misorientation threshold, but a choice of 12-15° is found to be useful in assessing structure-property relationships such as the dependence of the fracture facet size on that of the nodule, [33]. Many of the older data on nodule size were determined using optical microscopy alone and consistently show the nodules to be much smaller than austenite grains [e.g., 34]. However, the crystallographic grain-size is not just a less subjective measurement but is justified in terms of fracture mechanism in fully transformed samples.

The data illustrated in Figure 2.14, which represent eutectoid steels containing small concentrations of manganese and silicon but no microalloying additions, obey the classical grain coarsening law to a correlation coefficient of 0.94:

$$\overline{L}_\gamma^2 \approx 4.16 \times 10^{11} \times \underbrace{\exp\left\{-\frac{Q}{RT_\gamma}\right\} t_\gamma}_{\text{kinetic strength}} \qquad \mu\text{m}^2 \tag{2.5}$$

where $Q = 190\,\text{kJ}\,\text{mol}^{-1}$ [35], and the isothermal austenitisation time and temperature are expressed in units of hours and Kelvin, respectively. A more complex analysis is necessary when dealing with large samples to allow for the time taken to reach T_γ from the Ac_3 temperature and to account for the grain size immediately after austenitisation is complete and grain coarsening begins [35].

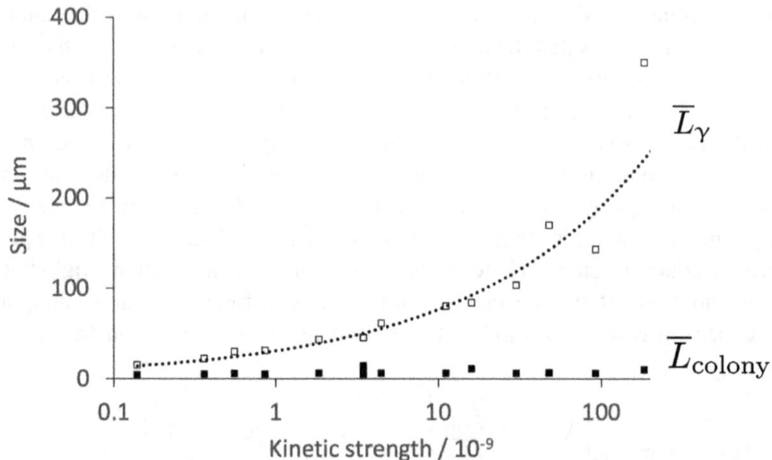

Figure 2.14 Plot of the austenite grain size (□) and the pearlite colony size (■) against the kinetic strength of the austenitisation heat treatment, given by $t_\gamma \exp\{-Q/RT_\gamma\}$, where t_γ is the austenitisation time in hours, T_γ is the austenitisation temperature in Kelvin, and the activation energy $Q \approx 190\,\mathrm{kJ\,mol^{-1}}$ [35]. The steels are all of eutectoid composition, data are selected from [36–38].

The interlamellar spacing is expected to decrease with the undercooling below the eutectoid temperature, because the larger free energy change is then able to support a greater α/θ interfacial area per unit volume, Figure 2.15a. The minimum achievable size of the interlamellar spacing, S_I^C, is when all of the available free energy change ΔG for the formation of pearlite from austenite is consumed in creating α/θ interfaces. Since from stereology the amount of interfacial area per unit volume is $2/S_I$, the minimum spacing is given by

$$S_I^C = \frac{2\sigma^{\alpha\theta}}{|\Delta G|}$$
$$\text{where } \Delta G = \Delta S(T_E - T), \tag{2.6}$$

T_E is the eutectoid temperature, T is the transformation temperature and the interfacial energy $\sigma^{\alpha\theta} \approx 1\,\mathrm{J\,m^{-2}}$. $\Delta G = \Delta H - T\Delta S$, where H and S are the enthalpy and entropy terms; $\Delta G = 0$ at T_E, $\Delta H = \Delta S T_E$ and assuming that ΔS does not vary much with temperature, $\therefore \Delta G = \Delta S(T_E - T)$. For a Fe-C alloy of eutectoid composition, it can be shown that $\Delta S = -862815\,\mathrm{J\,m^{-3}\,K}$, to a high degree independent of $(T_E - T) = 0 \rightarrow 300$. The dashed line in Figure 2.15a shows the critical interlamellar spacing S_I^C as a function of $(T_E - T)$ for Fe-C, illustrating the fact that the relationship is non-linear. Therefore, the actual spacings measured in low-alloy steels also follow a similar non-linear relation. The reason why $S_I > S_I^C$ is to do with factors that determine the growth rate, as discussed in Chapter 8.

The degree of scatter in S_I increases with undercooling, possibly because it is difficult to avoid some transformation during cooling to the intended transformation temperature. This is also apparent in the data for the pearlite colony size (Figure 2.15b). There is no clear story on why the colony size should decrease with S_I; indeed, there is no theoretical interpretation of colony size, it does not represent a nucleation event, but presumably, a branching event. Should branching become easier at greater undercoolings or as the thickness of the cementite lamellae decreases? There is some correlation between the colony size and S_I, Figure 2.15c. For a fixed nodule size, a larger interlamellar spacing must lead to fewer and bigger colonies.

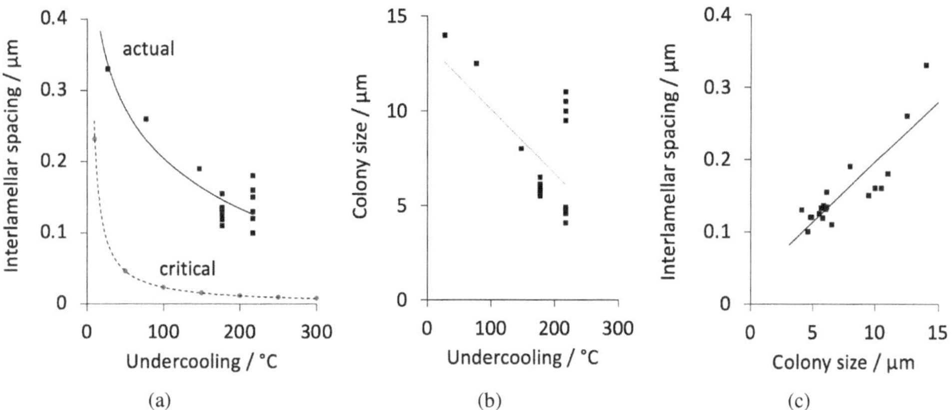

Figure 2.15 Eutectoid steels. (a) Measured variation in interlamellar spacing (labelled 'actual') with undercooling during isothermal transformation below the eutectoid temperature. Also presented is the calculated critical interlamellar-spacing for Fe-C. (b) Pearlite colony size versus the undercooling. (c) Correlation between S_I and $\overline{L}_{colony}$. The curves are all best fits. Selected data from [36–38].

As transformation progresses, physical impingement between nodules means that the spheroidal shapes evolve into more complex morphologies consistent with space-filling. Figure 2.11b,c shows low magnification images where large nodules of irregular shape are visible in the orientation image. It is probable that the crystallographic grain size defined by some *arbitrary* level of misorientation within, should better correlate with mechanical parameters such as toughness at low temperatures. The emphasis on the word arbitrary is because:

- A large misorientation needs to be specified with five degrees of freedom, including a unit vector to represent the axis of rotation, angle of rotation about this axis, and another unit vector that is normal to the interface between the misoriented regions. Each unit vector has just two independent components since the magnitude is known. [e.g., 39].
- The deflection of a crack on encountering a crystallographic discontinuity, depends on whether the cleavage planes on either side of a boundary are connected by a common line in the interface, or whether they have a relative twist, Figure 2.16. If the crack is diverted to a different path by the interface, the toughness should increase [40] because of the roughening of the crack path,

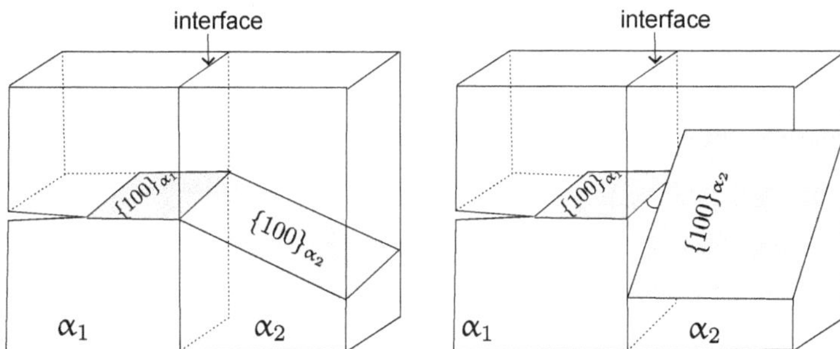

Figure 2.16 Schematic illustration of the propagation of a cleavage crack within the majority phase α within pearlite that is affected by the misorientation between α_1 and α_2, with their interface plane defined. In one case, the crack is tilted about a line common to both crystals, and in the other, the cleavage planes in the two crystals are relatively twisted. Diagram adapted from [40].

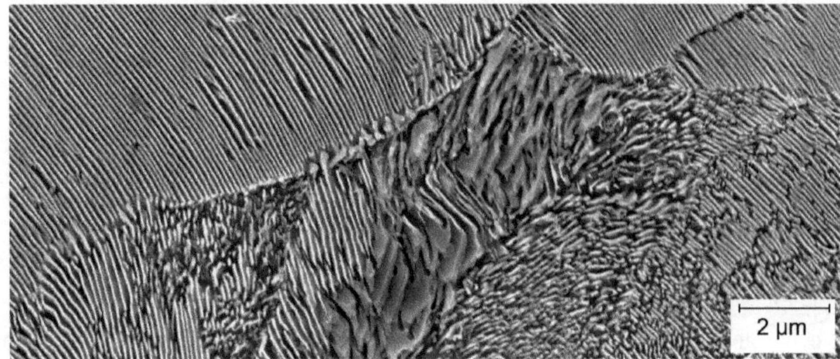

Figure 2.17 Pearlite in a eutectoid steel. Structure at junction between nodules of pearlite. Image courtesy of Apparao Chintha.

and change in stress experienced at the new crack orientation. An α_P/α_P misorientation of 15° between nodules is large, capable of hindering crack propagation in both the scenarios illustrated.

Given these mechanisms, the crystallographic grain-size appropriate for cleavage may be better defined by the continuity of $\{100\}_{\alpha_P}$ planes in neighbouring crystals [41].

- Nodule boundaries are known to retard crack propagation with the cleavage facet size correlating with $\overline{L}_{\text{nodule}}$, but each such facet contains many pearlite colonies [27, 42]. Not surprising since the nodule, irrespective of the direction of lamellae in colonies, is as a whole, a bi-crystal.

Impingement between nodules has other microstructural consequences. The carbon diffusion-field ahead of the γ/P interface extends a distance $\approx S_I$ so different pearlite nodules will interfere by the overlap of diffusion fields when they are within about $2S_I$. The intervening austenite, which should be richer in carbon than the average concentration, then tends to decompose into globular cementite or the ends of the cementite lamellae become bulbous, Figure 2.17. Whether or not this influences mechanical properties remains to be investigated.

In discussing Figure 2.13, it was observed that the pearlite nodule size becomes less sensitive to the austenite grain size when the latter is greater than about 100 μm. This is because the regions remote from γ/γ boundary do not contain easy nucleation sites. However, if a stress is applied during the transformation of coarse γ-grains, the resultant microplasticity stimulates the nucleation rate, resulting in a refinement of the nodule size, Figure 2.18a,b.

On the other hand, the stress has little effect on nodules produced when the austenite grain size is small, because as stated on page 16, each grain transforms to just one nodule because the distance available for growth is limited, Figure 2.18c,d.

2.4 COMPLEX CONFIGURATIONS

Even when the three-dimensional form of pearlite is accounted for, there are peculiarities that are oddities to understand but frequently ignored. The hot-rolled eutectoid steel illustrated in Figure 2.19 exhibits non-parallel formations of adjacent cementite-lamellae. The top right-hand corner shows structure in which coarse, parallel lamellae have finer, differently oriented and apparently discontinuous cementite lamellae in the space between the coarse lamellae. Quite unlike the common perception of pearlite consisting of parallel lamellae.

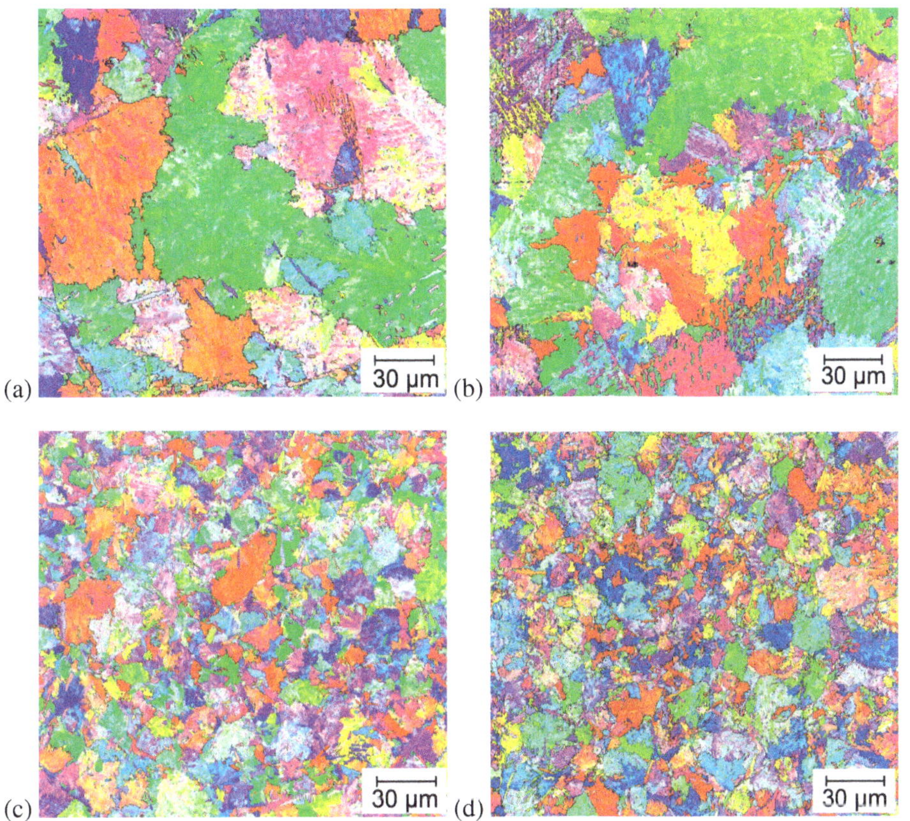

Figure 2.18 Crystallographic orientation images showing regions with a misorientation greater than 15°. The images show the influence of a uniaxial, compressive stress applied during the transformation of austenite into pearlite at 550 °C, on the size of pearlite nodules, for two different austenite grain sizes. (a) $\overline{L}_\gamma = 173\,\mu m$, $\sigma = 0\,MPa$. (b) $\overline{L}_\gamma = 173\,\mu m$, $\sigma = -85\,MPa$. (c) $\overline{L}_\gamma = 18\,\mu m$, $\sigma = 0\,MPa$. (d) $\overline{L}_\gamma = 18\,\mu m$, $\sigma = -100\,MPa$. Note that the yield strength of austenite at 550 °C in a typical eutectoid steel is 105 MPa [43]. Images courtesy of Rintaro Ueji.

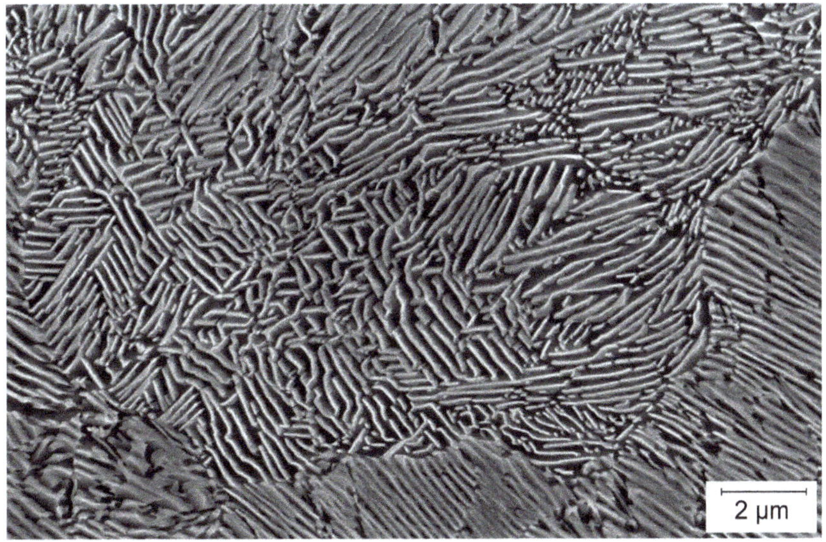

Figure 2.19 Complicated arrangement of cementite that does not resemble the common perception of pearlite. The image is from a fully transformed eutectoid steel, courtesy of Apparao Chintha.

2.5 ALIGNMENT OF UNDEFORMED PEARLITE

The collective directions of the cementite lamellae within pearlite colonies should be random relative to the sample axes. The γ-boundaries from which pearlite initiates also present a random network within the sample. Nevertheless, in unstrained wires, the direction of the lamellae aligns roughly parallel to the wire axis [44, 45], particularly in narrow wires and those transformed at low temperatures, Figure 2.20a,b.

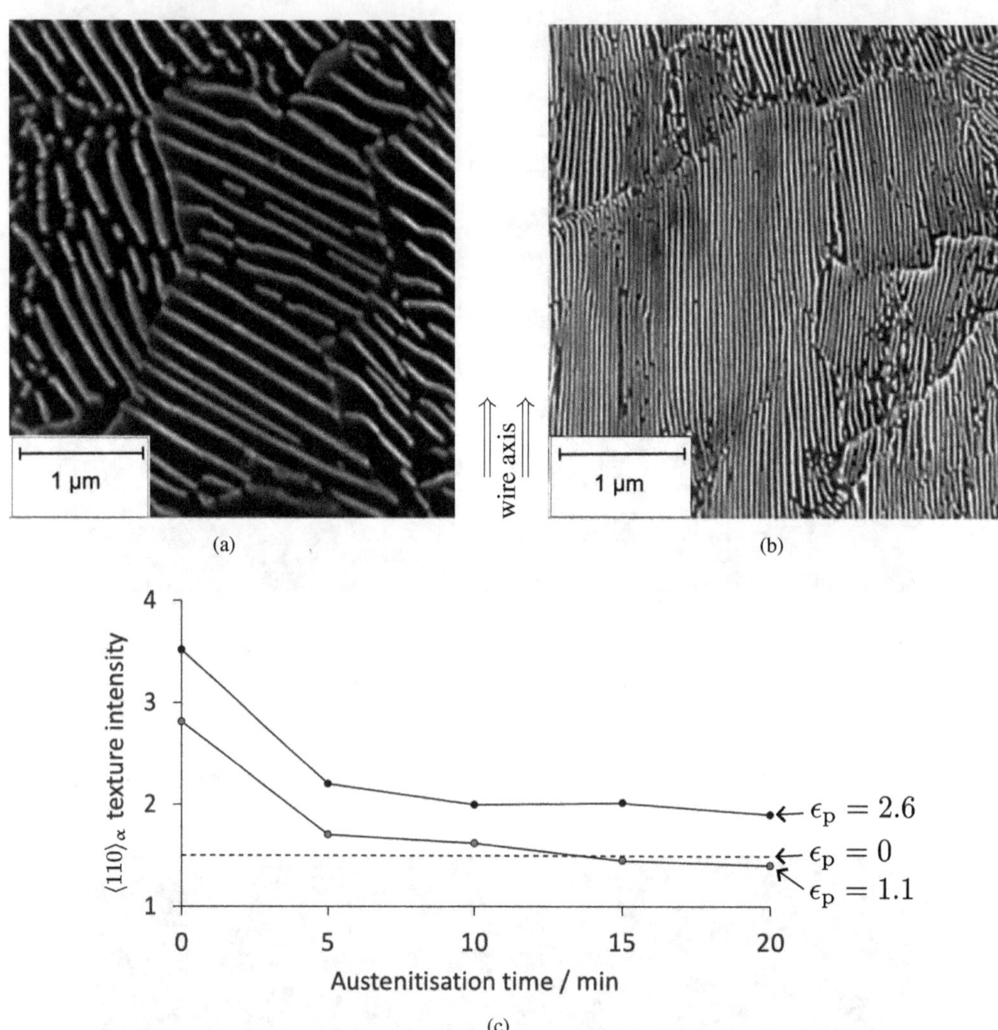

Figure 2.20 Pearlitic wires that are austenitised at 1000 °C for 60 s and transformed without deformation. The wire axis in the images is vertical. (a) Wire diameter 5.5 mm. (b) Wire diameter 1.6 mm. The interlamellar spacing of this sample is smaller because of the greater cooling rate experienced during cooling from austenite. Images reproduced from Durgaprasad et al. [44] with permission of Elsevier. (c) The texture intensity in wire drawn to the plastic strains indicated, followed by austenitisation at 820 °C for the time periods indicated. Selected data from Fang et al. [46].

The reason for the ordering of the lamellae directions along the wire axis is not well-understood. Wire drawing introduces a $\langle 110 \rangle_\alpha$ crystallographic texture that is recovered following austenitisation and transformation to pearlite. The wire in effect retains a *memory* of its original texture. The lamellar alignment is then attributed to the low elastic-modulus along $\langle 100 \rangle_\alpha$, which promotes the cementite to grow favourably along that direction more than any other. Prolonged austenitisation does weaken the original texture, the extent depending on the starting intensity, Figure 2.20c.

The texture memory is a feature of many reconstructive transformations that occur in steels. Ferrite nucleates at γ/γ boundaries in a crystal orientation that optimises its fit with both of the adjacent grains. The $\alpha \to \gamma$ transformation then leads to a recreation of the γ/γ boundaries and of the ferrite texture on cooling [47].

The thinner wire in Figure 2.20b exhibited stronger alignment, because of its greater cooling rate and the larger plastic strain (and hence, texture intensity) required to make the wire thin. This anisotropic distribution of cementite lamellae appears also in patented wire that has yet to be drawn.

Aligned cementite colonies will exhibit anisotropic properties – the oxidation rate, for example, is reduced because of the better fit at the α/θ interfaces [48]. When patented wire exhibits alignment prior to deformation, the fragmentation of cementite during wire drawing is reduced.

2.6 MAGNETIC PROPERTIES AND STRUCTURE

The Curie temperature of α-iron is 769 °C, with the magnetic moment per iron atom of 2.2 Bohr magnetons (μ_B) [49]. That of cementite is $\approx$ 186 °C with an average magnetisation per atom at 0 K of 1.86 μ_B [50]. When a magnetic field is applied to steel, a hysteresis loop is obtained showing how the magnetic flux density varies as the applied magnetic field is increased and then reduced, Figure 2.21a. The saturation magnetisation is illustrated for large H and the magnetic coercivity H_c is the amount of reverse field required to reduce the flux density to zero once the material has been fully magnetised. A low coercivity means that the material can easily be magnetised and de-magnetised, make it *magnetically soft*, associated with small hysteresis losses and hence suitable for applications such as electrical transformers and solenoids. On the other hand, a large coercivity helps make the material a 'permanent' magnet.

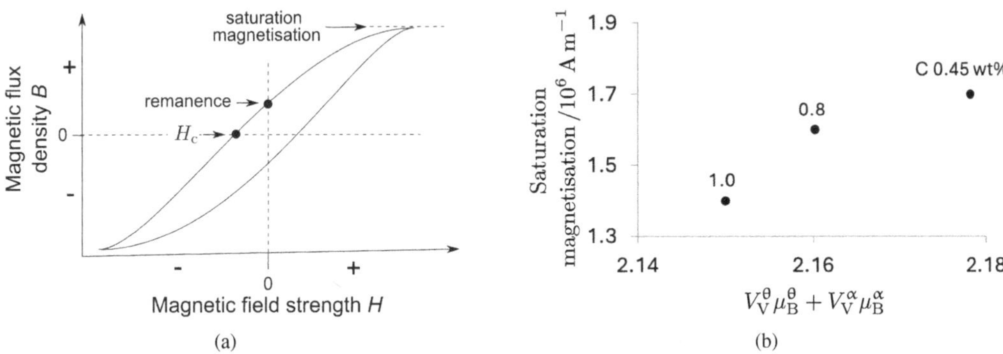

Figure 2.21 (a) Typical hysteresis curve showing how the magnetic flux density of a sample expressed (B) in Teslas varies as the magnetic field strength (H, A m^{-1}) is increased and then decreased. H_c is the magnetic coercivity, representing the ability of the ferromagnetic material to avoid demagnetisation due to an external field. Remanence is the residual magnetisation in the material when the field strength is zero. (b) A plot of the measured saturation magnetisation of mixtures of ferrite and cementite (selected data from [51]) plotted against the calculated number of Bohr magnetons per atom as the weighted average of the mixtures. The carbon concentrations of the steels ranged from 0.45 to 1 wt%, with the higher carbon steels were predominantly pearlitic.

The overall magnetic properties under ambient conditions should scale with the properties of each phase within pearlite; Figure 2.21b shows that the saturation magnetisation of mixtures of ferrite and cementite [51] scales with the average number of Bohr magnetons per atom calculated using the volume fractions of the two phases. The coercivity is a linear function of the carbon concentration of the steel in mixtures of α+P, the measurements covering structures with small fractions of pearlite to those that are fully pearlitic [52, 53]:

$$H_c\,/\,\mathrm{kA\,m^{-1}} = 1.186 w_C + 0.237 w_{Mn} \approx 8V_V^P \tag{2.7}$$

where the relationship with the volume fraction of pearlite has been derived assuming the equilibrium lever rule but neglecting the role of manganese.

The coercivity of pearlite is sensitive to temperature and affected by its lamellar structure. At $T > T_C^\theta$, the cementite acts as a nonmagnetic inclusion so the temperature dependence of the coercivity follows that of the ferrite, Figure 2.22, decreasing to zero on reaching T_E.

Below the Curie temperatures of both α and θ, complications arise due to differences in their crystal structures. The easy magnetisation direction for cementite is $[001]_\theta$, which is not parallel to the easy directions in ferrite [54]. For $T \ll T_C^\theta$, because of its strong anisotropy, the cementite maintains its easy direction whereas that of ferrite is pulled away from easy orientations by magnetostatic interactions with the cementite. With increasing temperature, the ferrite components rotate towards magnetostatically favoured direction while that of cementite is initially pulled away from $[001]_\theta$ but reverts to that as T_C^θ is approached, leading to a maximum in H_c. Beyond the maximum, the cementite is no longer ferromagnetic so the temperature dependence is due to ferrite alone.

The role of the lamellar shape is revealed by studying the same steel but with the cementite in a spheroidised condition, in which case, not only is the absolute value of H_c reduced but the behaviour just below T_C^θ becomes smoother. This presumably is because the proximity of cementite and ferrite is reduced in the spheroidised form.

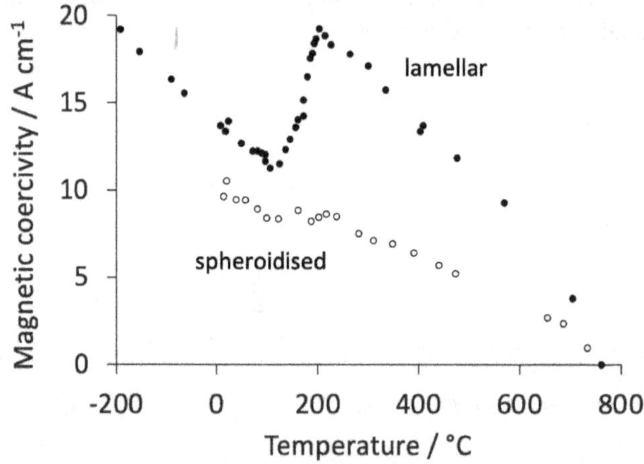

Figure 2.22 Magnetic coercivity of lamellar pearlite and a corresponding spheroidised sample as a function of temperature. The steel composition is Fe-0.9C-0.45Mn-0.23Si wt%. Selected data from Schwerer et al. [54]; ambient temperature datum from [51] for a predominantly pearlitic microstructure revealed $H_c = 10.7\,\mathrm{A\,cm^{-1}}$, consistent with the data plotted.

2.6.1 STRUCTURE RELATED TO MAGNETIC STRUCTURE + PROPERTIES

Ferromagnetic iron contains a domain structure that reduces the extent of its external magnetic field in order to minimise energy. The magnetic moments point in the same direction within a domain but not parallel to those in adjacent domains. Therefore, the direction of spin must change at a domain boundary, but the change is gradual over a distance of about $z_{ex} = 32$ nm in the case of iron [55]. This distance, which is the width of the transition between domains, is known as the 'exchange length' z_{ex}, that is dependent on the magnetic exchange and magnetocrystalline anisotropy. This latter term hinders the reorientation of magnetic moments from the easy orientation.

Refining the grain size ($\overline{L}$) leads to a greater coercivity because the domain boundaries then have to cope with a change in the easy magnetic moment direction across the grain boundaries. The coercivity H_c varies with $\overline{L}^{-1}$. This is why magnetically-soft iron-silicon alloys that are designed to enable huge ferrite grains have a low coercivity.

If the grain size becomes less than the exchange length z_{ex}, which therefore spans many grains, the coercivity decreases dramatically with $\overline{L}$, Figure 2.23. This is because the grain boundaries are so closely spaced that magnetocrystalline anisotropy of the small, randomly oriented grains is smoothed out over z_{ex} to a low average net-anisotropy [56, 57], sometimes designated the 'random anisotropy regime'.

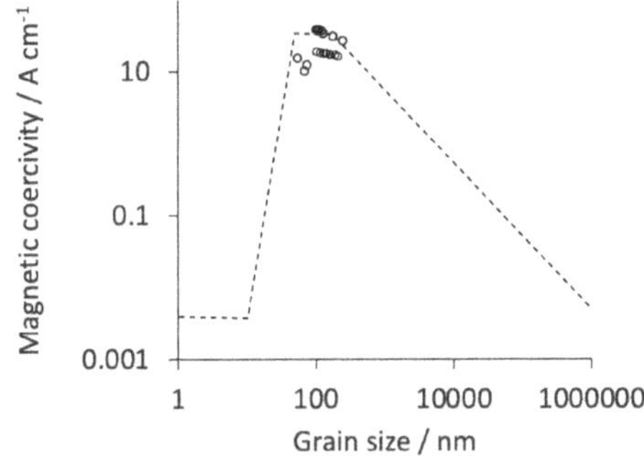

Figure 2.23 Coercivity as a function of grain size. The dashed line is from Herzer [56], based on a variety of amorphous, nanocrystalline, Fe-Si, Ni-Fe and Permalloy data (all omitted here for clarity). The points represent the measured coercivity values of fully pearlitic steels [58–61], where the interlamellar spacing has been taken here to represent the grain size.

Judging from the pearlite data in Figure 2.23, most common steel microstructures are not in the regime where $\overline{L} \leq z_{ex}$. Mechanically hard steels will therefore have a greater coercivity if the hardening mechanism is based on an increase in the number density of obstacles, which in turn interfere with domain-wall motion. Figure 2.24 shows that in general, H_c increases with hardness for a variety of microstructures generated in the same high-carbon steel.

There is an exception plotted in Figure 2.24, the open point representing the highest hardness and a particularly small coercivity – the fully martensitic sample has not been tempered but contains finely spaced (2-3 nm) transformation twins [62], that are an inherent feature of the crystallography of such martensite [63–65]. The easy magnetisation directions within adjacent twins are not parallel, thus reducing the effective grain size relative to z_{ex}, explaining the reduction in coercivity relative to the other data.

There are circumstances in which *severe*, mechanical deformation can change the scale of the structure into the random anisotropy regime, causing the coercivity to decrease. This goes against intuition because plastic deformation introduces obstructive defects, but a point is reached where the spacing between defects becomes small relative to z_{ex}. Coupled with other phenomena, this leads to some interesting outcomes for plastically deformed pearlite:

- Small plastic strains increase coercivity because of an increase in number density of defects that interfere with domain wall motion [55].

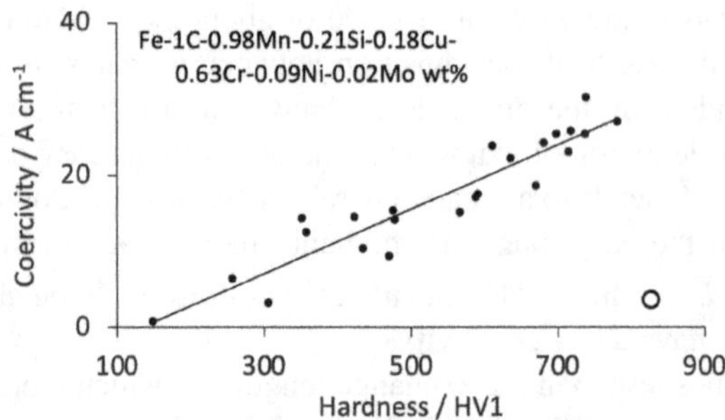

Figure 2.24 Increase in coercivity with the hardness for a variety of microstructures, P, α_b, α', γ_r, alone or in mixtures, in the same steel, austenitised at an identical temperature in all instances. The line does not represent the open circle which is for an untempered fully-martensitic condition. Selected data from Babasafari et al. [61].

- Deformation can induce the cementite in pearlite to fragment, causing a drop in coercivity; $[\varepsilon_p = 0, H_c = 12.8\,\mathrm{A\,cm^{-1}}]$, $[\varepsilon_p = 0.2, H_c = 16.4]$, $[\varepsilon_p = 3.6, H_c = 10.6]$ [66].
- At very large strains, the cementite dissolve into ferrite that now has an effective grain size of $\approx 10\,\mathrm{nm}$, which puts it into the random anisotropy regime. As a result the coercivity decreases to $[\varepsilon_p = 5.4, H_c = 4\,\mathrm{A\,cm^{-1}}]$.
- The saturation magnetisation of ferrite increases slightly when carbon dissolves in it, the magnetic moment of iron atoms is enhanced by $0.02\,\mu_B$ of magnetisation per atomic percent of dissolved carbon [67].
- Finally, it is useful that the saturation magnetisation of pearlite can be estimated from the weighted mean of those for ferrite and cementite, the weighting based on the phase fractions with the pearlite, as illustrated in Figure 2.21b and the data listed in Table 2.1.

Table 2.1
Saturation magnetisation data. That of pearlite can be obtained by a weighted average of the values for ferrite and cementite.

Structure	Saturation magnetisation / $\mathrm{A\,m^{-1}}$	Reference
α-ferrite	1.76×10^6	[p. 597, 68]
Cementite (θ)	0.48×10^6	[69]
Fe-0.77C wt% $\alpha + \theta$, pearlite	1.57×10^6	[55]

2.7 STRUCTURE INFLUENCED BY MAGNETIC FIELDS

The response of a material to a magnetic field $H(\mathrm{A\,m^{-1}})$ is its magnetisation J_m, representing the magnetic moment per unit volume of material in units of Teslas. The total flux of magnetic field lines per unit area is known as the magnetic flux density B with these quantities related as

$$B = \mu_\circ (H + J_m) \tag{2.8}$$

where $\mu_{\circ}$ is the permeability of free space.

The thermodynamic parameters of a transformation are expected to be influenced by an applied magnetic field if the parent and product phases have different magnetic characteristics. The quantity of ferrite increases transforming austenite under the influence of a mild magnetic field of just 1.2 T [70]. Suppose that the parent is paramagnetic and the product ferromagnetic, then the energy of the latter would be reduced by $H\Delta J_{\mathrm{m}}^{\mathrm{sat}}\,\mathrm{J\,m^{-3}}$, where $J_{\mathrm{m}}^{\mathrm{sat}}$ is the saturation magnetisation [71]. This is why the $\gamma \rightarrow \alpha'$ martensitic transformation in steels is promoted by applying a magnetic field. Our interest here is in pearlite, which is a mixture of two phases, each of which has a paramagnetic to ferromagnetic transition at a designated Curie temperature T_{C}.

One way of addressing this is to take the Bohr magneton $\mu_{\mathrm{B}} = 9.3 \times 10^{-24}\,\mathrm{J\,T^{-1}}$, multiply by Avogadro's number and the magnetic moment per iron atom in ferrite, to obtain a reduction in the free energy of ferrite $12\,\mathrm{J\,mol^{-1}\,T^{-1}}$ [72]. Figure 2.25 illustrates how the equilibrium phase diagram changes when a particular alloy is subjected to a large magnetic field of 30 T. The stability of the ferrite is dramatically increased with a tendency to grow in austenite even at temperatures exceeding 900 °C in alloyed steel; note the effect of the magnetic field on the free energy of cementite is not included in the calculations because the transformation temperatures concerned are well above T_{C}^{θ}.

The alloy, Fe-0.78C-1.6Si-2.02Mn-0.24Mo-1.01Cr-3.87Co-1.37Al wt%, ordinarily has such a large hardenability that after austenitisation at 1000 °C followed by cooling at $5\,°\mathrm{C\,s^{-1}}$, it transforms into a mixture of bainite and martensite, Figure 2.26a. Under identical cooling conditions but with an externally applied magnetic field of 30 T, it transforms instead into extremely fine pearlite, as illustrated in Figure 2.26b,c [72]. The magnetic field has accelerated the pearlite transformation because of its content of ferromagnetic ferrite. The minimum in the γ phase field has shifted to a greater carbon concentration, Figure 2.25b; this minimum is often identified as a eutectoid temperature, but the multicomponent nature of the alloy makes the eutectoid reaction occur over a range of temperatures within the $\alpha + \gamma + \theta$ phase field. In hypereutectoid steels, this should lead to a decrease in the proeutectoid cementite content in continuously cooled samples, as confirmed for a Fe-1C-0.22Si-0.2Mn wt% steel transformed under 12 T of magnetic field [73] and by calculations for the Fe-C system [74, 75].

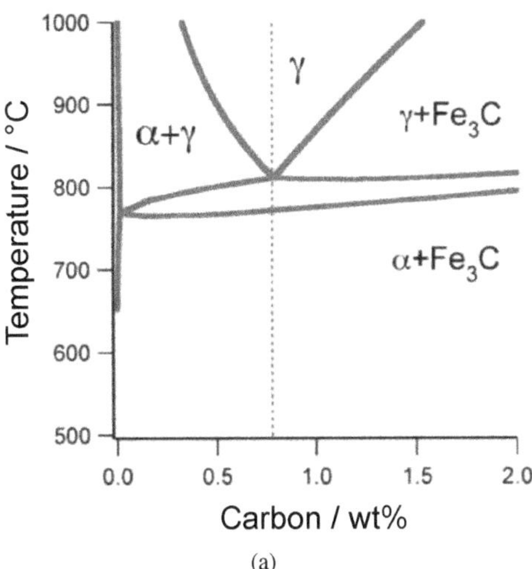

(a)

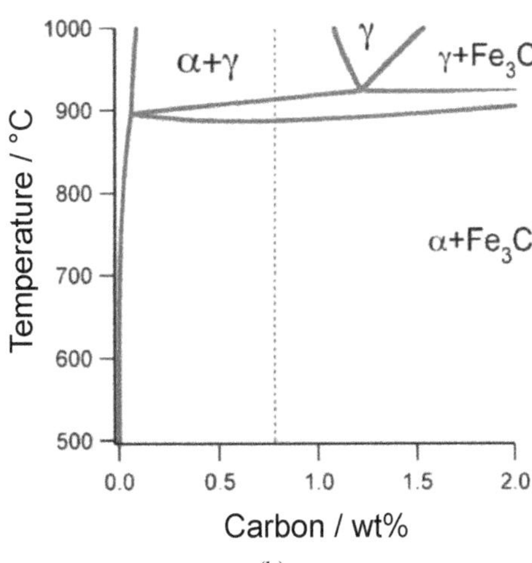

(b)

Figure 2.25 Calculated, vertical sections of Fe-0.78C-1.6Si-2.02Mn-0.24Mo-1.01Cr-3.87Co-1.37Al wt% phase diagram. (a) For transformation without the influence of an external magnetic field. (b) For transformation under the influence of an applied 30 T magnetic field, with the ferrite free energy reduced by $360\,\mathrm{J\,mol^{-1}}$ relative to the calculations in (a).

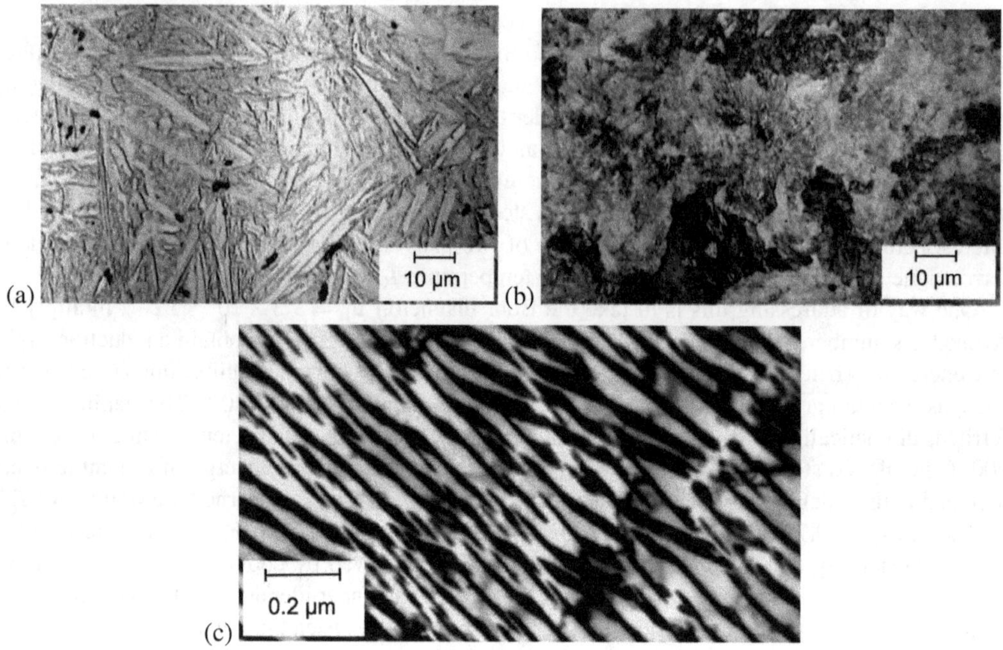

Figure 2.26 Fe-0.78C-1.6Si-2.02Mn-0.24Mo-1.01Cr-3.87Co-1.37Al wt%. (a) Transformed without magnetic field, showing a mixture of bainite and martensite plates. (b) Transformed under the influence of 30 T field, showing fine pearlite. (c) Higher resolution image of sample (b). After Jaramillo et al. [72].

2.7.1 FIELD-INDUCED ALIGNMENT IN α+P STRUCTURES

There are repeated observations that when a chemically homogeneous, hypoeutectoid steel is austenitised and then transformed into ferrite while under the influence of an applied magnetic field, the microstructure becomes *aligned*, Figure 2.27. The ferrite forms first with the carbon-enriched residual austenite then transforming into pearlite. This pearlite appears aligned because it forms in the spaces between the anisotropic ferrite. It is known that the aligned ferrite does not exhibit crystallographic texture [76]. The microstructure that develops in the absence of the applied field is isotropic with a dispersion of equiaxed ferrite and pearlite regions.

The columns of ferrite are in fact polycrystalline (Figure 2.27b) with the equiaxed grains stacked on top of each other. The structure therefore is not dependent on anisotropic growth. The phenomenon has been attributed to magnetic interactions between particles nucleated at independent locations; those that share the same direction of magnetisation are energetically favoured [76]. The potential energy U_m of interaction between two particles with magnetic dipole moments $\mathbf{m}_1$ and $\mathbf{m}_2$, with their centres located a vector $\mathbf{r}$ apart (unit vector $\hat{\mathbf{r}}$) is, with a term that vanishes everywhere except the origin not included, given by [77]

$$U_m = -\frac{\mu_\circ}{4\pi|\mathbf{r}|^3}[3(\mathbf{m}_1 \cdot \hat{\mathbf{r}})(\mathbf{m}_2 \cdot \hat{\mathbf{r}}) - \mathbf{m}_1 \cdot \mathbf{m}_2] \tag{2.9}$$

where $\mu_\circ$ is the magnetic permeability of free space. The reduction in energy is greatest if $\mathbf{r} \parallel \mathbf{m}_1 \parallel \mathbf{m}_2$. This would favour grains with a like magnetic moment to form with their centres on top of each other, consistent with the structure illustrated in Figure 2.27b.

In a highly alloyed steel (Fe-13Mn-1C wt%) where pearlite forms slowly, a 10 Tesla field led to a more than doubling of the number density of pearlite nodules without noticeably affecting the growth rate [78]. The enhanced driving force for transformation due to the magnetic field had a larger effect on nucleation, which in this alloy is slow. The maximum amount of transformation achieved was $V_V^P \approx 0.2$ with most nucleation events taking place at the austenite grain boundaries with particles isolated from each other, without perceptible alignment of the pearlite along the magnetic field.

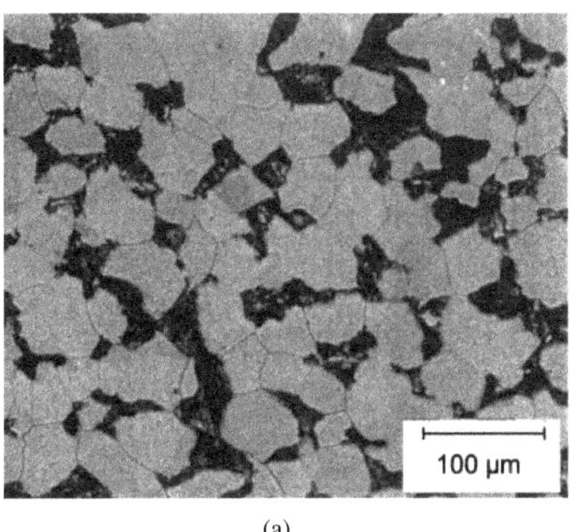

(a)

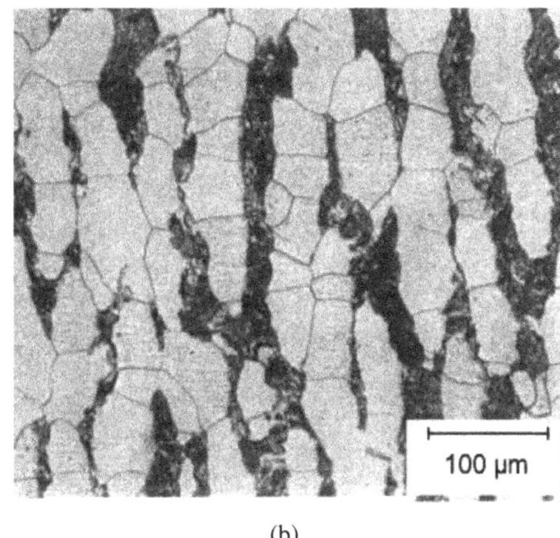

(b)

Figure 2.27 The transformation of Fe-0.4C wt% steel, austenitised at 950 °C, cooled rapidly to 850 °C, then at 5 °C min^{-1} from 700°C. (a) Without a magnetic field. (b) Under the influence of a 10 T magnetic field that is oriented vertically. Note that each column of ferrite is polycrystalline. Images courtesy of Hideyuki Ohtsuka, based on the work published in [79].

Finally, there is an unexplained increase in the hardness of pearlite ($\approx 384 \rightarrow 410\,\text{HV}$) during the transformation of austenite in a gradient of magnetic field, $\approx 50\,\text{T}\,\text{m}^{-1}$, speculated to be due to an increase in the carbon concentration within the pearlitic-ferrite [80].

REFERENCES

1. L. E. Samuels: Light Microscopy of Carbon Steels: Metals Park, Ohio, USA: ASM International, 1999.
2. W. Wright: 'Iridescence', *Leonardo*, 1974, **7**, 325–328.
3. G. E. Pellissier, M. F. Hawkes, W. A. Johnson, and R. F. Mehl: 'The interlamellar spacing of pearlite', *Transactions of ASM*, 1942, **30**, 1049–1086.
4. S. A. Saltykov: Stereometricheskaya Metallografiya: 2nd ed., Moscow: Metallurgizdat Publishers (English translation US Department of Defence AD267700), 1958.
5. E. E. Underwood: 'Stereology, or the quantitative evaluation of microstructures', *Journal of Microscopy*, 1969, **89**, 161–180.
6. G. F. V. Vroot, and A. Roósz: 'Measurement of the interlamellar spacing of pearlite', *Metallography*, 1984, **17**, 1–17.
7. J. F. Tilbury, T. D. Mottishaw, and G. D. W. Smith: 'Measurements of the spread in pearlite spacings', *Metallography*, 1986, **19**, 243–246.
8. G. Zhang, and M. Enomoto: 'Interlamellar spacing of pearlite in a near-eutectoid Fe-C alloy measured by serial sectioning', *ISIJ International*, 2009, **49**, 921–927.
9. E. B. Hawbolt, B. Chau, and J. K. Brimacombe: 'Kinetics of austenite-pearlite transformation in eutectoid carbon steel', *Metallurgical Transactions A*, 1983, **14**, 1803–1815.
10. A. Roósz, Z. Gácsi, and M. K. Baan: 'A simple method for determining the true interlamellar spacing', *Metallography*, 1980, **13**, 299–306.

11. C. Capdevila, F. G. Caballero, and C. García de Andrés: 'Neural network model for isothermal pearlite transformation. part I: Interlamellar spacing', *ISIJ International*, 2005, **45**, 229–237.
12. G. F. V. Vroot: 'The interlamellar spacing of pearlite', *Practical Metallography*, 2015, **52**, 419–436.
13. J. E. Hilliard: 'Stereology: an experimental viewpoint', *Advances in Applied Probability*, 1972, **4**, 92–111.
14. R. T. DeHoff, and F. N. Rhines: Quantitative Microscopy: New York, USA: McGraw Hill, 1968.
15. H. S. Fong: 'Determining true pearlite lamellar spacings from observed apparent spacings', *Metallography*, 1989, **23**, 173–188.
16. O. Forsman: 'Undersökning av rymdstrukturen hos ett kolstal av hypereutektoid sammansättning', *Jernkontorets Annaler*, 1918, **102**, 1–30.
17. L. S. Darken, and R. M. Fisher: 'Some observations on the growth of pearlite', In: V. F. Zackay, and H. I. Aaronson, eds. *Decomposition of austenite by diffusional processes, Interscience publishers*. 1962:249–294.
18. M. Hillert: 'The formation of pearlite', In: V. F. Zackay, and H. I. Aaronson, eds. *Decomposition of Austenite by Diffusional Processes*. New York, USA: Interscience, 1962:197–237.
19. Y. Adachi, S. Morooka, K. Nakajima, and Y. Sugimoto: 'Computer-aided three-dimensional visualization of twisted cementite lamellae in eutectoid steel', *Acta Materialia*, 2008, **56**, 5995–6002.
20. Y. T. Wang, Y. Adachi, K. Nakajima, and Y. Sugimoto: 'Quantitative three-dimensional characterization of pearlite spheroidization', *Acta Materialia*, 2010, **58**, 4849–4858.
21. Y. T. Wang, Y. Adachi, K. Nakajima, and Y. Sugimoto: 'Topology and differential geometry-based three-dimensional characterization of pearlite spheroidization', *ISIJ International*, 2012, **52**, 697–703.
22. R. J. Dippenaar, and R. W. K. Honeycombe: 'The crystallography and nucleation of pearlite', *Proceedings of the Royal Society A*, 1973, **333**, 455–467.
23. S. W. Thompson, and P. R. Howell: 'On the early stages of pearlite formation in hypoeutectoid steels', *Scripta Metallurgica*, 1988, **22**, 1775–1778.
24. N. Nakada, N. Koga, T. Tsuchiyama, and S. Takaki: 'Crystallographic orientation rotation and internal stress in pearlite colony', *Scripta Materialia*, 2009, **61**, 133–136.
25. M. Koga, E. C. Santos, T. Honda, K. Kida, and T. Shibukawa: 'Observation of wear in induction-heat-treated bearing steel bars under reciprocating motion', *Advanced Materials Research*, 2012, **457-458**, 504–510.
26. S. N. Doi, and H. J. Kestenbach: 'Determination of the pearlite nodule size in eutectoid steels', *Metallography*, 1989, **23**, 135–146.
27. S. Zhou, Y. Zuo, Z. Li, X. Wang, and Q. Yong: 'Microstructural analysis on cleavage fracture in pearlitic steels', *Materials Characterization*, 2016, **119**, 110–113.
28. P. Xu, Y. Liang, J. Li, and C. Meng: 'Further improvement in ductility induced by the refined hierarchical structures of pearlite', *Materials Science & Engineering A*, 2019, **745**, 176–184.
29. A. F. Gourgues, H. M. Flower, and T. C. Lindley: 'Electron backscattering diffraction study of acicular ferrite, bainite, and martensite steel microstructures', *Materials Science and Technology*, 2000, **16**, 26–40.
30. P. Yan, O. E. Güngör, P. Thibaux, M. Liebeherr, and H. K. D. H. Bhadeshia: 'Tackling the toughness of steel pipes produced by high frequency induction welding and heat-treatment', *Materials Science & Engineering A*, 2011, **528**, 8492–8499.
31. M. Masoumi, N. B. D. Lima, G. Tressia, A. Sinatora, and H. Goldenstein: 'Microstructure and crystallographic orientation evolutions below the superficial white layer of a used pearlitic rail', *Journal of Materials Research and Technology*, 2019, **8**, 6275–6288.

32. Z. Guo, N. Kimura, S. Tagashira, T. Furuhara, and T. Maki: 'Kinetics and crystallography of intragranular pearlite transformation nucleated at (mns+ vc) complex precipitates in hypereutectoid fe-mn-c alloys', *ISIJ international*, 2002, **42**, 1033–1041.
33. S. Behera, R. K. Barik, M. B. Sk, R. Mitra, and D. Chakrabarti: 'Recipe for improving the impact toughness of high-strength pearlitic steel by controlling the cleavage cracking mechanisms', *Materials Science & Engineering A*, 2019, **764**, 138256.
34. A. R. Marder, and B. L. Bramfitt: 'The effect of morphology on the strength of pearlite', *Metallurgical Transactions A*, 1976, **7**, 365–372.
35. H. Pous-Romero, I. Lonardelli, D. Cogswell, and H. K. D. H. Bhadeshia: 'Austenite grain growth in a nuclear pressure vessel steel', *Materials Science & Engineering A*, 2013, **567**, 72–79.
36. J. J. Lewandowski, and A. W. Thompson: 'Effects of the prior austenite grain size on the ductility of fully pearlitic eutectoid steel', *Metallurgical Transactions A*, 1986, **17**, 461–472.
37. F. P. L. Kavishe, and T. J. Baker: 'Effect of prior austenite grain size and pearlite interlamellar spacing on strength and fracture toughness of a eutectoid rail steel', *Materials Science and Technology*, 1986, **2**, 816–822.
38. F. Zhang, Y. Zhao, Y. Tan, X. Ji, and S. Xiang: 'Study on the nucleation and growth of pearlite colony and impact toughness of eutectoid steel', *Metals*, 2019, **9**, 1133.
39. H. K. D. H. Bhadeshia: Geometry of Crystals, Polycrystals and Phase Transformations: Florida, USA: CRC Press, Taylor and Francis, 2017.
40. Y. Wei, H. Gao, and A. F. Bower: 'Numerical simulations of crack deflection at a twist-misoriented grain boundary between two ideally brittle crystals', *Journal of the Mechanics and Physics of Solids*, 2009, **57**, 1865–1879.
41. A. Ghosh, S. Kundu, and D. Chakrabarti: 'Effect of crystallographic texture on the cleavage fracture mechanism and effective grain size of ferritic steel', *Scripta Materialia*, 2014, **81**, 8–11.
42. Y. J. Park, and I. M. Bernstein: 'The process of crack initiation and effective grain size for cleavage fracture in pearlitic eutectoid steel', *Metallurgical Transactions A*, 1979, **10**, 1653–1664.
43. A. Eres-Castellanos, I. Toda-Caraballo, A. Latz, F. G. Caballero, and C. Garcia-Mateo: 'An integrated-model for austenite yield strength considering the influence of temperature and strain rate in lean steels', *Materials and Design*, 2020, **188**, 108435.
44. A. Durgaprasad, S. Giri, S. Lenka, S. Kundu, S. Mishra, S. Chandra, R. D. Doherty, and I. Samajdar: 'Defining a relationship between pearlite morphology and ferrite crystallographic orientation', *Acta Materialia*, 2017, **129**, 278–289.
45. A. Durgaprasad, S. Giri, S. Lenka, S. Kundu, S. Chandra, S. Mishra, R. D. Doherty, and I. Samajdar: 'Microstructural engineering in eutectoid steel: A technological possibility?', *Metallurgical & Materials Transactions A*, 2018, **49**, 1520–1535.
46. F. Fang, Y. Zhao, L. Zhou, X. Hu, Z. Xie, and J. Jiang: 'Texture inheritance of cold drawn pearlite steel wires after austenitization', *Materials Science and Engineering: A*, 2014, **618**, 505–510.
47. T. Tomida, M. Wakita, M. Yasuyama, S. Sugaya, Y. Tomota, and S. Vogel: 'Memory effects of transformation textures in steel and its prediction by the double kurdjumov–sachs relation', *Acta materialia*, 2013, **61**, 2828–2839.
48. H. K. Mehtani, M. I. Khan, A. Durgaprasad, S. K. Deb, S. Parida, M. J. N. V. Prasad, and I. Samajdar: 'Oxidation kinetics in pearlite: The defining role of interface crystallography', *Scripta Materialia*, 2018, **152**, 44–48.
49. W. J. Moore: Physical Chemistry: 5th ed., New Jersey, USA: Prentice-Hall, 1972.
50. A. Dick, F. Körmann, T. Hickel, and J. Neugebauer: '*Ab initio* based determination of thermodynamic properties of cementite including vibronic, magnetic, and electronic excitations', *Physical Review B*, 2011, **84**, 125101.

51. Y. V. Khudorozhkova, and E. E. Chermitsina: 'The effect of a dispersed grain structure of hypoeutectoid, eutectoid and hypereutectoid carbon steels on their magnetic behavior', In: *Mechanics, Resource and Diagnostics of Materials and Structures , AIP Conference Proceedings*. New York, USA: AIP Publishing, 2018:040042.
52. B. Tanner, J. Szpunar, S. Willcock, L. Morgan, and P. Mundell: 'Magnetic and metallurgical properties of high-tensile steels', *Journal of Materials Science*, 1988, **23**, 4534–4540.
53. S. Thompson, and B. Tanner: 'The magnetic properties of pearlitic steels as a function of carbon content', *Journal of Magnetism and Magnetic Materials*, 1993, **123**, 283–298.
54. F. C. Schwerer, C. E. Spangler, and J. F. Kelly: 'Temperature dependence of the magnetic coercivity of pearlite', *Acta Metalllurgica*, 1978, **26**, 579–589.
55. S. Wurster, M. Stückler, L. Weissitsch, H. Krenn, A. Hohenwarter, R. Pippan, and A. Bachmaier: 'Soft magnetic properties of ultra-strong and nanocrystalline pearlitic wires', *Nanomaterials*, 2021, **12**, 23.
56. G. Herzer: 'Grain size dependence of coercivity and permeability in nanocrystalline ferromagnets', *IEEE Transactions on magnetics*, 1990, **26**, 1397–1402.
57. G. Herzer: 'Soft magnetic nanocrystalline materials', *Scripta Metallurgica et Materialia*, 1995, **33**, 1741–1756.
58. C. Lo, J. Jakubovics, and C. Scruby: 'Monitoring the microstructure of pearlitic steels by magnetoacoustic emission', *Journal of applied physics*, 1997, **81**, 4069–4071.
59. C. Lo, C. Scruby, and G. Smith: 'Dependences of magnetic barkhausen emission and magnetoacoustic emission on the microstructure of pearlitic steel', *Philosophical Magazine*, 2004, **84**, 1821–1839.
60. J. W. Byeon, and S. Kwun: 'Evaluation of microstructures of variously heat treated carbon steel by magnetic coercivity measurement', *Physica Status Solidi (b)*, 2004, **241**, 1697–1700.
61. Z. Babasafari, A. V. Pan, F. Pahlevani, C. Kong, M. Du Toit, and R. Dippenaar: 'Effect of microstructural features on magnetic properties of high-carbon steel', *Metallurgical and Materials Transactions A*, 2021, **52**, 5107–5122.
62. P. M. Kelly, and J. Nutting: 'The martensite transformation in carbon steels', *Proceedings of the Royal Society of London*, 1960, **A259**, 45–58.
63. J. S. Bowles, and J. K. Mackenzie: 'The crystallography of martensite transformations, part I', *Acta Metallurgica*, 1954, **2**, 129–137.
64. M. S. Wechsler, D. S. Lieberman, and T. A. Read: 'On the theory of the formation of martensite', *Transactions of the AIME Journal of Metals*, 1953, **197**, 1503–1515.
65. C. M. Wayman: 'The crystallography of martensitic transformations in alloys of iron': In: H. Hermans, ed. *Advances in Materials Research*, vol. 3. John Wiley & Sons, Inc., 1968:147–304.
66. E. S. Gorkunov, S. V. Grachev, S. V. Smirnov, V. M. Somova, S. M. Zadvorkin, and L. E. Kar'kina: 'Relation of physical-mechanical properties to the structural condition of severely deformed patented carbon steels at drawing', *Russian Journal of Nondestructive Testing*, 2005, **41**, 65–79.
67. M. Cadeville, C. Lerner, and J. Friedt: 'Electronic structure of interstitial carbon in ferromagnetic transition metals prepared by splat-quenching', *Physica B+ C*, 1977, **86**, 432–434.
68. J. M. Coey: Magnetism and magnetic materials: Cambridge, U. K.: Cambridge university press, 2010.
69. E. Duman, M. Acet, T. Hülser, E. Wassermann, B. Rellinghaus, J. Itié, and P. Munsch: 'Large spontaneous magnetostrictive softening below the Curie temperature of fe3c invar particles', *Journal of Applied Physics*, 2004, **96**, 5668–5672.
70. V. N. Pustovoit, Y. Dombrovskii, and S. Grishin: 'Structure features of carbon steel pearlite', *Metallovedenie i Termicheskaya Obrabotka Metallov*, 1979, **11**, 22–24.

71. K. R. Satyanarayan, W. Eliasz, and A. P. Miodownik: 'The effect of a magnetic field on the martensite transformation in steels', *Acta Materialia*, 1968, **16**, 877–887.
72. R. A. Jaramillo, S. S. Babu, G. M. Ludtka, R. A. Kisner, J. B. Wilgen, G. Mackiewicz-Ludtka, D. M. Nicholson, S. M. Kelly, M. Murugananth, and H. K. D. H. Bhadeshia: 'Effect of 30 Tesla magnetic field on transformations in a novel bainitic steel', *Scripta Materialia*, 2004, **52**, 461–466.
73. Y. D. Zhang, C. Esling, M. L. Gong, G. Vincent, X. Zhao, and L. Zuo: 'Microstructural features induced by a high magnetic field in a hypereutectoid steel during austenitic decomposition', *Scripta Materialia*, 2006, **54**, 1897–1900.
74. H. D. Joo, S. U. Kim, N. S. Shin, and Y. M. Koo: 'An effect of high magnetic field on phase transformation in Fe-C system', *Materials Letters*, 2000, **43**, 225–229.
75. J.-K. Choi, H. Ohtsuka, Y. Xu, and W.-Y.Choo: 'Effects of a strong magnetic field on the phase stability of plain carbon steels', *Scripta Materialia*, 2000, **43**, 221–226.
76. M. Shimotomai, K. Maruta, K. Mine, and M. Matsui: 'Formation of aligned two-phase microstructures by applying a magnetic field during the austenite to ferrite transformation in steels', *Acta Materialia*, 2003, **51**, 2921–2932.
77. Anonymous: 'Magnetic dipole-dipole interaction': https://en.wikipedia.org/wiki/Magnetic_dipole-dipole_interaction, 2022.
78. Y. Xu, H. Ohtsuka, and H. Wada: 'Effects of high magnetic field on pearlite transformation behavior and structure', *Transactions of the Materials Research Society of Japan*, 2000, **25**, 509–512.
79. H. Ohtsuka, Y. Xu, and H. Wada: 'Alignment of ferrite grains during austenite to ferrite transformation in a high magnetic field', *Materials Transactions, JIM*, 2000, **41**, 907–918.
80. M. Shimotomai: 'Influence of magnetic field gradients on pearlite', *Materials Transactions*, 2003, **44**, 2524–2528.

3 Crystallography

3.1 CRYSTAL STRUCTURE OF CEMENTITE

The cementite structure is discussed in some detail because of its low-symmetry. Cementite has an orthorhombic unit cell with lattice parameters as $a = 0.50837$ nm, $b = 0.67475$ nm and $c = 0.45165$ nm, corresponding to the space group *Pnma*. There are twelve atoms of iron in the unit cell and four of carbon, Figure 3.1. Four of the iron atoms are located on mirror planes whereas the other eight are at general positions (point symmetry 1).

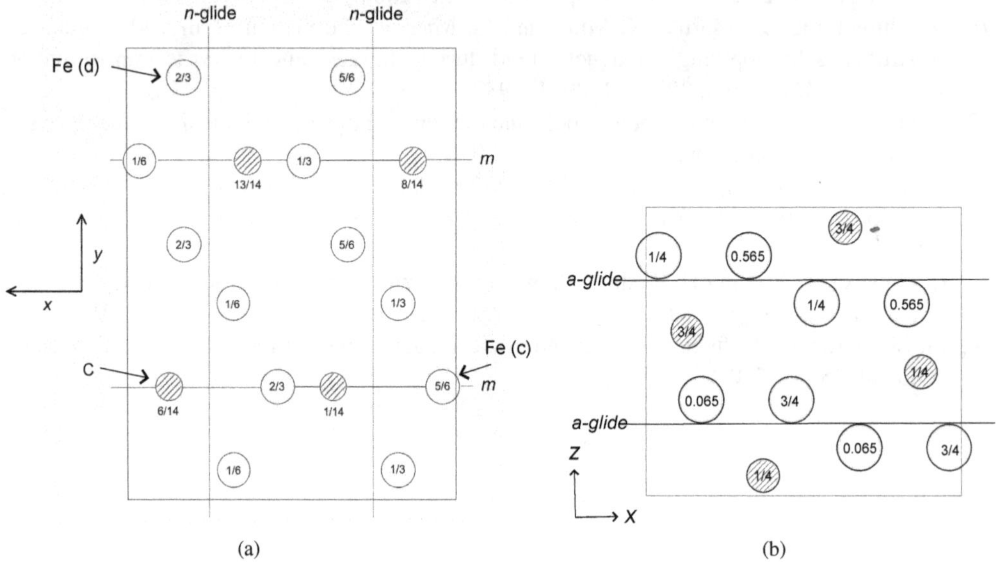

Figure 3.1 The crystal structure of cementite, consisting of twelve iron atoms and four carbon atoms, four Fe_3C formula units. (a) The fractional z coordinates of the atoms are marked. Four of the iron atoms (Wyckoff letter 'c') are located on mirror planes, whereas the others (Wyckoff letter 'd') are at general locations with point symmetry a monad. The pleated layers parallel to (100) are in ...*ABABAB*... stacking with carbon atoms occupying interstitial positions at the folds within the pleats, with all carbon atoms located on the mirror planes. (b) The fractional y coordinates are marked on this projection which illustrates the a-glide planes.

The lattice type is primitive (P). There are n-glide planes normal to the x-axis, at $\frac{1}{4}x$ and $\frac{3}{4}x$ involving translations of $\frac{b}{2} + \frac{c}{2}$. The mirror planes are normal to the y-axis, and a-glide planes normal to the z-axis, at heights $\frac{1}{4}z$ and $\frac{3}{4}z$ with fractional translations of $\frac{a}{2}$ parallel to the x-axis. The space group symbol is therefore *Pnma* [1].

3.1.1 MAGNETIC UNIT CELL

The space group of cementite is *Pnma* with point group *mmm*. There are eight magnetic point groups mmm, $m'mm$, $mm'm$, mmm', $\underline{m'm'm}$, $\underline{m'mm'}$, $\underline{mm'm'}$ and $m'm'm'$, where the primes signify the time reversal operations required to preserve the directions of magnetic spins. Only the three underlined are admissible in leaving one component of the magnetic spin invariant under operations of the point group. The component left invariant is that normal to the ordinary mirror plane. An

DOI: 10.1201/9781032631981-3

Table 3.1
Wyckoff positions for space group *Pnma* [2]. Each Wyckoff position in this space group is labelled with a letter (Table 3.1); thus, the eight iron atoms in general positions are labelled with the letter 'd', and the remaining four on mirror planes with the letter 'c'; the number preceding the letter, for example, the '8' in 8d, denotes the number of equivalent positions in the cell.

Multiplicity	Wyckoff letter	Site symmetry	Coordinates
8	d	1	(x,y,z) $(-x+\frac{1}{2},-y,z+\frac{1}{2})$ $(-x,y+\frac{1}{2},-z)$ $(x+\frac{1}{2},-y+\frac{1}{2},-z+\frac{1}{2})$ $(-x,-y,-z)$ $(x+\frac{1}{2},y,-z+\frac{1}{2})$ $(x,-y+\frac{1}{2},z)$ $(-x+\frac{1}{2},y+\frac{1}{2},z+\frac{1}{2})$
4	c	.m.	$(x,\frac{1}{4},z)$ $(-x+\frac{1}{2},\frac{3}{4},z+\frac{1}{2})$ $(-x,\frac{3}{4},-z)$ $(x+\frac{1}{2},\frac{1}{4},-z+\frac{1}{2})$
4	b	-1	$(0,0,\frac{1}{2})$ $(\frac{1}{2},0,0)$ $(0,\frac{1}{2},\frac{1}{2})$ $(\frac{1}{2},\frac{1}{2},0)$
4	a	-1	$(0,0,0)$ $(\frac{1}{2},0,\frac{1}{2})$ $(0,\frac{1}{2},0)$ $(\frac{1}{2},\frac{1}{2},\frac{1}{2})$

axial vector representing the magnetic moment parallel to that normal does not change direction on reflection through that mirror. But if that vector is parallel to a mirror plane, it switches to the opposite direction on reflection through that plane [3]. Therefore, time reversal is necessary to make it invariant to the other two mirrors with which it is parallel. The three underlined point groups thus are compatible with ferromagnetism.

Cementite is ferromagnetic below its Curie temperature of 186 °C. Neutron diffraction experiments indicate that the magnetic unit cell is not perceptibly different from the crystallographic unit cell [4]. The ferromagnetic state requires all the spins on the iron atoms to be aligned. The magnetic moments of the 8d iron atoms in general positions are not constrained by symmetry to point in a specific direction, but the 4c atoms that lie on mirror planes must have their moments normal to the mirror. The space groups corresponding to ferromagnetism are $Pn'ma'$, $Pnm'a'$ and $Pn'm'a$. The magnetic moments on the four iron atoms on mirror planes would need to lie parallel to $[010]_\theta$ for $Pn'ma'$ and in $(010)_\theta$ for the other two space groups. Based on small axial-ratio changes at the Curie temperature, it was concluded that the space group is likely to be $Pnm'a'$, in which case, the bulk magnetisation should be directed along $[100]_\theta$ [4]. However, experimental measurements made on a single-crystal of cementite do not appear consistent with this, because the easy magnetisation direction is $[001]_\theta$, Figure 3.2 [5, 6]; the magnetocrystalline anisotropy energy at 5 K was determined to be $334 \pm 20\,\text{kJ}\,\text{m}^{-3}$ [6].

Chromium is a common solute in steels and has a high solubility in cementite; because the magnetic moments of chromium and iron are aligned antiferromagnetically, the total magnetisation of cementite is reduced [7, 8].

3.2 STRUCTURAL DEFECTS AND DEFORMATION

The elastic moduli of cementite, like all crystals, are orientation-dependent. The shear modulus C_{44} is exceptionally small [9], some two times smaller than the corresponding term for aluminium. Nevertheless, the cementite has an exceptionally large ideal shear strength because elastic deformation

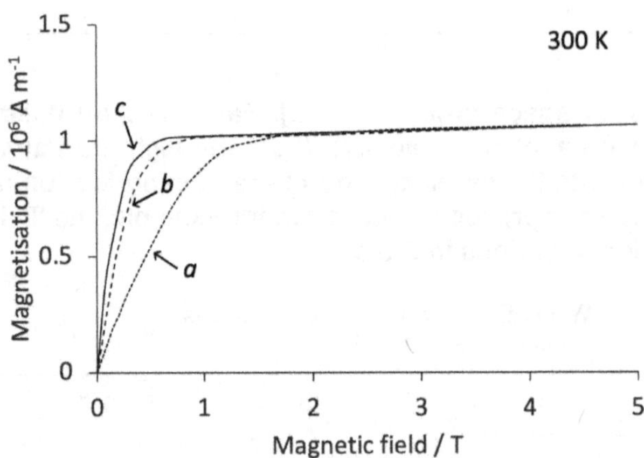

Figure 3.2 Magnetisation data for a single-crystal of cementite with lattice parameters *a*, *b* and *c* expressed in the *Pnma* space group. The measurements were made at 300 K but those at 5 K show a similar trend, albeit a greater saturation-magnetisation and magnetocrystalline anisotropy. Adapted from Yamamoto et al. [6]. The easy and hard magnetisation directions are along *c* and *a*, respectively.

reduces its symmetry from orthorhombic to monoclinic (space group $P2_1/c$), with an accompanying increase in three-dimensional covalent bonding that stiffens the material [10]. Thermal expansion too is a function of crystallographic orientation; when cementite in its polycrystalline state is subjected to a change in temperature, reversible strains develop due to the pronounced anisotropy in thermal expansion coefficients leading to the broadening of X-ray diffraction peaks [11, 12].

The experimentally observed slip systems in cementite include (001)[100], (100)[010], (100)[001], (010)[001] and (010)[100] [13]. Given the primitive nature of the lattice, it is assumed that the Burgers vectors of slip dislocations have magnitudes equal to the unit cell edges. These large vectors make slip difficult, rendering cementite a hard phase at ambient temperatures.

The common slip system appears to be (010)[001] [14]. The metal-metal bond is dominant between (010) planes so hardness depends on how solutes affect this bond strength [15]. Nickel weakens the metal-metal bond and hence reduces the hardness of cementite [16]. There may be other slip systems that operate when the cementite is forced to deform in a phase mixture such as pearlite [17]. There is a limited continuity between the slip planes and slip directions of the two lattices [18], but the Burgers vectors of dislocations in cementite are greater in magnitude than any in ferrite or in austenite. This makes the transfer of slip across cementite difficult.

Stress relaxation experiments on cementite at 1250 °C and 10 GPa pressure over a period of 8 h have revealed dislocation glide on (010)[001], tough operation of the (010)[100] slip system was more frequent. The [100] dislocations dissociate into $[\frac{1}{2}00]+[\frac{1}{2}00]$ [19]. Molecular dynamics simulations [20] based on the deformation of a cluster of 16 grains of nanocrystalline cementite have not reproduced the observed experimental data, but highlight the role of grain boundary sliding though at an unrealistic strain-rate of $5\times10^8\,s^{-1}$.

Planar striations can sometimes be observed in cementite, particularly when it precipitates in the solid-state. Nishiyama et al. [21] identified stacking faults on $(010)_\theta$ using transmission electron microscopy, involving translations by vectors parallel to $[100]_\theta$ that are not lattice vectors (Figure 3.3). It is known that dislocations with the Burgers vector equal to the lattice vector $[001]_\theta$, which lie in (010) slip plane, are not in general dissociated except when they lie in the $(130)_\theta$ plane [22]. More complex faults occur on other planes. Cementite that grows at low temperatures can contain planar defects that are identified as two-layer thick regions of transition carbide χ-Fe_5C_2 parallel to $(010)_\theta$; more complex faults occur due to the intercalation of iron into the cementite [23]. Partial

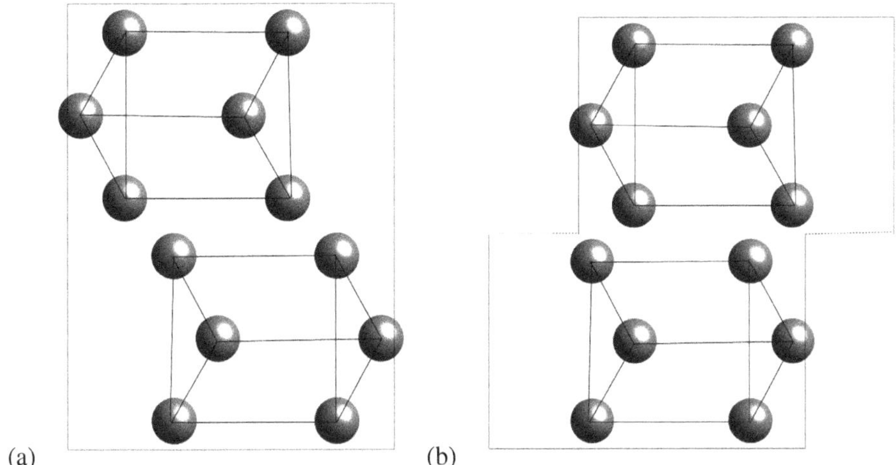

Figure 3.3 Creation of a stacking fault on $(010)_\theta$ by a partial displacement parallel to $[100]_\theta$, that does not recreate the lattice. Carbon atoms have been omitted for clarity. (a) Unfaulted structure. (b) Faulted structure. Adapted from [21].

dislocations, whose motion would leave trailing faults, have been proposed to exist in cementite [24]. Faults with varying levels of complexity have been hypothesised to exist in cementite but the evidence for the actual displacements involved is limited.

Point defects in cementite are mostly vacancies in the carbon atom sites or additional carbon atoms beyond the requirements of stoichiometry; the formation energy of vacancies at the prismatic sites is very large, some 66-69 kJ mol^{-1} at 0 K [25, 26]. Data from first principles calculations for a variety of point defects in cementite are listed in Table 3.2.

Simulations show that cascades produced by irradiation can lead to vacancies in the iron sites, and anti-site defects where iron atoms lodge in interstitial positions [28]. These point defect concentrations tend to be greater than those that occur in ferrite under the same conditions [29]. Similar

Table 3.2

Calculated formation energies at 0 K for point defects in cementite, referred to ferromagnetic α-iron and diamond as the reference states. Data from 128 atom simulations by Jiang et al. [27]. The mole fraction of carbon in cementite is denoted x^θ. For example, placing an iron atom in a prismatic interstice leads to a carbon concentration that is less than 25 at.%.

Defect	Formation energy per defect / kJ mol^{-1}	x^θ / mole fraction
Fe vacancy, 4c site	160	> 0.25
Fe vacancy, 8d site	140	> 0.25
C vacancy in prismatic interstice	61	< 0.25
C in Fe-4c site	358	> 0.25
C in Fe-8d site	273	> 0.25
Fe in prismatic interstice	256	< 0.25
C in octahedral interstice	71	> 0.25
Fe in octahedral interstice	511	< 0.25

simulations suggest that the moduli of cementite are affected by irradiation, but the lattice parameters used do not seem to be correct [30].

3.3 FERRITE AND AUSTENITE

The crystal structure of ferrite in pure iron above its Curie temperature (T_C) has a Cubic-I lattice with an iron atom at each lattice point, space group $Im\overline{3}m$. Its lattice parameter at ambient temperature when extrapolated to $T_C < 800\,°C$ is 0.2877 nm.[1]

Cubic symmetry is not compatible with ferromagnetism because of the alignment of magnetic moments along one of the cell edges. Below T_C, α-iron has its magnetic spins parallel to, say along the z axis, so rotations about the x or y axes must be combined with time reversal to preserve the directions of the spins. The magnetic point group therefore becomes tetragonal $4/mm'm'$. If the slight tetragonality is neglected, the cubic lattice parameter is 0.2867 nm at ambient temperature with an expansion coefficient of $\approx 1.3 \times 10^{-5}\,K^{-1}$.

Austenite in pure iron has a Cubic-F lattice with an iron atom at each lattice point, space group $Fm\overline{3}m$ and a lattice parameter at ambient temperature of 0.3477 nm and a thermal expansion coefficient of $\approx 2.1 \times 10^{-5}\,K^{-1}$.

3.4 WHY STUDY ORIENTATION RELATIONSHIPS?

Given the ease with which modern instruments permit the measurement of crystallography, it is common to report approximate orientation relationships between any combination of θ, α and γ. The information, however, is rarely interpreted in detail, and some theory requires more precise measurements that are not routinely available. One clear example is the requirement that as a minimum, an invariant-line must exist between the parent and product phases if the latter is to develop into a sizeable crystal by a displacive mechanism. The deformation that carries the parent into the product phase must therefore leave a line invariant and at the same time, defines the orientation relationship.

The crystallography of a transformation is important if the properties of interfaces are sensitive to the relative orientations and types of the crystals that they connect [33–35]. A particular orientation relationship may lead to an optimal fit at the interfaces that enclose newly formed crystals [36, 37]. This helps reduce the barrier to nucleation. If embryos evolve through thermally-activated fluctuations, those that by chance are oriented for best fit thrive as viable nuclei, whereas others would falter in their attempts to overcome larger activation barriers. This constitutes a mechanism for the generation of reproducible orientations [38, 39].

In some cases, nuclei are generated by the homogeneous deformation of a small region of the parent lattice [33], involving a collective movement of atoms, though this is not without a cost. The synchronised motion of a group of atoms does not occur in isolation, they push against the surrounding atoms that are not participants in their dance, causing strains to accumulate. Of all possible deformations, that which minimises strain energy would therefore be favoured. A mechanism like this results in strictly reproducible orientations, so it probably is not relevant for pearlite where the transformation is reconstructive with respect to both α and θ. It is likely that the plethora of observed relationships are not too energetically different.

Crystallographic orientations are today easy to measure, though accurate characterisation requires effort; we shall see in Section 3.5 that some of the orientations traditionally accepted are not found during precise measurements. It is important to appreciate the significance of such experiments – their role during nucleation has already been discussed, but they also can influence

[1] As Whewell points out, the ability of a magnet to attract iron was known to the ancients [31]. Pliny wrote in 77 CE that 'The hæmatites has not the same property of attracting iron that the ordinary magnet has' [32]. We know now that this is true below the Curie temperature when the iron becomes ferromagnetic.

mechanical properties when the polycrystalline steel is processed in a way that causes crystals to rotate into alignment.

The α/θ orientation within pearlite may be relevant to the development of crystallographic texture during plastic deformation. There are details in Chapter 13 but in cold-worked pearlitic wires, $\langle 011\rangle_\alpha$ tends to be parallel to the wire axis. Notice that the texture is always described in terms of the ferrite, simply because the texture of the cementite component of pearlite has never been characterised. It is an open question whether any of the reported α/θ relationships favour the conjoint deformation of the two phases and a combined crystallographic alignment. Texture does develop when polycrystalline cementite that begins with a random distribution of orientations, is deformed. The $(010)_\theta$ slip planes tend to align but this may not be the complete story since more than one slip system must operate in order to maintain continuity in a deforming polycrystalline aggregate [40]. There is no known α/θ orientation relationship that permits the $(010)_\theta$ plane to be aligned with a ferrite slip system.

There is more to crystallography than nucleation and deformation. Pitting corrosion involves intense and localised attack rather than the degradation of large areas of the metal surface. The localisation is induced when a passive surface-film is disturbed or contains a fault, by surface deposits of some kind, or mechanical damage. Unlike highly-alloyed steels [41], there are no simple formulae to estimate the pitting-corrosion resistance of pearlitic steels because they do not have a strong passive protective film. However, their microstructure is sufficiently heterogeneous to induce fine-scale corrosion that is not uniform. Both pearlitic and mixed α+P steels are susceptible to pitting corrosion in chloride environments, but the pits develop preferentially within the pearlitic ferrite α_P with anisotropic growth along the ferrite lamellae, Figure 3.4 [42].

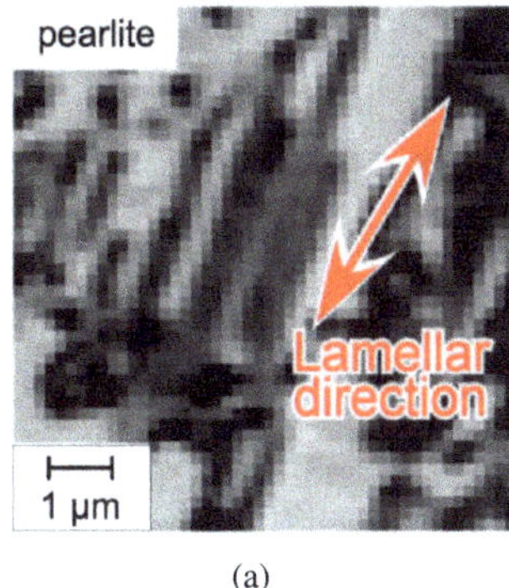

(a)

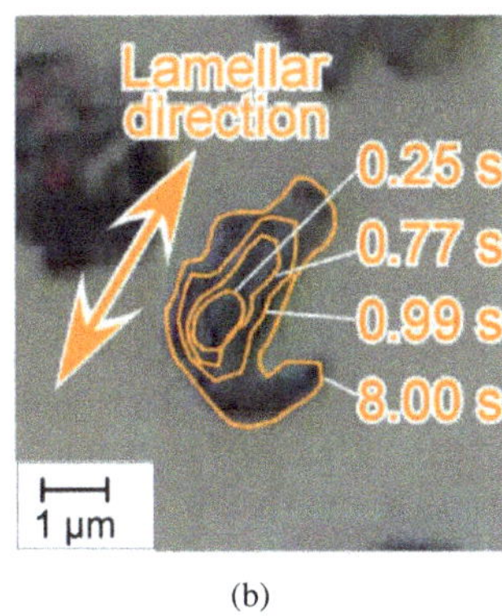

(b)

Figure 3.4 Real-time observations using confocal laser scanning microscopy, of pitting corrosion developing in pearlite in a hypoeutectoid steel containing a mixture of proeutectoid-α and pearlite. (a) Reflectance image of the lamellae. (b) Pit development in α_P along the lamellar orientation as a function of time, while subjected to a chloride solution. Images reproduced from Kadowaki et al. [42] under the CC BY 4.0 license.

Pitting is characterised using polarisation experiments in which an electrochemical cell is set up between the test-metal and a reference electrode. An electrical potential is applied such that a corrosion current is generated by the positive ions of the dissolving metal and the plot of potential versus this current density is known as the anodic polarisation curve. Figure 3.5 shows two such curves, one corresponding to the case where there is a large α_P/α_P orientation spread within the area studied, and the other where the spread is relatively small. The pitting potential is the most negative potential above which pits develop so the sample with a large orientation spread has a smaller resistance to the development of pits. Any role of the cementite in pearlite is neglected since the orientation data deal only with ferrite. The results are consistent with the role of gradients in surface structure stimulating pit formation [43].

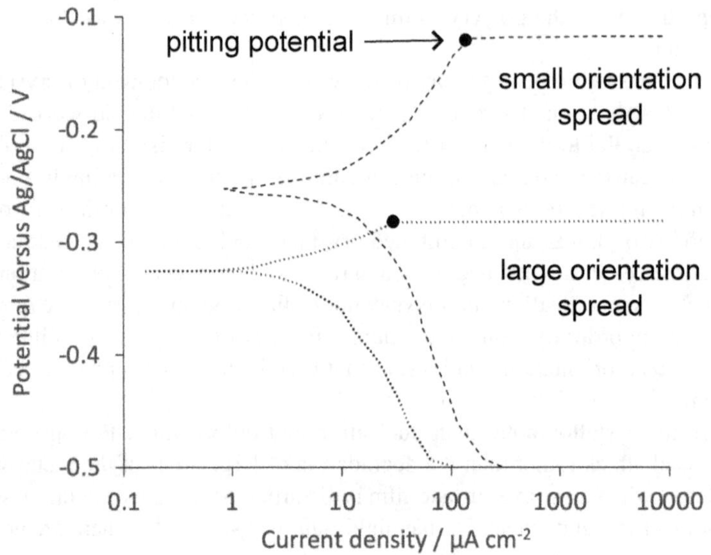

Figure 3.5 Anodic polarisation curves from local electrochemical experiments covering an area of about 320 μm^2, from a steel that has a fully pearlitic microstructure; the orientation spread refers to α/α crystallography within the pearlitic microstructure. The pitting potentials are indicated by dots. Adapted using selected data from Rault et al. [43], where substantial further measurements are reported.

It is well established that during corrosion, the ferrite is anodic with respect to cementite. But the presence of an α/θ interfaces also play a role, because they are defects with an excess energy relative to the lattices they connect. Therefore, corrosion within the ferrite is greatest in the vicinity of its interface with cementite, as shown in Figure 3.6a.

The magnitude of this interfacial energy per unit area depends on the α/θ orientation relationship. There exists a large variety of observed orientation relationships between cementite and ferrite in pearlite (Section 3.5), but the one that leads to an optimal fit is that due to Isaichev [44, 45]. Figure 3.6b shows that the depth of the local attack at the interface increases as a function of deviation from the Isaichev orientation relationship, presumably because the interfacial energy per unit area increases similarly. It also has been demonstrated that there is a correlation between oxidation rate and the α/θ orientation relationship [46]

The level of coherency at the α/θ interface has other consequences on the mechanical properties. It leads to the emission of dislocations in the ferrite during tensile loading, which in turn results in a smooth stress-strain curve rather than yield point and the propagation of a Lüders band. Both strain ageing and spheroidisation heat treatment that preserves the orientation relationship maintain the smooth σ/ε_p curve. It is only if the orientation relationship associated with the coherency is destroyed that a yield point appears, as discussed in detail on p. 52. Based on plasticity modelling where it is assumed that pearlite can be modelled by a slab of ferrite with cementite at its top and bottom surfaces, it is possible that the mechanical response depends on the α/θ orientation relationship [47] but the analysis does neglect the real complexity of pearlite.

A variety of properties evidently are affected by the crystallographic orientations. The only way in which the overall crystallography of pearlite can be manipulated is through plastic deformation, but that, in the context of pearlite, has the overriding function of producing a particular physical form. There is no example from industry, of the texture of pearlite being manipulated to achieve particular properties.

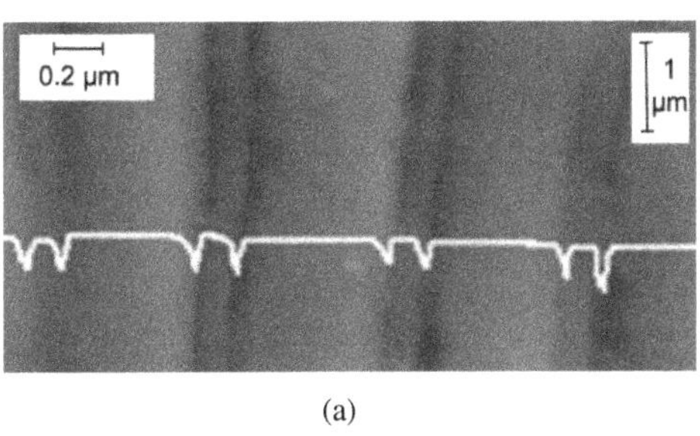

(a)

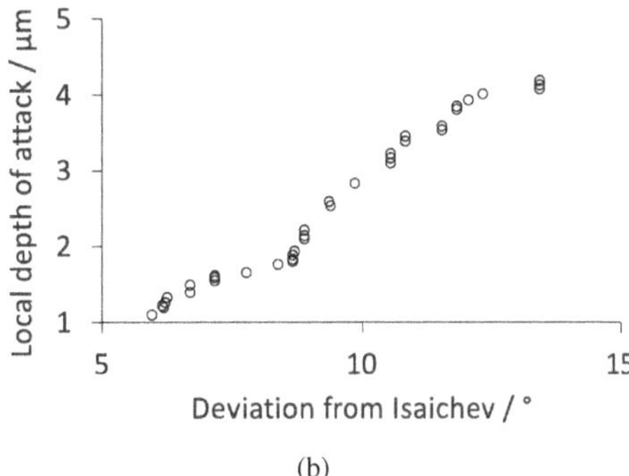

(b)

Figure 3.6 (a) Micrograph showing the depth profile after subjecting a flat sample to anodic polarisation attack. The vertical marker relates to the depth profile whereas the horizontal marker to the micrograph showing alternating lamellae of cementite and ferrite. Reproduced with permission of Taylor and Francis from [48]. (b) The depth of the local attack at α/θ interfaces as a function of the deviation of the orientation relationship from that due to Isaichev. Selected data from Khan et al. [48], with the small error bars omitted for clarity. Orientation relationships can be defined by axis-angle pairs but only the deviation angles were published because these represented the minimum disorientation angles, although the axis data were collected [49].

3.5 α_P/θ ORIENTATION RELATIONS

The crystallographic reproducibility that is a defining feature of displacive transformations is less relevant when the mechanism is reconstructive. The focus here is on precise data that enable differentiation between reported α_P/θ orientation relationships [50]:

'New-2'

$(0\,3\,\bar{1})_\theta$	$\parallel$	$(\bar{1}\,0\,1)_{\alpha_P}$
$[1\,0\,0]_\theta$	8.5° from	$[1\,3\,1]_{\alpha_P}$
$[\bar{1}\,1\,3]_\theta$	$\parallel$	$[1\,\bar{1}\,1]_{\alpha_P}$

'New-3'

$(0\,3\,\bar{1})_\theta$	$\parallel$	$(\bar{1}\,0\,1)_{\alpha_P}$
$[\bar{2}\,2\,0]_\theta$	2.4° from	$[1\,\bar{3}\,1]_{\alpha_P}$
$[1\,1\,3]_\theta$	$\parallel$	$[1\,\bar{1}\,1]_{\alpha_P}$

'New-4'

$(1\,0\,2)_\theta$	$\parallel$	$(\bar{1}\,0\,1)_{\alpha_P}$
$[1\,1\,3]_\theta$	5.95° from	$[1\,0\,1]_{\alpha_P}$
$[2\,1\,\bar{1}]_\theta$	$\parallel$	$[1\,3\,1]_{\alpha_P}$

'New-5'

$(0\,3\,\bar{1})_\theta$	$\parallel$	$(\bar{1}\,0\,1)_{\alpha_P}$
$[1\,1\,3]_\theta$	8.5° from	$[1\,\bar{1}\,1]_{\alpha_P}$
$[1\,0\,0]_\theta$	$\parallel$	$[1\,3\,1]_{\alpha_P}$

The New-2, New-3 and New-5 relations are close to but significantly different from the classical Pitsch-Petch [51, 52] orientation:

$(0\,1\,0)_\theta$	$\parallel$	$(5\,\bar{2}\,\bar{1})_{\alpha_P}$
$[0\,0\,1]_\theta$	2.6° from	$[1\,3\,\bar{1}]_{\alpha_P}$
$[1\,0\,0]_\theta$	2.6° from	$[1\,1\,3]_{\alpha_P}$

The original papers quoted in support of the Pitsch-Petch orientation relationship do not show the relations as stated above. Petch states the orientation, expressed here with cementite in the *Pnma* setting, to be $(010)_\theta \parallel (5\,\bar{2}\,\bar{1})_{\alpha_P}$ and $[0\,0\,1]_\theta \parallel [1\,3\,\bar{1}]_{\alpha_P}$. The Pitsch paper considers the θ/γ orientation and then assumes a conversion from austenite to ferrite to determine the θ/α_P relationship, which is not explicit in the original work. For the Pitsch-Petch orientation, using $a_{\alpha_P} = 0.0.2867$ nm, $a_\theta = 0.50837$ nm, $b_\theta = 0.67475$ nm, $c_\theta = 0.45165$ nm, [2]

$$(\alpha_P\,\mathrm{J}\,\theta) = \begin{pmatrix} 0.555629 & 2.148447 & 0.450241 \\ 0.598722 & -0.859379 & 1.399229 \\ 1.580702 & -0.429689 & -0.547252 \end{pmatrix}$$

$$(\theta\,\mathrm{J}\,\alpha_P) = \begin{pmatrix} 0.163668 & 0.150036 & 0.518271 \\ 0.387877 & -0.155151 & -0.077575 \\ 0.168194 & 0.555189 & -0.269411 \end{pmatrix}.$$

It is emphasised, as Zhang and Kelly [50] pointed out, that neither the Pitsch-Petch, nor the Bagaryatskii orientations were found, and suggested that they may not actually exist if accurate measurements are made; only the Isaichev relation was observed in addition to the new orientations relationships listed above and is known to lead to a greater α_P/θ coherency [45].

The mechanisms required to explain the orientation relationships observed in accurate determinations depend on the carbon concentration of the steel [50]. During transformation from austenite in hypoeutectoid steels, the ferrite precedes the cementite and therefore influences its nucleation, leading to the Isaichev orientation relationship. In hypereutectoid steels, the cementite nucleates first while in contact with austenite, leading to the New-5 orientation relationship with the ferrite; in this case the orientation of the ferrite within the pearlite is determined by its relationship with the austenite. The mechanisms determining the New-2, New-3 and New-4 relationships are not as clear, but apparently occur in eutectoid steel and are dominated by the relationship of the closest-packed planes within the cementite with those in austenite and ferrite.

3.6 PITSCH γ/θ ORIENTATION RELATIONSHIP

The Pitsch orientation relationship [52, 55] between austenite and cementite in the *Pnma* setting is given by:

$$\begin{aligned} [100]_\theta &\parallel [110]_\gamma \\ [010]_\theta &\parallel [\bar{2}25]_\gamma \\ [001]_\theta &\parallel [5\bar{5}4]_\gamma. \end{aligned}$$

The $[100]_\theta$ vector not only is parallel to $[110]_\gamma$, but its length is almost identical, making it an invariant line between the two lattices. For $a_\gamma = 0.3568$ nm, $a_\theta = 0.50837$ nm, $b_\theta = 0.67475$ nm, $c_\theta = 0.45165$ nm, the coordinate transformations become

$$(\gamma\,\mathrm{J}\,\theta) = \begin{pmatrix} 1.007488 & -0.658402 & 0.791147 \\ 1.007488 & 0.658402 & -0.791147 \\ 0 & 1.646005 & 0.632918 \end{pmatrix}$$

$$(\theta\,\mathrm{J}\,\gamma) = \begin{pmatrix} 0.496284 & 0.496284 & 0 \\ -0.184101 & 0.184101 & 0.460251 \\ 0.478783 & -0.478783 & 0.383027 \end{pmatrix}.$$

[2]The notation here is due to Mackenzie and Bowles [53]; its advantages have been described in [54]. Vectors **u** and plane normals **h** (referred to the reciprocal basis) can be converted using the coordinate transformation matrix with $(\alpha_P\,\mathrm{J}\,\theta)[\theta;\mathbf{u}] = [\alpha_P;\mathbf{u}]$ and $(\mathbf{h};\theta) = (\mathbf{h};\alpha_P)(\alpha_P\,\mathrm{J}\,\theta)$.

Using the correspondence matrix due to Sleeswyk [56], configured to the *Pnma* space-group setting, the deformation that converts the austenite lattice to that of cementite is given by:

$$\begin{aligned}
(\gamma\,\mathrm{S}\,\gamma) &= (\gamma\,\mathrm{J}\,\theta)(\theta\,\mathrm{C}\,\gamma)\\
&= \begin{pmatrix} 1.007488 & -0.658402 & 0.791147\\ 1.007488 & 0.658402 & -0.791147\\ 0 & 1.646005 & 0.632918 \end{pmatrix} \times \frac{1}{6}\begin{pmatrix} 4 & 2 & 0\\ 1 & -1 & -3\\ -3 & 3 & -3 \end{pmatrix}\\
&= \begin{pmatrix} 0.166352 & 0.841136 & -0.066373\\ 1.176970 & -0.169478 & 0.066373\\ -0.042124 & 0.042124 & -1.139460 \end{pmatrix}.
\end{aligned}$$

The eigenvectors and eigenvalues (λ_i) for $(\gamma\,\mathrm{S}\,\gamma)$ are given approximately by

$$\begin{aligned}
[0.344239\ 0.46953\ 1]_\gamma \quad & \lambda_1 = -1.17374\\
[1\ 1\ 0]_\gamma \quad & \lambda_2 = \ \ 1.00749\\
[-1.60845\ 2.26399\ 1]_\gamma \quad & \lambda_3 = -0.97634.
\end{aligned}$$

The alternative correspondence defined by Sleeswyk [56]:

$$(\theta\,\mathrm{C}\,\gamma) = \frac{1}{12}\begin{pmatrix} 7 & 5 & 3\\ 1 & -1 & -6\\ -6 & 6 & -6 \end{pmatrix}$$

leads to the eigenvectors and eigenvalues for the deformation matrix as follows:

$$\begin{aligned}
[0.26009\ -0.553681\ 1]_\gamma \quad & \lambda_1 = -1.17374\\
[1\ 1\ 0]_\gamma \quad & \lambda_2 = \ \ 1.00749\\
[-1.8993\ 1.97315\ 1]_\gamma \quad & \lambda_3 = -0.97634.
\end{aligned}$$

The deformations λ_i therefore are not different and it remains the case that the $[100]_\theta \parallel [110]_\gamma$ is unrotated and almost unchanged in length, an invariant line.

3.7 HABIT PLANE?

A variety of studies have indicated particular crystallographic indices of the plane connecting α_P and θ on the assumption that lamellae represent plates. Darken and Fisher indicated schematically the habit plane $(001)_\theta \parallel \{110\}_\alpha$; Bowden and Kelly [57] reported it to be $(001)_\theta \parallel \{215\}_{\alpha_P}$ or $(001)_\theta \parallel \{112\}_{\alpha_P}$, using single-surface trace analysis.[3] Zhou and Shiflet [60] find the plane to be orientation-relationship specific; for the Isaichev relation, they reported it to be $(101)_\theta \parallel \{112\}_{\alpha_P}$ based on transmission electron microscopy; others have deduced it from atom probe images. As a general principle, it is not appropriate to characterise habit planes at magnifications that do not respect the lattice-fit over the length scale of the structure.

The term *habit plane* refers to the broad interface between the parent and product phases. In crystallography, it implies that the indices of this plane are reproducible when determined on different particles of the product phase. It also implies a planar interface. None of these conditions are satisfied when considering the boundary between ferrite and cementite lamellae in a pearlite colony

[3]Baker et al. [58] have been quoted [59] to have found the habit plane to be $(001)_\theta$ but they do not in fact address this problem.

(Chapter 2). A cursory examination can show that the lamellae grow with ferrite, at a common front with the austenite, adopting growth directions and shapes that best accommodate growth normal to the transformation front.

There are, nevertheless, crystallographic aspects that are interesting. Both ferrite and cementite are crystalline and therefore anisotropic. Figure 3.7 shows the elastic anisotropy of single crystals of ferrite and cementite. The elastic modulus of ferrite has minima along the $\langle 100 \rangle_\alpha$ directions – there is a tendency for cementite lamellae to grow along these least stiff directions of ferrite [61] because there is limited coherency at the α_P/θ interfaces. It is this coherency that induces continuous yielding during tensile testing (p. 52). The extent of coherency must be a function of the specific orientation-relationship but judging from the behaviour of pearlite during tensile testing, there is sufficient fit associated with all observed orientations to enable continuous yielding.

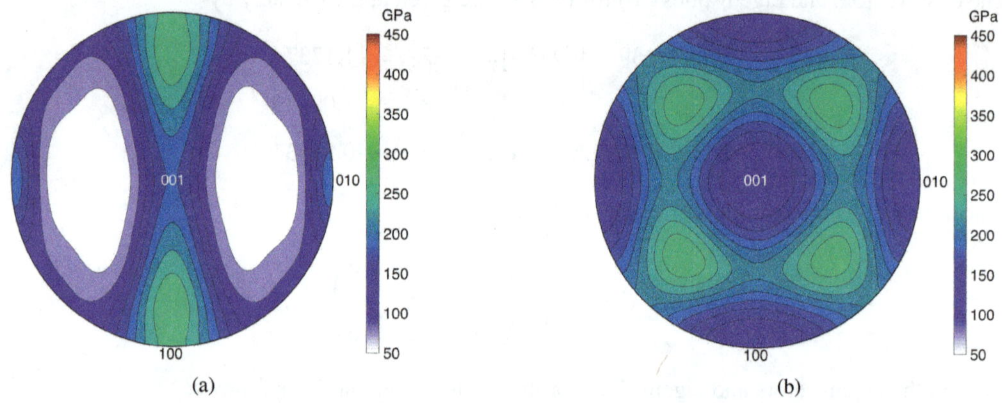

Figure 3.7 Stereographic projections showing the variation of calculated single-crystal elastic moduli as a function of orientation, for the setting *Pnma*. (a) Fe_3C [62]. (b) Corresponding data for body-centred cubic iron. Plots courtesy of Shaumik Lenka.

3.8 CRYSTALLOGRAPHIC TEXTURE

Pearlitic steel *plates* are not used in the cold-deformed state; the plate form is reserved for structural applications where safety, machinability and weldability are paramount. Cold rolling causes many mechanical properties to deteriorate, e.g., Figure 3.8. Comparisons are often made against the incredible success of cold-drawn wire, but as pointed out in Chapter 13, the twisting of wires into ropes provides certain safety features that cannot be incorporated in plates.[4]

There are, nevertheless, academic studies worthy of discussion. The texture is weak in the as-transformed condition, but on deformation, the $\langle 110 \rangle_\alpha$ directions tend to become parallel to the rolling direction, whereas $\langle 111 \rangle_\alpha$ aligns along the normal direction (i.e., $\{111\}_\alpha$ tends to become parallel to the rolling plane) [66]. Figure 3.9 shows the texture components that develop in a cold-rolled pearlitic steel with a 95% reduction in thickness from an initial thickness of 9.6 mm that had been reduced by rolling to a thickness of 0.43 mm ($\varepsilon_{vM} = 3.6$). This is the classic texture developed in cold-rolled ferrite [67, 68]. There are no similar data on the orientation of the cementite component of pearlite. However, in a study of eutectoid pearlitic-steel subjected to *in situ* shear

[4]Although the plates could be assembled from thin layers consisting of alternating layers of eutectoid steel and ferrite (Al-rich) bonded together by rolling; toughness improves depending on loading conditions [63]; many varieties have been studied with microstructures produced that are not ordinarily possible [64].

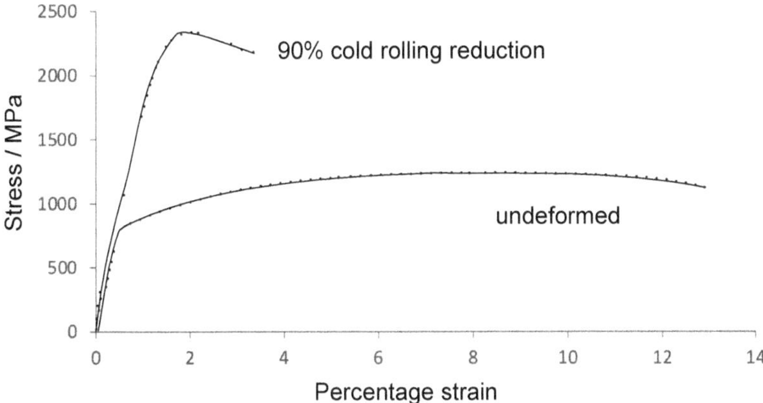

Figure 3.8 Tensile test data from pearlitic steel (Fe-0.82C-0.45Mn-0.2Si wt%) in a hot-rolled condition that is not cold-rolled prior to testing, and one where a 90% reduction is implemented by cold-rolling before testing in tension. Selected data from Liu et al. [65].

deformation ($\gamma \approx 0.6$) in a synchrotron X-ray source, the cementite orientation did not change much, while that of α_P did [69].

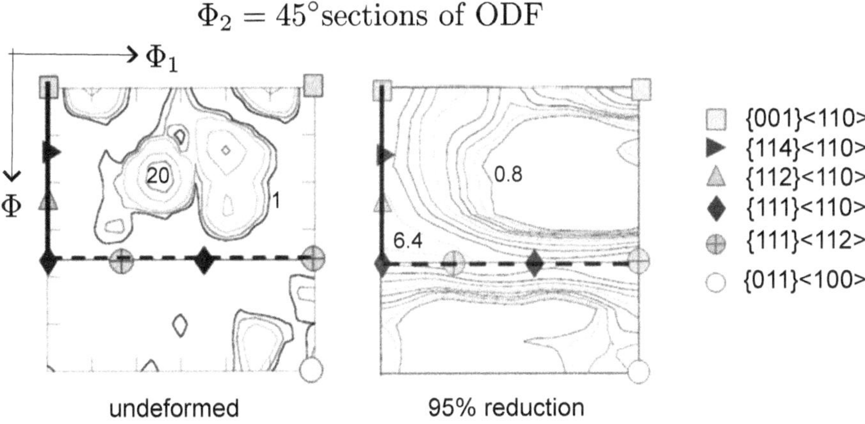

Figure 3.9 Sections of the orientation distribution functions (ODFs) of undeformed and deformed pearlite (cold-rolling reduction of 95%) in Fe-0.72C-0.75Mn-0.16Si wt% steel. The sections are at $\Phi_2 = 45^\circ$ because the main features can be represented there in a condensed manner; the numbers within the plots represent contour intensities. The angles $0 \le \Phi, \Phi_1 \le 90^\circ$. The pearlite was obtained by isothermal transformation at 650 °C. The bold line corresponds to the so-called α-fibre and the dashed one to the γ-fibre in ODF terminology. Images adapted from Narayanswami et al. [66] with permission of Elsevier.

Annealing the steel with a rolling reduction of 95%, at 700 °C for 60 min, produced little change in the cold-rolled texture shown in Figure 3.9. When the reduction was achieved by alternating the rolling direction in each pass by 90°, i.e., cross-rolling, the texture remains predominantly $\langle 110\rangle_\alpha \parallel$ rolling direction and $\langle 111\rangle_\alpha \parallel$ normal direction, but the $\{001\}_\alpha\langle 110\rangle_\alpha$ component was boosted [70].

The cross-rolling resulted in interesting microstructural changes in lamellar shape, illustrated in Figure 3.10. Studies on austenitic stainless steels [71] indicate that whereas the shear bands during unidirectional rolling are parallel, intersecting shear bands dominate a cross-rolled microstructure, which explains the doubly-bent shape of the θ-lamellae seen in Figure 3.10b.

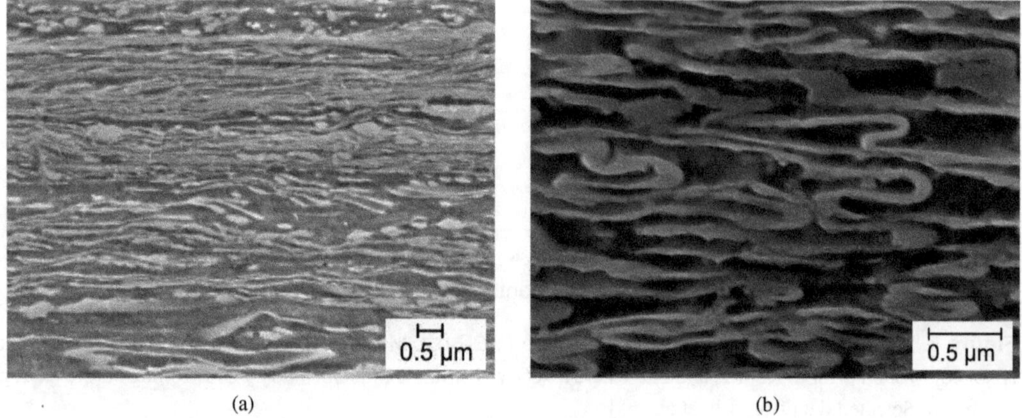

(a) (b)

Figure 3.10 Fe-0.72C-0.75Mn-0.16Si wt% steel transformed into pearlite at 650 °C. (a) Cold-rolled with the rolling direction maintained constant in each pass, to a 95% total reduction. (b) Structure resulting from multiple passes of cross-rolling, in which the direction of rolling is changed through 90° in alternating passes, until a 95% total reduction is achieved. Images courtesy of Pinaki Bhatacharjee, with more details in [66, 70].

Discussion of the development of texture in drawn pearlitic-wires is reserved for Chapter 13.

REFERENCES

1. E. J. Fasiska, and G. A. Jeffrey: 'On the cementite structure', *Acta Crystallographica*, 1965, **19**, 463–471.
2. 'Table of space group symbols': 2018: URL http://www.cryst.ehu.es/cgi-bin /cryst/programs/nph-wp-list.
3. M. D. Graef: 'Teaching crystallographic and magnetic point group symmetry using three-dimensional rendered visualizations': https://www.iucr.org/education/pamphlets/23/full-text, 2008.
4. I. G. Wood, L. Vočadlo, K. S. Knight, D. P. Dobson, W. G. Marshall, G. D. Price, and J. Brodholt: 'Thermal expansion and crystal structure of cementite, Fe_3C, between 4 and 600 K determined by time-of-flight neutron powder diffraction', *Journal of Applied Crystallography*, 2004, **37**, 82–90.
5. P. Blum, and R. Pauthenet: 'Étude ferromagnétique de la cémentite polycristalline et monocristalline', *Comptes Rendus Hebromadaires des Seances de L'Académie des Sciences*, 1953, **237**, 1501–1502.
6. S. Yamamoto, T. Terai, T. Fukuda, K. Sato, T. Kakeshita, S. Horii, M. Ito, and M. Yonemura: 'Magnetocrystalline anisotropy of cementite pseudo single crystal fabricated under a rotating magnetic field', *Journal of Magnetism and Magnetic Materials*, 2018, **451**, 1–4.
7. T. Shigematsu: 'Invar properties of cementite $(Fe_{1-x}Me_x)_3C$, Me= Cr, Mn, Ni', *Journal of the Physical Society of Japan*, 1975, **39**, 915–920.
8. N. I. Medvedeva, I. R. Shein, M. A. Konyaeva, and A. L. Ivanovskii: 'Effect of chromium on the electronic structure and magnetic properties of cementite', *Physics of Metals and Metallography*, 2008, **105**, 568–573.
9. C. Jiang, S. G. Srinivasan, A. Caro, and S. A. Maoly: 'Structural, elastic, and electronic properties of Fe_3C from first principles', *Journal of Applied Physics*, 2008, **103**, 043502.

10. C. Jiang, and S. G. Srinivasan: 'Unexpected strain-stiffening in crystalline solids', *Nature*, 2013, **496**, 339–342.
11. A. Leineweber: 'Anisotropic microstrain broadening in cementite, Fe_3C, caused by thermal microstress: comparison between prediction and results from diffraction-line profile analysis', *Journal of Applied Crystallography*, 2012, **45**, 944–949.
12. A. Leineweber: 'Thermal expansion anisotropy as source for microstrain broadening of polycrystalline cementite, Fe_3C', *Journal of Applied Crystallography*, 2016, **49**, 1632–1644.
13. A. Inoue, T. Ogura, and T. Muramatsu: 'Burgers vectors of dislocations in cementite crystal', *Scripta Metallurgica*, 1977, **11**, 1–5.
14. J. Alkorta, and J. G. Sevillano: 'Assessment of elastic anisotropy and incipient plasticity in Fe_3C d by nanoindentation', *Journal of Materials Research*, 2012, **27**, 45–52.
15. M. Mizuno, I. Tanaka, and H. Adachi: 'Effect of solute atoms on the chemical bonding of Fe_3C (cementite)', *Philosophical Magazine B*, 1997, **75**, 237–248.
16. A. Inoue, M. Ogura, and K. Masumoto: 'Lattice defects and deformation/fracture behavior of cementite in steel', *Bulletin of the Japan Institute of Metals*, 1974, **13**, 653–664.
17. J. Gil-Sevillano: 'Room temperature plastic deformation of pearlitic cementite', *Materials Science & Engineering*, 1975, **21**, 221–225.
18. L. E. Karkina, I. N. Karkin, I. G. Kabanova, and A. R. Kuznetov: 'Crystallographic analysis of slip transfer mechanisms across the ferrite/cementite interface in carbon steels with fine lamellar structure', *Journal of Applied Crystallography*, 2015, **48**, 97–106.
19. A. Mussi, P. Cordier, S. Ghosh, N. Garvik, B. C. Nzogang, P. Carrez, and S. Garruchet: 'Transmission electron microscopy of dislocations in cementite deformed at high pressure and high temperature', *Philosophical Magazine*, 2016, **96**, 1773–1789.
20. H. Ghaffarian, A. K. Taheri, K. Kang, and S. Ryu: 'Molecular dynamics simulation study of the effect of temperature and grain size on the deformation behavior of polycrystalline cementite', *Scripta Materialia*, 2015, **95**, 23–26.
21. A. Nishiyama, A. Kore'eda, and S. Katagiri: 'Study of plane defects in the cementite by transmission electron microscopy', *Transactions of JIM*, 1964, **5**, 115–121.
22. L. E. Kar'kina, I. N. Kar'kin, and T. A. Zubkova: 'Atomistic simulation of stacking faults in cementite: Planes containing vector [100]', *Physics of Metals and Metallography*, 2014, **115**, 814–829.
23. Y. Nakamura, T. Mikami, and S. Nagakura: '*In situ* high temperature electron microscopic study of the formation and growth of cementite particles in the third stage of tempering', *Transactions of the Japan Institute of Metals*, 1985, **26**, 876–885.
24. N. Garvik, P. Carrez, and P. Cordier: 'First-principles study of the ideal strength of Fe_3C cementite', *Materials Science & Engineering A*, 2013, **572**, 25–29.
25. G. A. Nematollahi, J. von Pezold, J. Neugebauer, and D. Raabe: 'Thermodynamics of carbon solubility in ferrite and vacancy formation in cementite in strained pearlite', *Acta Materialia*, 2013, **61**, 1773–1784.
26. H. Göhring, A. Leineweber, and E. J. Mittemeijer: 'A thermodynamic model for non-stoichiometric cementite; the Fe–C phase diagram', *CALPHAD*, 2016, **52**, 38–46.
27. C. Jiang, B. P. Uberuaga, and S. G. Srinivasan: 'Point defect thermodynamics and diffusion in Fe_3C: a first principles study', *Acta Materialia*, 2008, **56**, 3236–3244.
28. K. O. E. Henriksson: 'Cascades in model steels: The effect of cementite (Fe_3C) and $Cr_{23}C_6$ particles on short-term crystal damage', *Nuclear Instruments and Methods in Physics Research B*, 2015, **352**, 36–38.
29. K. O. E. Henriksson, and K. Nordlund: 'Irradiation cascades in cementite: 0.1-10 keV Fe recoils', *Nuclear Instruments and Methods in Physics Research B*, 2012, **277**, 136–139.
30. K. O. Henriksson, and K. Nordlund: 'Mechanical and elastic changes in cementite Fe_3C subjected to cumulative 1 keV Fe recoils', *Nuclear Instruments and Methods in Physics Research B*, 2014, **338**, 119–125.

31. W. Whewell: History of the inductive sciences, vol. 3: Cambridge, U. K.: Cambridge University Press, 1837.
32. Pliny the Elder: 'Natural history, book 36, chapter 25, 126-130': 77 CE.
33. P. L. Ryder, and W. Pitsch: 'The crystallographic analysis of grain boundary precipitation', *Acta Metallurgica*, 1966, **14**, 1437–1448.
34. P. L. Ryder, W. Pitsch, and R. F. Mehl: 'Crystallography of the precipitation of ferrite on austenite grain boundaries in a Co-20%Fe alloy', *Acta Metallurgica*, 1967, **15**, 1431–1440.
35. J. W. Christian: 'Martensitic transformations: a current assessment', In: *The Mechanism of Phase Transformations in Crystalline Solids, Monography 33*. London, U.K.: Institute of Metals, 1969:129–142.
36. W. C. Johnson, C. L. White, P. E. Marth, P. K. Ruf, S. M. Tuominen, K. D. Wade, K. C. Russell, and H. I. Aaronson: 'Influence of crystallography on aspects of solid-solid nucleation theory', *Metallurgical & Materials Transactions A*, 1975, **6**, 911–919.
37. W. Pitsch, and P. L. Ryder: 'Generation of an orientation relationship during nucleation in precipitation processes', *Scripta Metallurgica*, 1977, **11**, 431–434.
38. J. W. Christian: Theory of Transformations in Metals and Alloys, Part I: 2 ed., Oxford, U. K.: Pergamon Press, 1975.
39. H. K. D. H. Bhadeshia: 'Diffusional formation of ferrite in iron and its alloys', *Progress in Materials Science*, 1985, **29**, 321–386.
40. N. Adachi, H. Ueno, S. Morooka, P. Xu, and Y. Todaka: 'Deformation texture of bulk cementite investigated by neutron diffraction', *Materials*, 2022, **15**, 4485.
41. S. Ooi: 'A new index to estimate the corrosion resistance of aluminium-containing steel', *Materials and Corrosion*, 2024, **75**, 902–913.
42. M. Kadowaki, I. Muto, Y. Sugawara, T. Doi, K. Kawano, and N. Hara: 'Real-time microelectrochemical observations of very early stage pitting on ferrite-pearlite steel in chloride solutions', *Journal of The Electrochemical Society*, 2017, **164**, C261.
43. V. Rault, V. Vignal, H. Krawiec, and F. Dufour: 'Quantitative assessment of local misorientations and pitting corrosion behaviour of pearlitic steel using electron backscattered diffraction and microcapillary techniques', *Corrosion Science*, 2015, **100**, 667–671.
44. I. V. Isaichev: 'Orientation of cementite in tempered carbon steel (orientatsiya tsementita v otpushchennoi uglerodistoi stali)', *Zhurnal Tekhncheskoi Fiziki*, 1947, **17**, 835–838.
45. H. K. D. H. Bhadeshia: 'Solution to the Bagaryatskii and Isaichev ferrite-cementite orientation relationship problem', *Materials Science and Technology*, 2018, **34**, 1666–1668.
46. H. K. Mehtani, M. I. Khan, A. Durgaprasad, S. K. Deb, S. Parida, M. J. N. V. Prasad, and I. Samajdar: 'Oxidation kinetics in pearlite: The defining role of interface crystallography', *Scripta Materialia*, 2018, **152**, 44–48.
47. Y. Yasuda, and T. Ohashi: 'Crystal plasticity analyses of scale dependent mechanical properties of ferrite/cementite lamellar structure model in pearlite steel wire with bagaryatsky or pitsch-petch orientation relationship', *ISIJ International*, 2016, **56**, 2320–2326.
48. M. Khan, H. Mehtani, A. Durgaprasad, G. Goyal, M. Prasad, S. Parida, T. Dasgupta, N. Birbilis, and I. Samajdar: 'The defining role of interface crystallography in corrosion of a two-phase pearlitic steel', *Philosophical Magazine*, 2020, **100**, 1439–1453.
49. I. Samajdar: 'Personal communication', 2023: Deviation angles relative to Isaichev orientation relatioship.
50. M.-X. Zhang, and P. M. Kelly: 'Accurate orientation relationships between ferrite and cementite in pearlite', *Scripta Materialia*, 1997, **37**, 2009–2015.
51. N. J. Petch: 'The orientation relationships between cementite and α-iron.', *Acta Crystallographica*, 1953, **6**, 96–96.

52. W. Pitsch: 'Der orientierungszusammenhang zwischen zementit und austenit', *Acta Metallurgica*, 1962, **10**, 897–900.
53. J. K. Mackenzie, and J. S. Bowles: 'The crystallography of martensite transformations II', *Acta Metallurgica*, 1954, **2**, 138–147.
54. H. K. D. H. Bhadeshia: Geometry of Crystals, Polycrystals and Phase Transformations: Florida, USA: CRC Press, Taylor and Francis, 2017.
55. K. H. Yang, and W. K. Choo: 'Variants of the orientation relationship between austenite and cementite in Fe-30Mn-1C wt % alloy', *Acta Metallurgica et Materalia*, 1994, **42**, 263–269.
56. A. W. Sleeswyk: 'Crystallography of the austenite-cementite transition', *Philosophical Magazine*, 1966, **13**, 1223–1237.
57. H. G. Bowden, and P. M. Kelly: 'The crystallography of the pressure induced phase transformations in iron alloys', *Acta Metallurgica*, 1967, **15**, 1489–1500.
58. A. Baker, P. M. Kelly, and J. Nutting: 'Structures resulting from phase transformations in steels', In: J. W. G. Thomas, ed. *Electron Microscopy and the Strength of Crystals*. New York, USA: Interscience, 1963:899–915.
59. Y. Ohmori, A. Davenport, and R. Honeycombe: 'Crystallography of pearlite', *Transactions of the Iron and Steel Institute of Japan*, 1972, **12**, 128–137.
60. D. S. Zhou, and G. J. Shiflet: 'Ferrite: cementite crystallography in pearlite', *Metallurgical Transactions A*, 1992, **23**, 1259–1269.
61. A. Durgaprasad, S. Giri, S. Lenka, S. Kundu, S. Mishra, S. Chandra, R. D. Doherty, and I. Samajdar: 'Defining a relationship between pearlite morphology and ferrite crystallographic orientation', *Acta Materialia*, 2017, **129**, 278–289.
62. G. Ghosh: 'A first-principles study of cementite (Fe_3C) and its alloyed counterparts: Elastic constants, elastic anisotropies, and isotropic elastic moduli', *AIP Advances*, 2015, **5**, 087102.
63. J. A. Charles: 'Development and use of layered ferrous microstructure', *Materials Science and Technology*, 1998, **14**, 496–503.
64. D. K. Matlock, R. Johnson, E. Moor, and J. Speer: 'Microstructural simulations via thermal processing of roll bonded steel laminates', *International Journal of Metallurgical Engineering*, 2013, **2**, 10–17.
65. Y. Liu, C. Yang, M. Liu, C. Wang, Y. Dai, X. Li, A. Russell, C. Zhang, Z. Zhang, and G. Cao: 'Effects of microstructure and crystallography on mechanical properties of cold-rolled SAE1078 pearlitic steel', *Materials Science and Engineering: A*, 2018, **709**, 115–124.
66. S. Narayanswamy, R. Saha, and P. P. Bhattacharjee: 'Strain dependent evolution of microstructure and texture in severely cold-rolled and annealed ultrafine pearlite', *Materials Characterization*, 2020, **169**, 110583.
67. M. Hölscher, D. Raabe, and K. Lücke: 'Rolling and recrystallization textures of bcc steels', *Steel Research*, 1991, **62**, 567–575.
68. R. Ray, J. J. Jonas, and R. Hook: 'Cold rolling and annealing textures in low carbon and extra low carbon steels', *International Materials Reviews*, 1994, **39**, 129–172.
69. C. S. A. da Silva, B. R. C. Saraiva, L. Novotnỳ, P. W. C. Sarvezuk, M. Masoumi, C. C. Silva, L. F. G. Herculano, J. L. Cardoso, H. F. G. de Abreu, and M. Béreš: 'Texture and lattice strain evolution in a pearlitic steel during shear deformation: an in situ synchrotron X-ray diffraction study.', *International Journal of Plasticity*, 2024, **?**, https://doi.org/10.1016/j.ijplas.2024.104083.
70. S. Narayanswamy, R. Saha, and P. Bhattacharjee: 'Cross-rolling mediated microstructure and texture evolution in severely cold-rolled and annealed ultrafine pearlite', *Materials Characterization*, 2021, **171**, 110751.
71. S. Mohammadzehi, and H. Mirzadeh: 'Cold unidirectional/cross-rolling of austenitic stainless steels: a review', *Archives of Civil and Mechanical Engineering*, 2022, **22**, 129.

4 Strength

4.1 INTRODUCTION

Most structural steels are hypoeutectoid, with proeutectoid-ferrite that forms first during cooling, followed by pearlite as the carbon in the residual austenite reaches a concentration that allows both α_P and θ to grow simultaneously. They can be welded without worrying too much about what happens in the heat-affected zones. There is in addition, an important set of steels that consist only of pearlite. In considering the quantitative aspects of strength in both of these scenarios, the focus first is on hypoeutectoid steels.

Much of the theory involves approximations and simplifications, combined with a level of empiricism; mechanical properties are never easy to model given the plethora of variables that cannot be neglected. Some of the consequences of this will be described, if only to set the scene for a discussion of even more complex properties.

4.2 MIXTURES OF PROEUTECTOID FERRITE AND PEARLITE

The simplest way of dealing with the combined deformation of proeutectoid-α and pearlite (regarded here as an averaged contributor) is to assume a law of mixtures:

$$\overline{\sigma}\{\overline{\varepsilon}_p\} = (1 - V_V^P)\sigma^\alpha\{\overline{\varepsilon}_p\} + V_V^P\sigma^P\{\overline{\varepsilon}_p\} \tag{4.1}$$

where $\overline{\varepsilon}_p$ is the mean plastic strain measured on a macroscopic scale. The method works if the difference in the strengths of proeutectoid-α and pearlite[1] is not too large. A clever application of this equation is in finite element mechanical models that include microstructure, where the flow stresses of the individual components are needed. If it is assumed that the ratio $\sigma^\alpha/\sigma^P \approx \mathrm{HV}^\alpha/\mathrm{HV}^P$ is a constant independent of the magnitude of the plastic strain, then a measured tensile curve from a particular steel can, for that steel, be deconvoluted into the independent tensile curves for proeutectoid-α and pearlite. This is illustrated for the case where the hardness ratio is ≈ 0.7 in Figure 4.1. The method does not deal adequately with the occurrence of a yield point effect or Lüders strain, as described in Section 4.2.1, but this may not be too relevant in finite element modelling where the focus is on large plastic strains.

Constraint

The law of mixtures does in a sense, lack rigour. The overall deformation should occur in two stages, the first where the hard nodules of pearlite are only elastically strained while the proeutectoid-α flows in such a way that continuity is maintained. As the ferrite work hardens, it becomes possible to transfer sufficient stress onto the nodules to cause them to yield.

In this process, the elastic limits of the ferrite and pearlite are modified because of the heterogeneous nature of the deformation [2]. If the flow stress for a fully ferritic steel is σ_f^α and that of a fully pearlitic steel is σ_f^P, then at the early stages of deformation where only the ferrite yields while the nodules remain elastic, the applied uniaxial tensile stress σ can be expressed as a function of the average plastic strain $\overline{\varepsilon}_p$ as

$$\sigma = \sigma_f^\alpha\left\{\frac{\overline{\varepsilon}_p}{1-\overline{\varepsilon}_p}\right\} + \frac{V_V^P}{1-V_V^P}E'\overline{\varepsilon}_p \quad \text{where} \quad E' = \frac{E(7-5\nu)}{10(1-\nu^2)}. \tag{4.2a}$$

[1] Henceforth, a mixture of proeutectoid ferrite and pearlite is written α+P for brevity.

DOI: 10.1201/9781032631981-4

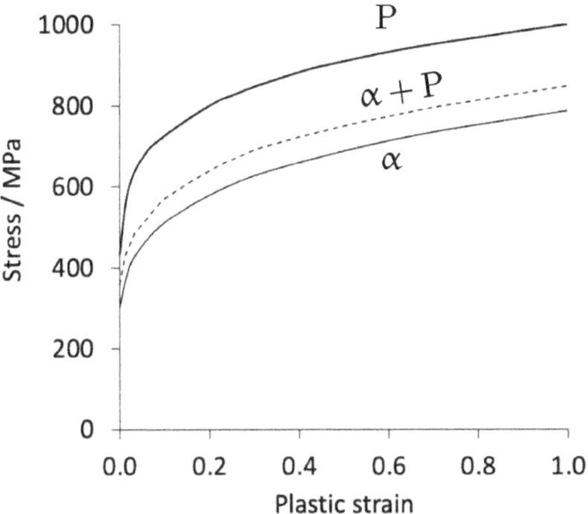

Figure 4.1 Stress-strain curves for the individual components of an α+P steel containing a volume fraction 0.3 of pearlite with $\overline{L}_\alpha \approx 25\,\mu$m, obtained by deconvoluting the measured tensile curve (dashed) for the α+P mixture using Equation 4.1. Adapted from Shoji et al. [1].

Therefore, if yielding is defined when $\overline{\varepsilon}_p = 0.002$, the ferrite will undergo plastic deformation at $\sigma > \sigma_y^\alpha$ because of the constraint provided by the rigid pearlite; σ_y^α represents the yield strength measured for a fully ferritic sample. At an average plastic strain $\overline{\varepsilon}_p = \overline{\varepsilon}_p^*$, the ferrite will have work hardened sufficiently to induce flow in the pearlite, given by the simultaneous equations

$$\sigma = \sigma_f^\alpha \left\{ \frac{\overline{\varepsilon}_p^*}{1 - \overline{\varepsilon}_p^*} \right\} + \frac{V_V^P}{1 - V_V^P} E' \overline{\varepsilon}_p^* \quad \text{and} \quad \sigma = \sigma_y^P - E' \overline{\varepsilon}_p^* \tag{4.2b}$$

so the pearlite yields at $\sigma < \sigma_y^P$. Beyond this, the two constituents $\alpha + P$ deform together while maintaining continuity, although it is possible to form voids in regions where the flows are not compatible (Figure 5.4).

When the soft region deforms first, the strain discontinuity at the interface with the hard region hinders its further flow, and aids the onset of flow in harder region.

Multiaxial stresses and strains

When the applied stress is not uniaxial, it is necessary to use a yield criterion. The von Mises yield criterion assumes that the yielding in a ductile material occurs when the equivalent stress

$$\sigma_{vM}^2 = \frac{1}{2}[(\sigma_{11} - \sigma_{22})^2 + (\sigma_{22} - \sigma_{33})^2 + (\sigma_{33} - \sigma_{11})^2 + 6(\sigma_{23}^2 + \sigma_{31}^2 + \sigma_{12}^2)] \tag{4.3}$$

reaches the yield stress $\sigma_y = \sqrt{3}\tau_y$ as measured in monotonic tension and τ_y is the shear yield strength similarly measured. The σ_{ij} are the components of a stress tensor. The von Mises equivalent strain is given by

$$\varepsilon_{vM}^2 = \frac{2}{3}[(\varepsilon_1 - \varepsilon_2)^2 + (\varepsilon_2 - \varepsilon_3)^2 + (\varepsilon_3 - \varepsilon_1)^2 + 6\varepsilon_{23}^2 + 6\varepsilon_{31}^2 + 6\varepsilon_{12}^2]$$

where ε_{ij} are the components of the relevant strain tensor. The concepts of equivalent stress and equivalent strain are useful in interpreting the behaviour of the steel when subjected to complex stress systems, for example, at the root of a notch.

4.2.1 TENSILE STRENGTH

Figure 4.2 shows the results of tensile tests conducted on a hot-rolled steel containing $V_V^P = 0.13$ of pearlite with the residue being proeutectoid ferrite. The curves are typical for mixtures of α+P, illustrating the reproducibility of the property, irrespective of the orientation of the test relative to the rolling direction within the rolling plane. The Lüders strain observed following yielding is typical of α+P steels and the strain hardening exponent (n) in $\sigma \propto \varepsilon_p^n$, following the propagation of the Lüders band across the gauge length of the sample, is approximately 0.16. Plastic instability occurs when $d\sigma_{engineering}/d\varepsilon_p$ changes from positive to zero, at which point the load divided by the original cross-sectional area defines the ultimate tensile strength.[2]

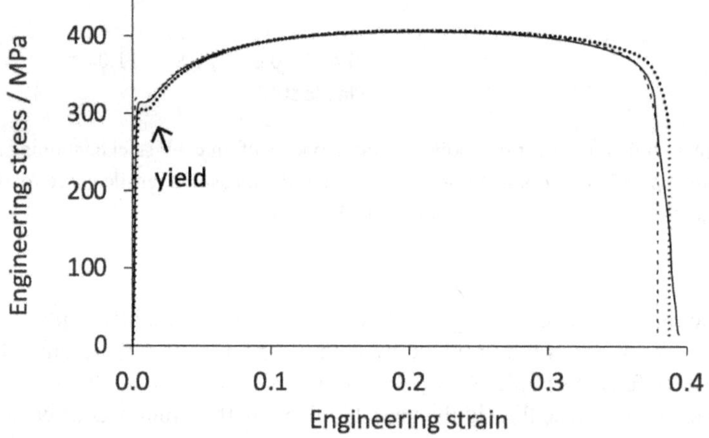

Figure 4.2 Three sets of tensile test data from a hot-rolled α+P steel (Fe-0.1C-1Mn-0.3Si wt%) on a 25 mm×1.6 mm rectangular section specimen with a 50 mm gauge length, with the crosshead-displacement rate of 5 mm min^{-1}. The test orientations are diagonal (continuous line), transverse (dashed) and longitudinal (dotted curve), with the longitudinal test parallel to the rolling direction. Data courtesy of M. Aamir, A. R. Chintha and S. Kundu.

Hypoeutectoid steels containing α+P exhibit discontinuous yielding. Fully pearlitic steels, in contrast, yield continuously, even though the major component of pearlite is ferrite; the other component, cementite, in single-crystal form is known to deform smoothly without abrupt yielding [3]. The stress-strain behaviour of fully pearlitic steels is also smooth because the α/θ interfaces within pearlite have a modicum of coherency, leading to an emission of dislocations under stress, Figure 4.3a [4]. Microscopy of deformed pearlite has shown that the density of dislocations increases at the α/θ interfaces due to the emitted dislocations [5].

The continuous yielding persists even when the pearlite is strain aged (5% strain in tension, followed by 150 °C for 1 h). The same steel, when spheroidised by annealing, persisted in exhibiting a smooth σ-ε curve even after strain ageing, because the spheroidised cementite retains the original orientation relationship and therefore, the coherency with the ferrite. On the other hand, severe deformation followed by recrystallisation of the microstructure led to Lüders extension in both the recrystallised and strain-aged conditions, because the new θ/α interfaces lost coherency. It is interesting that spheroidised bearing-steels also exhibit discontinuous yielding, because the spheroidal

[2]The true strain is given by $\ln\{\ell/\ell_o\}$ where ℓ and ℓ_o are the length and original length respectively, or alternatively, by $\ln\{A_o/A\}$ where A stands for the instantaneous cross-sectional area. The true and engineering strains are related by $\varepsilon_{true} = \ln\{1+\varepsilon_{engineering}\}$. Similarly, $\sigma_{true} = \sigma_{engineering}(1+\varepsilon_{engineering})$.

particles originate in the austenite which then transforms into ferrite and larger spheroidal-cementite via a divorced-eutectoid reaction, so there is no special θ/α orientation relationship.

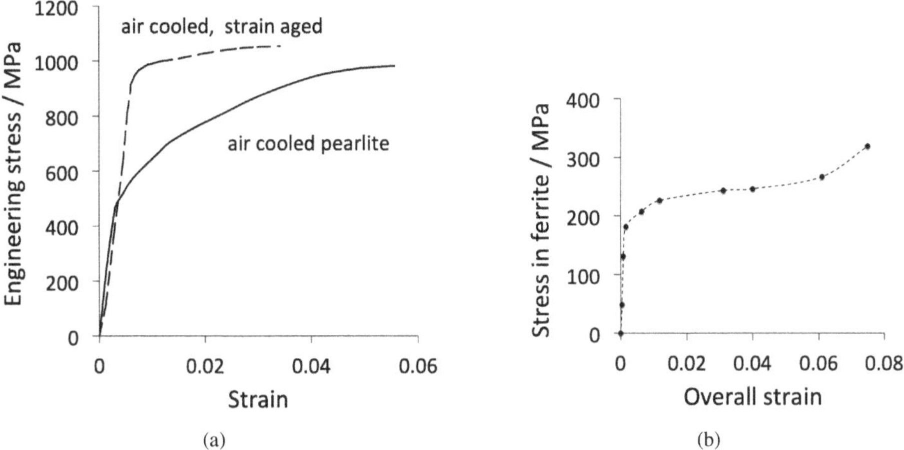

Figure 4.3 (a) Continuous yielding exhibited by pearlite in Fe-0.85C-0.4Mn-0.25Si wt% steel, both in the air-cooled and additionally strain-aged conditions. Selected data from Wang et al. [4]. (b) Deformation behaviour of the ferrite within pearlite in a hot rolled and air-cooled fully pearlitic steel, measured using X-ray diffraction. Selected data from Sidhom et al. [6].

Experiments in which the stress within the ferrite of fully pearlitic steel is monitored using X-ray diffraction confirm the continuous yielding behaviour described above. Figure 4.3b shows the evolution of stress within the ferritic component of pearlite during deformation [6]. The yielding is continuous with little work hardening, presumably because of the constraint provided by the surrounding cementite lamellae, until at large overall strain the cementite fractures and focusses strain into the ferrite.

4.2.2 MODELS OF TENSILE STRENGTH

In a structure containing dislocations, the average dislocation spacing $z_\perp \approx 1/\sqrt{\rho_\perp}$ where $\rho_\perp$ is the dislocation density. Since the resistance to shear is proportional to $1/z_\perp$, it will depend on $\sqrt{\rho_\perp}$ [7]. The shear stress τ required to translate dislocations in a single crystal, against both the resistance of the lattice (τ^{f}) and the presence of other dislocations, becomes

$$\tau = \tau^{\mathrm{f}} + b_{11} E_{\mathrm{s}} b \sqrt{\rho_\perp}$$

where b_{11} is a dimensionless parameter which for ferrite is about 0.33 [8], E_{s} is the shear modulus and b is the magnitude of the Burgers vector. A Taylor factor M can be used to relate the shear-induced flow in a single crystal to the uniaxial flow-stress [9] σ in a polycrystal

$$\sigma = \sigma^{\mathrm{f}} + b_{11} M b E_{\mathrm{s}} \left(\sqrt{\rho_\perp} - \sqrt{\rho_\perp^*}\right)$$

where $\rho_\perp^*$ is the dislocation density prior to the onset of plasticity. The strain ε_{p} caused by the movement of a cluster of dislocations of density $\rho_\perp$ over a free path $z_\perp$ is simply $M\rho_\perp b z_\perp$.

Table 4.1

Evolution of dislocation density in α_P, during tensile testing of a Fe-0.8C-0.84Mn-0.17Si wt% fully pearlitic steel with $\overline{L}_\gamma = 30\,\mu m$ [10]; a cold drawn 0.8C wt% pearlitic steel with other solutes unspecified [11]; a cold-drawn Fe-0.83C-0.38Si-0.92Mn-0.23Cr wt% pearlitic steel [12]. There are differences in the way the measurements were done, which must add uncertainties to the reported interlamellar spacings and dislocation densities. In the latter case, the X-ray diffraction data are likely to be the most reliable in the context of macroscopic properties.

ε_p	$\rho_\perp / m^{-2}$	$S_I / \mu m$	Technique	Reference
0	0.7×10^{14}	0.09	TEM	[11]
0	1.1×10^{14}	0.13	X-ray	[12]
0.008	$(2.35 \pm 0.7) \times 10^{13}$	0.32	TEM	[10]
0.013	6.6×10^{13}		TEM	[13]
0.023	$(1.02 \pm 0.2) \times 10^{14}$	0.32	TEM	[10]
0.046	$(1.73 \pm 0.3) \times 10^{14}$	0.13	TEM	[10]
0.048	$(1.91 \pm 0.3) \times 10^{14}$	0.32	TEM	[10]
0.071	$(2.27 \pm 0.2) \times 10^{14}$	0.32	TEM	[10]
0.073	$(1.98 \pm 0.3) \times 10^{14}$	0.13	TEM	[10]
0.63	2.3×10^{14}	0.10	X-ray	[12]
0.68	7×10^{14}	0.07	TEM	[11]
1.50	3.9×10^{15}		TEM	[13]
1.65	1.4×10^{15}	0.07	X-ray	[12]
2.67	8.8×10^{15}	0.03	TEM	[11]
3.68	2×10^{16}	0.02	TEM	[11]
4.2	2.8×10^{16}		TEM	[13]
5.4	5×10^{16}		TEM	[13]
0	4×10^{16}	0.03	HRTEM	[14]

Data determined using high-resolution transmission electron microscopy (HRTEM) cover incredibly small volumes ($\approx 10^{-25}\,m^3$) so it is unlikely that they are representative. Dislocations are known, at that level of resolution, to be heterogeneously distributed [13] and may even escape from the free surfaces of the necessarily thin sample used in transmission microscopy.

The dislocation density evolves during plastic deformation (Table 4.1, Figure 4.4)[3] as more defects are created and stored, while at the same time, some are lost in recovery processes:

$$\begin{aligned} \frac{d\rho_\perp}{d\varepsilon_p} &= \left.\frac{d\rho_\perp}{d\varepsilon_p}\right|_{\text{stored}} - \left.\frac{d\rho_\perp}{d\varepsilon_p}\right|_{\text{recovery}} \\ &= M\left(\frac{1}{bz_\perp} - b_{12}\rho_\perp\right) \end{aligned} \tag{4.4}$$

[3]It has been shown that $\rho_\perp$ does not vary significantly across the cross-section of a 2.74 mm diameter wire drawn to a strain of 1.39 [15].

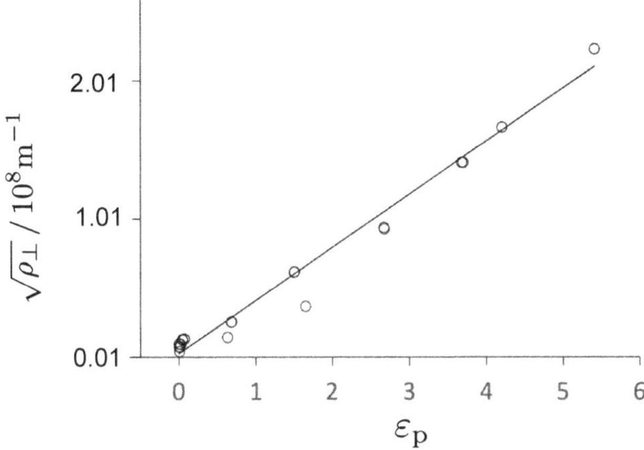

Figure 4.4 The square root of dislocation density in pearlitic ferrite versus plastic strain during wire drawing, using data from Table 4.1, with the datum from [14] excluded. The line is represented by the equation $\sqrt{\rho_\perp} = 4\times10^7\varepsilon_p + 4\times10^6$ to a correlation of 0.99. The finite dislocation density at zero strain is that which exists in the material prior to explicit deformation.

where $d\rho_\perp/d\varepsilon_p|_{stored}$ is set equal to $dL/b\,dA$, dL being the length of dislocation lines stored per area swept (dA), with dL/dA assumed to equal the mean free path $z_\perp$ [16]. The second term assumes that the recovery rate is proportional to the dislocation density. The dislocation density is obtained as a function of the plastic strain by directly integrating this equation with the assumption that $z_\perp$ is constant as follows:

$$\rho_\perp = \frac{1}{bz_\perp b_{12}}\left(1-\exp\{-b_{12}M\varepsilon_p\}\right) + \rho_\perp^*\exp\{-b_{12}M\varepsilon_p\} \tag{4.5}$$

which when incorporated into Equation 4.4 and assuming that $z_\perp$ is equal to the ferrite grain size defined as a mean lineal intercept $\overline{L}_\alpha$, gives [8, 17, 18]

$$\begin{aligned}\sigma_\alpha &= \sigma_\alpha^f + b_{11}ME_sb\left(\frac{1-\exp\{-Mb_{12}\varepsilon_p\}}{b_{12}\overline{L}_\alpha b} + \rho_\perp^*\exp\{-b_{12}M\varepsilon_p\}\right)^{\frac{1}{2}}\\ &\approx \sigma_\alpha^f + b_{11}ME_sb\left(\frac{1-\exp\{-Mb_{12}\varepsilon_p\}}{b_{12}\overline{L}_\alpha b}\right)^{\frac{1}{2}} \quad \text{if} \quad \rho_\perp^* \quad \text{can be neglected}\end{aligned} \tag{4.6}$$

the approximation being reasonable if the small dislocation-density of undeformed-ferrite ($\rho_\perp^*$) may be neglected. The magnitude of the Burgers vector in ferrite is $b \approx 2.5\times10^{-10}$ m, the Taylor factor $M \approx 3$ and E_s is the shear modulus. The fitted parameter $b_{11} = 0.33$ [17].

Taking $z_\perp = \overline{L}_\alpha$ is itself an approximation because the dislocation density increases with ε_p. To account for this, b_{12} has been expressed in terms of the ferrite grain size, i.e., $10^{-5}/\overline{L}_\alpha$ [17] with the lineal intercept in units of metres; this does diminish the Hall-Petch dependence of strength on the inverse square root of grain size in Equation 4.6 although a different dependence is retained.

The friction stress σ_α^f consists of a combination of the Peierls stress and solid solution strengthening components, represented empirically as

$$\sigma_\alpha^f/\mathrm{MPa} = 77 + 80w_{Mn} + 60w_{Si} + 45w_{Ni} + 60w_{Cr} + 11w_{Mo} + 750w_P + 5000(w_{C_{ss}} + w_{N_{ss}}) \tag{4.7}$$

where the solute concentrations are those in solid solution. Because the coefficients are derived by fitting to steels containing both ferrite and pearlite, they do not strictly represent solution strengthening. The true solution strengthening coefficients for Mn, Si, Ni and Cr are 37, 106, 40 and 11 MPa (wt%)$^{-1}$ respectively for 25 °C, based on binary alloys of polycrystalline ferrite [19].

Equation 4.6 can be used to estimate the yield strength by setting $\varepsilon_p = 0.002$, in which case it is found that the dependence of σ_y on $\overline{L}_\alpha$ is underestimated, possibly because of the problem described above in defining b_{12} with

$$\sigma_y = \sigma^f_\alpha + k_{HP}(\overline{L}_\alpha)^{-\frac{1}{2}} \quad \text{with} \quad k_{HP} = 0.1 + 0.12(10^6 x_{C,ss})^{0.31} \qquad \text{for} \quad x_{C,ss} \leq 1.23 \times 10^{-4} \tag{4.8}$$

where $x_{C,ss}$ is the atomic fraction of carbon in solution and the lineal intercept is in metres [20]. Equation 4.6 gives $k_{HP} \approx 0.095\,\text{MPa}\,\text{m}^{1/2}$ which is inconsistent with data on ferrite found in α+P steels where it reaches values of $0.74\,\text{MPa}\,\text{m}^{1/2}$ [21] or even in ferritic steels containing $x_{C,ss} = 1.23 \times 10^{-4}$ of carbon in solution where $k_{HP} \approx 0.6\,\text{MPa}\,\text{m}^{1/2}$ [20, 22].[4]

It may not be reasonable to assume that the Hall-Petch effect should be identical in single-phase ferritic steels and in α+P steels. Figure 4.5 shows the yield data for these microstructures. Focusing first on the data for a strain rate of 0.001 s^{-1}, the α-iron and the steel have similar values of σ^f, which increases on neutron irradiation, particularly in the case of α-iron, which also has a much greater grain-size dependence as characterised by the coefficient k_{HP}. The Hall-Petch effect is based on microscopic yielding in a particular grain leading to a pile-up of dislocations at its boundary. The resulting concentration of stress at the grain boundary triggers dislocation activity in the adjacent grain. The number of dislocations in the pile-up determines the magnitude of the stress concentration. A larger slip plane (i.e. a larger grain size) increases the number of dislocations that can participate in the pile-up for a given applied stress. Therefore, the stress concentration intensifies with increasing grain size, making the propagation of yield across crystals relatively easy. This description, however, applies to a single-phase microstructure. In a mixture of ferrite and pearlite, a fraction of the boundaries will be between ferrite and the harder pearlite. For the data illustrated in Figure 4.5, the fraction of α/P boundaries is estimated to be ≈ 0.6. Consequently, the sensitivity to $\overline{L}_\alpha$ must be reduced in α+P microstructures, making $k^{\alpha+P}_{HP} < k^{\alpha}_{HP}$. Slip in a ferrite crystal is less likely to induce significant dislocation activity in the harder pearlite colonies; instead, much of the general yielding should at first percolate through the ferrite grains that have reduced connectivity due to the presence of pearlite.

It is interesting that the grain size dependence of the yield strength of α-iron is reduced dramatically following hardening due to irradiation, with $k_{HP} \approx 0$ at the highest of strain rates, Figure 4.5b. The implication is that the deformation becomes more homogeneous, i.e., with multiple grains yielding at the same level of stress because of the pinning effects of irradiation-induced defects, thus diminishing the role of dislocation pile-ups at grain boundaries. The pile-ups that characterise the Hall-Petch mechanism may not occur at all if the intragranular defects lead to dislocation tangles.

In contrast, k_{HP} for the α+P steel is insensitive to irradiation and strain rate, but we have emphasised that the application of Equation 4.8 is essentially empirical. The continuity of the ferrite (Section 6.6.5) is limited by the α/P boundaries. In a single phase material with equiaxed grains, the grain boundary area per unit volume is given by [24]:

$$S_V = \frac{2}{\overline{L}} \tag{4.9}$$

where $\overline{L}$ is the mean lineal intercept. Consider an idealised mixture of ferrite and pearlite where both are equiaxed and isotropically distributed, with each equal-sized, spherical pearlite-nodule isolated from any other by ferrite. If the nodule radius is r, its surface area and volume would be $4\pi r^2$ and $4\pi r^3/3$ respectively, so for N_V particles per unit volume, the α/P interface per unit volume is

[4]The influence of dissolved carbon on k_{HP} has its origins in the segregation of the interstitial into grain boundaries [20].

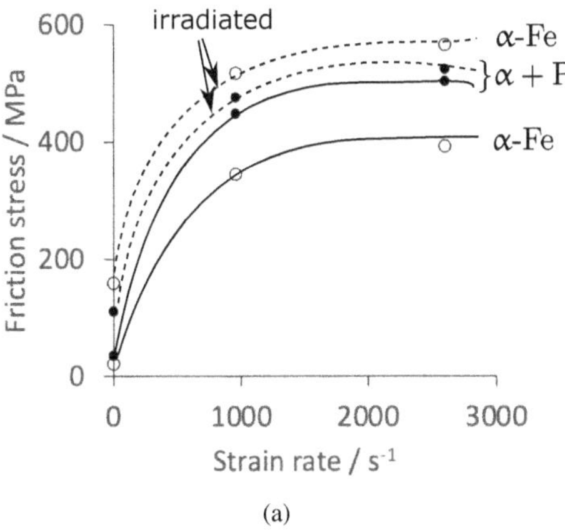

(a)

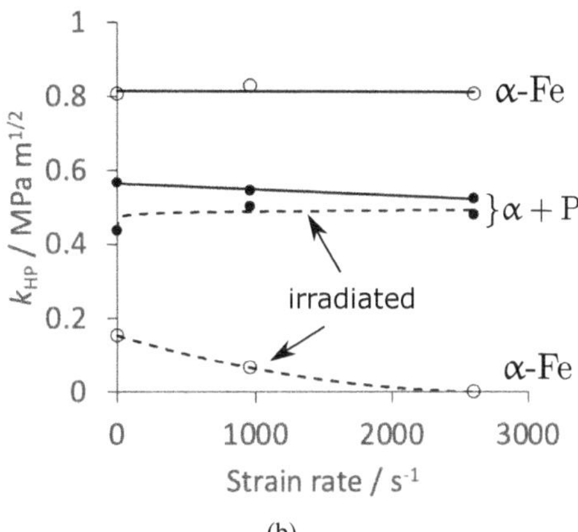

(b)

Figure 4.5 (a) The friction stress (σ^{f}) and (b) the coefficient k_{HP}, as a function of the strain rate. The open points refer to fully ferritic iron (0.0012C wt%) and the filled dots to Fe-0.21C wt% steel with $V_{\mathrm{V}}^{\mathrm{P}} \approx 0.28$. Selected data from Campbell and Harding [23]. The neutron irradiation dose was $5 \times 10^{22}\,\mathrm{m}^{-2}$.

$3V_{\mathrm{V}}^{\mathrm{P}}/r$. The mean lineal intercept presented by the nodules on random two-dimensional sections is $\overline{L}_{\mathrm{P}} = 4r/3$. It follows that the amount of ferrite-pearlite interface per unit volume is [p.94, 24, 25]

$$S_{\mathrm{V}}^{\mathrm{P}\alpha} = \frac{4V_{\mathrm{V}}^{\mathrm{P}}}{\overline{L}_{\mathrm{P}}}. \tag{4.10}$$

This relation does not apply at large $V_{\mathrm{V}}^{\mathrm{P}}$ because spheres cannot fill space; assuming a random distribution of equal spheres, the applicability of the equation should be limited to a volume fraction of pearlite that does not exceed 0.64. If it is assumed that $\overline{L} = V_{\mathrm{V}}^{\alpha}\overline{L}_{\alpha} + V_{\mathrm{V}}^{\mathrm{P}}\overline{L}_{\mathrm{P}}$ then it follows that

$$S_{\mathrm{V}}^{\alpha\alpha} = S_{\mathrm{V}} - S_{\mathrm{V}}^{\mathrm{P}\alpha} = \frac{2}{\overline{L}} - \frac{4V_{\mathrm{V}}^{\mathrm{P}}}{\overline{L}_{\mathrm{P}}} \quad \text{and} \quad \frac{S_{\mathrm{V}}^{\mathrm{P}\alpha}}{S_{\mathrm{V}}^{\alpha\alpha}} = \frac{4V_{\mathrm{V}}^{\mathrm{P}}}{2(\overline{L}_{\mathrm{P}}/\overline{L}) - 4V_{\mathrm{V}}^{\mathrm{P}}}. \tag{4.11}$$

Experimental measurements on a steel with $V_{\mathrm{V}}^{\mathrm{P}} = 0.18 \pm 0.03$, gave $S_{\mathrm{V}}^{\alpha\alpha}/S_{\mathrm{V}}^{\mathrm{P}\alpha} \approx 0.69$ [p.52, 26]; Equation 4.11 gives a value of 0.55 which is in reasonable agreement given the approximations involved; in particular, th assumption that the pearlite has no contiguity would lead to an underestimation of the contiguity of both α and pearlite in a banded microstructure, of the type on which the original experimental data were determined.[5] But the conclusion nevertheless stands, that it is unreasonable to apply the Hall-Petch equation given that the pearlite is essentially a rigid component of the microstructure at the point of yielding. Any values of σ_{f} and k_{HP} derived from experimental data on α+P steels should be regarded as empirical and not amenable to fundamental interpretation of the type applied to single-phase materials. $S_{\mathrm{V}}^{\alpha\alpha}$ becomes zero when $V_{\mathrm{V}}^{\mathrm{P}} = 0.5$ if it is assumed that $\overline{L}_{\mathrm{P}} = \overline{L}_{\alpha}$ in Equation 4.11.

The work-hardening behaviour following yielding is described adequately by an equation of the form

$$\sigma_{\alpha} = \sigma_{\alpha}^{\mathrm{Fe}} + \sum_i \sigma_{\mathrm{SS},i} + k_{\mathrm{HP}}(\overline{L}_{\alpha})^{-\frac{1}{2}} + f\{\varepsilon_{\mathrm{p}}\} \tag{4.12}$$

[5] A banded microstructure does not have alternating layers of ferrite and pearlite. The pearlitic regions have three-dimensional character such that ferrite grains and pearlite colonies mix within what might be called a ‘pearlite band’, Figure 6.1. The contiguity of pearlite within such a band is not, therefore, equal to one.

where $\sigma_\alpha^{\mathrm{Fe}} = 25\,\mathrm{MPa}$ is the strength of pure, annealed polycrystalline iron at 25 °C [27], representing the Peierls stress; $\sigma_{\mathrm{SS},i}$ is the solid solution strengthening due to component i dissolved in the ferrite. For typical values of $\overline{L}_\alpha = 20\,\mu\mathrm{m}$, 1.5Mn, 0.2Si wt% α+P steel, this equation gives a similar estimate of σ_y as the $\sigma_\alpha^{\mathrm{f}}$ of Equation 4.7 but is a better representation of the yield strength of ferrite. However, the representation does not allow for yield behaviour that is accompanied by a Lüders extension so $f\{\varepsilon_p\}$ applies at the stage where general, rather than localised work-hardening begins.

For pearlite, Equation 4.12 is adapted to allow for the interlamellar spacing S_I and the mean free path for dislocations in ferrite as follows:

$$\sigma_\mathrm{P} = \sigma^\mathrm{f} + \frac{3E_\mathrm{s}b}{S_\mathrm{I}} + b_{11}ME_\mathrm{s}b\left(\frac{1-\exp\{-Mb_{12}\varepsilon_\mathrm{p}\}}{b_{12}\overline{z}_\perp b}\right)^{\frac{1}{2}} \tag{4.13}$$

where the cementite is considered undeformable, its role being to obstruct the deformation of ferrite; this may be valid at small strains, but even at large strains, the equation performs well because of the empirical terms b_{11} and b_{12}. The initial dislocation density in the ferrite, $\rho_\perp^*$, is assumed to be insignificant. The second term in the equation has its origins in the *assumption* that the yield strength of pearlite depends not on the usual Hall-Petch relation with interlamellar spacing, but rather on dislocation sources being activated at the α/θ interfaces, so that $\sigma_y = \sigma^\mathrm{f} + b_{13}S_\mathrm{I}^{-1}$, with b_{13} deduced experimentally to be $\approx 3E_\mathrm{s}b\,\mathrm{N\,m^{-1}}$, where b is the magnitude of the Burgers vector of ferrite [10].[6] We shall see later (Section 4.3.1) that the $\sigma_y^\mathrm{P} \propto S_\mathrm{I}^{-1/2}$ is in fact the correct dependence [28], but since the parameters in Equation 4.13 are obtained by fitting empirically to experimental data, the importance of the Hall-Petch relation only becomes visible when applying the equation in a broader context. On implementing the fitting, $b_{12} = 7$ and $\overline{z}_\perp = 2\times10^{-6}\,\mathrm{m}$ [17]; since yielding first occurs in the interlamellar ferrite, which is the major constituent of pearlite, the stress $\sigma^\mathrm{f} \simeq \sigma_\alpha^\mathrm{Fe} + \sum_i \sigma_{\mathrm{SS},i}$, typically about 140 MPa [29].[7]

It is noticeable that in fully pearlitic steels, microstructural features such as the pearlite colony size and austenite grain size are not taken into account in calculating the yield strength. This is because both the yield strength and stress at which fracture occurs during tensile testing have been shown to be insensitive to anything other than the interlamellar spacing, which after all represents the most prominent barrier to the progress of plastic deformation [29].

Heterogeneous deformation

The equations described above cover the average plastic flow characteristics of α+P steels and can be implemented in finite element models for more complex scenarios than just uniaxial testing. They represent homogeneous deformation, not the intense shear bands that sometimes traverse colonies of pearlite (Figure 4.6), particularly when the pearlite spacing is fine or when the test is done in compression [30]. Similar features are observed during explosive deformation with the pearlite exhibiting a concatenated structure of kink bands [31]. This may explain why the dislocation density in the ferrite measured on tiny regions using transmission electron microscopy underestimates the average strain when compared with coarser pearlite (Table 4.1). Because plasticity is better at accommodating structural heterogeneities, the instabilities related to shear banding are less likely with harder materials, e.g., fine pearlite. The bands, on a microscopic scale, cause the fracture and shear displacement of the cementite, Figure 4.6b, but any resulting voids at the terminations of the cementite seem to be healed by the local and plastic deformation of the ferrite.

[6] The maximum number of dislocations on a glide plane that can pile up at a grain boundary depends on the grain size. The stress concentration at the boundary depends on this number of dislocations so a larger grain size would make it easier for slip to initiate in an adjacent grain. This in essence is the physical meaning behind the Hall-Petch equation.

[7] There are differences in the reported values of these parameters with a possible dependence on the average carbon concentration of the pearlite [8, 17, 18]. Note also that changes in S_I with plastic strain do not seem to be taken into account during the fitting of Equation 4.13 to data.

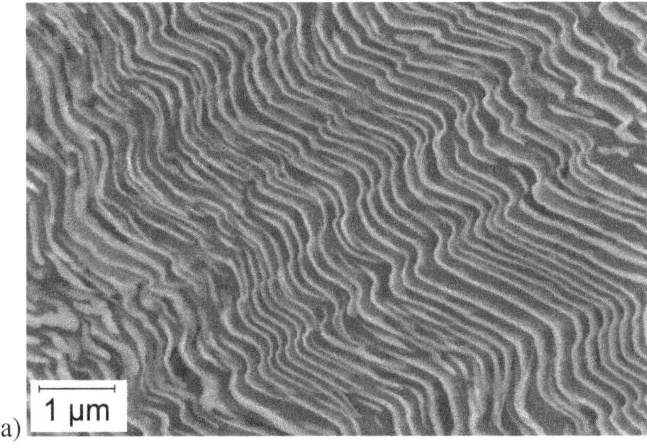

(a)

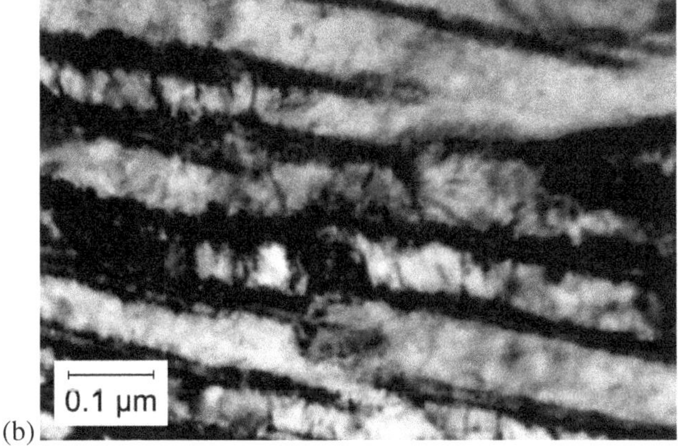

(b)

Figure 4.6 (a) Shear bands accompanied by the kinking of cementite, in deformed fully pearlitic steel that has a fine interlamellar spacing. Micrograph courtesy of Apparao Chintha.

(b) Transmission electron micrograph showing a shear band in deformed pearlite, whereby the cementite lamellae on the left- and right-hand sides of the image are no longer aligned. Micrograph courtesy of Professor Zhinan Yang, adapted from [32].

Equation 4.13 does not explicitly account for the work hardening rate when pearlite is deformed, other than by setting the mean free path for dislocations in the ferrite to a value $\overline{z}_{\perp}$ which is somewhat consistent with interlamellar spacing due to the barriers posed by cementite lamellae. This in part explains why the work-hardening rate of pearlite is so much greater than that of a fully ferritic microstructure. However, the transfer of load between ferrite and pearlite also has a role because the cementite constrains the ferrite which therefore, for a given plastic strain, deforms at a greater stress than would be the case for a single-phase ferritic microstructure [33].

More empirical approach

In his work on the plastic instability under conditions of plane stress, Swift [34] used an empirical $\sigma/\varepsilon_\mathrm{p}$ relationship that seems to work well for steels:

$$\sigma = b_{18}(b_{19} + \varepsilon_\mathrm{p})^{b_{20}} \tag{4.14}$$

It accounts for a yield point and Lüders strain so samples that exhibit discontinuous yielding are subjected to a small amount of plastic strain prior to testing, for example skin-pass rolling of sheet samples. Listings of the parameters b_i needed to apply Equation 4.14 are published (Table 4.2) [35, 36], but the empirical approach makes the interpretation of mechanisms uncertain.

Table 4.2
Empirical parameters in the Swift equation. $\overline{L}_{\alpha}^{\text{ASTM}}$ is the ASTM ferrite grain-size number, i.e., the number of grains per square inch when viewed at a magnification of $\times 100$. $w_{\text{C,ss}}$ is the carbon concentration that is dissolved in ferrite. $\overline{L}_{\alpha}$ and S_{I} are in micrometres. The range of applicability is included where available.

b_{18}^{α} / MPa	$109w_{\text{Si}} + 54.9w_{\text{P}} + 62.7w_{\text{Mn}} + 10.8(\overline{L}_{\alpha}^{\text{ASTM}})^{-\frac{1}{2}} + 373$	[35]
b_{19}^{α}	0.002	[35]
b_{20}^{α}	$0.35w_{\text{P}} - 0.037w_{\text{Si}} - 0.024w_{\text{Mn}} + \frac{5.35}{10+(\overline{L}_{\alpha}^{\text{ASTM}})^{-\frac{1}{2}}} - 0.12$	[35]
b_{18}^{P} / MPa	$189S_{\text{I}}^{-\frac{1}{2}} + 1140$	[35]
b_{19}^{P}	0.002	[35]
b_{20}^{P}	$0.202 - 2.798 \times 10^{-2}w_{\text{Si}} - 1.11 \times 10^{-2}w_{\text{Mn}} + 0.219(\overline{L}_{\alpha}^{\text{ASTM}})^{-\frac{1}{2}}$	[35]
b_{18}^{α} / MPa	$1550w_{\text{C}} - 31800w_{\text{C,ss}} + 142w_{\text{Si}} \ldots$ $\ldots + 51.3w_{\text{Mn}} + 2110w_{\text{P}} - 1320w_{\text{Nb}} + 428(\overline{L}_{\alpha})^{-\frac{1}{2}} + 457$	[37, 38]
b_{19}^{α}	0.002	[37, 38]
b_{20}^{α}	$4.25w_{\text{C}} - 52w_{\text{C,ss}} - 0.00204w_{\text{Si}} + 7.89 \times 10^{-3}w_{\text{Mn}} \ldots$ $\ldots + 1.16w_{\text{P}} - 3.38w_{\text{Nb}} - 0.459(\overline{L}_{\alpha})^{-\frac{1}{2}} + 0.307$	[37, 38]
b_{18}^{α} / MPa	$1446(\overline{L}_{\alpha})^{-\frac{1}{2}} + 140$	[39]
b_{19}^{α}	$0.003605(\overline{L}_{\alpha})^{\frac{1}{5}} - 0.004185$	[39]
b_{20}^{α}	$0.52721 - 0.2453(\overline{L}_{\alpha})^{\frac{1}{8}}$	[39]
b_{18}^{α} / MPa	$7190w_{\text{C}} - 27500w_{\text{C,ss}} + 2100w_{\text{P}} - 1010w_{\text{Nb}} + 230(\overline{L}_{\alpha})^{-\frac{1}{2}} + 459$	[40]
b_{19}^{α}	0.002	[40]
b_{20}^{α}	$20w_{\text{C}} - 42.4w_{\text{C,ss}} - 3.05 \times 10^{-3}4w_{\text{Si}} \ldots$ $\ldots + 1.24w_{\text{P}} - 2.54w_{\text{Nb}} - 0.934(\overline{L}_{\alpha})^{-\frac{1}{2}} + 0.299$	[40]
α range	$0.001 \leq \text{C} \leq 0.01,\ 0.02 \leq \text{Si} \leq 1.07,\ 0.04 \leq \text{Mn} \leq 0.08, \ldots$ $\ldots 0.003 \leq \text{P} \leq 0.048,\ 0 \leq \text{Nb}0.02$	
b_{18}^{P} / MPa	$423 + 97.95S_{\text{I}}^{-\frac{1}{2}} + 7909V_{\text{V}}^{\theta}$	[40]
b_{19}^{P}	0.001	[40]
b_{20}^{P}	$0.26 - 0.0517S_{\text{I}}^{-\frac{1}{2}} + 0.469V_{\text{V}}^{\theta}$	[40]
P range	$0.39 \leq \text{C} \leq 0.79,\ 0.75 \leq \text{Mn} \leq 1.95$	

4.3 FULLY PEARLITIC STEEL

Pearlite is not a phase, but a mixture of two phases. Once it starts to deform, the plastic strain distribution is initially somewhat uniform, with the cementite elongating along with the ferritic component. However, the deformation then becomes heterogeneous as shear bands form along with microscopic cracking. In fully pearlitic steels, *in situ* tests suggest that it is the cementite most aligned to the stress axis that cracks first (Figure 4.7). When the interlamellar spacing is small ($S_I \lesssim 60\,\text{nm}$,), there is a tendency for the collective cracking of parallel lamellae, whereas cracks can develop in a variety of lamellar orientations when the spacing is coarse [41]. This is because there is more room within the ferritic component in coarse pearlite for plasticity to evolve prior to cracking [42]. In all cases, the majority of damage is associated with strains beyond the average strain corresponding to the ultimate tensile strength. Many such detailed observations are at first neglected but will assume importance when discussing the ductility and toughness of pearlite (Chapter 5).

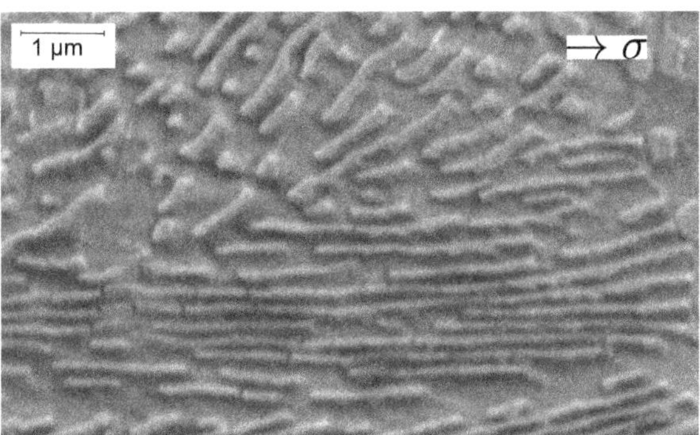

Figure 4.7 Cracks in pearlite with a fine interlamellar spacing, immediately after yielding. The fractured lamellae are aligned to the tensile stress axis. Micrograph courtesy of Professor Aparna Singh.

The proof strength σ_y of pearlite depends primarily on the interlamellar spacing S_I. The exact dependence has been the subject of debate for a number of reasons, the most common of which is the unphysical value of the strength when the interlamellar spacing is set to be infinite [e.g., 10]. A variety of relationships apparently can adequately represent measured strength data as a function of a length parameter [e.g., p. 230, 43]. It even has been suggested that the relationships should emulate the relaxation of an epitaxially deposited film, where the space available for dislocation sources to operate is the determining factor [44].

We know that pearlite is not a single phase, nor can it be described as consisting of alternating plates of cementite and ferrite [45]. Yielding occurs first in the softer ferrite which also is the predominant phase within pearlite. The cementite lamellae constrain this deformation and hence the role of the interlamellar spacing. It is useful therefore to examine the functional relationship between σ_y and S_I.

4.3.1 HALL-PETCH RELATIONSHIP

In a polycrystalline material, some crystals will be better oriented to react to the applied stress, leading to microscopic yielding in individual grains. Macroscopic yielding requires the propagation of plastic deformation across grains, but indirectly as stress concentrations at grain boundaries stimulate compatible dislocation sources in adjacent grains. The process does not involve dislocations traversing the boundaries because that would require a large degree of crystallographic continuity across grain interfaces [p.169, 28].

Figure 4.8 illustrates the mechanism, whereby the crystal on the left happens to be favourably oriented with respect to the applied shear stress τ_a, leading to a pile-up of like-dislocations at the grain boundary. Assuming that the length of the slip plane is equal to the mean lineal intercept $\overline{L}$ that defines the grain size, the number of dislocations involved in the pile-up, at equilibrium with the stress, is given for large n by

$$n = \frac{\overline{L}\tau_a}{2A} \quad \text{with} \quad A = \frac{E_s|\mathbf{b}|\,[\sin^2\phi + (1-\nu)\cos^2\phi]}{2\pi(1-\nu)} \tag{4.15a}$$

where ϕ is the angle between the dislocation line and its Burgers vector **b**, E_s is the shear modulus and ν is the Poisson's ratio [46, 47]. The pile-up has the effect of concentrating stress to an extent $n\tau_a$ at its tip so substituting for n using Equation 4.15a gives the stress there as

$$\tau_{tip} = n\tau_a = \frac{\overline{L}\tau_a^2}{2A} \quad \text{so that} \quad \tau_a = \sqrt{\frac{\tau_{tip}2A}{\overline{L}}}. \tag{4.15b}$$

Plastic deformation propagates when τ_{tip} reaches a constant critical value τ_c and after accounting for the friction stress τ^f which must also be overcome, the form of the classical Hall-Petch equation [48–50] is recovered:

$$\tau_y = \sqrt{\tau_c 2A} \times (\overline{L})^{-\frac{1}{2}} + \tau^f \tag{4.15c}$$

where τ_a has been replaced by the shear yield strength τ_y. If, in a polycrystalline material, the shear and uniaxial tensile stresses are related by the Taylor factor M [9], then it follows that

$$\sigma_y = \underbrace{M\sqrt{\tau_c 2A}}_{k_{HP}} \times (\overline{L})^{-\frac{1}{2}} + \sigma^f \tag{4.15d}$$

where k_{HP} is the Hall-Petch coefficient.

Dislocation pile-ups of the kind illustrated in Figure 4.8 are often not observed experimentally; instead, arrangements of dislocations described as *forests* within grain interiors are more common in iron. The general form of Equation 4.15d is nevertheless maintained when the forest dislocations are under stress from grain boundary steps [51, 52]. This is because the density ρ_f of such dislocations is proportional to the grain boundary area per unit volume, i.e., $1/\overline{L}$ and since the flow stress depends on $\sqrt{\rho_f}$, the net outcome remains the inverse square root dependence of flow stress on grain size.

Irrespective of the detailed mechanism, the friction stress σ^f is well defined. It probably does not matter that there may be a distribution of *true* S_I (p. 10), an average value of interlamellar spacing should suffice when dealing with macroscopic properties.

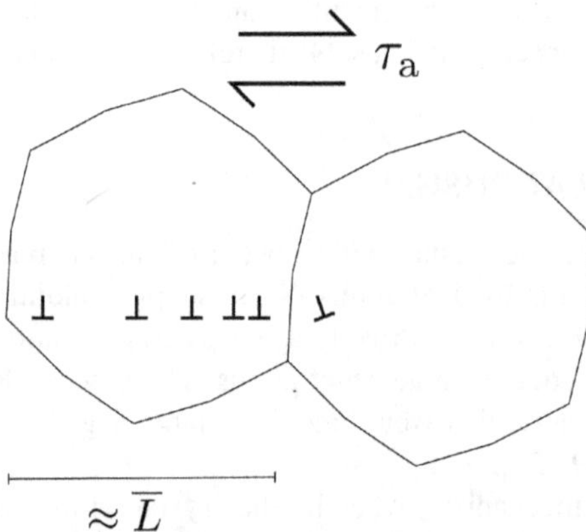

Figure 4.8 Pile-up of dislocations on a slip plane in the crystal on the left activates a dislocation source in an adjacent grain, thereby implementing general yielding in a polycrystalline material.

4.3.2 SLIP THROUGH CEMENTITE

Langford treated the deformation of pearlite in terms of the piled-up dislocations in ferrite slipping through the cementite, a model that led to a Hall-Petch type relationship.

Consider the deformation of cementite lamellae. We estimate the increment of shear strain caused by the translation of a dislocation of Burgers vector $\mathbf{b}_\theta$ across a cementite (θ) plate of length L_1, width L_2 and thickness L_θ. Assuming that $\mathbf{b}_\theta$ makes an angle of 45° with L_1, the shear strain is the magnitude of the displacement divided by the height over which that displacement occurs, i.e., [53]

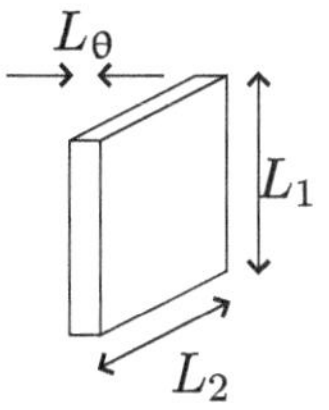

$$\Delta\gamma = \frac{|\mathbf{b}_\theta|}{\sqrt{2}L_1} \tag{4.16a}$$

and the work necessary to achieve this is the product of the shear stress and plastic shear-strain, which consists of components due to the friction shear stress, and the product of the line energy per unit length Γ_θ and the length of the dislocation, assuming that the latter has to be created:

$$\tau_\theta \Delta\gamma = \tau_\theta^{\mathrm{f}} \Delta\gamma + \frac{\Gamma_\theta L_2}{L_1 L_2 L_\theta}, \tag{4.16b}$$

noting that for pearlitic cementite, the thickness $L_\theta \ll L_1$ or L_2 and that the quantities $\Gamma_\theta L_2$ and $L_1 L_2 L_\theta$ represent the energy of the dislocation and the volume of the cementite plate respectively, so the term on the right has units of $\mathrm{J\,m^{-3}}$. Combining these equations gives

$$\tau_\theta = \tau_\theta^{\mathrm{f}} + \sqrt{2}\frac{\Gamma_\theta}{|\mathbf{b}_\theta| L_\theta}. \tag{4.16c}$$

Langford [53] considered that τ_θ calculated using Equation 4.16c is smaller than that required to deform pearlite but did not quote the values of τ_θ^{f} and Γ_θ used to reach this conclusion, but can be deduced as follows.

The ambient temperature Vickers hardness of almost pure cementite in cast iron is found to be just over 1000 HV corresponding to an ultimate strength in tension of about 3.3 GPa (≈hardness/3, [54]). Assuming that the friction stress in tension is smaller than this by about 10%, the *shear* friction strength $\tau_\theta^{\mathrm{f}} \approx 1.5$ GPa.

Cementite has a primitive lattice with space group *Pnma* and given its large lattice parameters, the magnitude of any Burgers vector of a slip dislocation will inevitably be large. Since the most commonly observed slip system is $(010)[001]$ [55, 56], $|\mathbf{b}_\theta| = 0.45165$ nm, which is much larger than that of any slip dislocation in austenite or ferrite.

The energy per unit length of a dislocation in an average population of screw and edge characters is given by [57]

$$\Gamma_\theta \approx \frac{E_{\mathrm{s}}^{\theta}|\mathbf{b}_\theta|^2}{4\pi}\frac{1-0.5\nu_\theta}{1-\nu_\theta}\ln\left\{\frac{L_\infty}{|\mathbf{b}_\theta|}\right\} + \Gamma_\theta^* \tag{4.16d}$$

where Γ_θ^* is the core energy per unit length; in the absence of information, this is taken here to be identical to that in ferrite, i.e., $\Gamma_\theta^* \approx 3 \times 10^{-10}\,\mathrm{J\,m^{-1}}$ [58]. The extent of the elastic-strain field L_∞ of the dislocation depends on stress screening by other dislocations; assuming that there are more dislocations present in the adjacent ferrite, this length could be approximated as the interlamellar spacing, typically less than a micrometre. The isotropic shear modulus and Poisson's ratio for cementite have been estimated using first principles calculations and averaging to be 75 GPa and 0.35 respectively [59, 60]. Using these values, $\Gamma_\theta \approx 1.2 \times 10^{-8}\,\mathrm{J\,m^{-1}}$ (an order of magnitude greater than that of dislocation in ferrite [61]). With these data and assuming that the cementite thickness $L_\theta = 0.2\,\mu\text{m}$, the value of the friction stress becomes $\tau^f_\theta \approx 1.7\,\text{GPa}$[8].

The shear yield strength of pearlite was then expressed using the Hall-Petch relation based on the work of Li and Chou [47]:[9]

$$\tau_P = \tau^f_\alpha + \sqrt{2E^\alpha_s|\mathbf{b}_\alpha|\tau_\theta \frac{2-\nu_\alpha}{4\pi(1-\nu_\alpha)}} \left(\frac{1}{L_\alpha}\right)^{\frac{1}{2}} \tag{4.16e}$$

where L_α is the thickness of ferrite, which at equilibrium is related to the cementite thickness L_θ by the fraction of cementite in a eutectoid steel. The term containing the Poisson's ratio is determined by an assumption of a mixed population of screw and edge dislocations orientations. Langford's model never seems to have been tested against data but Equation 4.16e leads to a gross overestimation of the yield strength of pearlite, even when τ_θ is reduced from the estimated 1.7 GPa to 0.2 GPa. For example, with $L_\theta = 0.02\,\mu\text{m}$, $L_\alpha = 0.118\,\mu\text{m}$ (giving $S_I = 0.138\,\mu\text{m}$), $\tau^f_\alpha = 77/2\,\text{MPa}$, $\tau^f_\theta = 1700\,\text{MPa}$, $|\mathbf{b}_\theta| = 0.452\,\text{nm}$, $|\mathbf{b}_\alpha| = 0.25\,\text{nm}$, the shear stress given by Equation 4.16e comes to $\tau_P = 2.5\,\text{GPa}$, i.e., a tensile strength twice that value whereas the actual tensile strength for that interlamellar spacing is 580 MPa. Reducing τ^f_θ to 200 MPa makes $\tau_P = 1.8\,\text{GPa}$.

It is known from direct observations using transmission electron microscopy that the yield strength of pearlite is that needed to move dislocations in ferrite between 'impenetrable' cementite walls; both yielding and work hardening are said to be controlled by dislocation activity within the ferrite [10]. Therefore, the assumption in the Langford model that a pile-up in ferrite will induce flow in the adjacent cementite, may not be justified given the tremendous disparity in the strengths of the two phases. Until the stress within the ferrite accumulates to a sufficient level, the cementite must remain elastically strained or strained to fracture as illustrated in Figure 4.7.

Bearing in mind that a pearlite colony is an interpenetrating bicrystal of α and θ [45], initial yielding may be determined by dislocation pile-ups at the α/θ interfaces stimulating other adjacent ferrite to yield, in which case a Hall-Petch relation would prevail with the cementite acting simply as a means to transmit stress. Given that the stress ahead of a dislocation pile-up diminishes with distance beyond its tip, there is a special strengthening effect because this intervening cementite makes it more difficult for yielding to be stimulated in the ferrite on the other side of the dislocation pile-up. This is separate from the usual strengthening due to the density of α/θ interfaces.

4.3.3 ASSESSMENT OF HALL-PETCH RELATION FOR PEARLITE

The dependence of the yield strength of pearlite on interlamellar spacing has been expressed in many different ways, all apparently with reasonable levels of empirical fit to experimental data. For example, it is stated in elementary metallurgical texts that equally linear dependence is observed between σ_y and the functions of grain size: $\overline{L}^{-1/2}$ and $\overline{L}^{-1}$ [p.230, 43]. It is worth delving into this in some detail in order decipher the physically correct mechanism of strengthening in the context of pearlite [28].

Figure 4.9 shows data from eutectoid steels containing similar, small concentrations of manganese and silicon. A comparison of Figures 4.9a & b shows that the correlation of the strength against $S_I^{-\frac{1}{2}}$ is better than against S_I^{-1}. It is known that in pearlite, yielding initiates in ferrite, so

[8]The magnitude of this stress contradicts the statement in the original work [53] that Equation 4.16c leads to a flow stress of cementite that is less than that of pearlite. It is possible that Langford neglected to consider τ^f_θ since its value was not stated.

[9]The ratio $|\mathbf{b}_\theta|/|\mathbf{b}_\alpha|$ in the original treatment is removed from this equation because it does not seem justified. Langford also assumed that the cementite deformation occurs by the propagation of a partial dislocation, which also is not justified so the Burgers vector corresponding to the common slip system, i.e., [001] is substituted. Similarly, Langford's $2L_\alpha$ term has been replaced by just L_α.

the intercept at $S_I = 0$ should correspond to the friction stress σ_α^f. Figure 4.9a has $\sigma^f = 128\,\text{MPa}$ for the Hall-Petch $S_I^{-\frac{1}{2}}$ dependence. Iron-containing 500 ppm of carbon has $\sigma^f = 100\,\text{MPa}$ [22, 62]; this concentration agrees with an atom probe determination of the carbon concentration of ferrite within pearlite [63]. Pearlitic steels typically contain 0.3Si and 0.4Mn wt%, which using Leslie's solution strengthening data [19] gives a combined contribution of 33 MPa so σ^f estimated in this way is 133 MPa, eminently consistent with the plot in Figure 4.9a represented by

$$\sigma_y \approx 174 S_I^{-\frac{1}{2}} + 128 \quad \text{MPa} \tag{4.17}$$

with S_I in micrometres.

In contrast, Figure 4.9b, with the dependence on S_I^{-1}, reveals $\sigma_\alpha^f =$ to be 397 MPa, a value that is too large and inconsistent with the known friction strength of ferrite.

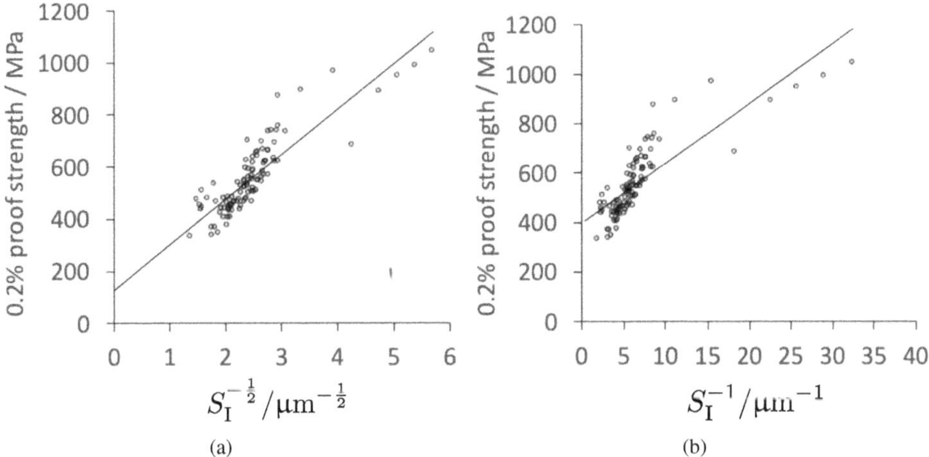

Figure 4.9 Yield strength of fully pearlitic, eutectoid, hypereutectoid and hypoeutectoid steels, none of which were in a plastically deformed state prior to testing. (a) Plotted against the inverse of the square root of the interlamellar spacing, with a correlation coefficient of 0.87. The intercept on the vertical axis for an infinite spacing, corresponding to the friction stress in the Hall-Petch equation, is $\sigma_\alpha^f = 128\,\text{MPa}$ with the slope of the regression line being $174\,\text{MPa}\,\mu\text{m}^{\frac{1}{2}}$. (b) Plotted against the inverse of the interlamellar spacing, with a correlation coefficient of 0.83. The intercept in this case is $\sigma_\alpha^f = 397\,\text{MPa}$ and the slope, $24\,\text{MPa}\,\mu\text{m}^{\frac{1}{2}}$.

The inverse square-root dependence has previously been criticised as leading to a negative friction-stress at infinite interlamellar spacing [e.g., 10]. This, however, is a consequence of analyses depending on limited data based on individual studies, a classic problem widely recognised in machine learning, of *overfitting* [64, 65], which leads to relationships then do not generalise well. This is illustrated in Figure 4.10 where data from individual experiments are plotted separately to reveal the large differences in slopes and intercepts when each plot covers a limited range of interlamellar spacing. Although the individual plots may exhibit high levels of correlation, they clearly would not generalise well on data from other experiments as is obvious by extrapolating any one of the straight lines. In some cases [e.g., 66], fitting to limited data leads to a negative friction stress!

It has to be concluded that one of the mistakes in fitting the Hall-Petch equation to limited data is to assume that σ_α^f can take on any value determined by best fit, whereas it is in fact easy to estimate its value independently. In regression analyses, the friction stress is not, and should not be treated as a fitting parameter.

It is instructive to compare the Hall-Petch coefficient k_{HP} from Equation 4.17 against data from interstitial-free ferrite, where the grain size of the recrystallised ferrite is comparable to the interlamellar spacing of pearlite. In Figure 4.11 the line fitted to the ferrite data had the intercept fixed at

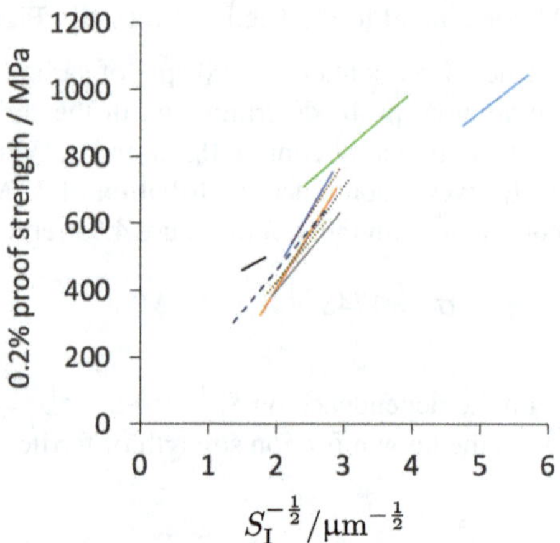

Figure 4.10 Hall-Petch best fits to individual datasets on the strength of pearlite. The totality of the data used here is identical to those in Figure 4.9.

$\sigma^f = 50\,\text{MPa}$ to be consistent with the value reported by Takaki [62] and can be represented by the equation:

$$\sigma_y \approx 413(\overline{L})^{-\frac{1}{2}} + 50 \quad \text{MPa} \tag{4.18}$$

where $\overline{L}$ is the grain size defined by the mean lineal intercept in units of μm. The astonishing result is that the ferrite grains are stronger than the ferrite associated with pearlite, at an identical length scale. Therefore, the sensitivity of the ferrite grains to grain size is much greater (larger k_{HP}) than is the case for pearlite as a function of the interlamellar spacing.

There are two reasons for this – first, that in a mixture of a soft and hard phase, the soft phase naturally yields first, but at a stress that is less than its yield strength in isolation (p. 50). Secondly, fully pearlitic steels yield continuously because the interfaces between ferrite and cementite have sufficient coherency to cause the emission of dislocations when stressed [4]. The continuous yielding persists even when the structure is subjected to low-temperature heat treatment of the kind associated with strain-ageing experiments. When annealed to induce a loss of coherency, discontinuous yielding and Lüders strains follow. This kind of a partly coherent interface that emits dislocations under applied stress would reduce the proof strength whereas the grain boundaries between the recrystallised-ferrite crystals in the interstitial-free steel produced using accumulative roll bonding (as in [67]) are incoherent, large misorientation boundaries [68].

4.3.4 THE INVERSE RELATIONSHIP

There is no theoretical justification for $\sigma_y \propto S_I^{-1}$ relationship. A composite model [52] in which cubic grains of size L_{soft} are surrounded by a hard phase of thickness L_{hard} predicts the strength to be a function of L_{hard}/L_{soft} provided $L_{hard} \ll L_{soft}$, i.e., an inverse dependence on the size of the soft phase. However, with the ratio for pearlite fixed by its chemical composition, the dependence of strength on interlamellar spacing vanishes.

A calculation of the lower-bound value of the critical stress necessary to push a dislocation in ferrite between two non-deforming cementite particles of planar separation L_α, leads to the relationship [69, 70]

$$\sigma_c = \sigma_\alpha^f + \frac{ME_s|\mathbf{b}_\alpha|}{2\pi L_\alpha}\left(\frac{1-0.5\nu_\alpha}{1-\nu_\alpha}\right)\ln\left\{\frac{L_\alpha}{|\mathbf{b}_\alpha|}\right\} \tag{4.19}$$

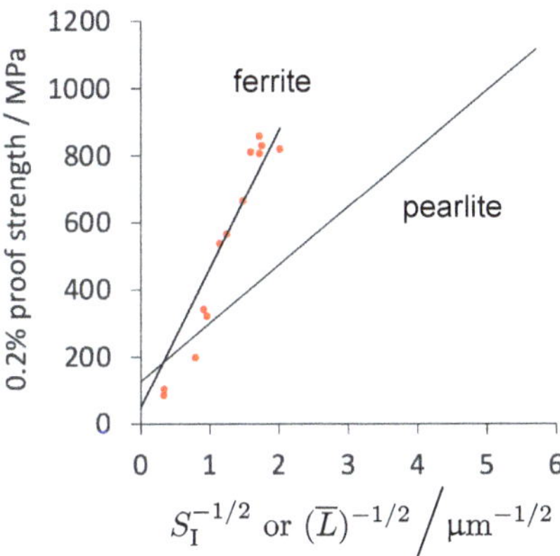

Figure 4.11 The ferrite data are from Tsuji et al. [67], representing interstitial free iron, and the line for pearlite is that from Figure 4.9a. The interstitial-free alloy has sufficient titanium to justify the assumption that the ferrite truly has a negligible carbon concentration in solution.

where the term $L_\alpha = S_\mathrm{I} V_\mathrm{V}^\alpha$ is the thickness of the ferrite layer. With σ_α^f given by Equation 4.7, the Taylor factor $M = 2.8$, $\nu_\alpha = 0.29$, Figure 4.12a shows that a reasonable *correlation* is obtained between the calculated and measured proof strengths. Nevertheless, Figure 4.12b shows the application of Equation 4.19 that for $S_\mathrm{I} < 50\,\mathrm{nm}$, the strength is grossly overestimated.

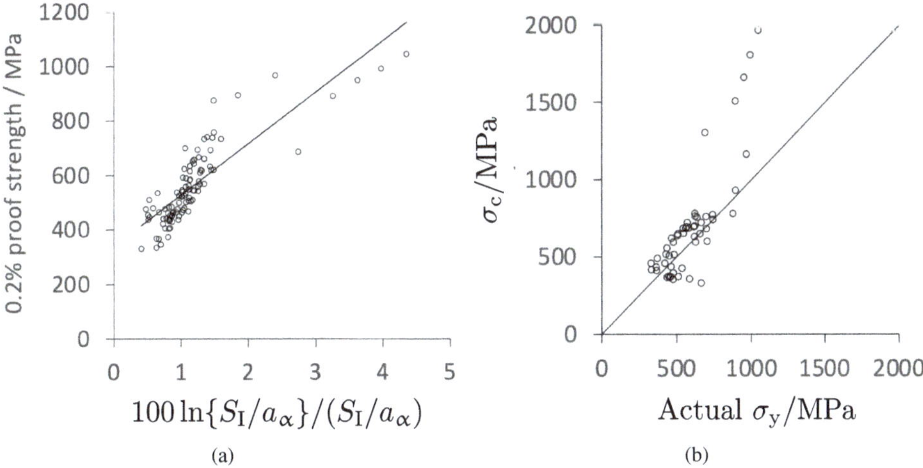

Figure 4.12 (a) Correlation of strength against a more complex function originating in thin-film theory; the units of both S_I and the lattice parameter of ferrite are in μm, correlation coefficient 0.85. The line has a unit slope and zero intercept. (b) Proof strength calculated using Equation 4.19 against measured data [10, 71–78].

An approach somewhat analogous to that of Langford has been claimed to pose a fundamental challenge to the Hall-Petch relationship. It has its origin in the semiconductor scenario, with an equation derived to estimate the elastic misfit strain that can be supported by a thin epitaxial layer deposited on a substrate of a different material. Beyond a critical thickness for a given misfit, the deposited layer should relax by the introduction of dislocations. With the assumption that a distance that is a multiple of this critical thickness must be used to account for the operation of dislocation sources, the strength is supposed to vary with grain size (here taken to be the interlamellar spacing) as [44]

$$\sigma_y \propto \frac{\ln\{S_I/a_\alpha\}}{S_I/a_\alpha} \tag{4.20}$$

where a_α is the lattice parameter of ferrite to make the grain size dimensionless. Although the original paper [44] uses grain size and dimensionless grain size interchangeably in describing the relationship with strength, this procedure in fact introduces two dependencies on interlamellar spacing because

$$\frac{\ln\{S_I/a_\alpha\}}{S_I/a_\alpha} \equiv a_\alpha \left(\frac{\ln\{S_I\}}{S_I} - \frac{\ln\{a_\alpha\}}{S_I} \right). \tag{4.21}$$

It isn't surprising, therefore, that the correlation of the data using the spacing term in Equation 4.20 is almost identical to that against S_I^{-1}, *cf.* Figures 4.12a,b because the inverse dependence is dominant in Equation 4.21 for $S_I \gtrsim 0.2\,\mu$m, which represents data for the majority of fully pearlitic steels, with small differences between these two relationships evident for the finest of spacings.

The proportionality constant in Equation 4.20 when converted into a dimensionless number by dividing with the Young's modulus (E) is expected to be of the order of unity [44]. The fitted line in Figure 4.12a can be written as:

$$\sigma_y = 188.9\left(\frac{100\ln\{S_I/a_\alpha\}}{S_I/a_\alpha} \right) + 339 \quad \text{MPa} \tag{4.22}$$

$$\text{so that } \frac{\sigma_y}{E} = \underbrace{\frac{1.889 \times 10^4}{E}}_{\text{coefficient}} \left(\frac{\ln\{S_I/a_\alpha\}}{S_I/a_\alpha} \right) + \frac{339}{E}. \tag{4.23}$$

Since $E = 2.1 \times 10^5$ MPa, the coefficient has a dimensionless value 0.09, which is about a tenth of that expected from Equation 1.3 of [44]. It follows that if the coefficient is close to unity then the theory should greatly overestimate the strength.

Notice that the form of the term $\ln\{S_I/a_\alpha\}/(S_I/a_\alpha)$ in Equation 4.21 is not similar to that of $\ln\{L_\alpha/|\mathbf{b}_\alpha|\}/(L_\alpha/|\mathbf{b}_\alpha|)$ of Equation 4.19 because $L_\alpha = S_I \times V_V^\alpha$ is not necessarily equal to S_I. At a constant S_I, the volume fraction of ferrite in pearlite depends on the average carbon concentration of the steel – both hypereutectoid and hypoeutectoid steel can be made fully pearlitic if the transformation is suppressed into the Hultgren-Belaiew extrapolation region (p. 193).

Equation 4.17 works rather well on independent data; Figure 4.13 illustrates careful measurements made on simple, model eutectoid-alloys: Fe-0.78C, Fe-0.8C-0.99Si, Fe-0.82C-0.97Mn, Fe-0.82C-2.01Si, Fe-0.77C-1.96Mn, Fe-0.79C-1Si-1Mn wt% [79]. Some of these alloys contain up to 3 wt% of substitutional solutes, and yet agreement is good even though σ_f is unchanged from the value in Equation 4.17. This is because the strength due to interlamellar spacing represents the largest proportion of total strength, more so at the smallest of interlamellar spacings. Therefore, the error due to the neglect of solid solution effects will be reflected in the noise, i.e., the scatter about the expectation. Solid-solution effects are in fact difficult to account for given that only average concentrations are available, whereas there is some partitioning expected between the ferrite and cementite at all transformation temperatures [80].

4.3.5 ANALYSIS IN BAYESIAN FRAMEWORK

The regression analysis of strength data results in unique values of the best-fit weights, which in linear analysis represent the slope and intercept, without commenting on how they depend, for example, on the size of the dataset. Figure 4.10 illustrates vividly how this can lead to misleading interpretations; not only are the weights different for each line, with no facility to indicate changes in uncertainty during extrapolation or interpolation.

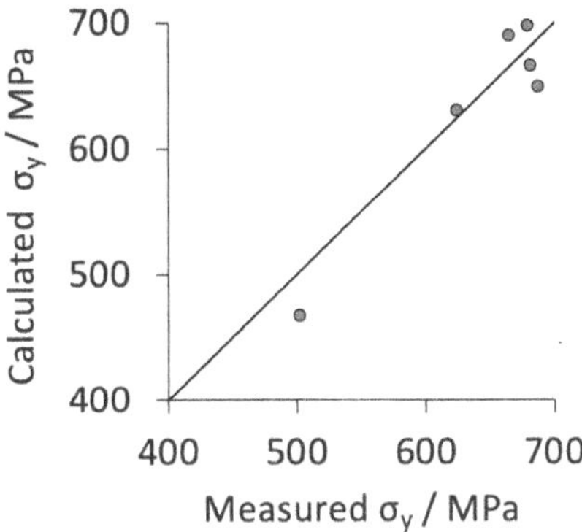

Figure 4.13 Comparison of the yield strength of pearlite as calculated using Equation 4.17 against experimental data due to Fu et al. [79]. The line has unit slope and zero intercept.

Bayesian inference on the other hand, captures how an estimate should depend on the availability of related evidence, i.e., the uncertainty should be greater in domains where data are sparse. Instead of providing the best-fit weights, each weight then has a *distribution* of possible values, reflecting the level of certainty about the weights *given the dataset.* The output is not estimated as a single value but as a probability distribution. A greater certainty in the prediction corresponds to a narrow distribution of weights in the context of the data. This is powerful because the uncertainty of any estimate can be quantified. This uncertainty is separate from the constant value of the perceived noise σ_v in the output, where noise refers to different outcomes during repeated experiments, attributed to missing variables or experimental error. The Bayesian uncertainty will vary with the location of the calculation in the input domain. The general principles have been described by MacKay [81–83].

To ensure that an empirical fit to data does not model noise (i.e., the overfitting problem), the data can be divided at random into a training and a test set, where the former is used to create the model and the latter to assess how it generalises on unseen experiments. Figure 4.14 shows two kinds of plots for fully pearlitic steels. It is evident that the Hall-Petch relationship is best suited to represent the propagation of yield across the structure of pearlite over a broad range of variation, with an optimum fit to both the training and test data. In contrast, the other two functions only show consistency between experiment and calculations over a narrow range of strength, with the consequence that there are very many outliers.

The same method can be applied to include all three functions in a multiple linear regression within the Bayesian framework. Figure 4.15a shows an interesting consequence of the fact that the S_{I}^{-1} and $\ln\{S_{\mathrm{I}}\}/S_{\mathrm{I}}$ represent only a narrow range of strength; the uncertainty becomes very large outside of this range because these two functions deviate so much from measured data. On the other hand, the Hall-Petch relation nicely follows the general trend of the multiple regression function.

Figure 4.15b shows the significance in the multiple regression analysis, of each of the three inputs S_{I}^{-1}, $S_{\mathrm{I}}^{-1/2}$ and $\ln\{S_{\mathrm{I}}\}/S_{\mathrm{I}}$. The term σ_{w} represents the values of the regularisation constants, expressed as standard deviations of the implicit Gaussians [81, 82]; in simple terms, it is related to the effectiveness of a particular input in explaining the variation in the output. The Hall-Petch parameter is again perceived to be the best at representing the variation of the strength of pearlite as a function of the interlamellar spacing.

It is noted that Occam's razor [84] is in the present case less useful as a method for distinguishing between the models because they all are linear functions with just the slope and intercept as coefficients.

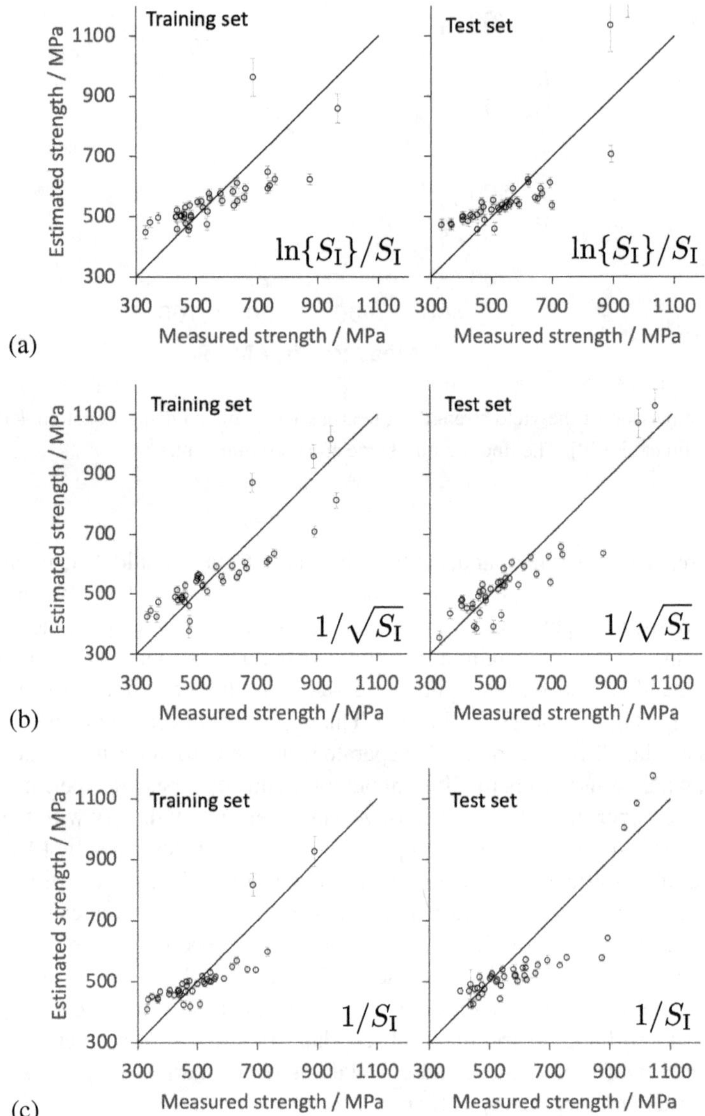

Figure 4.14 Linear regression in a Bayesian framework using the training data for model creation and the unseen test data to assess generalisation, using the method by MacKay [81, 82]. The uncertainty illustrated is due to the distribution of weights; the noise σ_ν has not been included in the error bars, for reasons of clarity. (a) Assuming that the strength varies with $\ln\{S_I\}/S_I$ (units μm), with $\sigma_\nu = \pm 0.13$. (b) The Hall-Petch relation, with $\sigma_\nu = \pm 0.11$. (c) The inverse relation, with $\sigma_\nu = \pm 0.11$.

In summary, the Hall-Petch equation is the correct representation of the proof strength of fully pearlitic steels, given that deformation initiates in the much softer ferrite. The friction stress deduced with this equation is physically reasonable whereas that is not the case with other interpretations.

Some of the problems perceived with the Hall-Petch equation in previous work are associated with overfitting of data. A discovery is that ferrite in interstitial-free iron is, at the same length scale,

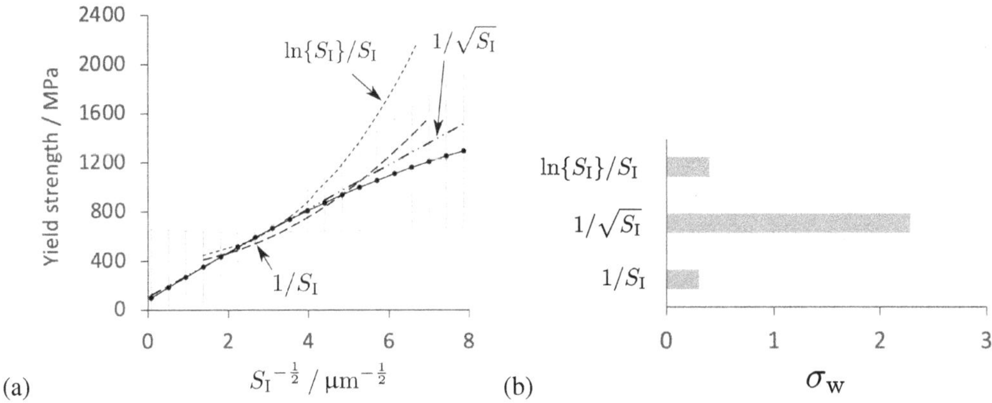

Figure 4.15 (a) The curve that includes uncertainties associated with the distribution of weights represents multiple linear regression, within a Bayesian framework, of predicted strength as a function of S_I^{-1}, $S_I^{-1/2}$ and $\ln\{S_I\}/S_I$. The perceived noise, not included in the plot, is $\sigma_\nu = \pm 0.09$. The other three curves are predictions using the individual models illustrated in Figure 4.14a-c. (b) The model perceived significance of each of the functions included in the multiple regression.

stronger in yield than the ferrite within pearlite; its sensitivity to grain size is also much greater than that of pearlite to interlamellar spacing. In all cases, irrespective of model, the perceived noise in the data is of the order ± 0.1, i.e., 10%, which might represent reasonable errors of measurement.

There are constitutive models available, for example [85], but they contain specific assumptions about the nature of plastic deformation and require at least ten 'calibration' parameters that are unlikely to have generic value. Unlike the Hall-Petch equation, elementary parameters such as the yield strength are not well modelled. The yield strength after all, is the critical engineering design parameter because the steel should not be loaded beyond σ_y.

4.3.6 HETEROGENEOUS STRAINS IN UNDEFORMED PEARLITIC-FERRITE

When a large piece of crystalline material that exhibits perfect long-range order is exposed to monochromatic X-rays, refections occur at specific Bragg angles, each exhibiting an imperceptibly narrow spread in a plot of the intensity against the angle of the reflected beam relative to the incident beam. Imperfections such as dislocations cause *heterogeneous* elastic strains (ε_h) that broaden the Bragg peaks, a phenomenon that can be used to estimate the strains [86]. Homogeneous strains would not cause peak broadening but would lead to a change in lattice parameter.

Figure 4.16 illustrates this for pearlite generated isothermally at the temperatures indicated over a period of just 60 s, followed by water cooling. The defect density is expected to increase at reduced transformation temperatures because the growth defects are more likely at large growth rates. Thus, massive ferrite in an interstitial-free alloy of iron has a dislocation density of $1.2 \pm 0.4 \times 10^{14}\,\mathrm{m}^{-2}$ whereas the proeutectoid variety which grows at a slower rate in the same alloy, has a dislocation density that is smaller by an order of magnitude [87].

The strain measurements illustrated in Figure 4.16 can be used to estimate the dislocation density of pearlitic-ferrite using the relationship $\rho_\perp \approx 14.4\varepsilon_h^2/|\mathbf{b}_\alpha|^2$ [88]; therefore, the dislocation density of pearlitic-ferrite would be in the range 2.5-31.5 $\times\, 10^{14}\,\mathrm{m}^{-2}$ for pearlite produced during a very short transformation time followed by water quenching. However, pearlitic steels that are cooled slowly have a much smaller $\rho_\perp \approx 6 \times 10^{13}\,\mathrm{m}^{-2}$ [89].[10]

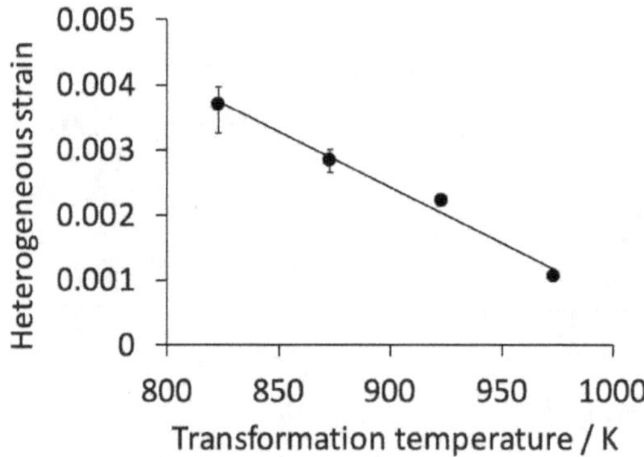

Figure 4.16 Heterogeneous elastic strains due to the defect structure within pearlitic ferrite, as a function of the isothermal transformation temperature. Selected data from a fully pearlitic Fe-0.9C-0.9Mn-0.4Si wt% steel, Nakada et al. [90].

4.3.7 HETEROGENEOUS STRAIN IN DEFORMED PEARLITE

Neutron diffraction can expose the elastic strains that develop in the cementite and ferrite within pearlite that is subjected to tension, Figure 4.17. The strains are deduced from changes in the spacings of particular crystallographic planes. Both ferrite and cementite are elastically anisotropic (Figure 3.7) so the magnitude of the strain depends on the crystallographic direction but it turns out that there is not much of a difference in the trend observed as a function of stress.

It is particularly interesting that the elastic strains in both θ and α are similar until the onset of plasticity in the ferrite. At that point, the cementite which has not yet yielded, continues alone to accommodate the stress by elastic deformation, so $|\varepsilon^\theta|$ increases while $|\varepsilon^\alpha|$ relaxes towards zero [92, 93]. The cementite therefore carries a greater proportion of the applied stress. It would be interesting to continue the measurements to the onset of plasticity in the cementite.

4.4 TEMPERATURE AND STRAIN RATE EFFECTS

The temperature and strain rate sensitivity of pearlite-containing steels are important, for example, in the modelling of machining and ballistic impact. A constitutive equation that is an adaptation of Equation 4.14, has the von Mises tensile flow stress given by [94]

$$\sigma_{\mathrm{vM}} = \sigma\{\varepsilon_{\mathrm{p}}, \dot{\varepsilon} \approx 10^{-3}\,\mathrm{s}^{-1}, T \approx 25^\circ\mathrm{C}\} \underbrace{\left[1 + b_{21} \ln\left\{\frac{\dot{\varepsilon}}{\dot{\varepsilon}_\circ}\right\}\right]}_{\sigma\{\dot{\varepsilon}\}/\sigma\{\dot{\varepsilon}_\circ\}} \underbrace{\left[1 - \left(\frac{T - T_\circ}{T_{\mathrm{m}} - T_\circ}\right)^{b_{22}}\right]}_{\sigma\{T\}/\sigma\{T_\circ\}} \tag{4.24}$$

[10]Nakada et al. [90] argued that the relationship between strength and interlamellar spacing may be fortuitous, the strain within the pearlitic-ferrite instead being the controlling feature. This is wrong because the measured lattice strain is assumed to be uniform whereas it must be interpreted as *heterogeneous* [86]. The formidable barriers to plastic flow provided by cementite are not to be neglected. In a separate study, Nakada and Kato [91] have attempted to calculate the strains at the α/θ interface for particular orientation relationships, incorrectly assuming that the interface plane can be deduced from the orientation relationship alone.

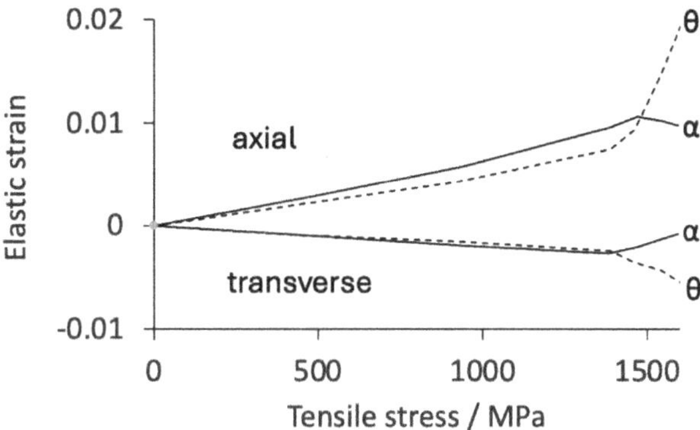

Figure 4.17 Elastic strains derived by measuring the $\{122\}_\theta$ and $\{200\}_\alpha$ interplanar spacings as a function of applied tensile-stress. The sharp change in gradient as a function of stress corresponds to the onset of plasticity in the ferrite. The steel composition is Fe-0.82C-0.23Si-0.73Mn wt%, with a fully pearlitic microstructure. Selected data from Kanie et al. – many more data are to be found in [92].

where $T_\circ$ is a reference temperature often taken to be $20\,^\circ$C and $\dot{\varepsilon}_\circ$ is used to normalise the plastic strain rate. The remaining empirical constants have to be determined by fitting to experimental data, but as seen later b_{22} can be set to unity because the stress is found experimentally to be linearly related to temperature below ambient. However, most data generated for finite element models of machining rely on properties at temperatures greater than ambient; data obtained by compression testing at $\dot{\varepsilon} < 1.1\,\mathrm{s}^{-1}$ and split Hopkinson bar experiments for $\dot{\varepsilon} > 1000\,\mathrm{s}^{-1}$ indicate the parameters of Equation 4.24 as follows, with $\dot{\varepsilon}_\circ = 0.002\,\mathrm{s}^{-1}$, $T_\mathrm{m} = 1500\,^\circ\mathrm{C}$ and $T_\circ = 25\,^\circ\mathrm{C}$ [95][11]:

C / wt%		b_{21}	b_{22}
0.75	$(0.99\mathrm{P}+0.01\alpha)$	0.011	1.10
0.45	$(0.60\mathrm{P}+0.40\alpha)$	0.015	1.22
0.05	$(0.01\mathrm{P}+0.99\alpha)$	0.034	1.86

A fully pearlitic microstructure therefore shows the smallest temperature and strain-rate sensitivity. Notice that the temperature and strain-rate effects are not affected by the magnitude of ε_p.

Temperature and strain rate are treated as independent effects in Equation 4.24 and strain softening which causes the localisation of deformation is not included. As a result, empirical modifications of the relationship are available in the literature [96].

There are no data for the hardness of cementite or of pearlite, except at temperature above ambient; it increases from 300-1000 HV as the temperature is reduced from 800-300 K [60]. Figure 4.18 compares the temperature sensitivity of strength of an α+P steel against that of ferrite in pure iron. Not surprisingly, the former is stronger at ambient temperature, but not so at the lowest of temperatures. This is because the ferrite strengthens rapidly as the test temperature is reduced; the screw dislocations within have a three-dimensional core structure that must change for the dislocations to move, leading to a large activation energy [27, 97]. In ferrite deformed at low temperatures or large strain rates, the edge components of dislocations are more mobile than the screw components, thus leaving elongated screws in the microstructure. Single crystals and polycrystalline samples of pure iron have identical temperature sensitivity of strength.

[11] Equation 1 of [95] is incorrectly presented, the second + sign should be a multiplication sign.

A three orders of magnitude increase in the strain rate leads to large increase in the yield strength of α+P steel, Figure 4.18 but the temperature dependence of flow stress is only slightly greater at the higher strain rate.

The role of mechanical twinning within the ferrite has been used to interpret the temperature dependence of its strength. Such twinning does occur at low temperatures but metallography [98] suggests that the volume fraction of twins is small, with some measurements indicating a fraction less than 0.05 for peak impact stresses up to 8.9 GPa [99]. The data interpreted originally to show a sharp change in slope of the σ_y-T curve for the Fe-0.19-0.51Mn steel, associated with the onset of twinning, have been replotted in Figure 4.18 using a polynomial fit demonstrating a continuous increase in strength as the temperature is reduced, not consistent with the role of twinning on the temperature dependence of strength. Each of the curves can be approximated by straight lines below 0 °C, consistent with $b_{22} = 1$ in Equation 4.24.

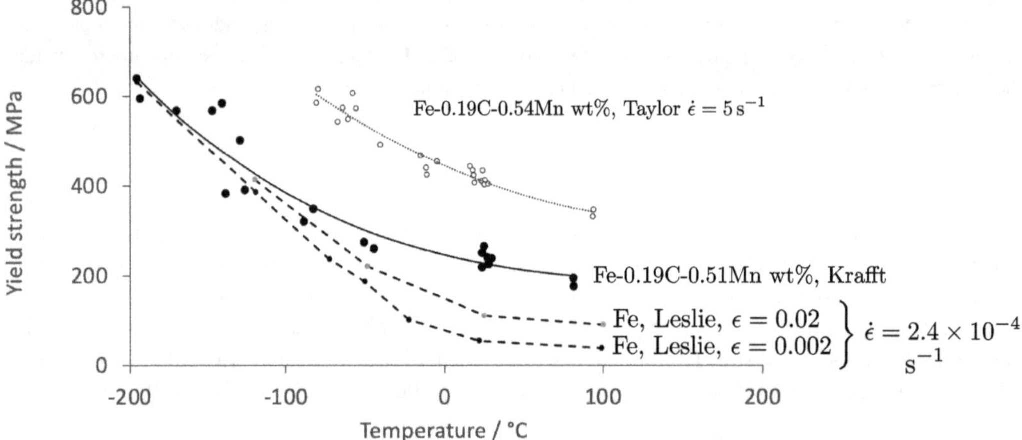

Figure 4.18 Temperature dependence of the compression yield-strength of pure iron and of an α+P steel containing about 25 vol.% of pearlite. The dashed curves represent data for pure iron where the yield is defined at either 0.002 or 0.02 plastic strain [19]. The data due to Krafft et al. [98] defined the yield strength at plastic strains in the range 0.01–0.03; although the strain rate was not stated, the tests were classified as static as opposed to dynamic, implying rates comparable to the Leslie data. The Taylor data [100] measure the lower yield-strength values.

Fully pearlitic steels also strengthen as the temperature is reduced, the largest dependence being for the coarsest S_I, Figure 4.19a,b, because the α_P is then least constrained by the cementite lamellae. It is interesting that the temperature dependencies of the yield strength of pearlite and pure ferrite match when the interlamellar spacing of the pearlite is large, indicative of the fact that the yielding of pearlite is defined by that of its ferrite, and that the constraint effect of cementite diminishes as the interlamellar spacing becomes coarse.

Independent measurements confirm these trends [101]. Figure 4.19c shows proof strength determined at $-80\,^\circ$C. The line is plotted such that the intercept on the vertical axis is forced to $\sigma_y^\alpha\{-80^\circ\text{C}\}$ using data from [19], and yet, it represents the pearlite data rather well. The equation of the line, together with data from ambient temperature measurements for comparison, is:

$$\sigma_y \approx 146 S_I^{-\frac{1}{2}} + 310 \quad \text{MPa} \qquad \text{data determined at } -80^\circ\text{C} \tag{4.25}$$

$$\sigma_y \approx 174 S_I^{-\frac{1}{2}} + 128 \quad \text{MPa} \qquad \text{ambient temperature data (Equation 4.17)}$$

which confirms that the sensitivity of strength to interlamellar spacing is smaller at low temperatures.

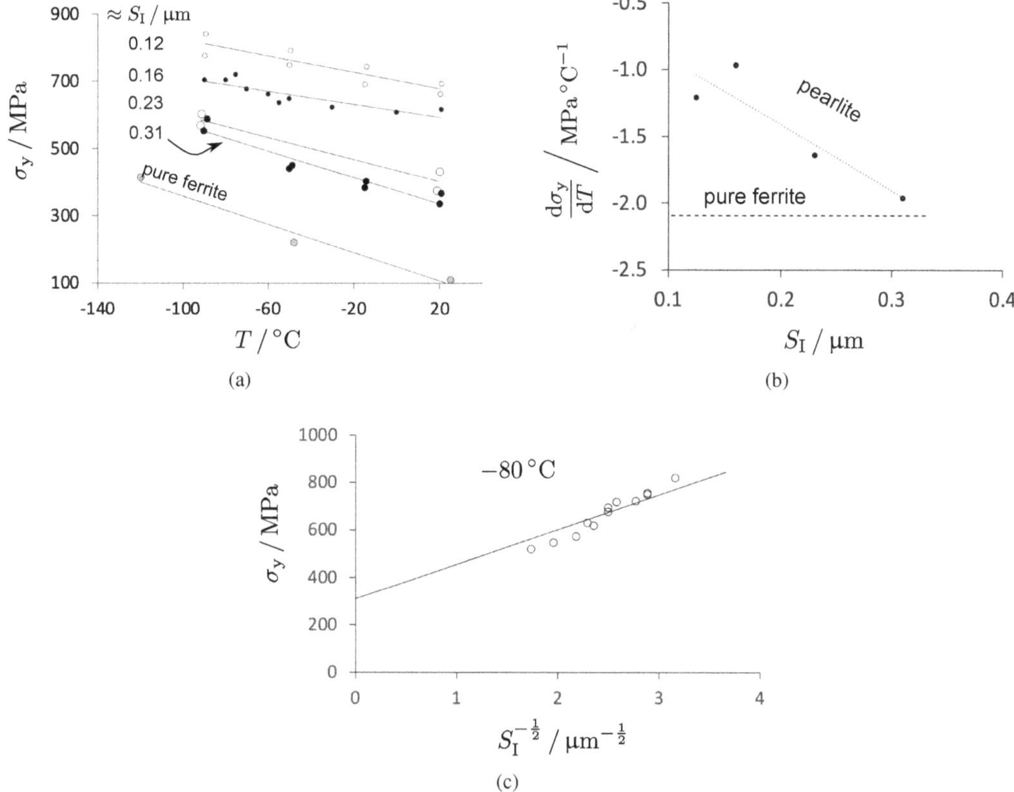

Figure 4.19 Temperature dependence of the yield strength of fully pearlitic steel (Fe-0.8C-0.17Si-0.84Mn wt%). (a) σ_y as a function of T, S_I. Also included are data for pure iron in the ferritic state. (b) The slope of the σ_y versus temperature line in (a), versus S_I. The pearlite data are selected from Alexander and Bernstein [74], the ferrite data selected from Leslie [19]. (c) Yield strength of fully pearlitic Fe-0.75C-0.23Si-1.12Mn wt% steel, measured at −80 °C. Selected data from Kavish and Baker [101].

Figure 4.20 shows that the strain rate sensitivity of ferrite in pure iron, over the range 4-500 × $10^{-6}\,s^{-1}$ is not large – all data fall on the same trend line. There are no similar data for pearlite. Over the range studied, the strain rate sensitivity $(\Delta\sigma/\Delta \ln\dot{\varepsilon})_T$ is independent of $\dot{\varepsilon}$ [27], although the strain rates studied are not particularly large.

The strength of pearlite as a function of very large strain rates has been measured using an adaptation of the Hopkinson bar technique. A high-speed hardened-steel projectile is impacted on the sample, one end of which touches a fixed bar instrumented with strain gauges. There is then a procedure to interpret the strain-time signal in order to convert it into a stress versus strain plot, Figure 4.21a. σ_y is taken as the initial peak in the σ-ε plot [102] which may or may not be appropriate since the peak is located at a large plastic strain. The subsequent decrease in stress is designated a 'yield drop' although this too is not analogous to conventional testing, given the magnitudes of the plastic strains involved.

As expected, the tests show that a large strain rate ($\dot{\varepsilon} \gtrsim 10^3\,s^{-1}$) causes an increase in the strength of pearlite. Figure 4.21b shows an open point at ambient temperature which represents strength measured using hardness, against the data measured at a strain rate some six orders of magnitude

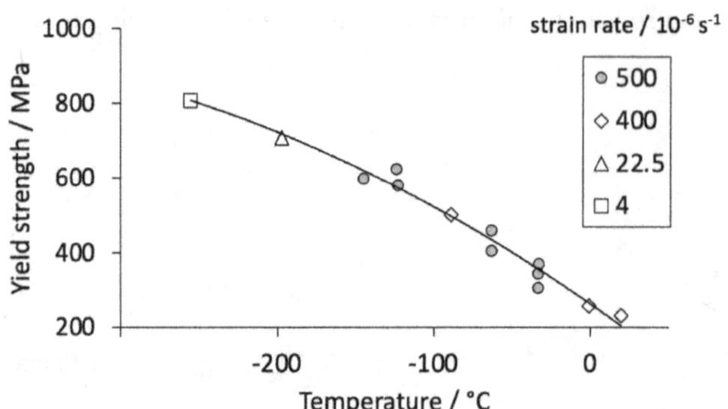

Figure 4.20 Strain-rate dependence of the upper yield-strength of body-centred cubic iron. Selected data, all of which seem to fall on the same curve, from Altshuler and Christian [27].

greater. There is a difference between those data by $\approx 350\,\text{MPa}$. Similar strain-rate sensitivity is found in dynamic tensile tests on predominantly pearlitic steel, with the additional observation that the fracture mode in tension becomes more brittle at high strain rates [103]. It is likely that this trend persists for elevated temperatures but comparative low-$\dot{\varepsilon}$ data do not seem to be available.

Finite element models of machinability where $\dot{\varepsilon}$ can be large enough to induce heating, incorporate the temperature sensitivity of strength in an efficient manner as follows [96, 104][12]:

$$\frac{\sigma\{T\}}{\sigma\{T_\circ\}} = \frac{1}{1+\exp\{b_{24}(T-b_{25})\}} \tag{4.26}$$

where the fitting constants, for α+P mixtures in steels strained at $\dot{\varepsilon} = 0.5\,\text{s}^{-1}$ are $b_{24} = 0.009$, $b_{25} = 650\,°\text{C}$ and the reference stress $\sigma\{T_\circ\}$ represents the strength at 20 °C. The equation covers the experimental data for fully pearlitic steel reasonably well (dashed line, Figure 4.21b) though not at the highest of temperatures. This might be because $\mathrm{d}\sigma/\mathrm{d}T$ is assumed to be identical for α and θ, whereas the latter may retain more of its strength at elevated temperatures [60]. A calculation using the form of temperature-dependence in Equation 4.24, shows the same trend with experimental data (continuous line, Figure 4.21b).

The second term in Equation 4.24 apparently does not adequately represent the strain-rate induced hardening of α+P steels when modelling machining behaviour [104]. The $\dot{\varepsilon}$ dependence is found not to be linear and there is a temperature dependence of the hardening mechanism, expressed empirically for $\dot{\varepsilon} \lesssim 4000\,\text{s}^{-1}$ and $T \lesssim 680\,°\text{C}$ as follows:

$$\frac{\sigma\{\dot{\varepsilon}\}}{\sigma\{\dot{\varepsilon}_\circ\}} = 1 + \left(0.076 - \frac{0.0696}{1+(T/1108)^{6.4}}\right) \times \left(\ln\left\{\frac{\dot{\varepsilon}}{\dot{\varepsilon}_\circ}\right\}\right)^{1.848} \tag{4.27}$$

where the temperature is expressed as °C and $\dot{\varepsilon}_\circ = 0.5\,\text{s}^{-1}$ [104].

4.4.1 ADIABATIC SHEAR

Adiabatic bands are narrow layers of intensely sheared material [105–108], which develop when the heat generated by rapid deformation cannot be dissipated fast enough by diffusion. Deformation in practice is not homogeneous so a local rise in temperature causes a region to soften and

[12]These publications [96, 104] both contain sign errors in the exponential term of Equation 4.26, and the constants b_i derived in [104] for ferrite-pearlite mixtures are not correct and have been re-derived here using their data.

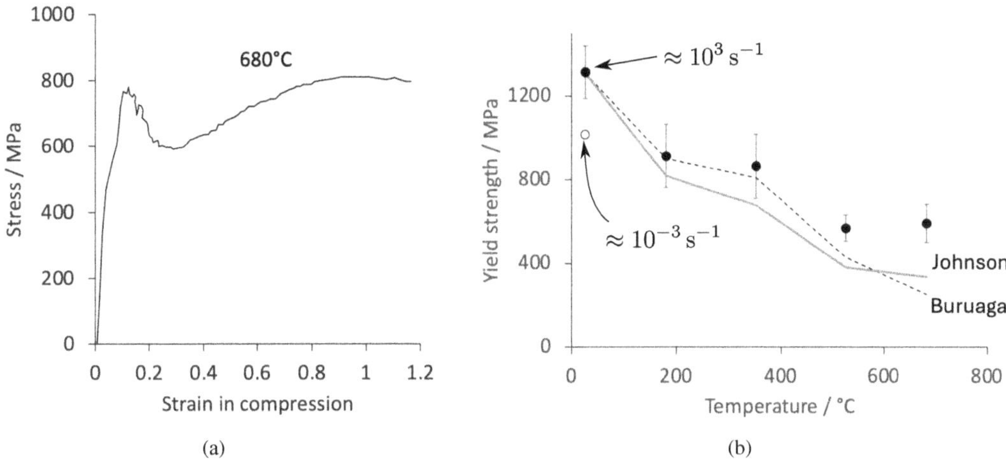

Figure 4.21 (a) A stress-strain curve for a pearlitic steel tested at 680 °C, derived from a Hopkinson bar experiment. The steel has the chemical composition Fe-0.77C-0.85Mn-0.56Si-0.41Cr wt%. (b) The yield strength of the same steel determined in compression at very large strain rates, using the Hopkinson bar technique. The open point is an estimate based on the hardness of the steel. Selected data from Nakkalil et al. [102]. The dashed and continuous lines are estimates using Equations 4.26 and 4.24 respectively.

therefore focuses the plasticity. The resulting bands typically are 10-100 μm in width [109]. They are not crystallographically oriented, extending over large distances relative to the microstructural scale. They tend not to form in tension because defects such as cracks and voids can terminate the deformation. The history of the development of adiabatic bands is nicely documented by Dodd et al. [110].

Assuming that the effect of strain rate on stress is less than that of temperature, the mechanical equation of state that expresses how the stress is raised by strain hardening but lowered by temperature increments becomes [111, 112]:

$$\left(\frac{\mathrm{d}\sigma}{\mathrm{d}\varepsilon_\mathrm{p}}\right) = \left(\frac{\partial\sigma}{\partial\varepsilon_\mathrm{p}}\right)_T + \left(\frac{\partial\sigma}{\partial T}\right)_{\varepsilon_\mathrm{p}} \frac{\mathrm{d}T}{\mathrm{d}\varepsilon_\mathrm{p}} \tag{4.28a}$$

where the first term is the slope of the dynamic stress-strain curve, with shear instability when

$$\left(\frac{\partial\sigma}{\partial\varepsilon_\mathrm{p}}\right)_T = -\left(\frac{\partial\sigma}{\partial T}\right)_{\varepsilon_\mathrm{p}} \frac{\mathrm{d}T}{\mathrm{d}\varepsilon_\mathrm{p}}. \tag{4.28b}$$

Representing the stress versus plastic-strain component of the Swift Equation 4.14 as $\sigma = b_{18}\varepsilon_\mathrm{p}^{b_{20}}$, the change in temperature caused by plastic work, assuming that a proportion β is dissipated as heat, is given by

$$\frac{\mathrm{d}T}{\mathrm{d}\varepsilon_\mathrm{p}} = \frac{\beta}{\rho C_\mathrm{P}} b_{18}\varepsilon_\mathrm{p}^{b_{20}} \tag{4.28c}$$

where C_P is the specific heat capacity at constant pressure and ρ the density. Since

$$\left(\frac{\partial\sigma}{\partial\varepsilon_\mathrm{p}}\right)_T = b_{20}\frac{b_{18}\varepsilon_\mathrm{p}^{b_{20}}}{\varepsilon_\mathrm{p}} \quad \text{and} \quad -\left(\frac{\partial\sigma}{\partial T}\right)_{\varepsilon_\mathrm{p}} \frac{\mathrm{d}T}{\mathrm{d}\varepsilon_\mathrm{p}} = -\left(\frac{\partial\sigma}{\partial T}\right)_{\varepsilon_\mathrm{p}} \frac{\beta}{\rho C_\mathrm{P}} b_{18}\varepsilon_\mathrm{p}^{b_{20}} \tag{4.28d}$$

it follows by setting these two terms equal, that the critical value of the strain associated with the onset of adiabatic shear is given by [112]

$$\varepsilon_{\mathrm{p}}^{\mathrm{c}} = \frac{b_{20}\rho C_{\mathrm{P}}}{\beta|\partial\sigma/\partial T|_{\varepsilon,\dot{\varepsilon}}}. \tag{4.28e}$$

For mild steel, the slow $\dot{\varepsilon}$ *static* parameters are approximately $b_{18} = 174\,\mathrm{MPa}$, $b_{20} \approx 0.17$ whereas during *dynamic* testing conducted in torsion with a shear strain rate of $\approx 180\,\mathrm{s}^{-1}$, $b_{18} = 198\,\mathrm{MPa}$, $b_{20} \approx 0.13$ [112].

Adiabatic shear-strain is large so the small strain equality that $\gamma_{\mathrm{p}} = 2\varepsilon_{\mathrm{p}}$ is not appropriate because of the neglect of second-order terms that become significant at large strains [112]. Consider Figure 4.22, where the vector $\mathbf{u}=[11]$ is sheared by strain s into vector $\mathbf{v}$ [p.153, 113]:

$$\mathbf{v} = \begin{pmatrix} 1 & 0 \\ 0 & 1+s \end{pmatrix} \begin{bmatrix} 1 \\ 1 \end{bmatrix} = \begin{bmatrix} 1 \\ 1+s \end{bmatrix}$$

$$\therefore \frac{|\mathbf{v}|}{|\mathbf{u}|} = \frac{\sqrt{2+2s+s^2}}{\sqrt{2}} = \sqrt{1+s+\frac{1}{2}s^2}$$

and $\ln\{|\mathbf{v}|/|\mathbf{u}|\}$ is the true longitudinal strain ε.

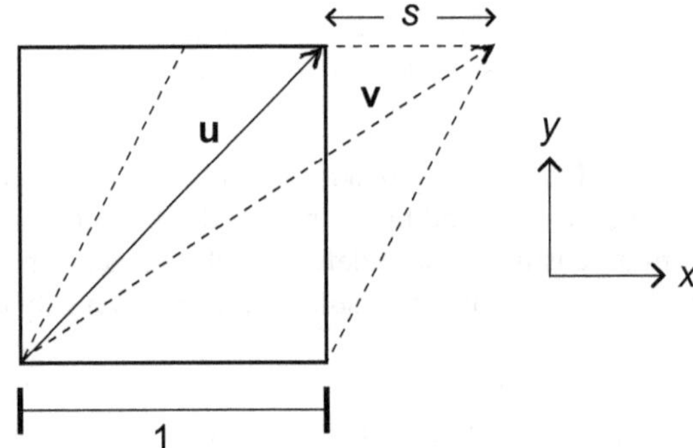

Figure 4.22 A unit square sheared to illustrate the initial and final vectors **u** and **v**.

The onset of adiabatic shear causes a discontinuity in the stress-strain curve; measurements reveal that $\varepsilon_{\mathrm{p}}^{\mathrm{c}}$ decreases as the interlamellar spacing of the pearlite is refined [114]. This may be explained by a reduction in the strain hardening coefficient as the pearlite spacing is refined, although the measurements showing this trend are conducted at much smaller strain rates than associated with adiabatic processes [74]. The temperature dependence of strength for $T = 25\text{-}700\,^{\circ}\mathrm{C}$ is found to be similar for fine and coarse interlamellar spacings when tests are conducted at large strain rates [102].

Figure 4.23a illustrates an adiabatic shear band in a pearlitic steel; the band etches light relative to the surrounding pearlite because of its much finer structure; a crack also has formed due to the increase in hardness (365→1040 HV) [114]. The structure within can be martensitic or simply severely deformed, depending on whether the transient rise in local temperature permits austenite to form and then undergo martensitic transformation during cooling. In a hypereutectoid steel (1.3C wt%), compressive deformation at $4\times10^3\,\mathrm{s}^{-1}$ causes adiabatic shear, but cementite in the microstructure fails to dissolve during the deformation and local heating. Therefore, the relatively low-carbon austenite generated in the heating stage subsequently forms low-carbon martensite [115], the hardness of which is attributed to the refined state of the final structure.[13]

[13] This difference in behaviour between hypereutectoid and eutectoid steels is not explained.

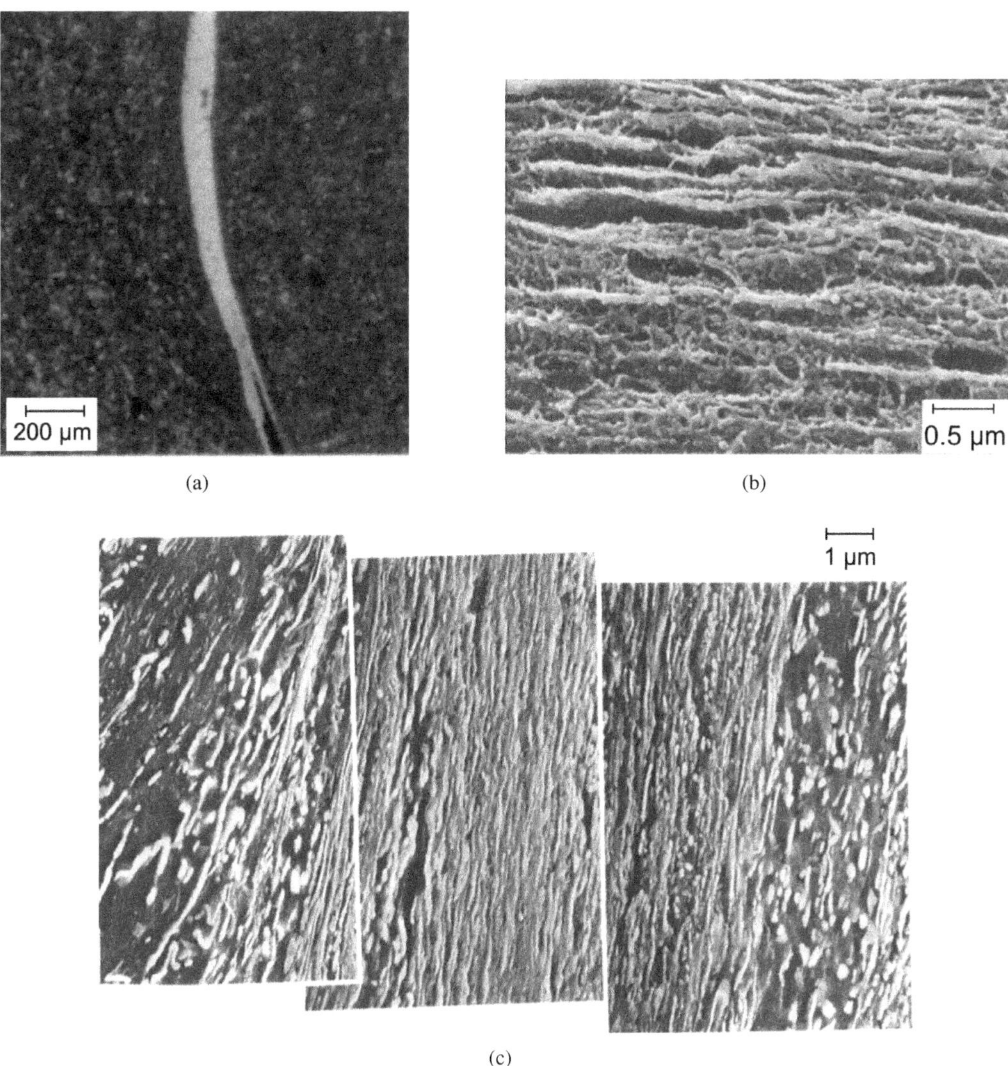

(a) (b) (c)

Figure 4.23 (a) Adiabatic shear band in an otherwise fully pearlitic steel (Fe-0.77C-0.85Mn-0.56Si-0.41Cr wt%) tested in compression at ambient temperature, at $\dot{\varepsilon} = 2\text{-}3 \times 10^3\,\mathrm{s}^{-1}$. Image reproduced with permission of Elsevier from Nakkalil et al. [102]. (b,c) Steel 4340 of approximate composition Fe-0.4C-0.7Mn-0.8Cr-0.25Mo wt% subjected to impact deformation at $\approx 18000\,\mathrm{s}^{-1}$ – the initial microstructure was pearlitic. (b) The structure across the band. (c) High-magnification image from within the shear band. Images reproduced from Zurek [116] with permission from Springer SNCSC.

Adiabatic shear bands are not always associated with phase transformation or even the refinement of the microstructure to a level where the band stands out as being different from its surroundings. When α+P steels are impacted by high-velocity projectiles, the thin bands that form, simply represent intensely sheared regions that still have a recognisable mixture of pearlitic and ferritic regions [117], although the rate of deformation within is large enough to cause the lamellae to fragment with the local heating inducing a degree of spheroidisation [116], Figure 4.23b,c.

Some general features of adiabatic shear bands can be emphasised:

- the paths followed by the bands in a polycrystalline metal are not crystallographic; their trajectory complies with the impact event, often leading to a perforation or fragmentation of the steel.
- There are initiation points such as inclusions, beyond which the bands propagate rather like running cracks, without separation of the material, just localised shear [117]. Separation may follow given the momentum of any ballistic impact, aided by the compromised structure within the adiabatic shear band.
- Thermal softening occurs at the advancing tip.
- Large strain rates can stimulate hard shear bands that etch lightly because of their fine structure and so appear white in an optical microscope. However, deformation can also localise without causing severe microstructural change at moderately large strain rates, and high deformation temperatures, as mapped in Figure 4.24.

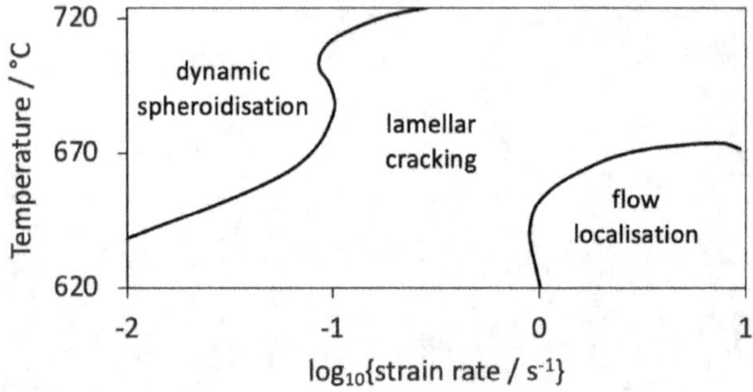

Figure 4.24 Map showing the combination of temperature and strain rate that can lead to strain localisation in a fully pearlitic Fe-0.82C-0.66Mn-0.18Si wt% steel. Selected data from Rastegari et al. [118].

4.4.2 SHOCK LOADING OF α+P

The impact of a flyer plate onto a target plate can create a free surface inside the metal and sometimes cause the detachment of segments, known as *spalls*. The sort of damage that occurs when a flyer plate impacts a steel containing a mixture of ferrite and pearlite, with an impact velocity of 237 $m\,s^{-1}$, is illustrated in Figure 4.25. There is a significant opening up of the target steel along its central plane. The plane shock wave caused by the impact is reflected from the opposite free surface of the plate as a tensile wave responsible for the spalling event.[14] The flyer and target are in contact longer than required for the shock wave to propagate through the material. Because of this, the flyer and target release waves are able to interact within the target to create a region of tension (spall plane). This superposition becomes possible because the speed of sound in the metal is much greater than that of the impact. The parting surfaces may not be continuous, as illustrated in Figure 4.25.

[14]When the elastic displacements are all in the direction of the propagation of a wave, the wave is said to be *plane*. An elastic wave in iron can travel at 5000-5950 $m\,s^{-1}$ depending on it character with respect to transverse stresses, the upper limit corresponding to a plane elastic-wave. If the pressure at a detonation wave front is 15 GPa, this leads to a shock wave velocity of 5240 $m\,s^{-1}$ which is less than that of a plane elastic wave. When the stress is at the yield of the iron, the corresponding plastic wave velocity would be about 4600 $m\,s^{-1}$ [119]. Both the shock and plastic wave velocities are therefore smaller than that of the plane elastic wave.

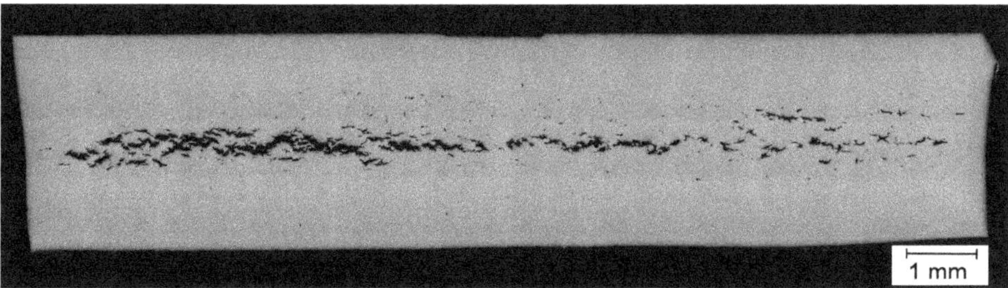

Figure 4.25 Damage caused in Fe-0.17C-0.46Mn-0.21Si wt% steel impacted by a flyer plate at 237 m s^{-1} generating a peak stress of 3.8 GPa. Image courtesy of V. K. Euser.

Studies have been conducted on α+P steels to examine damage evolution during dynamic loading involving the impact with flyer plates at 216-293 m s^{-1} [120]. Two steels were studied, one containing Fe-0.45C-0.71Mn-0.18Si wt% and the other, Fe-0.17C-0.46Mn-0.21Si wt%, along with small concentrations of other solutes. The former obviously contained more pearlite, so damage on the central plane of the target plate was macroscopically more brittle than with the steel containing less pearlite, Figure 4.26. The cementite lamellae appeared to crack periodically when oriented normal to the plate plane, with most of the damage initiated in the pearlite in the case of the higher carbon steel, whereas there was clear evidence of damage in both ferrite and pearlite in the steel containing less carbon. Cementite therefore reduces the spall resistance of the steel. Because the peak stresses reached during these experiments were less than 5 GPa, mechanical twinning was not observed as a deformation mechanism.

In some shock experiments, the pressures reached can exceed the 13 GPa required to cause the pure α-ferrite in iron to transform into the hexagonal close-packed ϵ-martensite, which then transforms back into α'-martensite on relief of the pressure [p. 2 121]. It is known, however, that when the steel contains carbides, the time required for the $\alpha \rightarrow \epsilon$ transformation to occur increases so a greater peak pressure is necessary before the change to ϵ is detected [122]; the specific role of the carbides is not known but a similar trend has been noted in prior work [123].

4.4.3 DEFORMATION TWINNING ASSOCIATED WITH PEARLITE

Deformation twinning has never been observed in cementite, which is an obstacle to the continuous propagation of $\{112\}_\alpha\langle\bar{1}\bar{1}1\rangle_\alpha$ twins with a shear strain of $1/\sqrt{2}$ within the ferrite. In grains that are fully ferritic, the twins usually are halted at the α/α interfaces that present substantial crystallographic discontinuities, or α/P boundaries where cementite interferes with twin propagation. The ferrite does not exhibit much mechanical twinning unless the impact pressure exceeds about 8 GPa at a strain rate of the order of 1000 s^{-1}, Figure 4.27. The fraction of twinning scales with the shock pressure, until the $\alpha \rightarrow \epsilon$ transition becomes possible, the transition in pure iron being at 13 GPa [124]. After that, both twinning and vestiges of the transformation appear in the microstructure that is examined *post-mortem*, i.e., after the ϵ has transformed back into ferrite. The vestiges are in the form of markings representing defects created during the displacive transformation. However, above ≈ 33 GPa, the pressure wave travels faster than the transformation wave, leaving only twins in the microstructure [124]. If the pressure exceeds ≈ 67 GPa, the pressure wave velocity becomes greater than that of an elastic wave, so twinning is absent. These numbers relate to pure iron and are known to be modified if the iron contains other solutes.

Figure 4.28 shows regions of explosively deformed steel [126] that exhibit mechanical twinning, confined primarily to the ferrite but with some penetration into the pearlite colonies. The lenticular shape of individual twins (Figure 4.28c) is because that shape minimises the strain energy associated

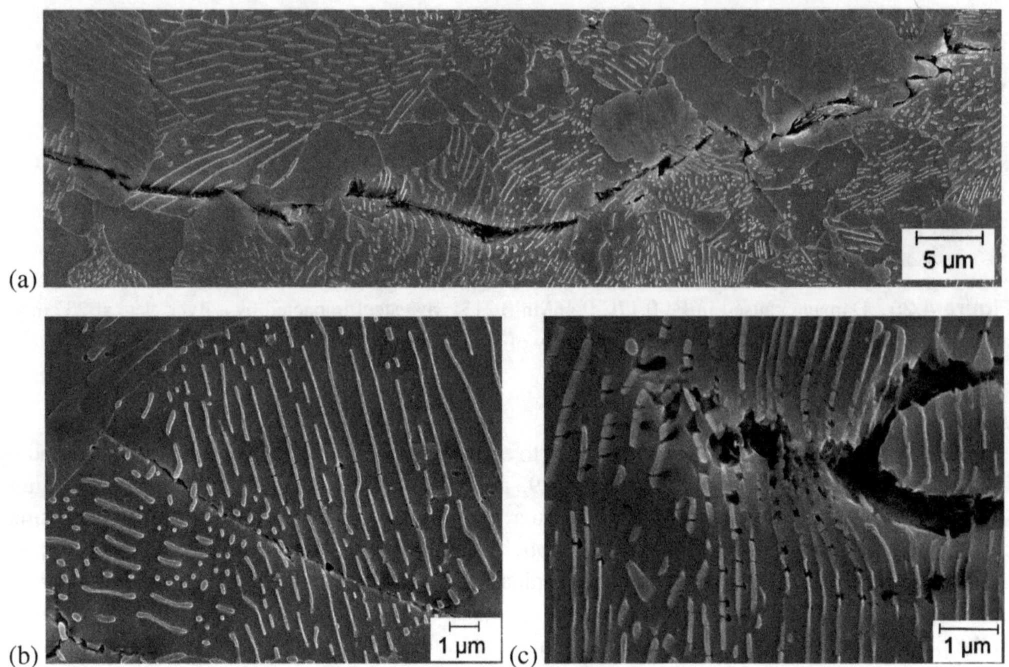

Figure 4.26 Spall damage caused on α+P Fe-0.45C-0.71Mn-0.18Si wt% steel impacted by a flyer plate at $216\,\mathrm{m\,s^{-1}}$. The loading direction is in all cases, vertical. (a) Macroscopic cracks across pearlitic regions, which then may propagate into the adjacent ferrite. (b) Higher magnification image showing cracks across the pearlite colony, and within the cementite lamellae. (c) Void formation and extensive cementite lamellar cracking. Images courtesy of V. K. Euser. Reproduced from Euser et al. [120] with the permission of AIP publishing.

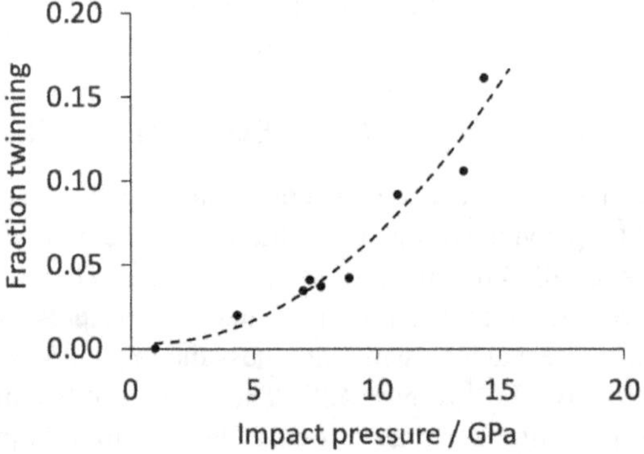

Figure 4.27 The influence of impact pressure in experiments where a projectile hits a target at a strain rate in excess of $1000\,\mathrm{s^{-1}}$, on the volume fraction of mechanical twins in the ferrite. The steel has an approximate chemical composition Fe-0.23C-1.35Mn-0.25Si wt%, with a microstructure of ferrite and pearlite. Selected data from Visser and Ghonem [125].

with the large shear deformation [127, 128]. The coordinated motion of atoms during the twinning deformation cannot be sustained across grain boundaries, which are barriers to twin propagation. There are large stresses generated in the vicinity of a mechanical twin in a polycrystalline material, which may be accommodated by plastic relaxation in the surrounding material. Figure 4.28c shows some blunting of the lenticular-twin tip at the α/α boundary, due to accommodation of the shear by dislocation plasticity in the adjacent grain. The twins sometimes appear to be curved macroscopically due to plastic deformation following twinning.

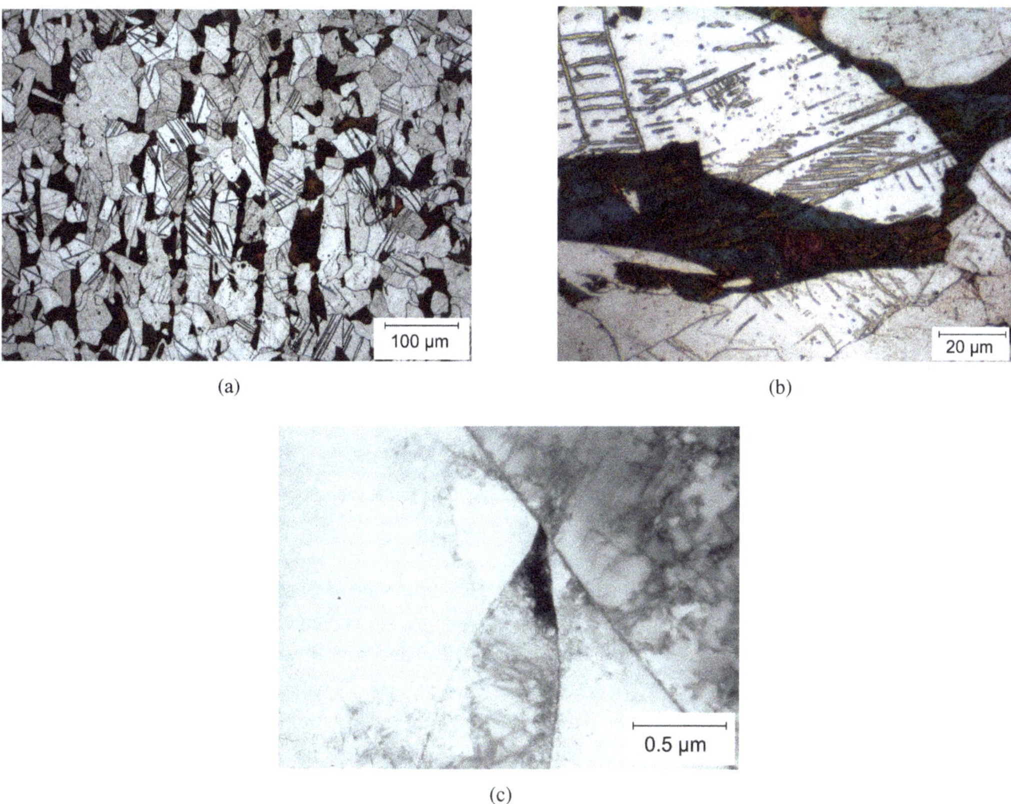

(a) (b) (c)

Figure 4.28 Explosively deformed 0.18C-0.16Mn wt% steel, α+P steel. (a) Optical micrograph showing mechanical twins. (b) Higher magnification optical micrograph showing twin distribution in ferrite and apparently also in pearlite, courtesy of Ivan Bataev of Novosibirsk State Technical University. (c) Transmission electron micrograph from a different sample, a pipe-bomb from an α+P steel showing lenticular twin halted by α/α boundary.

Explosive deformation

During explosive deformation of α+P steels, in some cases, twins in the ferrite continue apparently undisturbed into the pearlitic regions, implying some sort of crystallographic continuity between the proeutectoid and pearlitic ferrite [129]. In spite of appearances (Figure 4.28b), mechanical twinning in shock deformed pearlite is confined to its ferritic phase, but the twins then interact with the cementite lamellae, in one of two ways, Figure 4.29. If the twinning system in the ferrite corresponds to a slip system within the cementite, than the latter is sheared. Alternatively, the cementite lamellae retain their original shapes when the correspondence between the deformation systems of the two phases is not appropriate [126, 130]. In either case, the ferrite twin has crystallographic continuity between the lamellae of the cementite; whether this is by repeated nucleation and propagation of the twin along the same system, or because there is three-dimensional connectivity with the ferrite in the pearlite colony, has not been investigated.

There is metallographic evidence of a third scenario, where a mechanical twin in ferrite can induce a region of the pearlite to fracture at $-196\,^\circ$C, with the crack propagating across both the cementite and ferrite within the pearlite [131].

Figure 4.29 shows also that the cementite, which often is regarded as a brittle phase, has sufficient structural integrity to avoid cracking and even exhibit plasticity during the course of impingement with propagating ferrite twins, under conditions of shock loading.[15] Given the tenacity of cementite when present within the lamellar composite structure, fully pearlitic steels harden less than mixtures of proeutectoid-ferrite and pearlite following shock loading, consistent with the ferrite accommodating the majority of the resulting plastic strain, Figure 4.30.

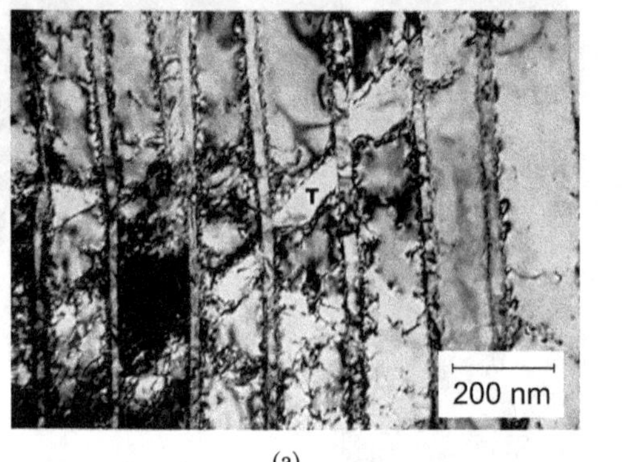

(a)

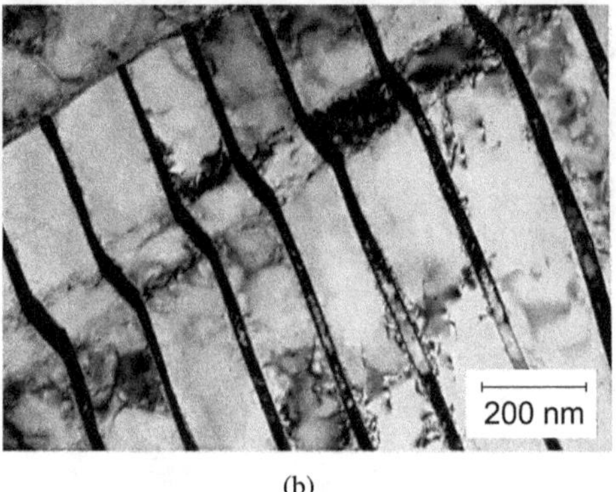

(b)

Figure 4.29 Mechanical twins confined within the ferritic component of pearlite, in shock loaded 0.99C wt% steel. (a) Twin traversing the colony of pearlite without shearing the cementite. (b) Ferrite twin shearing the cementite. Micrographs reproduced with the permission of Elsevier, from Bowden and Kelly [130].

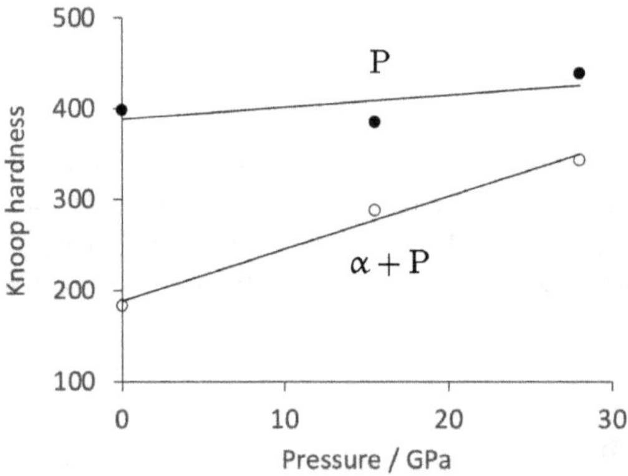

Figure 4.30 Hardness following the shock loading of samples of normalised steels. The carbon concentrations of the steels were ≈ 0.4C wt% and ≈ 1C wt% for the α+P and fully pearlitic samples, respectively. Selected data from Koepke et al. [132].

Rapid compression tests conducted at impact pressures in the range 14-16.3 GPa on α+P steel have revealed an additional ferrite twinning mode on $\{33\bar{2}\}_\alpha\langle 113\rangle_\alpha$ [133]. This mode is predicted

[15] In fibre composites where the minor component is firmly bonded to the matrix, shear causes brittle fibres to crack but ductile fibres kink in order to accommodate the shear.

to have half the twinning shear of the much more common $\{112\}_\alpha\langle\overline{1}\overline{1}1\rangle_\alpha$ system, but half the atoms must then shuffle during the process in order to recover the exact body-centred cubic structure [128, 134]. Other experiments conducted over a range 10-20 GPa [135] also reported the $\{33\overline{2}\}_\alpha\langle113\rangle_\alpha$ twinning mode when the pressure exceeded that where ϵ-iron forms. A possible explanation for this has been proposed, that the formation of the hexagonal iron reduces the magnitudes of the shuffles required to achieve the correct twinned structure [134].

A reduction in the grain size of a wide range of polycrystalline materials tends to suppress mechanical twinning. The underlying mechanism is not known, but it could be argued that the suppression is a natural consequence of the enhanced strength accompanying the refined grain structure, so for a given applied stress, the tendency for any kind of deformation is reduced. Grains of proeutectoid-ferrite readily twin whereas the ferrite within the pearlite does not, consistent with the reduced mean free path of the pearlitic-ferrite.

4.5 STRENGTH OF AUSTENITE

The strength of austenite at high temperatures is used, for example, to calculate rolling-mill loads, to design experiments that reveal how stress influences the rate of austenite transformation, during which $\sigma < \sigma_y$ to avoid conflating the effects of stress and plastic strain. Figure 4.31a shows approximately, how the strength of austenite in near eutectoid steels depends on temperature; the σ_{UTS} data will obviously be an overestimate of the yield strength.

There are data on the strength of austenite, including eutectoid steels, as a function of temperature, strain and strain rate, determined using torsion tests; these extend into regimes of large strains typical in hot-rolling processes. The von Mises equivalent-stress/equivalent-strain curves obtained from such tests fall below those measured using axisymmetric tensile testing [136]. The σ_y is expected to be smaller in torsion by $\approx 6.5\%$ compared with that measured in tension. This is because the critical resolved shear stress at yield is converted into one appropriate for a polycrystalline substance using a mean Taylor factor, which is different in tension and torsion. At large strains, the development of crystallographic texture, and hence the effective Taylor factor, also is different in tension and torsion. There may be corresponding differences in the work-hardening behaviour.

A recent empirical equation represents the yield strength of austenite as follows [141]:

$$\begin{aligned}\sigma_y^\gamma/\text{MPa} &= (105.3 + 47a_\text{C} + 31.3a_\text{Si} + a_\text{Mn} + a_\text{Cr} \\ &\quad +31.3a_\text{Mo} + 1.3a_\text{Ni} + 10.7a_\text{V} + 3.8a_\text{Al} + 16.8a_\text{Cu} + 15.2a_\text{Ti}\ [\pm 34]) \\ &\quad \times\left(1 + 0.001\ln\left\{\frac{\dot{\varepsilon}}{0.001}\right\}\right) \times \left(1 - \left[\frac{T-25}{1487-25}\right]^{0.658}\right)\end{aligned} \tag{4.29}$$

where the temperature is expressed in °C, $\dot{\varepsilon}$ is the strain rate in units of s^{-1} and a_X is the atomic percent of the solute identified in the subscript. The data included in the analysis covered the ranges listed below, using these same units; the dataset consisted of 845 points, predominantly involving temperatures associated with the hot-rolling of low-alloy steels. The austenite grain size is not included in the analysis due to a lack of data.

C	Si	Mn	Cr	Mo	Ni	V	Al	Cu	Ti	$T/°\text{C}$	$\dot{\varepsilon}/\text{s}^{-1}$	$\sigma_y^\gamma/\text{MPa}$
0.12	0.00	0.00	0.01	0.00	0.00	0.00	0.02	0.00	0.00	200	0.003	7
4.42	5.80	2.62	1.70	0.37	1.90	0.22	3.63	0.23	0.19	1100	11	298
at% (C–Ti)												

Figure 4.32 illustrates the performance of Equation 4.29 against experimental data for near-eutectoid steels [137, 138]. Low-temperature data are not well-predicted. It is possible that this is a reflection of limited data in the domain of low temperatures.

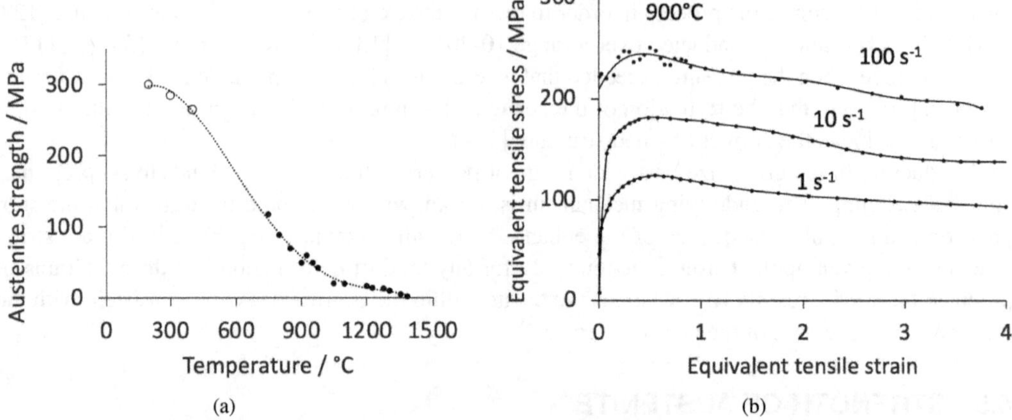

Figure 4.31 (a) Compilation of the strength of austenite as a function of temperature. The open points are from Fe-0.79C-1.56Si-1.98Mn-1.01Cr-1.51Co-0.24Mo wt% steel and represent the 0.2% proof strength determined in uniaxial compression [137]. The filled points are from Fe-0.80C-0.21Si-0.79Mn wt% steel and represent σ_{UTS} [138]. (b) Equivalent tensile strength of austenite 900 °C versus equivalent tensile strain, as measured using torsion testing at a variety of strain rates. The torsion data are converted into equivalent tensile values using the Fields and Backofen [139] von Mises criterion. Selected data from Kong et al. [140].

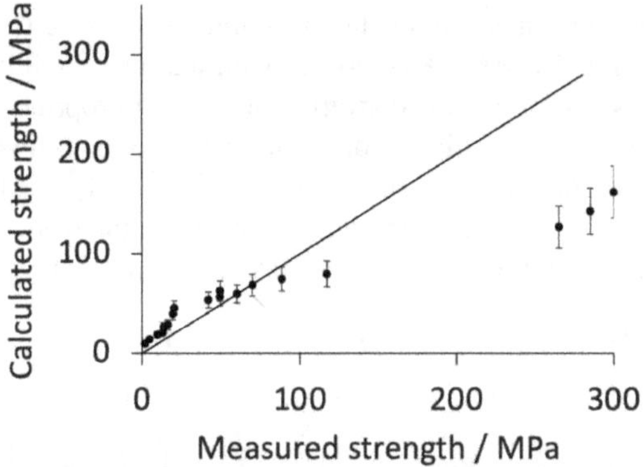

Figure 4.32 A comparison of austenite yield strength calculated using Equation 4.29 against the data plotted in Figure 4.31a.

REFERENCES

1. H. Shoji, M. Ohata, and F. Minami: 'Simulation-based method for hierarchal material design to improve ductile crack growth resistance of structural component', *International Journal of Fracture*, 2015, **192**, 167–178.
2. Y. Tomota, K. Kuroki, T. Mori, and I. Tamura: 'Tensile deformation of two ductile phase alloys: flow curves of α/γ Fe-Cr-Ni alloys', *Materials Science & Engineering*, 1976, **24**, 85–94.
3. W. W. Webb, and W. J. Foregeng: 'Mechanical behavior of microcrystals', *Acta Metallurgica*, 1958, **6**, 462–469.

4. Y. Wang, Y. Tomota, T. Ohmura, W. Gong, S. Harjo, and M. Tanaka: 'Continuous and discontinuous yielding behaviors in ferrite-cementite steels', *Acta Materialia*, 2020, **196**, 565–575.
5. T. Takahashi, and M. Nagumo: 'Flow stress and work-hardening of pearlitic steel', *Transactions of the JIM*, 1970, **11**, 113–119.
6. H. Sidhom, H. Yahyaoui, C. Braham, and G. Gonzalez: 'Analysis of the deformation and damage mechanisms of pearlitic steel by EBSD and "in-situ" SEM tensile tests', *Journal of Materials Processing and Manufacturing Science*, 2015, **24**, 2586–2596.
7. D. Kuhlmann-Wilsdorf: 'Theory of workhardening 1934-1984', *Metallurgical Transactions A*, 1985, **16**, 2091–2108.
8. R. Rodriguez, and I. Gutierrez: 'Unified formulation to predict the tensile curves of steels with different microstructures', *Materials Science Forum*, 2003, **426-432**, 4525–4530.
9. G. I. Taylor: 'Plastic strain in metals', *Journal of the Institute of Metals*, 1938, **62**, 307–324.
10. M. Dollar, I. M. Bernstein, and A. W. Thompson: 'Influence of deformation substructure on flow and fracture of fully pearlitic steel', *Acta Metallurgica*, 1988, **36**, 311–320.
11. X. Zhang, A. Godfrey, X. Huang, N. Hansen, and Q. Liu: 'Microstructure and strengthening mechanisms in cold-drawn pearlitic steel wire', *Acta Materialia*, 2011, **59**, 3422–3430.
12. N. Guo, T. Liu, B. Luan, B. Wang, and Q. Liu: 'Dislocation density and configuration in fully pearlitic steel during wire drawing', *Materials Research Innovations*, 2014, **18**, S4:249–254.
13. X. Zhang, N. Hansen, A. Godfrey, and X. Huang: 'Dislocation-based plasticity and strengthening mechanisms in sub-20 nm lamellar structures in pearlitic steel wire', *Acta Materialia*, 2016, **114**, 176–183.
14. C. Borchers, T. Al-Kassab, S. Goto, and R. Kirchheim: 'Partially amorphous nanocomposite obtained from heavily deformed pearlitic steel', *Materials Science & Engineering A*, 2009, **502**, 131–138.
15. S. Sato, T. Shobu, K. Satoh, H. Ogawa, K. Wagatsuma, M. Kumagai, M. Imafuku, H. Tashiro, and S. Suzuki: 'Distribution and anisotropy of dislocations in cold-drawn pearlitic steel wires analyzed using micro-beam X-ray diffraction', *ISIJ International*, 2015, **55**, 1432–1438.
16. U. Kocks, and H. Mecking: 'Physics and phenomenology of strain hardening: the FCC case', *Progress in materials science*, 2003, **48**, 171–273.
17. I. Gutierrez: 'AME modellling the mechanical behaviour of steels with mixed microstructures', *Metalurgija*, 2005, **11**, 201–214.
18. B. Berisha, C. Raemy, C. Becker, M. Gorji, and P. Hora: 'Multiscale modeling of failure initiation in a ferritic–pearlitic steel', *Acta Materialia*, 2015, **100**, 191–201.
19. W. C. Leslie: 'Iron and its dilute substitutional solid solutions', *Metallurgical Transactions A*, 1972, **3**, 5–23.
20. K. Takeda, N. Nakada, T. Tsuchiyama, and S. Takaki: 'Effect of interstitial elements on Hall–Petch coefficient of ferritic iron', *ISIJ International*, 2008, **48**, 1122–1125.
21. K. Shibanuma, Y. Nemoto, T. Hiraide, K. Suzuki, S. Sadamatsu, Y. Adachi, and S. Aihara: 'A strategy to predict the fracture toughness of steels with a banded ferrite pearlite structure based on the micromechanics of brittle fracture initiation', *Acta Materialia*, 2018, **144**, 386–399.
22. S. Takaki, K. Kawasaki, and Y. Kimura: 'Mechanical properties of ultra fine grained steels', *Journal of Materials Processing Technology*, 2001, **117**, 359–363.
23. J. D. Campbell, and J. Harding: 'The effect of grain size, rate of strain, and neutron irradiation on the tensile strength of α-iron', In: P. G. Shewmon, and V. F. Zackay, eds. *Response of Metals to High Velocity Deformation*. New York, USA: Interscience, 1961:51–76.
24. S. A. Saltykov: Stereometricheskaya Metallografiya: 2nd ed., Moscow: Metallurgizdat Publishers (English translation US Department of Defence AD267700), 1958.
25. E. E. Underwood: Quantitative Microscopy, eds R. T. DeHoff and F. N. Rhines, chap. 4: New York, USA: McGraw Hill, 1968:77–125.
26. R. L. Higginson, and C. M. Sellars: Worked examples in quantitative metallography: London, U.K.: Maney Publishing, 2003.

27. T. L. Altshuler, and J. W. Christian: 'Mechanical properties of pure iron tested in compression over the temperature range 2-293 K', *Proceedings of the Royal Society A*, 1967, **261**, 253–287.
28. H. K. D. H. Bhadeshia, and A. R. Chintha: 'Critical assessment: the strength of undeformed pearlite', *Materials Science and Technology*, 2022, **38**, 1291–1299.
29. A. R. Marder, and B. L. Bramfitt: 'The effect of morphology on the strength of pearlite', *Metallurgical Transactions A*, 1976, **7**, 365–372.
30. L. E. Miller, and G. C. Smith: 'Tensile fractures in carbon steels', *Journal of the Iron and Steel Institute*, 1970, **11**, 998–1005.
31. E. W. L. Rocca: 'Unique structure of pearlite deformed by explosive load', *Acta Metallurgica*, 1957, **5**, 408–410.
32. S. Liu, F. Zhang, Z. Yang, M. Wang, and C. Zheng: 'Effects of Al and Mn on the formation and properties of nanostructured pearlite in high-carbon steels', *Materials and Design*, 2016, **93**, 73–80.
33. B. Karlsson, and G. Linden: 'Plastic deformation of ferrite–pearlite structures in steel', *Materials Science and Engineering*, 1975, **17**, 209–219.
34. H. W. Swift: 'Plastic instability under plane stress', *Journal of the Mechanics and Physics of Solids*, 1952, **1**, 1–8.
35. Y. Tomota, W. Nakagawara, K. Tsuzaki, and T. Maki: 'Reversion of stress-induced ε martensite and two-way shape memory in Fe-24Mn and Fe-24Mn-6Si alloys', *Scripta Materialia*, 1992, **26**, 1571–1574.
36. X. Wang, X. Chen, H. Ding, and H. Yang: 'Synthesis and magnetic properties of Fe_3C doped with Mn or Ni for applications as adsorbents', *Dyes and Pigments*, 2017, **144**, 76–79.
37. N. Furukawa, G. Anan, and S. Nanba: 'Microstructure and tensile properties of pearlite single structure steels', In: H. Yoshinaga, and H. Yada, eds. *Prediction and Control of Deformation Property*. Tokyo, Japan: Iron and Steel Institute of Japan, 1994:269–276.
38. A. Hiramatsu, and T. Yamada: 'Prediction of stress-strain curves in ferrite single structure steels', In: H. Yoshinaga, and H. Yada, eds. *Prediction and Control of Deformation Property*. Tokyo, Japan: Iron and Steel Institute of Japan, 1994:263–268.
39. L. Wang, D. Tang, and Y. Song: 'Prediction of mechanical behaviour of ferrite-pearlite steels', *Journal of Iron and Steel Research International*, 2017, **24**, 321–327.
40. M. Umemoto: 'Relationships between microstructure and mechanical properties in steels', *Tetsu-to-Hagané*, 1995, **81**, 157–166.
41. K. Mishra, V. N. Khiratkar, and A. Singh: 'Change of deformation mechanism through nanostructuring of pearlite: An in-situ study', *Materials Characterization*, 2020, **167**, 110487.
42. D. Porter, K. Easterling, and G. Smith: 'Dynamic studies of the tensile deformation and fracture of pearlite', *Acta Metallurgica*, 1978, **26**, 1405–1422.
43. R. E. Reed-Hill: Physical Metallurgy Principles: 2nd ed., Boston, USA: PWS Publishers, 1973.
44. Y. Li, A. J. Bushby, and D. J. Dunstan: 'The Hall-Petch effect as a manifestation of the general size effect', *Proceedings of the Royal Society A*, 2016, **472**, 20150890.
45. M. Hillert: 'The formation of pearlite', In: V. F. Zackay, and H. I. Aaronson, eds. *Decomposition of Austenite by Diffusional Processes*. New York, USA: Interscience, 1962:197–237.
46. J. D. Eshelby, F. C. Frank, and F. R. N. Nabarro: 'XLI. The equilibrium of linear arrays of dislocations.', *The London, Edinburgh, and Dublin Philosophical Magazine and Journal of Science*, 1951, **42**, 351–364.
47. J. C. M. Li, and Y. T. Chou: 'The role of dislocations in the flow stress grain size relationships', *Metallurgical Transactions*, 1970, **1**, 1145–1159.
48. E. O. Hall: 'The deformation and ageing of mild steel: III discussion of results.', *Proceedings of the Physical Society B*, 1951, **64**, 747–753.

49. N. J. Petch: 'The cleavage strength of polycrystals', *Journal of the Iron and Steel Institute*, 1953, **174**, 25–28.
50. E. O. Hall: Yield point phenomena in metals and alloys: New York, USA: Plenum Press, 1970.
51. J. C. M. Li: 'Petch relation and grain boundary sources', *Transactions of the Metallurgical Society of the AIME*, 1963, **227**, 239–247.
52. M. Kato: 'Hall-Petch relationship and dislocation model for deformation of ultrafine-grained and nanocrystalline metals', *Materials Transactions, JIM*, 2014, **55**, 19–24.
53. G. Langford: 'Deformation of pearlite', *Metallurgical Transactions A*, 1977, **8**, 861 – 875.
54. D. Tabor: 'The hardness and strength of metals', *Journal of the Institute of Metals*, 1951, **79**, 1–12.
55. A. Inoue, T. Ogura, and T. Muramatsu: 'Burgers vectors of dislocations in cementite crystal', *Scripta Metallurgica*, 1977, **11**, 1–5.
56. J. Alkorta, and J. G. Sevillano: 'Assessment of elastic anisotropy and incipient plasticity in Fe_3C d by nanoindentation', *Journal of Materials Research*, 2012, **27**, 45–52.
57. N. Hansen, and D. Kuhlmann-Wilsdorf: 'Low energy dislocation structures due to unidirectional deformation at low temperatures', *Materials Science & Engineering*, 1986, **81**, 141–161.
58. E. Clouet, L. Ventelon, and F. Wilaime: 'Dislocation core energies and core fields from first principles', *Physical Review Letters*, 2009, **102**, 055502.
59. C. Jiang, S. G. Srinivasan, A. Caro, and S. A. Maoly: 'Structural, elastic, and electronic properties of Fe_3C from first principles', *Journal of Applied Physics*, 2008, **103**, 043502.
60. H. K. D. H. Bhadeshia: 'Cementite', *International Materials Reviews*, 2020, **65**, 1–27.
61. D. Dingley, and D. McLean: 'Components of the flow stress of iron', *Acta metallurgica*, 1967, **15**, 885–901.
62. S. Takaki: 'Influence of alloying elements on the Hall-Petch coefficient in ferritic steel', *Materials Science Forum*, 2012, **706-709**, 181–185.
63. N. Maruyama, T. Tarui, and H. Tashiro: 'Atom probe study on the ductility of drawn pearlitic steels', *Scripta Materialia*, 2002, **46**, 599–603.
64. H. K. D. H. Bhadeshia: 'Neural networks in materials science', *ISIJ International*, 1999, **39**, 966–979.
65. H. Pous-Romero, I. Lonardelli, D. Cogswell, and H. K. D. H. Bhadeshia: 'Austenite grain growth in a nuclear pressure vessel steel', *Materials Science & Engineering A*, 2013, **567**, 72–79.
66. H. Sunwoo, M. Fine, M. Meshii, and D. Stone: 'Cyclic deformation of pearlitic eutectoid rail steel', *Metallurgical Transactions A*, 1982, **13**, 2035–2047.
67. N. Tsuji, Y. Ito, Y. Saito, and Y. Minamino: 'Strength and ductility of ultrafine grained aluminium and iron produced by ARB and annealing', *Scripta Materialia*, 2002, **47**, 893–899.
68. A. C. C. Reis, and L. Kestens: 'Cross-sectional texture gradients in interstitial free steels processed by accumulated roll bonding', *Solid State Phenomena*, 2005, **105**, 233–238.
69. A. J. E. Foreman: 'The bowing of a dislocation segment', *Philosophical Magazine*, 1967, **15**, 1011–1021.
70. J. G. Sevillano: 'On the yield and flow stress of lamellar pearlite', In: *Strength of Metals and Alloys*. Oxford, U. K.: Pergamon Press, 1979:819–824.
71. T. Gladman, I. D. McIvor, and F. B. Pickering: 'Some aspects of the structure-property relationships in high-C ferrite-pearlite steels', *Journal of the Iron and Steel Institute*, 1972, **210**, 916–930.
72. J. M. Hyzak, and I. M. Bernstein: 'The role of microstructure on the strength and toughness of fully pearlitic steels', *Metallurgical Transactions A*, 1976, **7**, 1217–1221.

73. J. P. Houin, A. Simon, and G. Beck: 'Relationship between structure and mechanical properties of pearlite between 0.2% and 0.8%C', *Transactions of the Iron and Steel Institute of Japan*, 1981, **21**, 726–731.
74. D. J. Alexander, and I. M. Bernstein: 'Cleavage fracture in pearlitic eutectoid steel', *Metallurgical Transactions A*, 1989, **20**, 2321–2335.
75. C. M. Bae, C. S. Lee, and W. J. Nam: 'Effect of carbon content on mechanical properties of fully pearlitic steels', *Materials Science and Technology*, 2002, **18**, 1317–1321.
76. J. Toribio: 'Relationship between microstructure and strength in eutectoid steels', *Materials Science & Engineering A*, 2004, **387-389**, 227–230.
77. Y. S. Jang, M. P. Phaniraj, D. I. Kim, J. H. Shim, and M. Y. Huh: 'Effect of aluminum content on the microstructure and mechanical properties of hypereutectoid steels', *Metallurgical & Materials Transactions A*, 2010, **41**, 2078–2084.
78. J. A. Mohandesi, and M. Saadatmand: 'The optimization of interlamellar spacing in a nanopearlitic lead-patented hypoeutectoid steel wire', *Journal of Materials Engineering and Performance*, 2011, **20**, 1467–1473.
79. W. Fu, T. Furuhara, and T. Maki: 'Effect of Mn and Si addition on microstructure and tensile properties of cold-rolled and annealed pearlite in eutectoid Fe–C alloys', *ISIJ International*, 2004, **44**, 171–178.
80. J. Chance, and N. Ridley: 'Chromium partitioning during isothermal transformation of a eutectoid steel', *Metallurgical Transactions A*, 1981, **12A**, 1205–1213.
81. D. J. C. MacKay: 'Bayesian interpolation', *Neural Computation*, 1992, **4**, 415–447.
82. D. J. C. MacKay: 'Bayesian non-linear modelling for the energy prediction competition', *ASHRAE Transactions*, 1994, **100**, 1053–1062.
83. D. J. C. MacKay: Information Theory, Inference, and Learning Algorithms: Cambridge University Press, 2003.
84. D. J. C. MacKay: 'Probable networks and plausible predictions – a review of practical Bayesian methods for supervised neural networks', *Computation in Neural Systems*, 1995, **6**, 469–505.
85. J. Páez, A. Dorronsoro, J. Martínez-Esnaola, J. G. Sevillano, and J. Alkorta: 'A microstructure-based constitutive model for eutectoid steels', *Acta Materialia*, 2024, **263**, 119533.
86. A. R. Stokes, and A. J. C. Wilson: 'The diffraction of X rays by distorted crystal aggregates-I', *Proceedings of the Physical Society*, 1944, **56**, 174–181.
87. B. P. Liu, T. F. Chung, J. R. Yang, J. Fu, C. Y. Chen, S. H. Wang, M. C. Tsai, and C. Y. Huang: 'Microstructure characterization of massive ferrite in laser-weldments of interstitial-free steels', *Metals*, 2020, **10**, 898.
88. G. K. Williamson, and R. E. Smallman: 'Dislocation densities in some annealed and cold-worked metals from measurements on the X-ray Debye-Scherrer spectrum', *Philosophical Magazine*, 1956, **1**, 34–46.
89. D. Nikas, X. Zhang, and J. Ahlström: 'Evaluation of local strength via microstructural quantification in a pearlitic rail steel deformed by simultaneous compression and torsion', *Materials Science & Engineering A*, 2018, **737**, 341–347.
90. N. Nakada, N. Koga, Y. Tanaka, T. Tsuchiyama, S. Takaki, and M. Ueda: 'Strengthening of pearlitic steel by ferrite/cementite elastic misfit strain', *ISIJ International*, 2015, **55**, 2036–2038.
91. N. Nakada, and M. Kato: 'Internal stress and elastic strain energy in pearlite and their accommodation by misfit dislocations', *ISIJ International*, 2016, **56**, 1866–1873.
92. A. Kanie, Y. Tomota, S. Torii, and T. Kamiyama: 'Elastic strains of cementite in a pearlite steel during tensile deformation measured by neutron diffraction', *ISIJ international*, 2004, **44**, 1952–1956.

93. Y. Wang, T. Ohnuki, Y. Tomota, S. Harjo, and T. Ohmura: 'Multi-scaled heterogeneous deformation behavior of pearlite steel studied by in situ neutron diffraction', *Scripta Materialia*, 2017, **140**, 45–49.
94. K. L. Johnson: Contact Mechanics: Cambridge, U. K.: Cambridge University Press, 1985.
95. M. Abouridouane, F. Klocke, and D. Lung: 'Microstructure-based 3D finite element model for micro drilling carbon steels', *Procedia CIRP*, 2013, **8**, 94–99.
96. A. Iturbe, E. Giraud, E. Hormaetxe, A. Garay, G. Germain, K. Ostolaza, and P. J. Arrazole: 'Mechanical characterization and modelling of Inconel 718 material behavior for machining process assessment', *Materials Science & Engineering A*, 2017, **682**, 441–453.
97. J. W. Christian: 'Some surprising features of the plastic deformation of body-centered cubic metals and alloys', *Metallurgical Transactions A*, 1983, **14A**, 1237–1256.
98. J. M. Krafft, A. M. Sullivan, and C. F. Tipper: 'The effect of static and dynamic loading and temperature on the yield stress of iron and mild steel in compression', *Proceedings of the Royal Society of London A: Mathematical, Physical and Engineering Sciences*, 1954, **221**, 114–127.
99. W. Visser, Y. Sun, O. Gregory, G. Plume, C.-E. Rousseau, and H. Ghonem: 'Deformation characteristics of low carbon steel subjected to dynamic impact loading', *Materials Science & Engineering A*, 2011, **528**, 7857–7866.
100. D. B. C. Taylor, and L. E. Malvern: 'Dynamic stress and deformation in a mild steel at normal and low temperatures', In: P. G. Shewmon, and V. F. Zackay, eds. *Response of Metals to High Velocity Deformation*. New York, USA: Interscience, 1961:77–91.
101. F. P. L. Kavishe, and T. J. Baker: 'Effect of prior austenite grain size and pearlite interlamellar spacing on strength and fracture toughness of a eutectoid rail steel', *Materials Science and Technology*, 1986, **2**, 816–822.
102. R. Nakkalil, J. R. Haonaday Jr, and M. Nabil Bassim: 'Characterization of the compression properties of rail steels at high temperatures and strain rates', *Materials Science & Engineering A*, 1991, **141**, 247–260.
103. L. Jing, L. Han, L. Zhao, and Y. Zhang: 'The dynamic tensile behavior of railway wheel steel at high strain rates', *Journal of Materials Engineering and Performance*, 2016, **25**, 4959–4966.
104. M. S. de Buruaga, P. Aristimuño, D. Soler, E. D'Eramo, A. Roth, and P. J. Arrazola: 'Microstructure based flow stress model to predict machinability in ferrite–pearlite steels', *CIRP Annals – Manufacturing Technology*, 2019, **68**, 49–52.
105. V. P. Kravz-Tarnavskii: 'A peculiar band discovered in steel', *Zeitschrift der Russischen Metallurgischen Gesellschaft*, 1928, **3**, 162–167.
106. N. Davidenkov, and I. Mirolubov: 'Eine besondere art der stauchdeformation von stahl: Der krawz-tarnawskij effekt', *Tech. Phys. USSR*, 1935, **2**, 281–298.
107. C. Zener, and J. H. Hollomon: 'Effect of strain rate upon plastic flow of steel', *Journal of Applied Physics*, 1944, **15**, 22–32.
108. T. W. Wright: The physics and mathematics of adiabatic shear bands: Cambridge, U. K.: Cambridge University Press, 2002.
109. S. M. Walley: 'Shear localisation: a historical overview', *Metallurgical & Materials Transactions A*, 2007, **38A**, 2629–2654.
110. B. Dodd, S. M. Walley, R. Yang, and V. F. Nesterenko: 'Major steps in the discover of adiabatic shear bands', *Metallurgical & Materials Transactions A*, 2015, **46**, 4454–4458.
111. H. G. Baron: 'Stress-strain curves of some metals and alloys at low temperatures and high rates of strain', *Journal of the Iron and Steel Institute*, 1956, **182**, 354–365.
112. R. S. Culver: 'Thermal instability strain in dynamic plastic deformation', In: R. W. Rhode, B. M. Butcher, J. R. Holland, and C. H. Karnes, eds. *Metallurgical effects at high strain rates*. New York, USA: Plenum Press, 1973:519–530.

113. H. K. D. H. Bhadeshia: Geometry of Crystals, Polycrystals, and Phase Transformations: Florida, USA: CRC Press, ISBN 9781138070783, freely downloadable from www.phase-trans.msm.cam.ac.uk, 2017.
114. R. Nakkalil: 'Formation of adiabatic shear bands in eutectoid steels in high strain rate compression', *Acta Metallurgica et Materialia*, 1991, **39**, 2553–2563.
115. C. K. Syn, D. R. Lesuer, and O. D. Sherby: 'Microstructure in adiabatic shear bands in a pearlitic ultrahigh carbon steel', *Materials Science and Technology*, 2005, **21**, 317–324.
116. A. K. Zurek: 'The study of adiabatic shear band instability in a pearlitic 4340 steel using a dynamic punch test', *Metallurgical and Materials Transactions A*, 1994, **25**, 2483–2489.
117. M. E. Backman, and S. A. Finnegan: 'The propagation of adiabatic shear', In: R. W. Rhode, B. M. Butcher, J. R. Holland, and C. H. Karnes, eds. *Metallurgical effects at high strain rates*. New York, USA: Plenum Press, 1973:531–543.
118. H. Rastegari, A. Kermanpur, A. Najafizadeh, D. Porter, and M. Somani: 'Warm deformation processing maps for the plain eutectoid steels', *Journal of Alloys and Compounds*, 2015, **626**, 136–144.
119. D. C. Pack, W. M. Evans, and H. J. James: 'The propagation of shock waves in steel and lead', *Proceedings of the Physical Society*, 1948, **60**, 1–8.
120. V. K. Euser, D. T. Martinez, J. A. Valdez, C. P. Trujillo, C. M. Cady, D. R. Jones, and S. J. Fensin: 'The influence of pearlite fraction on the shock properties of ferrite–pearlite steel microstructures: Insight into the effect of second-phase particles', *Journal of Applied Physics*, 2022, **131**, 115902.
121. H. K. D. H. Bhadeshia: Theory of Transformations in Steels: London, U.K.: CRC Press, Taylor and Francis Group, 2021.
122. B. G. Koepke, R. P. Jewett, W. T. Chandler, and T. E. Scott: 'Effects of initial microstructure and shock method on the shock-induced transformation strengthening of carbon steels', *Metallurgical Transactions*, 1971, **2**, 2043–2049.
123. H. G. Bowden, and P. M. Kelly: 'The structure and mechanical properties of shock-loaded steels', *Metal Science*, 1967, **1**, 75–77.
124. E. G. Zukas: 'Metallurgical results from shock-loaded iron alloys applied to a meteorite', *Journal of Geophysical Research*, 1969, **74**, 1993–2001.
125. W. Visser, and H. Ghonem: 'Dynamic flow stress of shock loaded low carbon steel', *Materials Science & Engineering A*, 2019, **753**, 317–330.
126. I. A. Bataev, I. A. Balagansky, A. Bataev, and K. Hokamoto: 'Transformation of structure in carbon steel specimen under loading by Mach stem, formed in preliminary compressed high explosive charge TG-40', *Materials Science Forum*, 2011, **673**, 89–94.
127. J. W. Christian: 'Accommodation strains in martensite formation, the use of the dilatation parameter', *Acta Metallurgica*, 1958, **6**, 377–379.
128. J. W. Christian, and S. Mahajan: 'Deformation twinning', *Progress in Materials Science*, 1995, **39**, 1–157.
129. H. P. Tardif: 'Mechanical twinning in pearlite', *Journal of the Australian Institute of Metals*, 1963, **8**, 415–416.
130. H. G. Bowden, and P. M. Kelly: 'The crystallography of the pressure induced phase transformations in iron alloys', *Acta Metallurgica*, 1967, **15**, 1489–1500.
131. Y. Ohmori, and F. Terasaki: 'Effects of nickel and manganese on cleavage fracture of ferritic-pearlitic steels', *Transactions of the Iron and Steel Institute of Japan*, 1976, **16**, 561–568.
132. B. G. Koepke, R. P. Jewett, and W. T. Chandler: 'An investigation of shock-hardened steels by electron microscopy', *Transactions of ASM*, 1965, **58**, 510–519.
133. L. M. Dougherty, G. T. Gray III, E. K. Cerreta, R. J. McCabe, R. D. Field, and J. F. Bingert: 'Rare twin linked to high-pressure phase transition in iron', *Scripta Materialia*, 2009, **60**, 772–775.

134. A. G. Crocker: 'Twinned martensite', *Acta Metallurgica*, 1962, **10**, 113–122.
135. K. Yang, C. Li, X. J. Zhao, Y. X. Xiang, S. N. Luo, and Y. Cai: 'Impact-induced twinning and phase transition in a medium carbon steel', *Journal of Alloys and Compounds*, 2021, **881**, 160–421.
136. S. C. Shrivastava, J. J. Jonas, and G. Canova: 'Equivalent strain in large deformation torsion testing : Theoretical and practical considerations', *Journal of the Mechanics and Physics of Solids*, 1982, **30**, 75–90.
137. K. Hase, C. G. Mateo, and H. K. D. H. Bhadeshia: 'Bainite formation influenced by large stress', *Materials Science and Technology*, 2004, **20**, 1499–1505.
138. J. Fu, F.-M. Wang, F. Hao, and G.-X. Jin: 'High-temperature mechanical properties of near-eutectoid steel', *International Journal of Minerals, Metallurgy and Materials*, 2013, **20**, 829–834.
139. D. S. Fields, and W. A. Backofen: 'Determination of strain hardening characteristics by torsion testing', *Proceedings of ASTM*, 1957, **57**, 1259–1272.
140. L. X. Kong, P. D. Hodgson, and D. C. Collinson: 'Modelling the effect of carbon content on hot strength of steels using a modified artificial neural network', *ISIJ International*, 1998, **38**, 1121–1130.
141. A. Eres-Castellanos, I. Toda-Caraballo, A. Latz, F. G. Caballero, and C. Garcia-Mateo: 'An integrated-model for austenite yield strength considering the influence of temperature and strain rate in lean steels', *Materials and Design*, 2020, **188**, 108435.

5 Ductility, toughness and fatigue

5.1 INTRODUCTION

The metallic bond is a huge asset. It allows atoms to move into entirely new positions, *without* compromising the integrity of the material, because the bond itself is abiding. There is no other class of materials where this is possible. There are simple crystals, such as solid argon, where each atom is bound to its twelve near neighbours by short-range, central, van der Waals forces [1], which enable some of the normal mechanical phenomena associated with metallic materials to manifest. Solid argon can deform plastically, it can work harden and recrystallise [2]. But van der Waals bonds are weak relative to metallic bonding. Graphite has a layered structure held together by van der Waals bonds so it too can accomplish displacements, but the weak bonding is unable to support much shear, hence its use in pencils or solid-lubricants.

Metallic bonding facilitates plasticity, that gifts to the material, an ability to irreversibly absorb energy and to do this while in three-dimensional form. When metals are scaled to large sizes, the metallic bond remains tolerant to inevitable defects, whether these imperfections are a thermodynamic necessity accompanying the increase in the number of participating entities, or macroscopic manufacturing-defects. This is not so for materials that rely on perfection, graphene to wit; the mechanical properties of such substances deteriorate dramatically on scaling. Graphene is not a 'wonder material' by any stretch of the imagination.

The lesson to be learnt from modern, exaggerated science is that the safe application of a material in structural engineering requires a combination of properties. Excluding elevated temperature service for which there exist better options than pearlite, ductility, toughness and fatigue feature in many of its applications. These are complex properties for which there is no theory that is predictive as a function of the abundance of variables involved.

There are, therefore, two objectives in attempting to decipher sophisticated properties. It is necessary to understand *mechanisms* and to express some aspects quantitatively in order to engage in educated design. A reduction in the size of inclusions can enhance ductility, but taking this to extremes at constant volume fraction can cause the steel to unzip in an apparently brittle manner. This would require a theoretical framework that incorporates both the nucleation, growth and linking of ductile voids as a function of the number density, distribution, size of inclusions, and of the features that determine the extension and connection of voids.

The second is to attempt the quantitative expression of vast experimental datasets in order to recognise patterns and uncertainties within the domain of a large number of variables. This is necessary because theory of the type we are familiar with, is so simple as to be incapable of representing reality. Machine learning, on the other hand, can handle complexity to a level never dreamt of before. The precursor to machine learning was linear or non-linear regression and contributed much to the development of steels some half a century ago, nicely documented by Pickering [3]; the method normally is applied when scientific understanding is limited. Regression equations, however, necessitate a prior assumption of the form of the relationship between the inputs and output. Machine learning does not require that presumption – it discovers the empirical relationship, which can be arbitrarily non-linear, making it possible to capture behaviours that depend on position in the input domain. Thus, the effect of carbon on steel is quite different from its influence on cast iron – both can be captured in a single model using machine learning. The use and limitations of machine learning in materials science are discussed elsewhere [4].

DOI: 10.1201/9781032631981-5

5.2 DUCTILITY

5.2.1 DURING TENSION

A tensile-test specimen has a parallel length over which the sample has a uniform cross-sectional area (A). Elongation is measured along a set distance within that length, the distance is known as the 'gauge length', equal to $5.65\sqrt{A}$. This procedure has been shown by experience to minimise the effect of specimen size or shape, thus enabling comparisons between different kinds of test samples. Figure 5.1 illustrates a broken sample which before deformation satisfied the $5.65\sqrt{A}$ gauge length criterion.

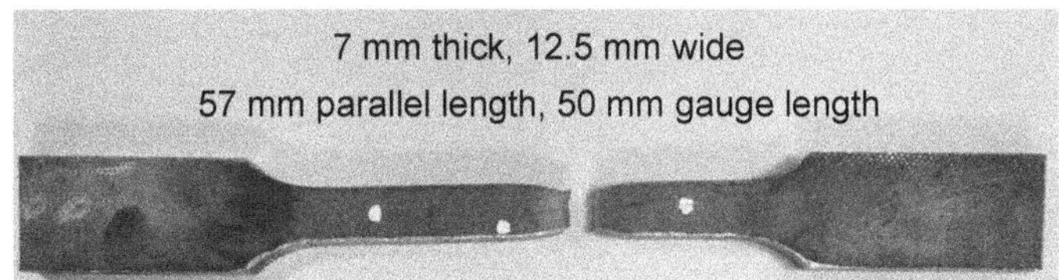

Figure 5.1 A broken tensile-test specimen from steel; the dimensions quoted are prior to deformation. The test at first leads to uniform strain but a plastic instability eventually ensues, leading to the necking illustrated. The total elongation which includes homogeneous and heterogeneous deformation, is measured by putting the two halves together such that they touch at their fracture positions. Image and data courtesy of Apparao Chintha.

Ductile failure during tension is determined by the nucleation of voids which then grow and coalesce, or link by tearing between voids, leading ultimately to overall fracture. If the number density of voids nucleated is large, they can link together at a smaller overall strain to fracture, and vice versa. Void generation of this kind is associated with stress triaxiality at the neck, but shear failure can manifest when uniaxial stress is predominant, for example during the testing of thin sheets or of pure iron. Triaxiality can be defined as the ratio $\overline{\sigma}/\sigma_{\mathrm{vM}}$ of the hydrostatic stress (mean of normal-stresses) to the equivalent stress calculated using Equation 4.3. As an example, for the plane stress condition,

$$\text{hydrostatic } \overline{\sigma} = \frac{\sigma_1 + \sigma_2 + 0}{3}, \quad \text{von Mises equivalent } \sigma_{\mathrm{vM}} = \sqrt{\sigma_1^2 + \sigma_2^2 - \sigma_1\sigma_2} \tag{5.1}$$

where the σ_1, σ_2 are the principal stresses. During uniaxial testing, the other components of stress arise because of the local change in geometry at the neck. It is assumed in Equation 5.1 that the steel is isotropic, which may not hold true for a sample with crystallographic texture.

Ductility during a tensile test has two components – at first, the elongation is uniform, with an identical stress experienced on all cross-sections of the gauge length. This morphs into the development of a neck so the deformation in its vicinity becomes heterogeneous. The necking is plastic instability that occurs when the ability of the material to strain harden cannot keep up with the increase in stress experienced as the cross-sectional area diminishes. It is stimulated in a local region where the sample may contain a non-uniformity such as a machining defect.

Whereas the evolution of damage in the form of microscopic fractures or voids can occur throughout the sample before the instability sets in, the damage intensifies at the minimum cross-section of the tensile specimen. Figure 5.2 shows the extent of damage just prior to overall fracture, in a tensile specimen characterised using X-ray tomography [5]. In this particular test, the damage at the centre of the sample was ductile but following internal fracture, the peripheries failed by a brittle, quasi-cleavage mechanism. This kind of cleavage is not flat over large areas but is interrupted by other features such as steps – the term *quasi* cleavage has long been used in geology to refer to slates or other formations that do not present flat cleavage but have roughening features [6, 7].

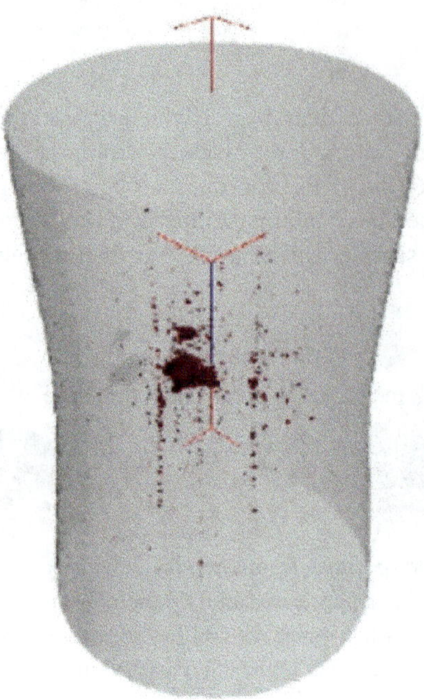

Figure 5.2 X-ray tomograph of the necked region of a tensile specimen, tested to an engineering strain of 0.128, which is close to the failure strain. Much of the gross damage is focused at the minimum cross-section of the neck. The composition of the steel is Fe-0.83C-0.25Si-0.72Mn-0.24Cr wt%, with a fully pearlitic microstructure. Image reproduced from Suárez et al. [5] under the CC BY 4.0 licence, https://creativecommons.org/licenses/by/4.0/.

X-ray tomography does not have sufficient resolution, but using other techniques, it is established that during the deformation of pearlite, damage begins with the fracture of cementite lamellae. The lamellae are oriented to experience the greatest tension break first. The consequent voids then propagate into the adjacent ferrite, eventually link, causing fracture. Figure 5.3 shows the fracture surface from a eutectoid steel; there are regions where the lamellar structure of the pearlite is visible due to the cylindrical voids that have formed in the ferrite between the cementite. The fracture is mixed in character with some large, spherical voids that may be associated with non-metallic inclusions, together with some brittle facets. The fracture facet revealing the lamellar structure is flat is because the pearlite nodule presents a homogeneous crystallography. The work of fracture is predominantly due to the elongation of the ferrite voids through a distance approximately equal to the interlamellar spacing, which also means that *if this fracture mode is dominant*, the ductility must decrease as the interlamellar spacing is reduced.

5.2.2 MIXED α+P STEELS

Necking *per se* is not necessarily associated with internal decohesion within the sample; tensile specimens made from pure metals can neck to a point, shear on a single macroscopic plane, or the shear can be localised on multiple planes [8]. In 'impure' alloys, any localisation of plastic strain leads to a geometry change that generates triaxial stresses within the neck, which in turn promote void generation, with or without the presence of inclusions, for example at locations where slip alternates on intersecting planes. Damage may not accumulate homogeneously in mixed microstructures – softer ferrite experiences greater plastic strain than the pearlite (Figure 4.1). Much

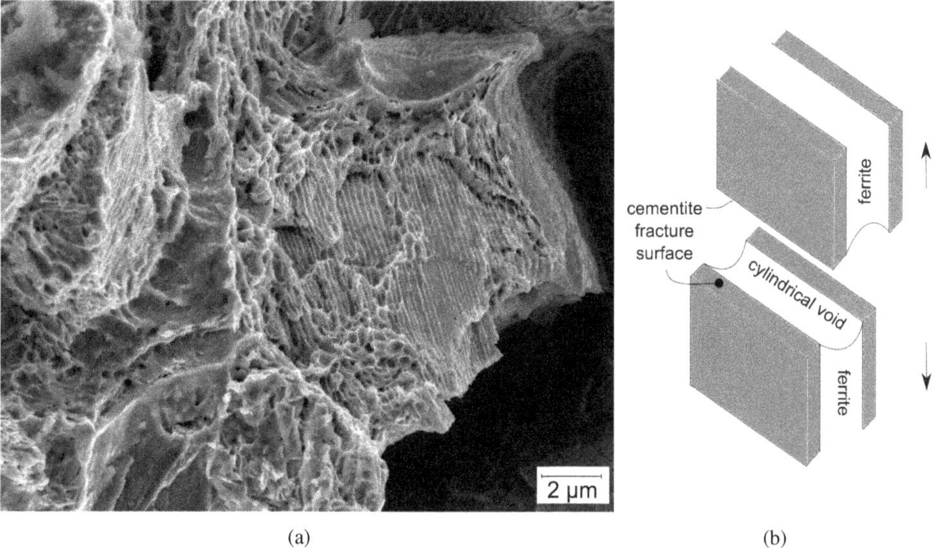

(a) (b)

Figure 5.3 (a) Essentially ductile fracture surface from a eutectoid steel, showing vestiges of the lamellar structure of pearlite. The lamellar structure is visible also on the inclined surface at the top left of the image. Fractograph courtesy of Apparao Chintha. (b) Schematic representation of the development of cylindrical voids parallel to the lamellae.

of the damage during the early stages of deformation is associated with this ferrite, with voids initiating in regions of strain discontinuities, such as at the interfaces between the ferrite and pearlite colonies. In steels that have small impurity concentrations (S, P ≤ 0.02 wt%), voids nucleate in the ferrite at the α/P interfaces located at the centres of both parallel-sided and notched tensile specimens [9]. Any inclusions that are weakly bonded to the steel (e.g., MnS, [Al,Mg]O, CaS, complex oxy-sulphides), or undeformable precipitates such as titanium nitride where the matrix must flow around the precipitate, or precipitates that fracture leaving gaps, can all contribute to void initiation. Plastic strain causes the cementite within coarse pearlite to crack, particularly when the cementite lamellae happen to be aligned along the tensile axis; fine pearlite is more resistant to this kind of microscopic cracking [10]. Spheroidal particles of cementite do not similarly crack because the ferrite is able to flow around them, though voids can form in the wake of the flow around large particles, Figure 5.4 [11].

Triaxiality and local strain

The ductility measured on a smooth, round-sectioned tensile specimen will be different from that determined on circumferentially-notched samples of identical minimum cross-sectional area, and indeed, for other shapes. It is possible, however, to rationalise ductile fracture if parameters such as the strain at the known location where a ductile crack might initiate and the stress triaxiality at the location can be calculated. The data can then be used alongside calibration curves to estimate ductile failure, as follows, but noting that the aim is to associate α+P microstructures where the hardness of the two components is not identical, with ductile crack-growth resistance.

Changes in geometry within the stressed region cause stress-triaxiality that alter the response of the steel [12]. Ductile failure begins at the surface of the V-shaped notch that has a root radius much less than a millimetre, by a shear-slip mechanism, at the maximum in the load-displacement curve. Subsequent propagation of the crack occurs as non-uniform strain accumulates, Figure 5.5a. This applies also during compression testing of steel where the onset of plastic instability defines

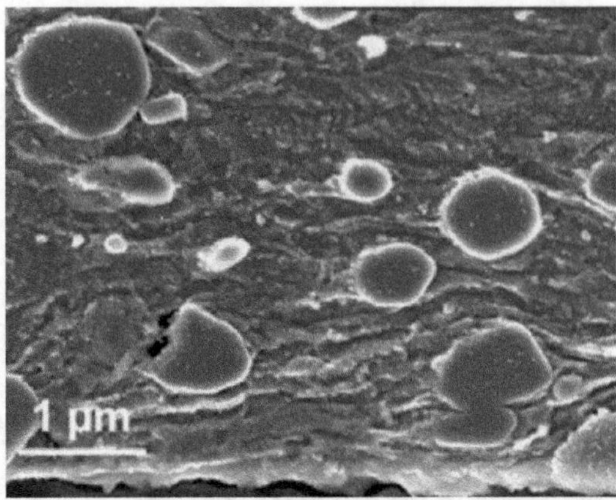

Figure 5.4 Void developing between a large cementite particle and the ferrite matrix during ring–rolling of 52100 steel in its spheroidised condition. Micrograph reproduced from [11], with the permission of Elsevier.

the beginning of cavity growth and coalescence by internal necking [13]. For smooth or U-shaped notches, the ductile crack initiates in the centre of the sample cross-section when the maximum in the load-displacement curve has been exceeded; overall failure quickly follows, Figure 5.5b. This difference arises because the volume of the material that is triaxially stressed is small at the surface of a sharp notch, when compared against that at the centre of a smoothly-notched tensile specimen.

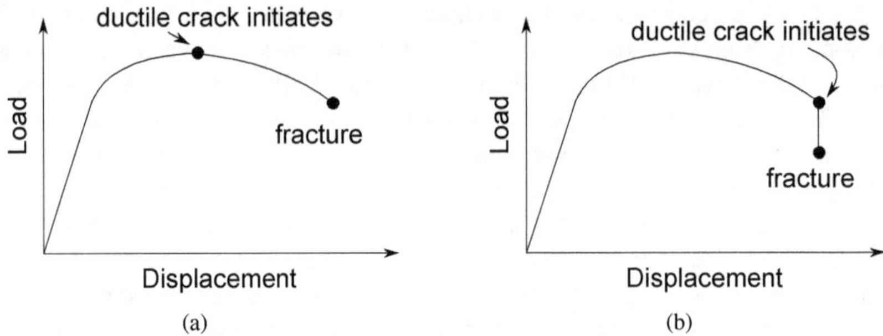

Figure 5.5 Schematic load-extension curves for α+P steel tested in tension. (a) Sharp circumferential notch on round-sectioned tensile specimen. (b) Corresponding behaviour when the notch is smooth with a large root radius. Adapted from Shoji et al. [12].

A crack might be said to initiate at a critical *local* value of the equivalent strain ε^{c}_{vM} [12], which can be related to stress triaxiality $\overline{\sigma}/\sigma_{vM}$ as illustrated in Figure 5.6, where $\overline{\sigma}$ is the mean of the normal-stresses. The local strains and stresses at the point of initiation have to be estimated using finite element analysis. Experimental results show that ductile crack initiation occurs from the surface at the root of sharp-notched samples but at the centre of the tensile specimen that has smooth circumferential notches. Since crack initiation is at a smaller critical strain when the stress triaxiality is large, it is expected that the overall ductility decreases with the triaxiality.

The results illustrated in Figure 5.6 can form the basis of finite element schemes to calculate overall ductility as a function of the shape of α+P steels; this would involve implementation of a damage parameter within each homogenised element. The three-dimensional representation of the ferrite and pearlite regions is expressed as Voronoi tessellations distributed according to experimental observations, with the flow properties of each phase represented by relations such as that in Equation 4.1. To deal with ductile damage, volume elements are defined in which the properties are assumed to be homogeneous. The strength of the element is reduced as an empirical damage parameter evolves, with the element deleted to leave a void in the finite element scheme when the damage parameter reaches a unit value.

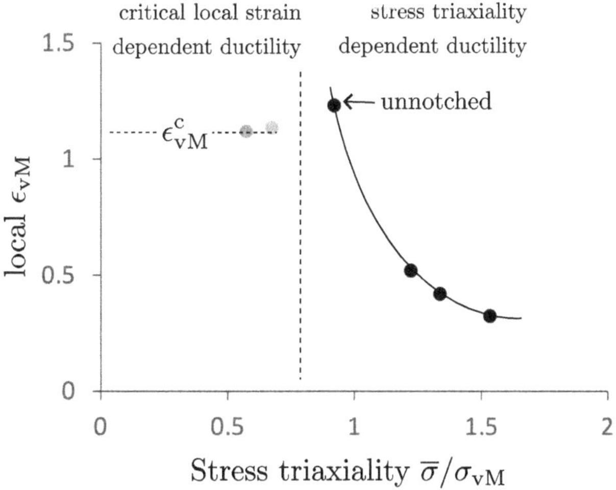

Figure 5.6 Local value of equivalent strain versus the degree of stress triaxiality, based on experimental observations on a ferrite-pearlite steel. The triaxiality is varied by altering the radius and shape (V- or U-shaped) of the notch root on round-sectioned tensile specimens. The local ε_{vM} is calculated at the observed crack initiation point using finite element analysis. Adapted from Shoji et al. [12].

There are contradictions in the literature about the initiation of damage during a tensile test on an un-notched cylindrical sample. Whereas the description above [12] considers initiation at the centre of the specimen, others base their simulations on surface observations assuming that the failure occurs first in the pearlite and at some critical value of strain measured at a somewhat arbitrary distance from the fracture surface [14]. This latter approach does not seem reasonable given that the ultimate fracture surface shows a cup-and-cone morphology, which is convincing evidence for ductile crack initiation at the centre of the sample.

Finally, it is worth pointing out that while the local strain and stress triaxiality considerations are useful in the basic understanding of the ductile behaviour of α+P steels, their implementation into numerical models requires a considerable number of fitting parameters. Neglecting certain assumptions of how these parameters could be obtained for the α and pearlite regions, Shoji et al. listed seven empirical values that are needed to model damage evolution.

The models described do not seem to have been used to illustrate how changes in phase fractions or how modifying the pearlite internal-structure and morphology might influence ductility. The quote by Anderson [15] comes to mind:

> One of my strongest stylistic prejudices in science is that many of the facts nature confronts us with are so implausible ..., that the mere demonstration of a reasonable mechanism leaves no doubt of the correct explanation. This is so especially if it also correctly predicts unexpected facts.... Very often such, a simplified model throws more light on the real workings of nature than any number of 'ab initio' calculations of individual situations, which even where correct often contain so much detail as to conceal rather than reveal reality. It can be a disadvantage rather than an advantage to be able to compute or to measure too accurately, since often what one measures or computes is irrelevant in terms of mechanism. After all, the perfect computation simply reproduces Nature, does not explain her.

... unless the models are used to create [16].

5.2.3 HYDROSTATIC PRESSURE

Hydrostatic pressure leads only to a change in volume, the magnitude of which can be estimated using the bulk modulus. For the typical pressures achievable in experiments, it cannot cause plastic deformation because there are no shear components associated with such pressure. Extreme pressures ($\gtrapprox$ 13 GPa) induce $\alpha \rightarrow \epsilon$ phase transformation [17–19] that is martensitic and hence accompanied by a shape deformation that involves a large shear strain.

Subjecting a steel to hydrostatic pressure has little effect on the tensile-stress-strain curve measured at ambient pressure [20], unless the intention is to mitigate defects. Additively-manufactured steel components might be pressurised to 100 MPa, at 1150 °C for 3 h in order to seal pores or heal cracks [p.415, 21]. But there is an independent effect, when a tension test is conducted with superimposed hydrostatic pressure, the strain to failure increases with the magnitude of the pressure. Voids, as they form, cause the volume of the sample to increase, so it is easy to imagine that hydrostatic pressure must retard the process. There is considerable evidence that large hydrostatic pressure can suppress ductile fracture [22]. The fraction of voids increases dramatically with strain in the necked region of a tensile sample [23] so the interaction with the pressure becomes prominent in delaying failure.

This phenomenon is observed in many materials but there are special effects associated with pearlite or pearlite-containing microstructures [24, 25]. Hydrostatic pressure applied during tensile testing suppresses the growth of voids in ferrite, which improves elongation (Figure 5.7) in hypo- and hypereutectoid steels [24, 25]. The fracture surface at the highest of pressures tends to become flat as planar shear becomes favoured over the void mechanism, with an absence of the dimples that characterise ordinary ductile fracture.

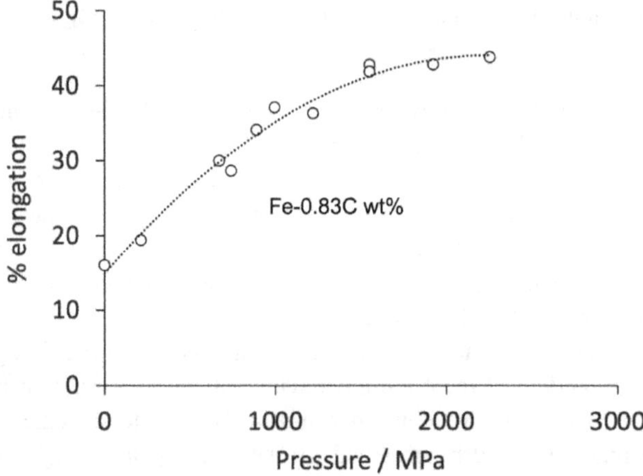

Figure 5.7 Elongation measured on a fully pearlitic steel following uniaxial tension under the influence of hydrostatic pressure. Selected data from Davidson and Ansell [24].

Hydrostatic pressure also enhances the ductility of cementite; its crystal structure remains intact to pressures up to 73 GPa although the volume of the cell decreases by some 20% at the highest of pressures [26].

5.2.4 DURING UNIAXIAL COMPRESSION

The yield strength of pearlite is the same in compression or tension, but there is a greater tendency to fracture by shearing across a plane oblique to the compression axis. This assumes that deformation is heterogeneous because uniform deformation cannot cause ductile fracture during compression. It is the barrelling of the sample caused by friction at the sample/platen interfaces that limits the deformation there, causing non-uniform deformation.

Plastic instability in uniaxial tension is defined by $\mathrm{d}\sigma/\mathrm{d}\varepsilon_\mathrm{p} = \sigma$, where σ is the true stress, which forms the basis of the Considère construction [27]. This defines the strain where the rate of increase of tensile stress on a decreasing cross-sectional area exceeds the load-carrying capacity of the section.

When a cylinder is deformed by uniaxial compression with the help of platens, the circumferential stress on the equatorial free-surface becomes *tensile*, causing barrelling [13]. The coordinate system is defined with z (compression axis), r and θ (radial and circumferential axes, respectively). Deformation of a thin ring element of thickness t and radius r on the free surface at the equator occurs due to changes in σ_z and σ_r. By analogy with tensile testing, the instability condition is:

$$\frac{\mathrm{d}\sigma_\theta}{\mathrm{d}\varepsilon_\theta - \mathrm{d}\varepsilon_r} = \sigma_\theta \tag{5.2a}$$

where $\mathrm{d}\varepsilon_\theta = \mathrm{d}r/r$ and $\mathrm{d}\varepsilon_r = \mathrm{d}t/t$. Using the fact that volume is conserved, $\mathrm{d}\varepsilon_z + \mathrm{d}\varepsilon_\theta + \mathrm{d}\varepsilon_r = 0$ so Equation 5.2a becomes

$$\frac{\mathrm{d}\sigma_\theta}{\mathrm{d}\varepsilon_\theta} = \sigma_\theta\left(2 + \frac{\mathrm{d}\varepsilon_z}{\mathrm{d}\varepsilon_\theta}\right). \tag{5.2b}$$

If the second term in the bracket is > 1, the instability will occur first under the action of σ_θ

$$\frac{\mathrm{d}\sigma_\theta}{\mathrm{d}\varepsilon_\theta} = \sigma_\theta. \tag{5.2c}$$

The work hardening rate therefore controls the onset of plastic instability during compression testing. If the sample thickness to height ratio is small, then buckling may precede other phenomena that occur during uniaxial compression.

We have emphasised that the plastic instability during the uniaxial compression of a cylinder is really a tensile plastic-instability along the equatorial plane due to circumferential stress and strain. However, as pointed out by Thomason [13], localised external necking is not obvious on the barrelled surface of the sample. The point of instability can instead be measured by a Considère type construction in which σ_θ is plotted against the circumferential strain ε_θ, and finding the point where the tangent to the σ_θ curve has the gradient $\sigma_\theta(2 + \mathrm{d}\varepsilon_z/\mathrm{d}\varepsilon_\theta)$. Damage that leads to ductile failure evolves rapidly within the sample at the point of plastic instability.

Figure 5.8 compares the stress-strain curves obtained during uniaxial tension and uniaxial compression of a α+P steel containing a volume fraction 0.25 of pearlite [28]. [1] The compression test naturally exhibits a much greater ductility given that plastic instability relies on the development of barrelling that is a function of friction at the sample/platen surfaces, the details of which depend on the lubrication there. Furthermore, the upper and lower yield points and subsequent Lúders extension are sometimes absent during compression (Figure 5.8a), though this may be an artefact because the strain in compression was derived from machine displacement due to the small size of the cylindrical samples (10 mm diameter, 15 mm length). More accurate measurements on slender specimens covered in a brittle coating and instrumented with strain gauges reveal the two yield points and the propagation of Lúders bands across the sample as the stress is increased [29]. Similarly, when a flat, instrumented sample 76 mm long and 12.7×9.5 mm in cross section, is tested in uniaxial compression, all of the yield features including the Lúders extension are clearly observed (Figure 5.8b).

[1] Laschet et al. presented four values for V_V^P, claiming a dependence on the orientation of the surface on which the measurements were made. However, the volume fraction is independent of the plane of observation, so the value quoted here is an average.

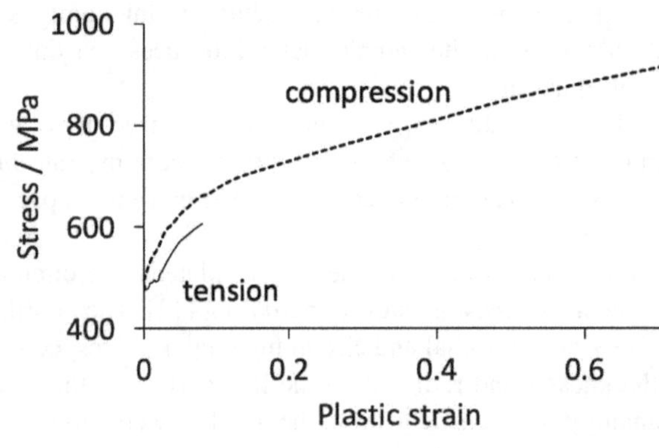

(a)

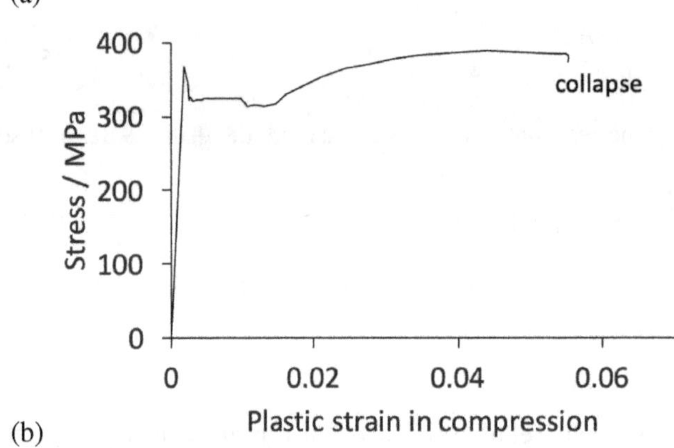

(b)

Figure 5.8 (a) Tensile and compression tests on a steel (Fe-0.11C-0.34Si-1.52Mn wt%) containing a banded microstructure of ferrite and pearlite, with $V_V^P = 0.25$. The stress axis was in each case parallel to the rolling direction. Selected data from Laschet et al. [28].

(b) Stress-strain curve from a slender sample of α+P steel containing $V_V^P \approx 0.22$, tested in compression, showing an upper and lower yield point and a Lúders extension. The test was terminated when the sample buckled and collapsed. Adapted using data from Corona et al. [29].

5.2.5 FULLY PEARLITIC STEELS

Fully pearlitic steels are ductile but at constant composition, exhibit an *increase* in elongation to fracture (ε_u) as the strength increases, Figure 5.9. The strength, in the case illustrated, is doubled by reducing the interlamellar spacing. The uniform elongation is found to be almost identical at about $\varepsilon_u = 0.08$ [30] so the non-uniform component, $\varepsilon_f - \varepsilon_u$ can be assumed to correspond to the onset of damage in the form of cracks and voids, leading ultimately to failure when these features link.

It is interesting that the microstructure can sustain a uniform strain of 0.08 without showing macroscopic signs of damage, but this presumably is because ferrite is the dominant phase. Whereas it is likely that some cementite particles that are appropriately oriented relative to the stress, may fracture at an early stage for $\varepsilon_p < \varepsilon_u$, the extent of damage does not appear sufficient to disrupt uniform elongation.

The critical event in the damage process seems to be the widespread cracking of cementite. The stress to break the cementite due to a dislocation pile-up in the adjacent ferrite is estimated as a function of interlamellar spacing (μm) as [30]

$$\sigma_\theta = \frac{0.403}{\sqrt{S_I}} \quad \text{MPa}. \tag{5.3}$$

The origin of this equation is from the work of Smith and Barnby [31], who considered the stress concentration due to the pile-up of dislocations, triggering the nucleation of a crack at a grain boundary. If $z_\perp$ is the length of the pile-up against a barrier of cementite of thickness L_θ, then the shear stress τ_* to nucleate a crack is given by [30–32]

$$\tau_* = \left(\frac{2\sigma_s E_s}{\pi(1-\nu)z_\perp}\right)^{\frac{1}{2}} \left(\frac{L_\theta}{z_\perp + L_\theta}\right)^{\frac{1}{2}} \tag{5.4a}$$

where σ_s is the surface energy per unit area of the fractured region and the other terms have their usual meanings. Taking $z_\perp = L_\alpha$ and noting that $L_\alpha + L_\theta = S_I$ with $L_\theta = V_V^\theta S_I$ at equilibrium, Equation 5.4a simplifies to

$$\tau_* = 0.34\left(\frac{2\sigma_s E_s}{\pi(1-\nu)z_\perp}\right)^{\frac{1}{2}}. \tag{5.4b}$$

Dollar et al. assumed that work of creating new surfaces $\sigma_s = 2.33\,\mathrm{J\,m^{-2}}$ and allowing for the Taylor factor in converting the shear stress into a tensile stress, Equation 5.3 is obtained on making the appropriate substitutions for the elastic constants.[2]

The application of Equation 5.3 to literature data on pearlitic steels indicates that the cementite cracks at a stress often significantly less than the ultimate tensile strength of the material. This means that although voids are formed, they may not have grown to a sufficient extent to link, or that the number density of voids needs to increase sufficiently for the linking to occur. An important outcome is that the stress required to crack cementite increases with $1/\sqrt{S_I}$; it has been known for some time that thin lamellae of cementite are more resistant to brittle fracture and may even exhibit plasticity before fracture [10, 33, 34]. This does not, however, mean that ductility increases at the same time because a full theory for ductility would require a quantitative treatment of factors such as the damage evolution rate and the work-hardening ability of the microstructure. Whereas studies on individual steels do indeed show an increase in ductility as S_I is reduced, there is little or no correlation between the reduction of area and the interlamellar spacing when a large dataset is examined, Figure 5.10, and this remains the case when the ductility is plotted against other functions of S_I. It would be interesting in this context to conduct tensile tests while monitoring acoustic emissions in order to (a) justify the form of Equation 5.3, and (b) the extent of damage at the point where the macroscopic elongation ceases to be uniform.

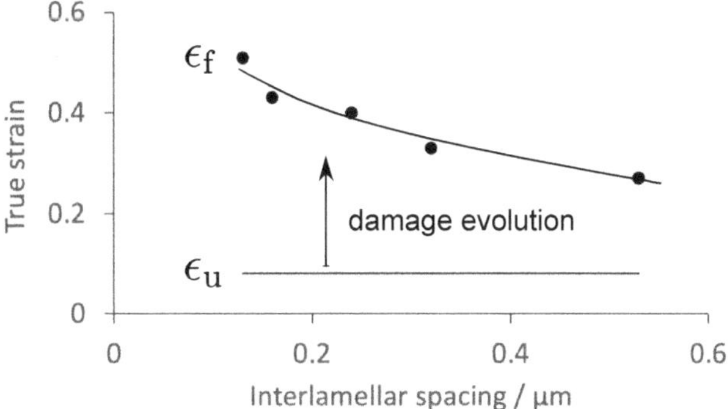

Figure 5.9 True fracture-strain ε_f in tension at a strain rate of $5 \times 10^{-4}\,\mathrm{s^{-1}}$, as a function of the interlamellar spacing, Fe-0.8C-0.84Mn-0.17Si-0.013S-0.018P wt% fully pearlitic steels. The 0.2% proof strength doubled from 331 to 662 MPa over the same range of interlamellar spacings. The true uniform-strain ε_u was about 0.08 for all the spacings, represented by the horizontal line. Selected data from Dollar et al. [30].

[2] In estimating E_s, Dollar et al. used data for ferrite rather than cementite, but this may not matter since E_s is essentially empirical because of the unknown plasticity associated with the creation of cementite surfaces during cleavage. It was also assumed that $z_\perp = S_I$ whereas it should be smaller at L_α.

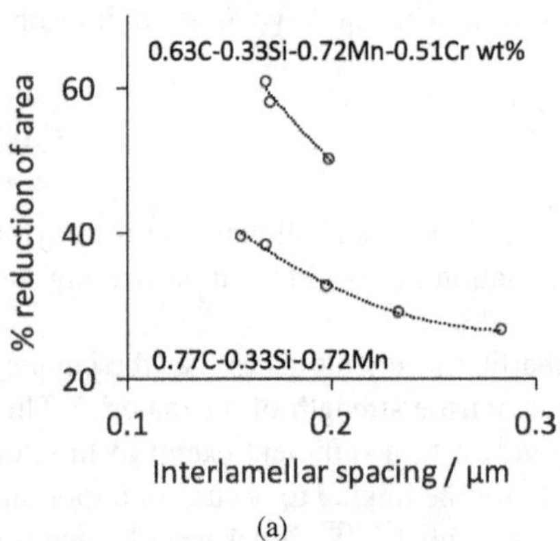

(a)

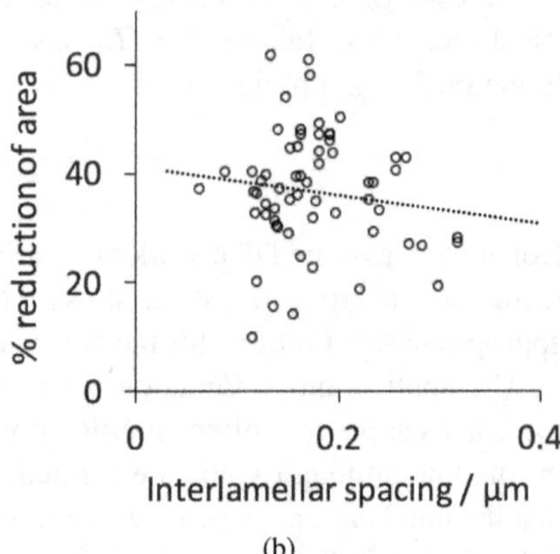

(b)

Figure 5.10 Reduction of area data measured on tensile specimens from fully pearlitic steels, as a function of the interlamellar spacing. (a) Selected data from Nakase and Bernstein [35]. (b) Data compiled from [30, 35–39], showing little dependence on S_I with a correlation coefficient of 0.17.

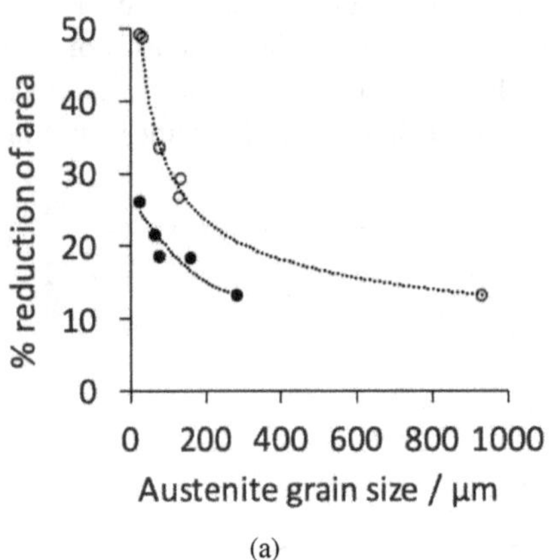

(a)

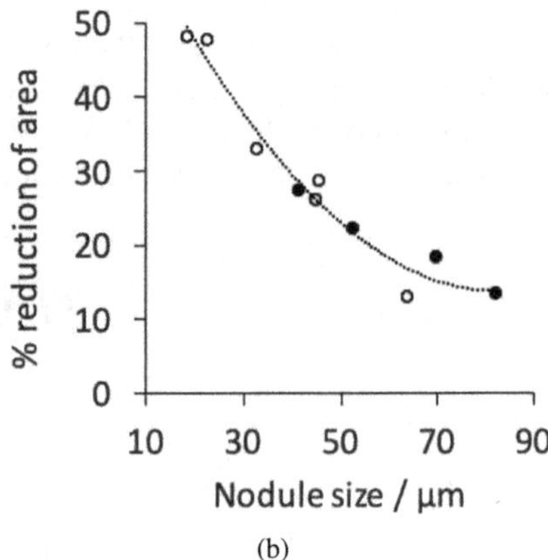

(b)

Figure 5.11 Ductility expressed as the reduction of area in a tensile test conducted to fracture, as a function of (a) the austenite grain size, (b) the pearlite nodule size. The data represent fully pearlitic steels, with the filled points referring to Fe-0.83C-0.25Si-0.97Mn-0.98Cr wt% steel transformed isothermally at 680 °C; the open points represent Fe-0.85C-0.25Si-1.51Mn wt% steel transformed isothermally at 600 °C. Selected data from Takahashi et al. [40].

Once the cementite begins to crack during tensile testing, the crystallography of the microstructure becomes an important factor in determining the fracture path [40]. Figure 5.11 shows that although the austenite grain size might be construed to influence ductility, the data can better be rationalised when plotted against the nodule size. Shear bands and cleavage cracks are found to propagate with little deviation across pearlite nodules that do not contain large orientation differences.[3] Therefore, the nodule size becomes an important factor in determining macroscopic ductility irrespective of whether the fracture is microscopically ductile or brittle. The refinement of austenite grain size is useful because it leads to smaller pearlite nodules (Section 2.3).

[3]There is an elegant montage of micrographs in [40], of slip bands observed on the surface of a sheet-tensile specimen, showing clearly that the bands are not macroscopically deviated within pearlite nodules.

5.2.6 α+P+GRAPHITE CAST IRON

One form of ductile cast-iron contains spheroidal-graphite nodules occupying typically ≈ 10% of the volume, embedded in a mixture of pearlite and ferrite, Figure 5.12. The proportion of ferrite depends both on the composition and the heat treatment because annealing causes the cementite in the pearlite to decompose and deposit on to the existing graphite, leaving behind a ferritic matrix. The structure of a graphite nodule in such irons is complex with a thin (≈ 2μm) shell on the surface of an otherwise larger core that grows by a different mechanism to that associated with solidification [41]. It is believed that the shell forms through the decomposition of pearlitic cementite.

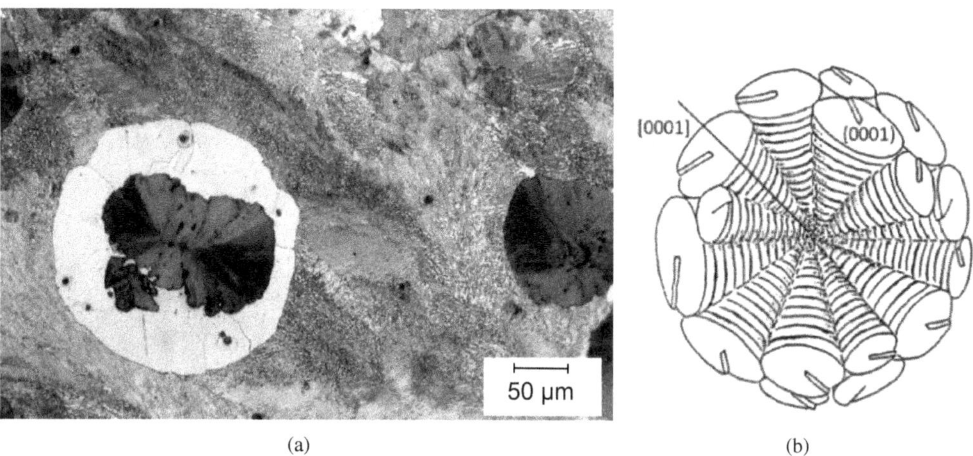

Figure 5.12 (a) Spheroidal graphite cast iron, Fe-3.2C-2.5Si-0.05Mg wt%, containing graphite nodules in a matrix that is predominantly pearlitic, with some nodules surrounded by ferrite. It is the decomposition of cementite that deposits carbon on to the graphite that causes the ferrite to form. Note the cone-like structure within the graphite nodule. (b) Schematic representation of the crystallography of graphite ($P6_3/mmc$) in a spheroid, with cones of stacked (0001) planes radiating from an origin. Each of these cones has a growth step at the surface. Diagram reproduced from Ştefănescu et al. [42] under the CC BY 4.0 license, https://creativecommons.org/licenses/by/4.0/

Graphite on its own is soft and friable, a weak inclusion that can easily be debonded from the matrix; for some purposes, its presence is equivalent to that of voids. Its actual behaviour does depend on the properties of the surrounding matrix [43].

When the matrix is soft, with a fully ferritic microstructure, damage is not observed during elastic loading, but plastic flow in the matrix induces the peeling of the surface layers of the graphite (Figure 5.13a). This *onion peeling* that begins at the poles of the graphite [43], is a consequence of the shell-core structure of graphite as described earlier. Continued plastic deformation opens up the parted regions, but there are no macroscopic cracks that can be attributed to de-bonding or trans-graphite fracture.

The lack of visible signs of damage during elastic loading persists when the ferrite fraction in the matrix is reduced to 0.7 with the remainder being pearlite (Figure 5.13b), with partial de-bonding at the graphite/ferrite interface visible in the plastic regime. The damage then penetrates the ferrite during extended plastic deformation. If the pearlite fraction is increased further at the expense of ferrite, the stronger matrix then causes brittle cracking within the graphite, in part because the ferrite at that point has lost continuity, isolated around individual nodules of graphite.

With a strong, fully pearlitic matrix, damage occurs at all stages of deformation, with both trans-graphite cracking and limited de-bonding (Figure 5.13c). The fact that the matrix is more rigid means that the trans-graphite cracks develop early in the deformation process and open up in an

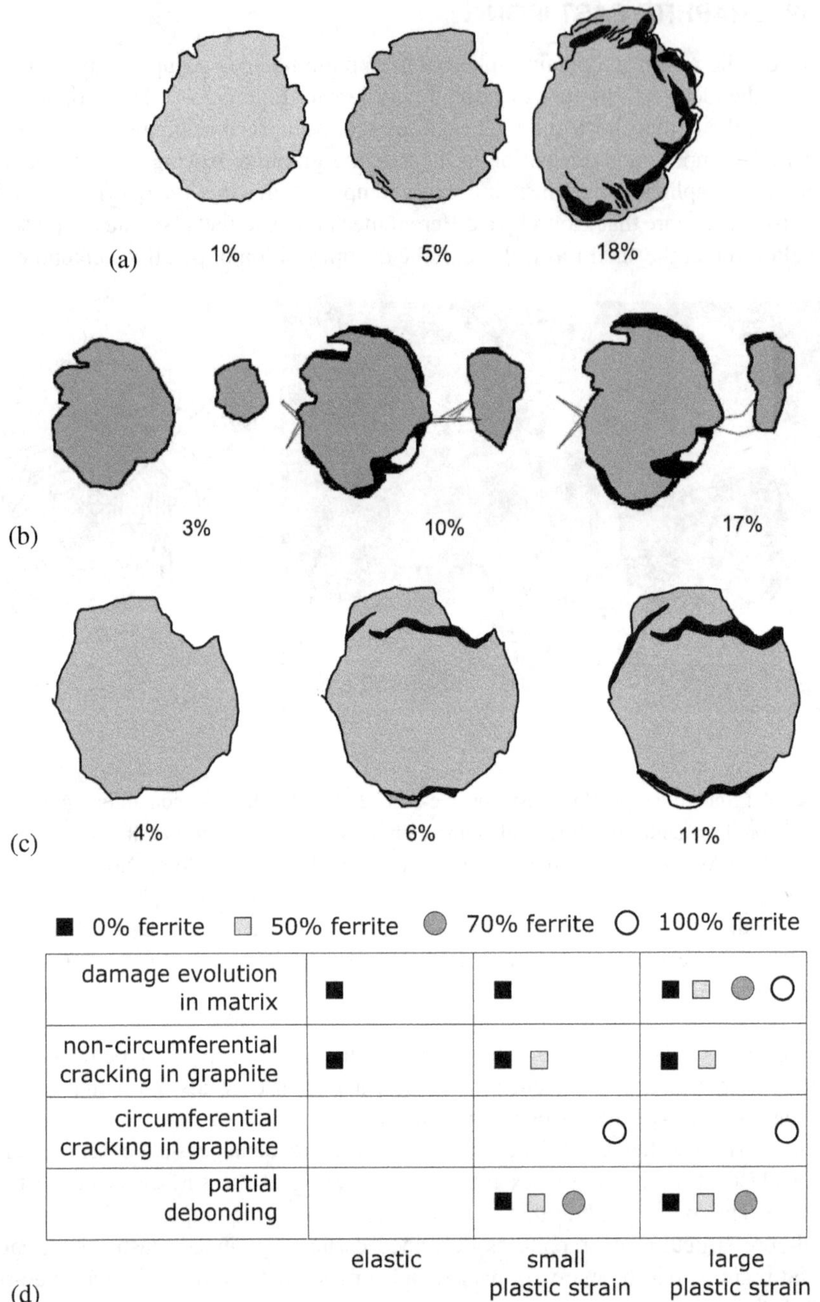

Figure 5.13 Evolution of damage during the tensile testing of ductile, spheroidal cast-iron. The sketches are based on *in situ* observations by Iacoviello et al. [43] during the tensile testing of cast iron containing about 10% of graphite. The direction of tension in each case is vertical and the percentages represent the tensile elongation at which the observation was recorded. (a) Completely ferritic matrix. (b) Matrix containing 70% ferrite with the remainder being pearlite. There are two graphite particles in each sub-image. (c) Fully pearlitic matrix. (d) Summary of the consequences of tensile loading of ductile iron; the terms *small* and *large* in describing the plastic strain should be taken in the context of parts a-c.

orientation normal to the tensile stress.

The results are summarised in Figure 5.13d, but they emphasise that models of ductility that assume graphite particles to be equivalent to voids may not be a good representation of what actually happens [43]. It would be interesting to see how the deformation links between the graphite nodules prior to the onset of gross fracture.

5.2.7 SUPERPLASTICITY

Exceptionally large tensile-elongations can sometimes be achieved when fine pearlite is deformed at a temperature just below Ae_1, by a mechanism that involves diffusion and associated structural changes. During the isothermal flow of a viscous liquid, the shear stress is proportional to the shear-strain *rate*, with the proportionality constant defining the viscosity. A similar sort of relationship applies during superplasticity with $\sigma \propto \dot{\varepsilon}^m$, where the strain-rate sensitivity $m < 1$ varies with temperature and even with $\dot{\varepsilon}$ so the magnitude of m should be defined over a narrow strain-rate range. A large strain-rate sensitivity can hinder the development of necking during tension, because the local strain rate and hence the hardening within the neck, will be greater than in uniformly deformed regions. This is illustrated in Figure 5.14a for tests done on an initially pearlitic microstructure, tested at just below the Ae_1 temperature [44].

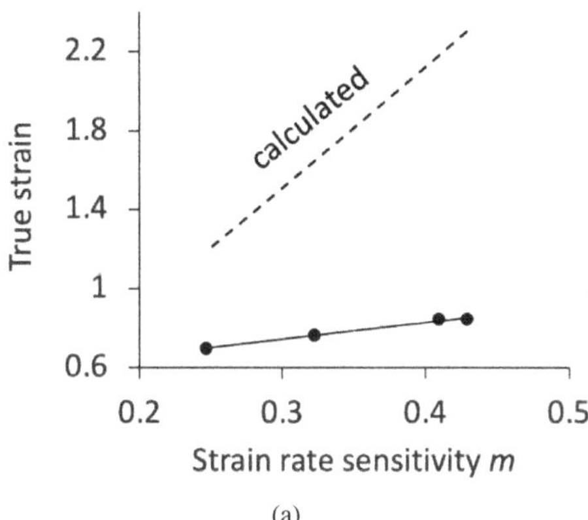

(a)

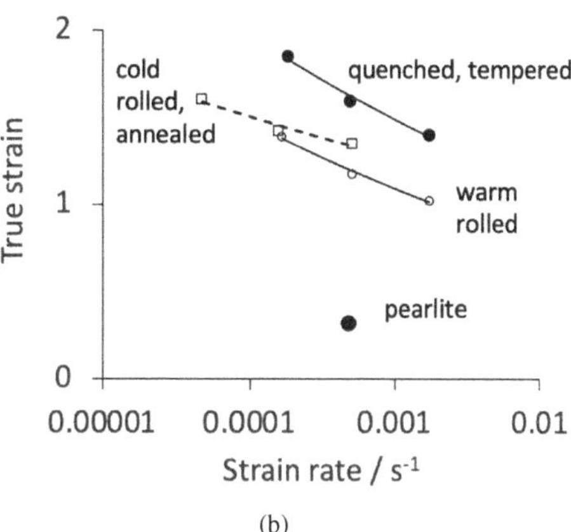

(b)

Figure 5.14 (a) Tensile elongation as a function of the strain rate sensitivity m, determined over the strain-rate range $10^{-3} \rightarrow 1\,\text{min}^{-1}$ during testing at 716 °C. The fully pearlitic microstructure of Fe-0.91C-0.12Si-0.45Mn wt% steel had an interlamellar spacing of about 65 nm but spheroidised shortly after the test began. Selected data from Yoder and Weiss [44]. (b) Fe-0.98C-0.27Si-0.41Mn-1.44Cr wt% steel tested in tension at 700 °C, with a variety of initial microstructures. The warm rolling at 650 °C, with a 90% reduction ($m \approx 0.27$); cold rolling with similar reduction but annealed at 700 °C ($m \approx 0.25$); quenched from 770 °C followed by tempering at 650 °C ($m \approx 0.54$). Selected data from Furuhara et al. [45].

Consider a tensile bar that is smooth other than a region of reduced cross-section by a factor f_A that is only slightly less than unity. Assume that the sample deforms according to $\sigma \propto \dot{\varepsilon}^m$. The ratio of the instantaneous to original cross-sectional area is given by $\exp\{-\varepsilon_p\}$. Load equilibrium between the perfect and reduced cross-section requires that [46]:

$$\exp\{-\varepsilon_u\}\,\dot{\varepsilon}_u^m = f_A \exp\{-\varepsilon_n\}\,\dot{\varepsilon}_n^m \tag{5.5a}$$

where ε_u is the true uniform-strain and ε_n the corresponding quantity in the region with smaller cross section. It follows that

$$\exp\{-\varepsilon_u/m\}\,d\varepsilon_u = f_A^{\frac{1}{m}} \exp\{-\varepsilon_n/m\}\,d\varepsilon_n \tag{5.5b}$$

which on integrating from zero initial strain to $\varepsilon_{u,max}$ at fracture, gives:

$$\varepsilon_{u,max} = -m \ln\{1 - f_A^{\frac{1}{m}}\{1 - \exp\{-\varepsilon_n/m\}\} \approx -m \ln\{1 - f_A^{\frac{1}{m}}\}. \quad (5.5c)$$

The approximation follows from the assumption that ε_n at fracture will be very large due to superplasticity, much greater than ε_u [46]. The total elongation in the necked region must in superplasticity be much less that the total elongation measured, so experimental data from tensile tests can be approximated to the maximum uniform strain $\varepsilon_{u,max}$. The relationship between elongation and the strain hardening rate based on Equation 5.5c seems to estimate the observed elongation rather well for a range or materials subjected to superplastic deformation [46]. However, in the case of pearlite, Figure 5.14a shows that the elongation expected should be much larger than observed for any given value of m. This is difficult to explain but the discrepancy indicates premature failure, perhaps due to the presence of considerable quantities of cementite particles; the materials compared successfully against theory in [46] are either pure metals or alloys with more benign second phases. If this interpretation is correct, then a reduction of the cementite particle size should enhance closure between theory and experiments on pearlite. The results described in Figure 5.14b seem to confirm this hypothesis because the processes used to generate the spheroidised microstructure prior to testing under superplastic conditions led to a much more refined cementite particle size (≈ 0.06-$0.2\,\mu$m compared with several micrometres in [44]). There may be other contributing factors as discussed below.

Superplasticity occurs at homologous temperatures in excess of $T/T_m \gtrsim 0.5$ because the mechanism of plastic deformation is the sliding of grains past each other combined with grain rotation. These diffusion-assisted processes are facilitated by fine-grained structures where atomic mobility along grain boundaries becomes an important component of the overall flux of atoms. The grain interiors, during superplastic deformation, do not therefore exhibit the clusters of defects that typify cold deformation. The fine structure that facilitates the diffusion flux must be maintained during superplastic extension. Two-phase mixtures are ideal in this context, because coarsening is hindered by the interphase interfaces across which the phases differ in chemical composition *and* crystal structure.

It is hard to imagine grain rotations or sliding within the convoluted three-dimensional form of the lamellar structure of pearlite. The structure must therefore spheroidise in the early stages of deformation at elevated temperatures, as observed by Yoder and Weiss [44]. Thus, *pearlite* does not exhibit superplasticity, rather, it is the mixture of ferrite and spheroidised cementite that does. Indeed, it is possible to generate the spheroidised structure from a variety of initial microstructures, which when tested, exhibit superplasticity [45].

Since grain boundary diffusion and mobility are important factors in the mechanism of superplasticity, the large-misorientation grain α/α grain boundaries are needed. When pearlite is rolled at a temperature close to Ae_1, it spheroidises but the majority of α/α boundaries are of a low-misorientation character. Cold-rolling, followed by annealing leads to a heterogeneous structure consisting of fine and coarse-grained regions, with the former containing predominantly low-misorientation boundaries. Full austenitisation, followed by quenching into the martensitic state and tempering leads to the greatest fraction of large-misorientation boundaries, thus enhancing superplastic elongation, Figure 5.14b [45].

Bulk, polycrystalline cementite is brittle even when deformed at elevated temperatures. However, when uniaxially compressed at 700 °C under the influence of a constant stress of 100 MPa, a strain of -1.94 has been achieved without the sample showing signs of damage [47]. Transmission electron microscopy of the structure revealed that the deformation did not lead to dislocated cementite grains with altered shapes, rather, they remained equiaxed and essentially dislocation-free, albeit with some coarsening, Figure 5.15. The strain rate involved was not reported but the observed characteristics, when combined with the fact that bulk cementite is brittle at high strain-rates, are consistent with superplastic flow involving grain rotation. This assumes that the structure did not recrystallise during the course of the experiment.

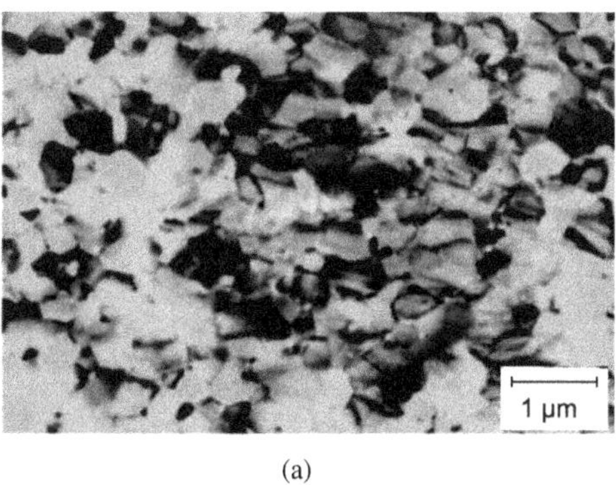

(a)

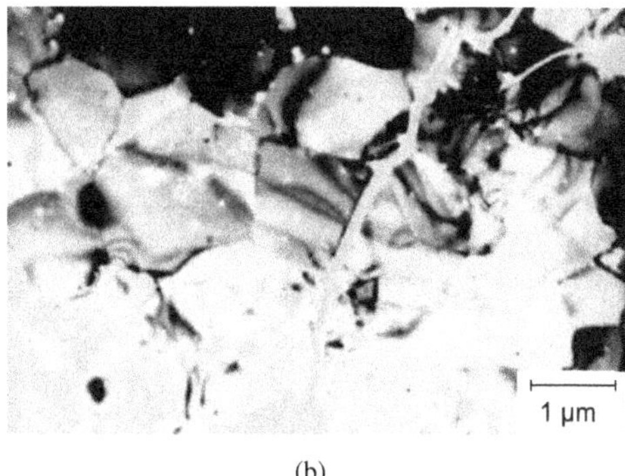

(b)

Figure 5.15 Transmission electron micrographs of $(Fe_{0.95}Mn_{0.05})_3C$ compressed uniaxially at 700 °C. (a) Before deformation. (b) After deformation. Images reproduced with the permission of Elsevier from Umemoto et al. [47].

5.3 TOUGHNESS

The ability of a material to irreversibly *absorb* energy when deformed is an indication of its toughness. The deformation can manifest in many ways: sudden impact as in a car crash, local plasticity at a crack tip resulting in an increased work of fracture, free energy absorbed in the form of an induced phase change, by intense delamination, as so on.

Toughness depends also on engineering design – it goes through a maximum as a function of sample thickness before it becomes a 'material property' [p.116, 48]. When a component contains stress concentrators, such as a notch, the lines of force have to meander around that notch, giving rise to triaxial stresses. Figure 5.16 illustrates that the maximum shear stress τ_{max} experienced, for the same magnitude of normal stress σ_1, is much smaller when both σ_1 and σ_2 are > 0 – this makes plastic deformation more difficult unless σ_1 is increased. This constraint to plastic deformation reduces the ability of the material to absorb energy. This is why a notched impact test is used to empirically characterise toughness or for quality control purposes.

Particularly in the case of α-iron but less so for γ-iron, the strength increases as the temperature is reduced or the strain-rate increased (Chapter 4); since plastic deformation then becomes more difficult, the toughness deteriorates and there is a change in fracture mode to cleavage. This change is not sudden and pure brittle-failure is in any case rare because the energy per unit area of cleavage plane is in almost all measurements, larger than the surface energy per unit area. Indeed, dislocation activity is often thought to be necessary to initiate cleavage, for example by the pile-up of dislocations at obstacles causing stress concentrations, or the combination of dislocations on two non-parallel slip planes $(101)_\alpha + (10\bar{1})_\alpha$ leading to a resultant $(100)_\alpha$ fracture. Alternatively, cleavage may initiate during the fracture under load of a brittle particle such as cementite or a non-metallic inclusion.

5.3.1 BRITTLE FRACTURE: FULLY PEARLITIC STEELS

During cleavage across a pearlite colony in a eutectoid steel at low temperatures, the crack propagates on the $\{100\}_\alpha$ planes inside the colony, with the cementite playing a less significant role in the energetics of the process [50]. After all, the toughness of cementite is just 2-4 MPa m$^{1/2}$ at ambient temperature [51–53]. Figure 5.17 shows such a crack and emphasises the fact that the defect density in the pearlite near the fracture surface is unaffected by the brittle mode of failure. Cleavage may

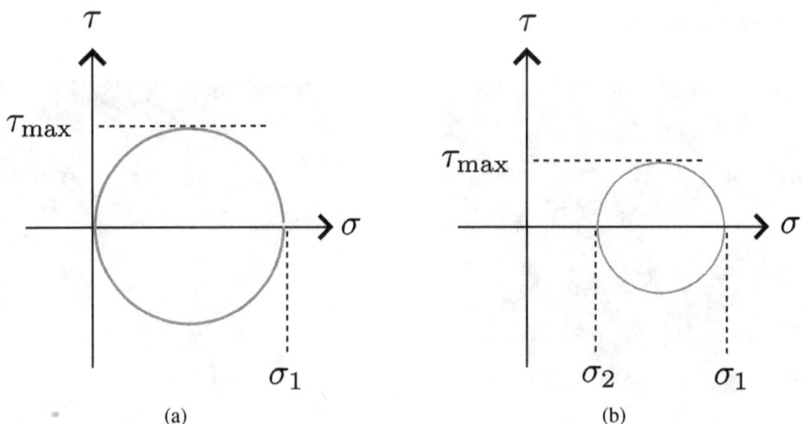

Figure 5.16 Mohr's circle [49] constructions to illustrate the maximum shear stress as a function of the normal stresses. The stress σ_1 is identical for both cases; the horizontal and vertical stress-scales also are intended to be identical for both diagrams. (a) Uniaxial normal-stress σ_1. (b) Biaxial normal-stresses σ_1 and σ_2, resulting in a smaller τ_{max}. Therefore, a circularly notched sample with the same cross-sectional area as a smooth tensile specimen, would fail at a greater tensile stress.

additionally occur on $\{110\}_\alpha$ in pearlite [54], but the evidence is not strong because the samples were fractured at ambient temperature with evidence for localised plasticity.

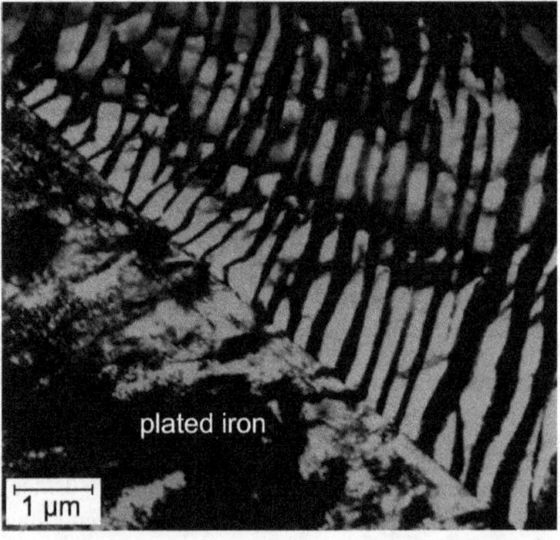

Figure 5.17 Cleavage front in pearlite fractured at $-196°C$, Fe-0.89C-0.84Mn-0.17Si wt%, examined in a transmission electron microscope after plating the fracture surface with iron. The trace of the fracture facet is precisely a straight line, seemingly unaffected even locally by the cementite. Reprinted with permission from Springer Nature Customer Service Centre GmbH, copyright 2008, from Bernstein [50], published by Springer.

This behaviour is a reflection of two phenomena: (i) cementite is brittle at cryogenic temperatures [55]; (ii) a colony is an interpenetrating bicrystal of cementite and ferrite [56, 57] containing only small orientation gradients within each phase (Chapter 2). Cleavage is therefore able to maintain its path within the ferrite, with little resistance from the cementite. This is why the toughness is sensitive to the austenite grain size rather than S_I, because $\overline{L}_\gamma$ determines the variety of nodule orientations present in each γ-grain [36] and therefore, the nodule size. The brittle-fracture facets should be smaller than L_γ but for grain sizes less than about 50 μm, $\overline{L}_{\text{nodule}} \approx \overline{L}_\gamma$, Figure 2.13.

The crystallographic size of nodules and fracture facets on Charpy specimens broken at ambient temperature have been measured [58]. Their data were presented as equivalent sphere diameters, but are converted here to the mean projected-height $\overline{H}$ of space-filling tetrakaidecahedra, Figure 5.18a.[4] It is evident that there is a strong correlation between the fracture-facet size and the measure of the crystallographic size of pearlite nodules.

Although the mechanistic link between cleavage and the crystallographic nodule-size is clear, Figure 5.18b shows that the Charpy energy does not improve much as the nodules are refined. The range is just 4-7 J across the whole span of length scales. Varying $\overline{L}_\gamma$ from 43 $\rightarrow$ 227 μm hardly had an effect on the 11 J recorded for a eutectoid steel transformed isothermally at 550 °C [61].[5] The inevitable conclusion is that the roughness of the fracture surface caused by traversing nodule boundaries does not contribute much to the work of fracture. It is probable that the connectivity of the brittle cementite within pearlite, that controls overall toughness. Tempered martensitic or bainitic steels that are much stronger than pearlite have better Charpy properties. Figure 5.18c shows that pearlite at ambient temperature falls well below the ductile-brittle transition temperature (T_{DB}).

It is claimed that there are two ductile-brittle transition temperatures associated with pearlite, the first at about −140 °C below which cementite lamellae crack readily; the second at 0 °C, attributed to plasticity in the cementite [64]. The experiments, however, were conducted on unacceptably small impact-samples, 1.2 × 0.8 mm in cross section and some of the data ignored in reaching the interpretation. Only one ductile-brittle transition temperature is observed when proper Charpy specimens of pearlite are tested; e.g., Figure 5.18c and [65, 66].

The Charpy test is empirical though useful for quality control and in some cases for comparisons between different steels. A fracture mechanics approach is more useful in that the data facilitate the calculation of a tolerable crack size after which fast fracture follows. It can help assess fatigue life by estimating the time taken for a slowly growing fatigue crack to reach a critical size. Fracture-toughness samples incorporate a sharp crack before testing begins; in the linear elastic fracture mechanics approach, the application of an external stress then leads to a distribution of stress ahead of the initial crack tip. When the sample is loaded in tension normal to the crack faces, the stress as a function of distance z ahead of the crack tip, is given by

$$\sigma = \frac{K_\text{I}}{\sqrt{2\pi z}} \tag{5.6a}$$

where K_I is a stress intensification factor, in this case for mode I (tensile) loading for a through crack of length $2z_\text{c}$. The stress asymptotically reaches the applied value σ_{app} far away from the crack tip. The stress intensity is a function of the applied stress and the specimen geometry:

$$K_\text{I} = \sigma_{\text{app}} \times Y\{z_\text{c}/W\} \tag{5.6b}$$

where Y is a compliance function that depends on the crack length z_c and the sample width W. For a large enough sample containing a central through-thickness crack of length $2z_\text{c}$, normal to

[4]The conversion from equivalent sphere diameter to stereologically sound measures can be achieved by noting that a tetrakaidecahedron (truncated octahedron) [59] with an edge length a has a volume $11.314a^3$, a mean lineal intercept $\overline{L} = 1.69a$, a mean projected height $\overline{H} = 3a$ and a mean areal intercept $\overline{A} = 3.77a^2$ [60].

[5]A normal Charpy test sample has a notch radius of 0.25 mm. Sometimes, non-standard samples are used, for example with a larger notch radius of 1 mm, resulting in much greater recorded impact-energies, approaching 40 J at ambient temperature [62].

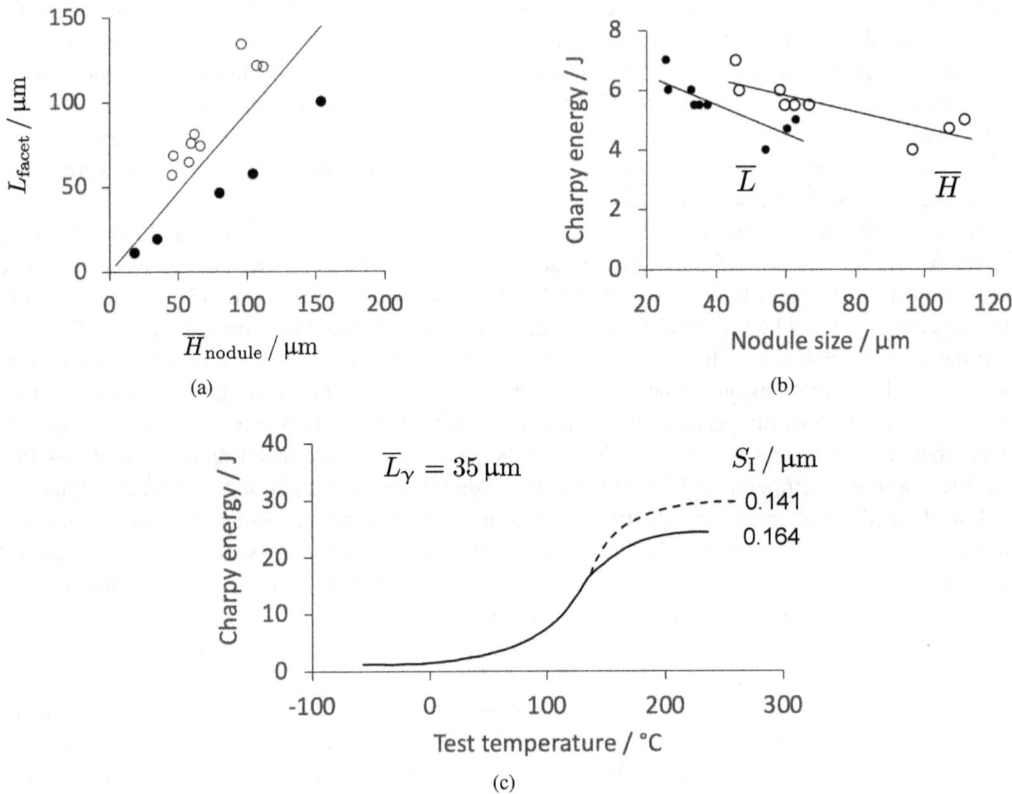

Figure 5.18 (a) Brittle-fracture facet size as measured on the broken surfaces of Charpy specimens, against a measure of the crystallographic size of pearlite nodules in fully pearlitic eutectoid steels. The best-fit line has a slope close to unity at 0.94, with an intercept of zero. Open circles are data using a 12° orientation threshold adapted from Behera et al. [58] and filled circles represent 15° threshold data from Zhou et al. [63]. (b) Correlation of the Charpy impact toughness measured at $\approx 25\,°C$, against the nodule size expressed either as the mean lineal intercept of the mean projected nodule height. Selected data from [58]. (c) Charpy impact curves for a pearlitic steel (Fe-0.81C-0.17Si-0.87Mn wt%). Selected data from [36].

σ, $Y = \sqrt{\pi z_c}$. Under plane strain conditions, K_I takes on a unique critical value K_{IC} that can be regarded as a material parameter such that fast fracture follows when $K_I \geq K_{IC}$.

The toughness K_{IC} is typically 30-40 MPa m$^{\frac{1}{2}}$ for a fully pearlitic steel at ambient temperature. Like the Charpy energy, it is insensitive to interlamellar spacing over the range 0.1-0.35 μm [67]. Strength does depend on S_I. At coarse interlamellar spacings, the ferrite is less constrained by the adjacent cementite, so the toughness increases as the yield strength decreases, Figure 5.19a; the fracture mechanism during K_{IC} testing involves ductility. Ferrite is much stronger at low temperatures (Figure 4.19) and hence more susceptible to cleavage fracture, with K_{IC} at low-T remaining constant at about 30 MPa m$^{\frac{1}{2}}$, irrespective of the overall yield strength. Nevertheless, a modicum of α-plasticity persists even at cryogenic temperatures, so the fracture toughness measured at $-196\,°C$ increases as the strength decreases, Figure 5.19b. It is noticeable, though, that the cryogenic data don't match well across the two figures.

In cobalt-containing (3.8 wt%) steel, the fracture toughness of pearlite has been reported to reach 42 MPa m$^{\frac{1}{2}}$ [71]; cobalt is known to enhance the toughness of ferrite when a comparison is made between pure iron and Fe-Co alloys, although the exact toughening mechanism remains speculative [74].

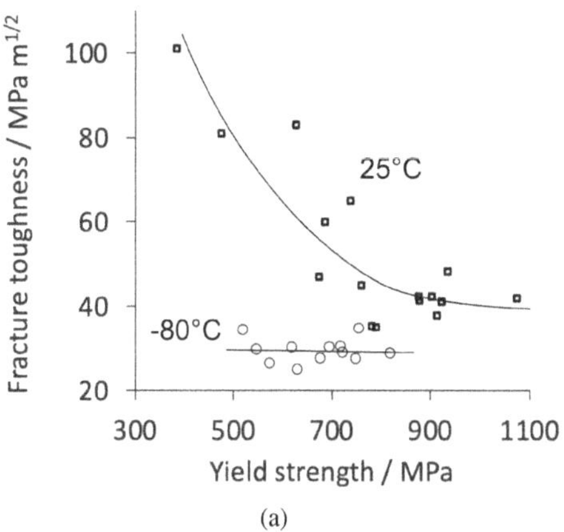

(a)

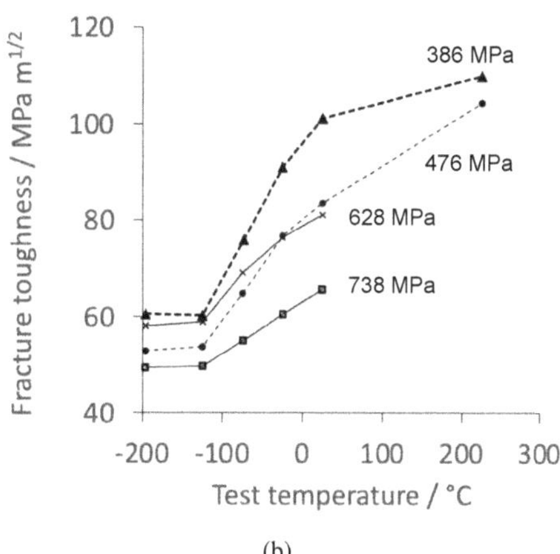

(b)

Figure 5.19 (a) Fracture toughness as a function of the yield strength. It is assumed that ambient temperature corresponds to about 25 °C. Data compiled from [67–73]. (b) Fracture toughness of fully pearlitic steel 'AISI 1080' as a function of test temperature. Adapted using data from El-Shabasy and Lewandowski [69].

5.3.2 BRITTLE FRACTURE: FERRITE+PEARLITE MIXTURES

Tensile fracture of α+P steels at $T < T_{\mathrm{DB}}$ begins with the development of damage within pearlite colonies, where T_{DB} is the ductile-brittle transition temperature. Any crack then advances into the proeutectoid ferrite [75]. During tensile testing at ambient temperature, cracks form first within the pearlite at modest overall-strains [76] but the sample elongates unabated, even during testing at cryogenic temperatures, Figure 5.20. This is because the cementite lamellae fracture first, with the interlamellar ferrite then failing by a ductile process; the resulting 'composite crack' may then penetrate into the adjacent α. The number density of cracked pearlite colonies increases with the total plastic-strain. Other observations confirm this general behaviour; in α+P structural steels with $V_V^P \approx 0.25$, tested at $-130\,^\circ$C, brittle failure is observed first in pearlite [77], although in fatigue tests (smooth samples, ambient temperature), crack initiation begins in ferrite [78]. It is well-established that the fatigue-limit scales with hardness (HV $\lessapprox 400$) [79, 80].

The lamellar ferrite does not cleave on $\{100\}_\alpha$ but shears on the $\{110\}_\alpha$ with harder adjacent cementite constraining this deformation. The crack thus has an element of roughness associated with plastic work and shear fracture. The crack path does not follow the lamellar direction, presumably because of the convoluted three-dimensional structure of pearlite (Chapter 2). The crack in this first stage of damage is arrested at the boundary between the pearlite and the α that surrounds the colony, Figure 5.21a. If the pearlite fraction is large then the different crystallographic orientations between pearlite nodules may provide some resistance to crack propagation.

The number of cracks per unit length of pearlite observed on a two-dimensional section, N_L, is expressed empirically as [81]

$$N_L \approx 5.5\varepsilon_{vM}^3 + \varepsilon_{vM} \qquad \text{for} \qquad \varepsilon_{vM} \leq 0.8, \qquad -80^\circ\mathrm{C} \geq T \geq -160^\circ\mathrm{C} \tag{5.7}$$

although a finer colony size is known to be more resistant to fracture [75].

Cracks in pearlite are able to advance into the adjacent α, when the normal stress at the junction exceeds the fracture stress of the ferrite (stage 2, Figure 5.21). This stress, for a circular microcrack,

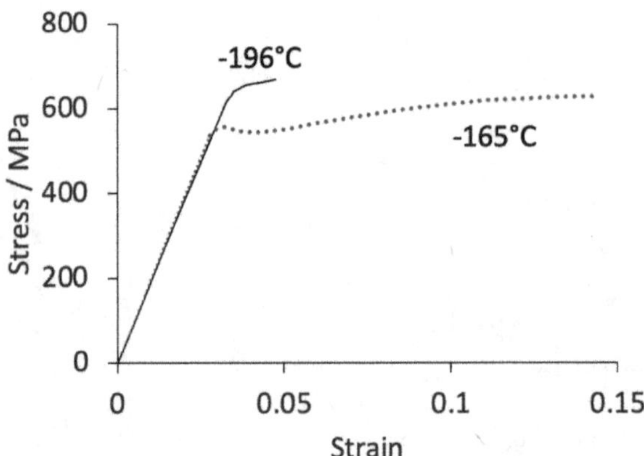

Figure 5.20 Tensile stress-strain curves for Fe-0.1C-0.62Mn-0.26Si wt%, with a microstructure of ferrite and pearlite ($V_V^P \approx 0.13$). The temperatures at which the tests were conducted are indicated. Selected data from Ohmori and Terasaki [75]

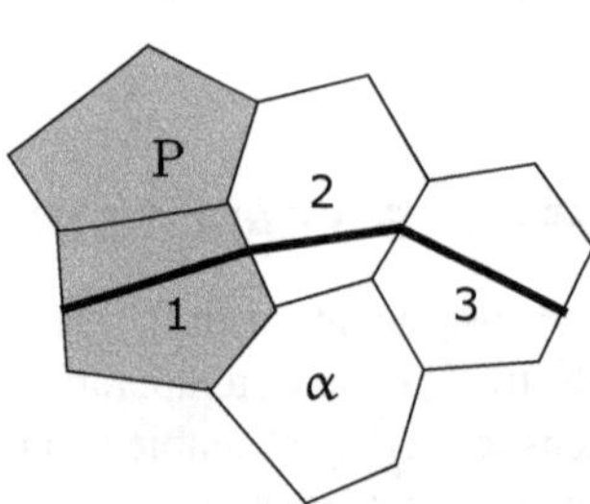

Figure 5.21 Stages during the brittle failure of an α+P steel. Stage 1 refers to initiation by the shear fracture of the pearlite colony, followed by the propagation of the crack into the adjacent α and then the advance across α/α grain boundaries. Adapted from Shibanuma et al. [77].

under plane-strain conditions and assuming linear elastic fracture mechanics, is [82]:

$$\sigma_f\{\text{stage 2}\} = \sqrt{\frac{\pi E \sigma_s^{\alpha\alpha}}{2(1-\nu^2)z_c}} \tag{5.8}$$

where z_c is the radius of the microcrack, $\sigma_s^{\alpha\alpha}$ is the work done per unit area in creating the ferrite crack-surfaces. Figure 5.22 shows how σ_s depends on temperature; the values are much greater than associated with, for example, pure cleavage on $\{100\}_\alpha$ of $2.5\,\mathrm{J\,m^{-2}}$ [83]. Plasticity evidently contributes to σ_s at the temperatures concerned; the scatter in the data at the highest test-temperature illustrated in Figure 5.22, is related to the proximity of T_{DB}. The σ_s values in fully pearlitic steels are smaller than in α+P alloys, because the cementite has a toughness of only 2-3 $\mathrm{MPa\,m^{1/2}}$ [52, 84]).

Cracks traverse α/α boundaries during stage 3. If the normal stress on the crack faces of the first ferrite grain exceeds $\sigma_f\{\text{stage 3}\}$, then Equation 5.8 can be applied again by substituting $z_c \approx \overline{L}_\alpha$ to represent the α/α cleavage plane created at stage 2 [86].

The scheme in Figure 5.21 can be implemented in a finite-element algorithm that contains microstructure and crystallography, given the constitutive equations for ferrite and pearlite [81]. Such analysis indicates that although thinner pearlite bands increase $\sigma_f\{\text{stage 2}\}$, the number density of pearlite cracks during stage 1 also increases, making the toughness insensitive to the band thickness over the range 6-30 μm. Refining $\overline{L}_\alpha$ increases σ_f beyond expectation from the Hall-Petch effect.

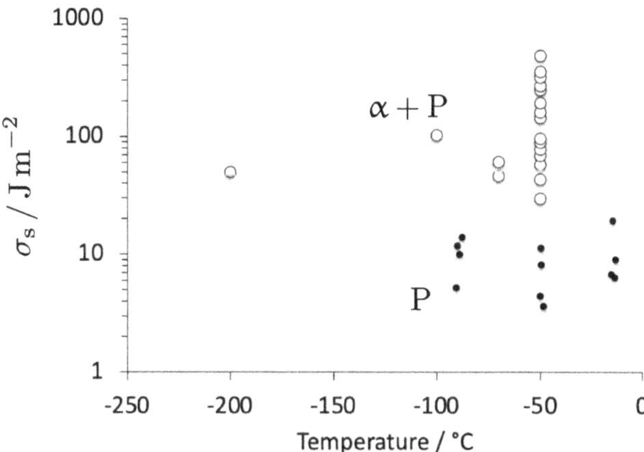

Figure 5.22 The work done per unit area in creating surfaces during fracture. Data for α+P steels from San Martin et al. [85] and on fully pearlitic steel from Alexander and Bernstein [38], covering a range of interlamellar spacings.

Charpy toughness

A small quantity of α in an otherwise pearlitic microstructure increases Charpy toughness; with $V_V^\alpha \approx 0.2$, tested at ambient temperature, the Charpy energy was in the range 14–23 J, which exceeds that of fully pearlitic samples (Figure 5.18) [87, 88]. The strength decreases as V_V^α increases and a yield point may appear in the stress-strain curve. Fully pearlitic samples have poor Charpy toughness because it takes just one cleavage crack to initiate general failure at the large strain-rates involved in impact testing. Crack Initiation can occur at multiple sites during more gentle toughness tests, leading to tortuous crack paths [87].

5.3.3 TOUGHNESS OF BANDED MICROSTRUCTURES

A banded structure might exhibit properties that are orientation-dependent provided that the scale of the variations is consistent with the controlling features of the mechanical test. The plastic zone of a tensile-test sample usually exceeds the length scale of banding, rendering the strength insensitive to orientation. Toughness is more sensitive to the microstructural scale because it depends on the small plastic zone at a crack or notch tip. Figure 5.23 shows the orientation-dependent Charpy energy of a steel containing sulphides that align within the bands, making them elongated and flattened parallel to the rolling plane (Alloy [89], Table 5.1). The steel is susceptible to *delamination*, which describes ductile-separation parallel to the rolling plane, sometimes along several parallel planes. The Charpy energy absorbed for $T > T_{DB}$ is therefore large when the test specimen delaminates. The effect can also lead to a decrease in the ductile-brittle transition temperature; in some elegant experiments on brazed steel laminates, Embury et al. [90] showed that T_{DB} decrease dramatically because of the effective thickness of the steel is reduced to that of the laminates, thus reducing plastic constraint; delamination ‘enables a thick specimen to fracture as the sum of a number of thinner specimens’.

The type of delamination cracks illustrated in Figure 5.24 occur parallel to the rolling plane and are observed in a variety of materials, ranging from artificial composites to steels, aluminium alloys, and even on some steels following irradiation [91]. They can be observed both during tensile and toughness testing.

Inclusions such as the elongated sulphides or arrays of aligned inclusions constitute an obvious mechanism that promotes delamination, but so does crystallographic texture in the α within banded steels [92–94]. This makes it difficult to decipher the actual cause of the orientation dependence of

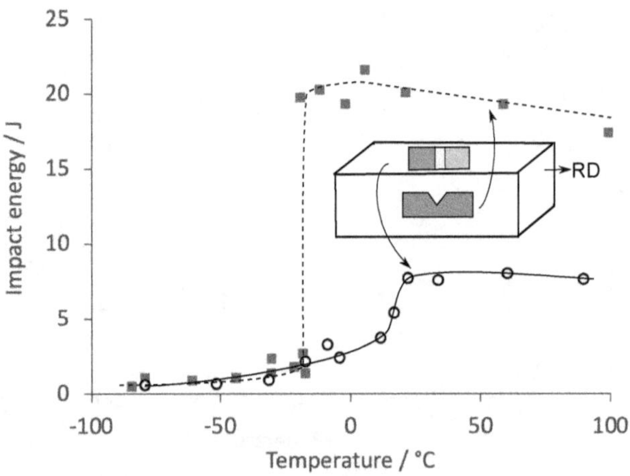

Figure 5.23 Huge anisotropy in Charpy impact energy of a hot-rolled, banded ferrite-pearlite steel (Table 5.1) as a function of notch orientation relative to the bands. Adapted using data from Almond [89].

Table 5.1
Chemical compositions, wt%, of hot-rolled steels with a ferrite-pearlite microstructure.

Reference	C	Si	Mn	P	S
[89]	0.25	0.01	0.65	0.019	0.04
[95]	0.21	0.20	0.85	0.016	0.028
[100]	0.14	0.27	1.45	0.01	0.00076

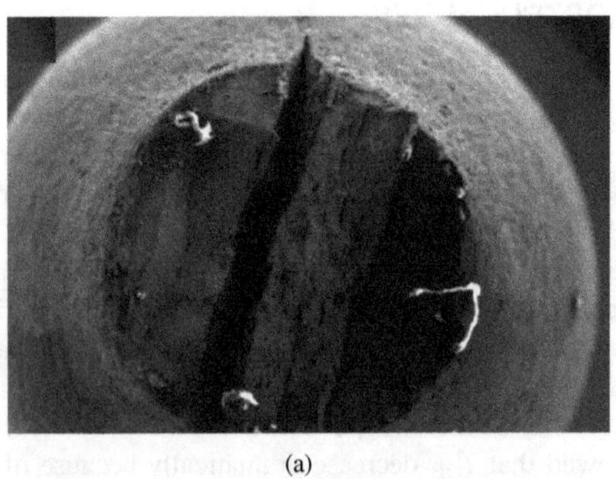
(a)

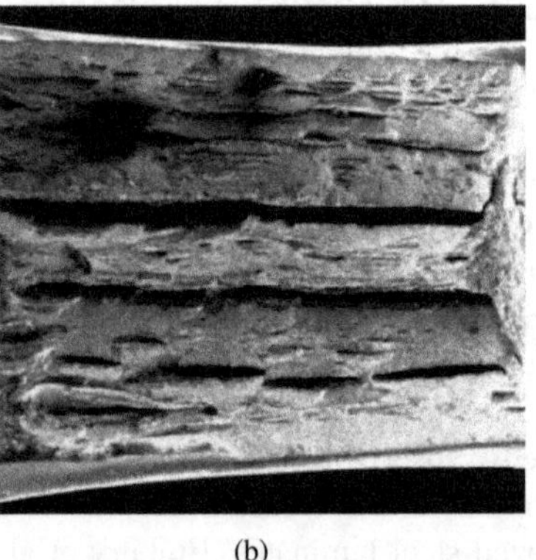
(b)

Figure 5.24 (a) Delamination observed on a broken half of a tensile test sample. (b) Similar observation on the broken half of a Charpy specimen. Images adapted from Arnoult et al. [91], under the CC BY 4.0 licence, https://creativecommons.org/licenses/by/4.0/.

toughness, a phenomenon of particular importance in steels used in the production of pipes for the transmission of crude oil and natural gas. The problem becomes worse as the strength of the steel is increased, which is one of the reasons why the stronger pipe steels (e.g., X80, X100) have specifications requiring sulphur concentrations less than 0.018 wt%, with actual concentrations much smaller than that specified.

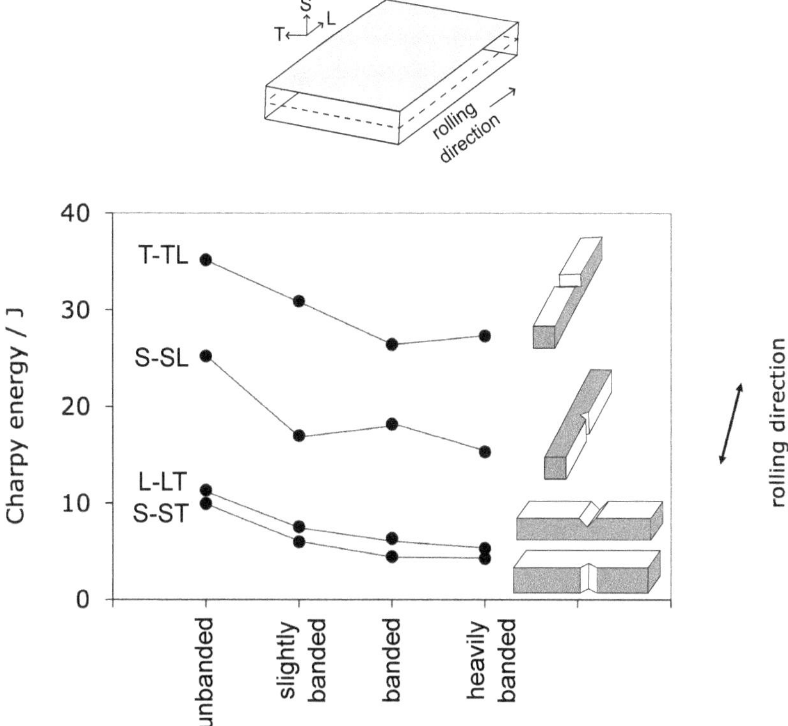

Figure 5.25 The ambient-temperature Charpy impact energy measured on sub-size 5 × 5 mm cross-section samples at room temperature, as a function of the orientation of the notch relative to the banded microstructure hot-rolled steel (Alloy [95] Table 5.1). The first letter in notation such as T-TL is the notch direction, the final two letters define the plane in which the notch lies. The yield and ultimate tensile strengths measured parallel to the rolling direction were found to be 378±24 and 550±15 MPa respectively. Data from [95].

The toughness is found to be similarly orientation-dependent when comparing banded and un-banded varieties of the same steel, though there was an enhancement of toughness in the un-banded samples, Figure 5.25 [95]. This indicates the controlling role of strings of manganese sulphide inclusions aligned during the hot-rolling process. Other experiments, where both the inclusion characteristics and pearlite banding were controlled, indicated no independent effect of banding on the Charpy toughness, with most variations attributed to changes in inclusion morphology, number density and size [96].

Apart from orientation dependence due to inclusion distributions, delamination of the steel is an important aspect controlling how the toughness varies with direction. Crystallographic texture can cause clusters of ferrite grains to have different Taylor factors, reported in the range 2.2-4.8, causing strength incompatibilities that initiate delamination at the α/α boundaries. Delamination is more prominent at high test-temperatures, so much so that the upper-shelf energy decreases, Figure 5.26 [95, 97, 98]. At low test-temperatures where brittle fracture dominates, the mild delamination contributes to the work of fracture, leading to an increase in the energy absorbed. But it is emphasised that this relationship between the extent of delamination and temperature is not generic [e.g., 99].

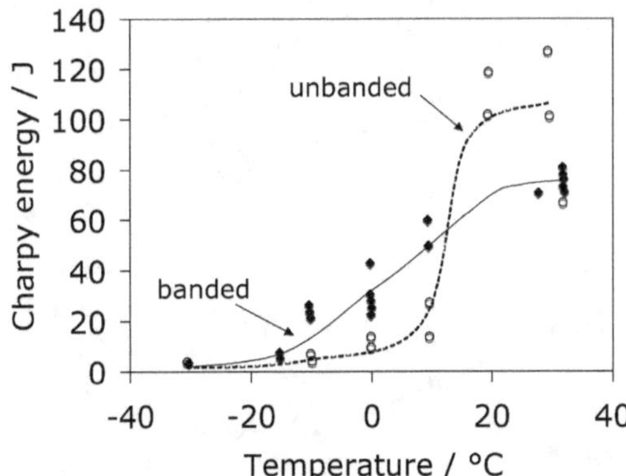

Figure 5.26 The Charpy impact energy as a function of temperature in Fe-0.2C-1.26Mn-0.02Si-0.014P-0.018S wt% hot-rolled steel. The extent of banding was varied by heat-treatment. Data from [98].

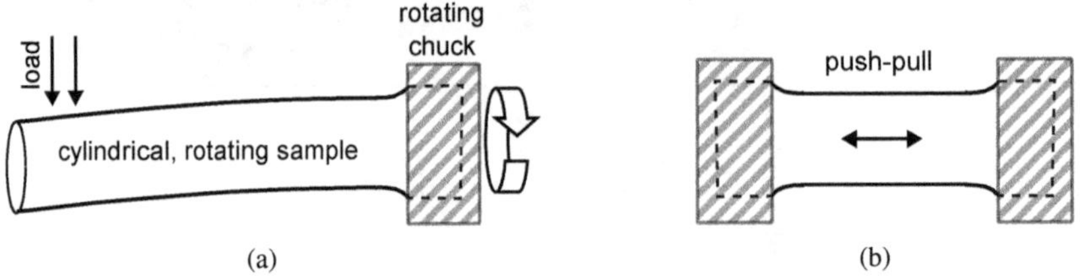

Figure 5.27 Schematic diagram of (a) a rotating-bending fatigue test, (b) a push-pull fatigue test.

5.4 FATIGUE - SMOOTH SAMPLES, CRACK INITIATION

The initiation of fatigue is best studied using smooth samples with a geometry that avoids explicit stress concentrations. In one such test, a cylindrical sample has an end clamped to a rotating chuck with the opposite end being dead-loaded, Figure 5.27 [101]. It bends while it rotates, experiencing cyclic tension and compression with each revolution with the magnitude of stress being largest at the surface. The distribution of stress through the sample is therefore not uniform and there is a neutral axis where the stress is zero. Alternatively, a sample may be loaded axially so that the gauge length is uniformly stressed, exposing a larger volume of material to the stress, making for example, the detection of inclusion-initiated fatigue is more likely.

The results from such tests are expressed as plots of the applied stress amplitude σ_a versus the logarithm of the number of cycles to failure (N_f), sometimes referred to as *S-N* curves, Figure 5.28. Materials that strain-age exhibit a *fatigue limit* corresponding to the value of σ_a below which failure in principle does not ever occur during cyclic loading. In steels, this limit often is attributed to the pinning of mobile dislocations by interstitials with the applied stress being insufficient to liberate the dislocations. Another view is that the limit should be identified with the need for plasticity to spread across grain boundaries for the successful propagation of cracks [102–104].

For materials that do not exhibit a fatigue limit, an *endurance limit* is defined to represent σ_a below which failure by fatigue does not occur after a number of cycles that is greater than the expected service life, typically 10^7-10^8 cycles. The definition of the endurance limit σ_e is clear in tests involving uniaxial loading (axial or rotating bending), but it also is used in the context of complex stress systems such as that associated with rolling contact, to describe the number of loading cycles N_e below which failure may be avoided [105],

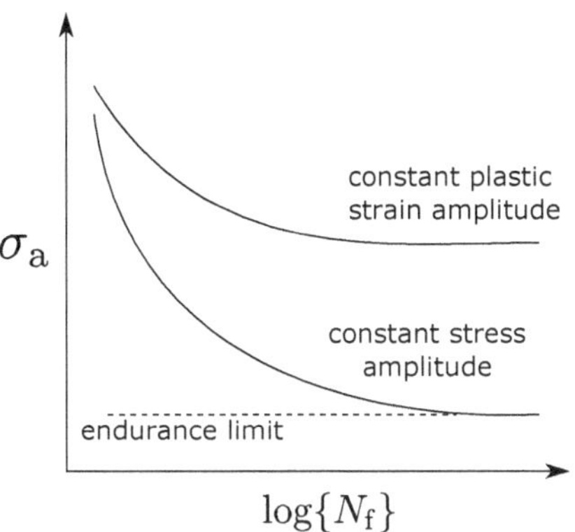

Figure 5.28 Two kinds of fatigue tests on smooth samples, constant stress or constant strain amplitude.

$$N = \begin{cases} f\{\sigma_a\} & \text{for} \quad \sigma_a > \sigma_e \\ \geq N_e & \text{for} \quad \sigma_a \leq \sigma_e. \end{cases}$$

In practice, the estimation of the endurance limit is based on the largest defect within the volume at risk. The potency of a surface defect in determining σ_e is related approximately to the square root of its area A projected on to the plane normal to the applied stress [106, 107]. An analysis of fatigue data indicates that for large defects (10-1000 μm), when the ratio of the maximum to minimum applied stress amplitude of $R_\sigma = -1$, the endurance stress is given empirically by [106, 107]

$$\begin{aligned} \sigma_e\,/\,\text{MPa} &\approx f\{\text{HV}\}(\sqrt{A})^{-1/6} \\ &= \frac{1.43(\text{HV}+120)}{A^{\frac{1}{12}}} \qquad \text{for surface cracks} \\ &= \frac{1.56(\text{HV}+120)}{A^{\frac{1}{12}}} \qquad \text{for internal cracks} \end{aligned} \tag{5.9}$$

where the units of A are in μm^2, those of Vickers hardness, $\text{kgf}\,\text{mm}^{-2}$. In a eutectoid steel heat-treated to obtain a variety of length scales, $\sqrt{A}$ has been associated with the pearlite nodule size to estimate the endurance strength [108], so a reduction in L_{nodule} should raise σ_e.

Non-metallic inclusions commonly present in commercial steels are likely to play a significant role in the crack initiation process. To deal with this, an analysis based on commercial eutectoid-steels revealed reasonable correlations between the endurance limit and the yield strength, for both smooth-bar specimens and notched samples regardless of $\overline{L}_\gamma$ [109]:

$$\sigma_e\,/\,\text{MPa} = 0.35\sigma_y + 141 \qquad \text{Rotating bend, } R_\sigma = -1,\ 160\,\text{Hz} \tag{5.10a}$$
$$\sigma_{se} = \pm 10\,\text{MPa},\ \sigma_y \text{ range 400-650 MPa}$$

$$\sigma_e\,/\,\text{MPa} = 0.09\sigma_y + 204 \qquad \text{Charpy 3-point bend, } R_\sigma = 0.1,\ 30\,\text{Hz} \tag{5.10b}$$
$$\sigma_{se} = \pm 6\,\text{MPa},\ \sigma_y \text{ range 400-650 MPa}$$

with the interlamellar spacing contributing only via its influence on strength (σ_{se} is the standard error of regression). It is odd that the defect size appears explicitly ($\sqrt{A}$) in the first set of equations but not in the second, where the only variable is σ_y. It must be assumed, therefore, that the commercial steels represented in Equations 5.10a all have a similar initiating-defect size. This may not be unrealistic given modern mass-production methods and the degree of scatter in experimental data.

The analysis above does not apply to cold-deformed pearlite because factors such as compressive residual-stresses can enhance the endurance strength [110], although as we shall see, the estimation is not bad for cold-rolled pearlite. We note that if deformation prior to fatigue testing introduces microstructural damage such as cracks or voids at inclusions/matrix interfaces, then there may be a consequential reduction in σ_e [111].

The fatigue limit clearly depends on the shape of the sample and the nature of the test. Figure 5.29a shows *S-N* curves for pearlitic steel, representing cantilever plane-bending fatigue tests[6] for both the undeformed and 75% cold-rolled states [112]. Based on Wohler's data [101], Basquin proposed an exponential law for *S-N* curves [113]:

$$\sigma_a = b_{26}(2N_f)^{b_{27}} \tag{5.11}$$

which fits the data in Figure 5.29 rather well with $b_{26} = 1138\,\text{MPa}$ and $b_{27} = -0.12$ for the undeformed pearlite, and $b_{26} = 2750\,\text{MPa}$ and $b_{27} = -0.14$ for the cold-worked pearlite. The last points in each of these curves were neglected in the derivation of the empirical constants because they lie on the asymptotic parts of the curves, well beyond the point at which the endurance limit is reached. Equation 5.10a predicts the endurance limits reasonably well, at 258 ± 10 ($\sigma_y = 335\,\text{MPa}$) and $525 \pm 10\,\text{MPa}$ ($\sigma_y = 1099\,\text{MPa}$) for the undeformed and deformed pearlite respectively, which compare with measurements at 219 and 526 MPa, respectively.[7]

The endurance limit in Figure 5.29a is, as expected, greater in the harder condition. Data from round-sectioned specimens [79, 115] are plotted in Figure 5.29b for a variety of steels and steel-microstructures, showing the positive correlation of fatigue limit with hardness, Because fatigue relies on microscopic-plasticity at stresses less than the overall yield strength, it is more difficult to initiate cracks in hard samples [116]. The cantilever data, although following the same trend, exhibit reduced fatigue limits due to sample-shape effects [117].

Crack initiation

Cracks develop at the free surface of a homogeneous sample either at stress concentrations arising from machining marks, the intersection of grain boundaries with the surface, slip steps at the free surface, from surface-breaking inclusions [119] or even unfavourable crystallographic orientations [120]. During fatigue testing, the maximum stress applied is less than σ_y. The surface seems to be the favoured initiation site at large applied stresses but internal cracks develop instead from inclusions or brittle phases at lower σ_a. This change in the location of initiation sites is a general observation [121] but the details must depend on sample integrity and test geometries – the surface/interior locations are not exclusive in that both sites can in principle operate simultaneously. Figure 5.30 shows the different regimes of initiation tendencies. A large stress-range is required for a surface-initiated crack to exceed the threshold stress intensity range $\Delta K_\circ$, before crack growth can occur at a reasonable pace (Section 5.5). Surface cracks are now able to grow at low stress-ranges, so fatigue life becomes dependent on defects within the sample that are large enough to allow $\Delta K_\circ$ to be exceeded [122].

During the fatigue loading of smooth samples of hypoeutectoid steels, cracks initiate most frequently at α/pearlite interfaces due to the incompatible plastic strains between these two regions of the microstructure [124]. Metallographic evidence indicates that this damage mechanism operates also during the creep deformation of α+P steels [125]. The cracks then propagate first into

[6]Vertical, planar sample, fixed rigidly at one end, with the other end subjected to repeated bending along the horizontal axis normal to the plane.

[7]In another study on undeformed, eutectoid pearlitic-steel, $b_{26} = 1461\,\text{MPa}$ and $b_{27} = -0.13$, $\sigma_y = 436\,\text{MPa}$ [114]. The endurance limit estimated using Equation 5.10a is $\sigma_e = 281 \pm 10\,\text{MPa}$ but a measured value is not available.

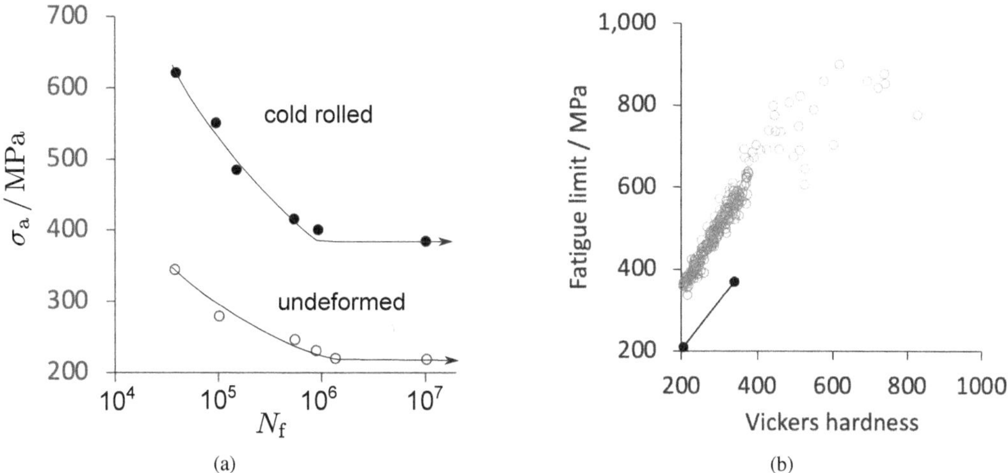

Figure 5.29 (a) *S-N* curves from plane bending cantilever fatigue tests on pearlitic steel (Fe-0.85C-0.77Mn-0.17Si wt%), before and after 75% reduction by cold-rolling. Selected data from Querales and Byrne [112]. The mean stress was zero, i.e., fully reversed alternating stresses. (b) The open points represent fatigue limit versus the hardness data from a variety of steels and steel microstructures, compiled by Peet et al. [118] from listings by [79, 115]. The filled points represent the two curves plotted in (a).

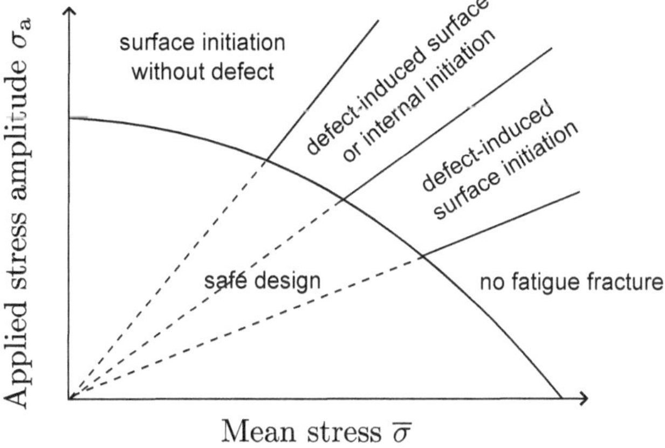

Figure 5.30 Schematic map of the mechanisms of fatigue failure during smooth sample testing, as a function of the applied stress amplitude and mean stress $\overline{\sigma}$. The curve, Gerber's parabola [123], defines the locus of a constant fatigue life for various combinations of σ_a and $\overline{\sigma}$. Adapted from [122].

the α until encountering a barrier such as a pearlite colony, making it possible to observe cracks confined within the ferrite prior to gross failure [126]. In eutectoid steels, it is the localised shear bands in ferrite that induce the cementite to crack, which sets the scene for ductile tearing across the pearlite colony [127]. Figure 5.31 illustrates schematically that the crack path is undeviated across the colony, consistent with observations that fatigue cracks do not change direction within a colony of pearlite, only at intersections with other colonies. It follows that fatigue life can be extended by refining the colony size [128]. In fully pearlitic samples, fatigue cracks form in the ferrite in close proximity of α/θ interfaces [129].

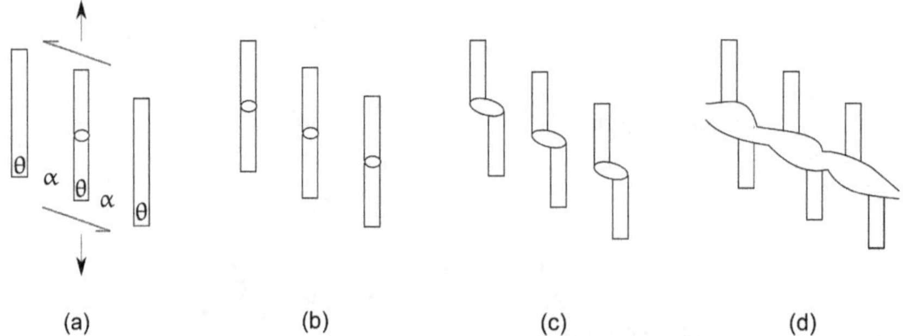

Figure 5.31 Schematic representation of the initiation of fatigue. (a) Crack develops in cementite lamella under the action of shear and tensile stresses. (b) As a result, adjacent cementite lamellae in the shear-zone crack. (c) Void expansion. (d) Void coalescence. Adapted from Miller and Smith [10].

Strain controlled

During a constant-stress fatigue test, the material work-hardens with each cycle, leading to an ever decreasing strain, Figure 5.32. In contrast, if the strain-range is maintained constant with each fatigue cycle, then cyclic hardening or cyclic softening occur during which the stress varies with each cycle until some asymptotic value is reached. Most practical assessments are conducted using the constant σ_a test.

Whereas cyclic hardening is easy to understand based on the usual work-hardening mechanisms during monotonic deformation, the mechanism for cyclic softening is less well-defined. In α+P steels, cyclic softening is associated with plastic strain amplitudes that are close to the elastic limit ($\Delta\varepsilon_p = 0.003$, Figure 5.32d) [130]. In fully pearlitic steel the cyclic softening occurs when $\Delta\varepsilon_p \approx 0.007$ [131]; the softening is attributed to the ferrite with diffraction data indicating reversible plasticity within the α-lamellae [132]. The dislocation creation and annihilation rates eventually balance such that $\rho_\perp$ does not change.

The detailed mechanism of cyclic hardening and softening of pearlite is revealed using transmission electron microscopy, Figure 5.33 [129]. Before deformation, pearlite has a small dislocation density and sharply defined α/θ interfaces. At low cyclic strains, the dislocation density increases but is uniform throughout the ferrite, with cyclic softening as the density of mobile dislocations increases. At larger cyclic strains, dislocation cell structures develop in the ferrite and the spacing between the cell-walls decreases as the number of fatigue cycles increases. This stage corresponds to cyclic hardening because the presence of cell boundaries provides barriers that enhance work hardening.

The onset of cyclic hardening depends on the interlamellar spacing [129]. Fine pearlite ($S_I \approx 0.08\,\mu$m) readily exhibits cyclic softening because dislocation interactions are less probable due to the limited space available, and cell formation also is limited in the same way. Only cyclic hardening is observed with coarse pearlite ($S_I \approx 3.3\,\mu$m) where dislocations are generated at relatively low stresses with opportunities for work hardening through mobile-dislocation interactions. The micrographs illustrated in Figure 5.33 represent an intermediate S_I so cyclic softening is observed at low strain amplitudes but hardening with the onset of cell formation.

We have seen that fatigue tests on smooth samples reveal phenomena such as the *initiation* of cracks, and how short cracks may surmount barriers such as grain boundaries [78]. For most of the fatigue life, there is little difference between stress and strain-controlled experiments until the fatigue crack becomes large [133]. To accelerate testing, ultrasonic vibrations at 20-100 kHz frequencies can be used to apply the load cycles. The samples resonate to the stimulus with the maximum stress experienced at a displacement node located half way along its length [134, 135]. Hour-glass

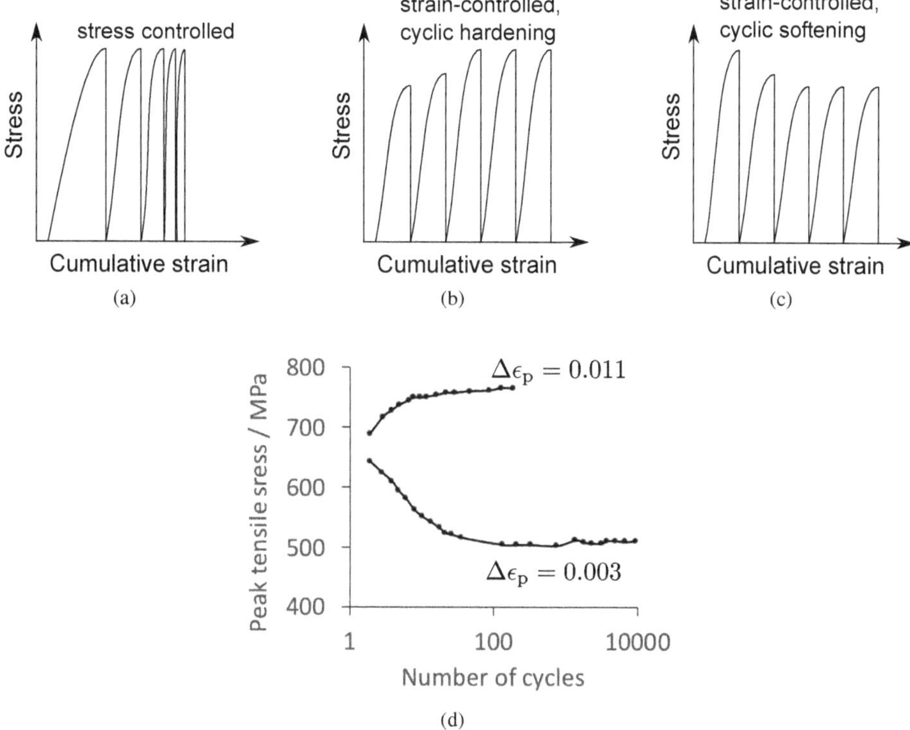

Figure 5.32 Push-pull fatigue tests. (a) Stress controlled. (b) Strain controlled with cyclic hardening. (c) Strain controlled with cyclic softening. (d) Strain controlled fatigue tests showing cyclic softening and hardening as a function of plastic strain amplitude. Selected data from Sankaran et al. [130] for α+P steel.

shaped samples are normally used (Figure 5.34a) but parallel gauge lengths can also be introduced to increase the volume at risk, although the distribution of stress within that length will not be uniform. To deal with this, parallel-sided samples can be tested using push-pull hydraulic testing-machines, at 0.4-1 kHz frequencies, but necessitating longer tests. The two kinds of tests do not give identical results, in part because the response of the material to ultrasonic loading may not be the same as during conventional loading, assuming that the sample temperatures are maintained constant. It is possible during ultrasonic fatigue testing for the temperature of the steel to increase by almost 100 °C and the effect becomes exaggerated as the rate of damage evolution increases [136].

In rotating-bending tests typically at 50 Hz, parallel-sided samples with smooth shoulders experience maximum stress amplitudes at the surface of the test specimens (Figure 5.34b). The volume of material exposed is therefore small when compared with push-pull tests.

Coffin and Manson observed that during low-cycle fatigue, which usually is accompanied by cyclic hardening,[8] the number of cycles to failure is related to the plastic-strain amplitude $\Delta\varepsilon_p/2$ per cycle

$$N_f = b_{15}\left(\frac{\Delta\varepsilon_p}{2}\right)^{-b_{14}} \tag{5.12}$$

[8] Low-cycle fatigue is associated with high stress-amplitudes ($\sigma_a > \sigma_e$) and low frequencies, with a plastic strain per cycle in excess of $\approx 10^{-5}$. N_f is typically less than 10^5.

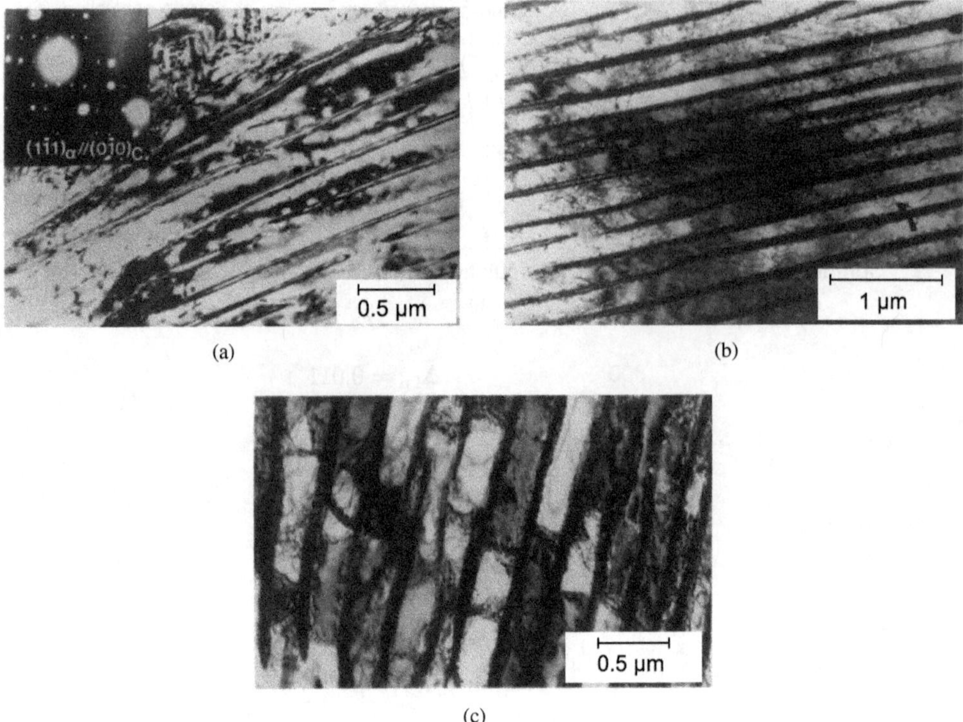

(a) (b) (c)

Figure 5.33 A fully pearlitic steel of composition Fe-0.71C-0.85Mn-0.17Si wt%, $S_\mathrm{I} \approx 0.16\,\mu\mathrm{m}$, subjected to fatigue loading under strain control. (a) Sample prior to testing. The diffraction pattern shows the crystallographic relationship between the α and θ. (b) Subjected to $\approx$ 15000 cycles at 0.3% strain amplitude. (c) Subjected to $\approx$ 150 cycles at 1.5% strain amplitude, exhibiting a dislocation cell structure. Images reproduced from Sunwoo et al. [129] with permission of Springer Nature.

where the empirical exponent $b_{14} \approx 2$ for single-phase metals [137] but can be much smaller (0.67) for α+P steels; the proportionality constant $b_{15} = 0.54$ [130]. Stronger steels will have smaller $\Delta\varepsilon_\mathrm{p}$ for a given stress, so it is reasonable that the fatigue life increases with hardness, as illustrated in Figure 5.29b.

The full curve of the type illustrated in Figure 5.28 can be estimated if $\Delta\varepsilon_\mathrm{p}$ is known as a function of the stress range. Ramberg and Osgood [138] proposed an analytical expression relating uniaxial-stress and strain, which using their notation for the empirical constants K and n is:

$$\varepsilon_{\mathrm{total}} = \underbrace{\frac{\sigma}{E}}_{\text{elastic strain}} + \underbrace{K\left(\frac{\sigma}{E}\right)^n}_{\text{plastic strain}}. \tag{5.13}$$

This has been adapted for cyclic loading,[9]

$$\frac{\Delta\varepsilon_\mathrm{p}}{2} = \left(\frac{\Delta\sigma}{2b_{16}}\right)^{1/b_{17}} \quad \text{or} \quad \frac{\Delta\sigma}{2} = b_{16}\left(\frac{\Delta\varepsilon_\mathrm{p}}{2}\right)^{b_{17}} \tag{5.14}$$

[9]Equation 5.14 seems to be obtained by the substitution of $\Delta\sigma$ for σ in Equation 5.13, but this is difficult to justify.

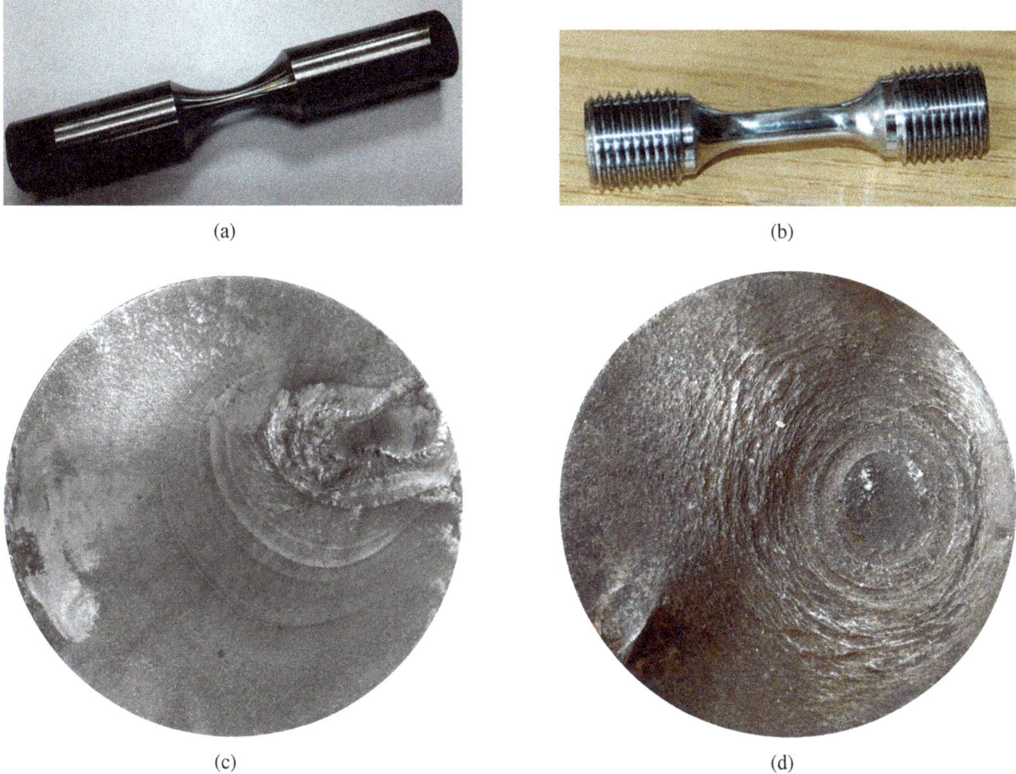

(a) (b)

(c) (d)

Figure 5.34 (a) Sample for high-frequency push-pull fatigue testing. (b) Sample for rotating-bending fatigue tests. (c) Push-pull sample fracture topography showing the surface-initiation of the fatigue crack. (d) Rotating-bending sample fracture surface. The crack in this case has started within the sample.

where $\Delta\sigma/2$ is the cyclic-stress range and for α+P steels, b_{17} can be set to the strain hardening coefficient measured in a monotonic tensile test and [139]

$$b_{16} \approx 1.21\sigma_{\mathrm{UTS}} + 555 \qquad \mathrm{MPa}. \tag{5.15}$$

To estimate the plastic strain increment per cycle, it is necessary to consider the cyclic stress-strain relationship which differs from the monotonic relationship because load reversals can lead to a different dislocation structure. The cyclic yield strength often is suggested or found to define the endurance limit since dislocation activity should be minimal at low stresses. Its value for α+P steels is, in terms of the monotonic properties given approximately by [126, 140]

$$\sigma_{\mathrm{y}}^{\mathrm{c}} = (1+\varepsilon_{\mathrm{RA}})\sigma_{\mathrm{UTS}}\left(-\frac{0.002}{\ln\{1-\varepsilon_{\mathrm{RA}}\}}\right)^{0.16} \qquad \mathrm{MPa} \tag{5.16}$$

where $\varepsilon_{\mathrm{RA}}$ is the fractional reduction of area in a tensile test and 0.002 is the plastic strain used to define yielding. So for typical α+P steels with $V_{\mathrm{V}}^{\mathrm{P}} < 0.3$, $\sigma_{\mathrm{y}}^{\mathrm{c}}$ would typically be about 0.4 of the ultimate tensile strength, which is consistent with the factor of ≈ 0.48 that has been measured [126, 141]; at large fractions of pearlite in α+P steels, the equation correctly estimates a larger ratio, as is observed experimentally [130]. The ratio is closer to 1 in strong steels that have bainitic or martensitic microstructures [141].[10]

[10]Hussain and Rios [126] reported their measurement as the cyclic shear-yield strength, which has been converted into a cyclic tensile strength for comparison against Equation 5.16.

5.5 FATIGUE: CRACK GROWTH

If it is assumed that defects exist in manufactured steels, then the problem of defining life as a function of cyclic loading is determined by the time taken for the defects to reach a critical size that promotes fast fracture. When Paris, Gomez and Anderson wrote their 'rational analytic theory of fatigue' [142], the elastic-stress field at the tip of a crack was used to understand crack extension under static loading, the processes leading to fast fracture and the definition of stress intensity [143–145]. Fatigue in its early stages involves slow crack growth so it was assumed that the crack extension per cycle is some function of the minimum and maximum stress intensities:

$$\frac{\mathrm{d}z_\ell}{\mathrm{d}N} = f\left\{K_\mathrm{I}^{\max}, \frac{K_\mathrm{I}^{\max}}{K_\mathrm{I}^{\min}}\right\} \tag{5.17}$$

where z_ℓ is the crack length, N the number of loading cycles and K_I is the stress intensity in mode I loading.[11] Figure 5.35 shows that there are three regimes represented by the function f, the first involving the threshold where subcritical cracks do not grow below a stress intensity range ΔK_o and grow rather slowly beyond that; the majority of the service life of a component is therefore spent in that domain, where the microstructure plays a prominent role.

The second regime is where the 'Paris law' applies, with the crack advance per cycle proportional to $(\Delta K)^m$, where m is the Paris exponent. The plastic zone at the crack tip can be large relative to microstructure, in which case the latter plays a less significant role. Finally, in the third regime, the stress intensity range is so large that the crack reaches the critical crack size defined by K_c and fast fracture follows.

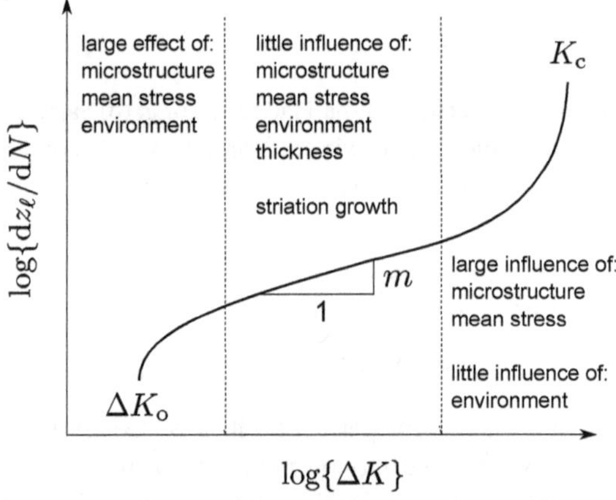

Figure 5.35 The function f of Equation 5.17, illustrated schematically with key regimes of crack growth marked. Adapted from Ritchie [146].

In the context of α+P normalised steels designed for structural applications, the crack advance per cycle within the Paris law regime is found to be [147]

$$\frac{\mathrm{d}z_\ell}{\mathrm{d}N} = 6.41\times10^{-12}\begin{cases}\text{range}\\ 11.2\times10^{-12}\\ 3.67\times10^{-12}\end{cases} \times \Delta K^{3.06} \quad \mathrm{m\,cycle^{-1}} \tag{5.18}$$

[11]Mode I loading is normal to the crack plane; mode II is a shear on the plane of the crack in a direction normal to the crack front; mode III is a shear on the crack plane but in a direction parallel to the crack front.

where the range in the Paris constant represents plus or minus two standard deviations about the mean, the tests are in air, covering a frequency range 0.005-50 Hz, stress ratio 0.05-0.85, $\sigma_y =$ 248-653 MPa, $\sigma_{UTS} =$ 434-766 MPa with the Paris regime defined to extend over the range $20 \leq \Delta K \leq 40\,\text{MPa}\,\text{m}^{1/2}$.

The fatigue crack growth per cycle in a fully pearlitic steel of composition Fe-0.79C-0.43Mn-0.24Si-1.18Ni-0.43Cr wt% with $S_I \approx 0.11\,\mu\text{m}$ and $\sigma_y = 650\,\text{MPa}$ is given by [148]

$$\frac{dz_\ell}{dN} = 7.2 \times 10^{-12} \Delta K^{2.97} \quad \text{m cycle}^{-1}, \tag{5.19}$$

similar to the data for the α+P steels (Equation 5.18). It might be speculated that this is connected with a dominant role of pearlite during the fatigue of the mixed microstructure.

The crack growth rate increases when the tests are conducted in seawater or under a high pressure of gaseous hydrogen, Figure 5.36. Environmental effects can present formidable challenges to design against fatigue.

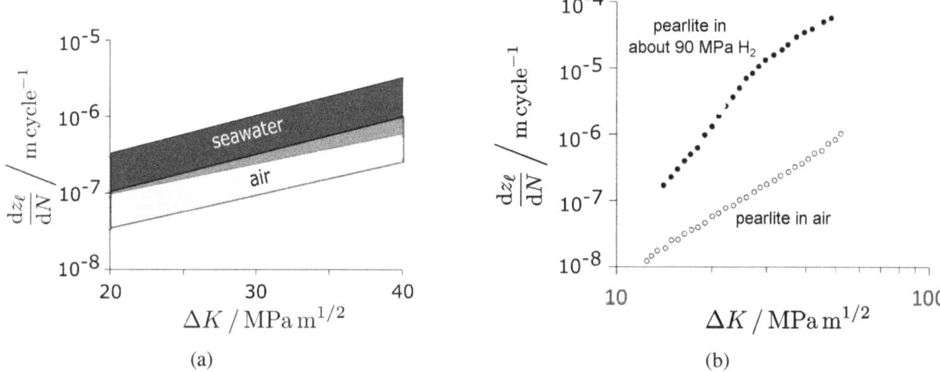

Figure 5.36 (a) Fatigue crack growth rate in seawater, compared against that in air, covering the Paris Law region of stress-intensity range for hot-rolled and normalised α+P steels. Adapted using data from Igwemezie et al. [147]. (b) Selected data from Ogawa and Iwata [148] showing the effect of hydrogen at high pressure on the fatigue crack advance per cycle in a fully-pearlitic steel.

There are caveats to the use of the raw stress intensity range $\Delta K = K_{max} - K_{min}$ in crack growth analysis [149]. Crack closure, i.e., when the crack faces are pushed against each other, can lead to a stress intensity $K_{closure} \geq K_{min}$. In ductile materials such as α+P steels, the opposing faces of the crack can be rough which prevents them from a neat closure, in which case the effective stress intensity range at the crack tip becomes $\Delta K_{effective} = K_{max} - K_{closure}$. Another mechanism of processes operating behind the crack tip is 'grain bridging' whereby roughness is introduced by grains that elongate before fracture and hence protrude beyond the crack faces. The accumulation of oxide or corrosion patches can also contribute to crack closure.

Figure 5.37 illustrates two different α+P microstructures, one of which is 'networked' with pearlite enclosing large regions of ferrite, whereas in the 'distributed' form, the pearlite is dispersed uniformly throughout the microstructure. The growth rate of fatigue cracks is smaller for the distributed microstructure because of a greater degree of crack branching and roughness, even when the comparison is made at a constant effective stress intensity range, Figure 5.37c [150]. Microstructural detail affects fatigue crack growth, even within the Paris law regime [147], because the pearlite-dispersion length-scale is larger than the expected plastic-zone size at the tip of the crack.

So, it seems that variations in structure that have a greater span than the crack-tip plastic zone influence fatigue. Pearlite colonies have a capacity to deflect cracks, as evident in the erratic variation

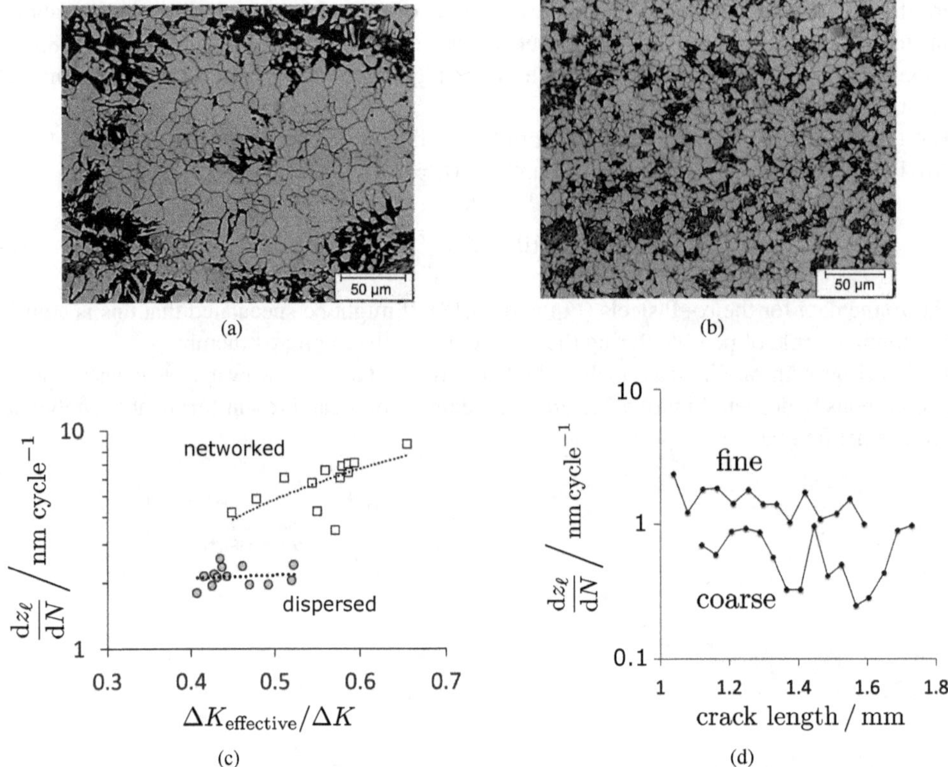

Figure 5.37 Fatigue resistance of two α+P microstructural variants in hot-rolled steels. (a) Networked pearlite in Fe-0.15C-0.16Si-1.07Mn wt% steel, with $V_V^P = 0.25$. (b) Distributed pearlite in Fe-0.11C-0.31Si-1.31Mn wt% steel, with $V_V^P = 0.20$. Micrographs reproduced with the permission of Elsevier, from Mutoh et al. [150]. (c) Fatigue crack growth rate in the Paris regime, as a function of the normalised value of the effective stress intensity range, with $\Delta K = 15.3\,\text{MPa}\,\text{m}^{1/2}$. Adapted using data from [150]. (d) Fatigue crack growth in coarse and fine α+P microstructures. Adapted using data measured by direct observation during crack propagation, from Mustapa and Mutoh [151].

in dz_ℓ/dN shown in Figure 5.37d, which compares structures with 70 μm and 5 μm colony sizes, with an estimated crack-tip plastic zone size of 150 μm during fatigue at $\Delta K = 12\,\text{MPa}\,\text{m}^{1/2}$. Large colonies are associated with greater variations in crack growth per cycle, but only slightly deflected by finer colonies [151]. As a result, the finer structure exhibits a greater crack propagation rate.

5.6 HIGH-CYCLE FATIGUE: CRACK GROWTH

High-cycle fatigue means that failure is expected after a large number of cycles because the plastic strain increment per cycle is less than about 10^{-6}; as a rough guide, N_f should exceed 10^4, bearing in mind that the endurance limit corresponds to about 10^7 cycles. The stress amplitude should not be a large fraction of the yield strength.

Modern α+P structural steels have less impurity content (0.002 wt%) than the open hearth steel of old, which contained about 0.15 wt% of sulphur and phosphorus [152, 153]. Fatigue cracks therefore begin at the steel surface but the initiation process may constitute just 1% of the total fatigue life so it is appropriate to focus on crack growth as the controlling factor [154]. The fatigue life of an engineering structure can be estimated assuming that subcritical cracks already exist when the steel begins service.

Cracks that are short, such as the size of a grain, have characteristics different from long ones that extend over many such structural features. Short cracks in general grow at a greater rate, are subject to crystallography and orientation relative to the applied stress, whereas long cracks can be treated in a continuum manner based on the stress intensity range ΔK [155]. Nevertheless, it is possible to treat the entire problem as the stepwise progress of cracks between grains using a finite element framework to represent the structure and the assumed or measured crystallographic texture.

5.6.1 CRACK-OPENING STRESS

A crack loaded such that it recovers its original configuration on unloading would not advance; the loading would be elastic. In a ductile material, the tip of the crack opens irreversibly through localised plastic deformation – if this opening displacement reaches a critical value then the material may fracture, forming the basis of the CTOD test [156]. In fatigue, the advance of the crack is driven by crack tip sliding displacements driven by shear stresses. It is this incremental plasticity that leads to the formation of striations on the surface of a fatigue-induced fracture. A tensile-loaded crack will tend to open, the extent of opening at the crack tip depending on the ability of the material to blunt the crack via mechanisms such as plasticity. The crack advance per cycle during fatigue is proportional directly to the cyclic crack tip opening displacement with the advance of the crack orthogonal to the maximum principal stress.

In modelling the process for the evolution of the cracks, the opening displacement usually is measured at some distance behind the tip. When a crack is loaded or unloaded, the response initially is elastic but followed by opening at the tip, manifested as a smaller slope in the plot of nominal stress versus displacement normal to the crack face. The nominal stress at the point where the slope changes is designated the crack opening stress σ_{op}, which is a function of the crack length, given by direct observations during the fatigue of ferrite-pearlite steels by [153]

$$\sigma_{\text{op}} / \text{MPa} = \sigma_{\text{op}}^{\infty}(1 - \exp\{-0.028 z_c\}) - \sigma_a \tag{5.20}$$

where z_c is half the width of the crack as measured in micrometres on the sample surface. σ_a is the nominal stress amplitude, $\sigma_{\text{op}}^{\infty}$ is the *crack opening stress* of a long crack, i.e., if σ_{op} is plotted against z_c, it asymptotically reaches the limit $\sigma_{\text{op}}^{\infty}$ as the crack lengthens, Figure 5.38.

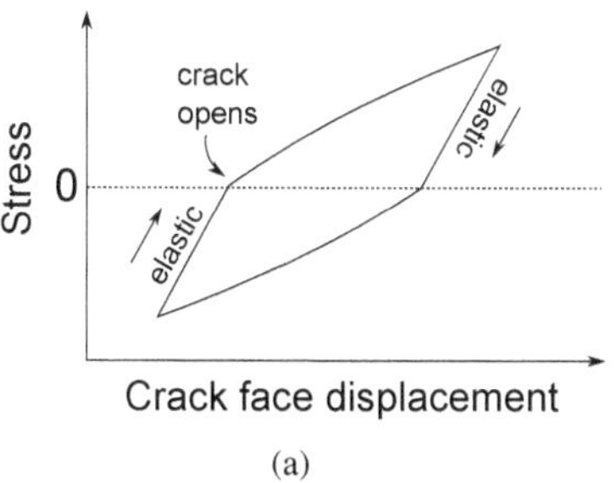

(a)

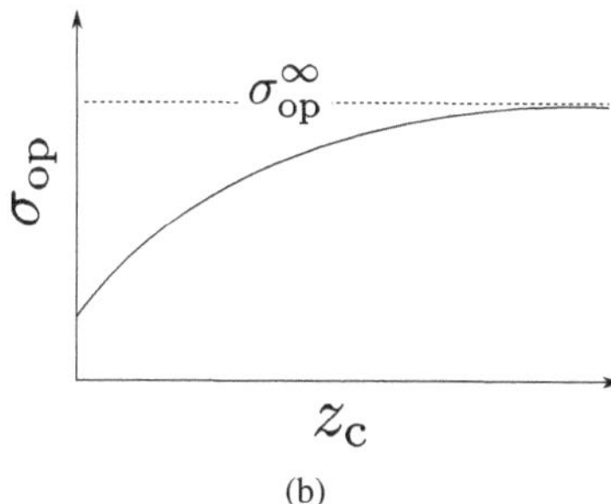

(b)

Figure 5.38 (a) Hysteresis plot during cyclic fatigue loading. The stress plotted is the overall value σ rather than any local magnitude. The crack opening stress is defined on the loading side of the hysteresis loop. (b) Crack opening stress as a function of crack length.

Using numerical analysis, Neuman [157] expressed $\sigma_{\text{op}}^{\infty}$ as follows:

$$\begin{aligned} \sigma_{\text{op}}^{\infty} &= (b_7 + b_8 R_{\text{F}} + b_9 R_{\text{F}}^2 + b_{10} R_{\text{F}}^3)\sigma_{\max} \qquad \text{for} \quad R_{\text{F}} \geq 0 \\ \sigma_{\text{op}}^{\infty} &= (b_7 + b_8 R_{\text{F}})\sigma_{\max} \qquad \text{for} \quad -1 \leq R_{\text{F}} < 0 \end{aligned} \tag{5.21}$$

$$\begin{aligned} \text{with} \quad b_7 &= (0.825 - 0.34\zeta + 0.05\zeta^2)\left[\cos\left\{\frac{\pi\sigma_{\max}}{\sigma_{\text{y}} + \sigma_{\text{UTS}}}\right\}\right]^{1/\zeta} \\ b_8 &= (0.415 - 0.071\zeta)\frac{2\sigma_{\max}}{\sigma_{\text{y}} + \sigma_{\text{UTS}}} \\ b_9 &= 1 - b_7 - b_8 - b_{10} \\ b_{10} &= 2b_7 + b_8 - 1 \end{aligned} \tag{5.22}$$

where ζ is a constraint factor depending on the geometry of the sample and crack, with $\zeta = 3$ for plane strain conditions and $\zeta = 1$ for plane stress; R_{F} is the ratio of the maximum ($\sigma_{\max}$) to minimum stress amplitude. The flow stress is taken to be the mean of the uniaxial quantity $(\sigma_{\text{y}} + \sigma_{\text{UTS}})/2$; the effect of constraint would be to scale this quantity by ζ. Figure 5.39 shows how the crack opening stress varies with the stress ratio and the degree of constraint. The opening stress decreases as the constraint increases and as R_{F} decreases.

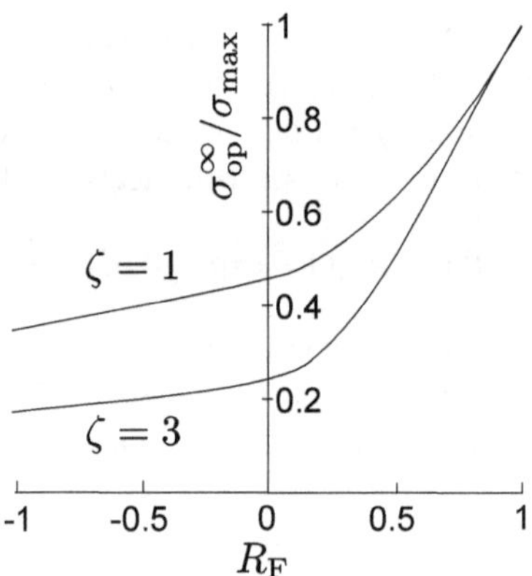

Figure 5.39 Plot of the normalised crack opening stress versus the stress ratio, for $\sigma_{\max} = \sigma_{\text{y}}/3$. Adapted using selected data from Neuman [157], consistent with Equation 5.21.

5.6.2 STRAIN RESPONSE

The cyclic stress-strain relationship illustrated schematically in Figure 5.38a can be represented by [139, 153]

$$\varepsilon_{\text{vM}}^{\text{a}} = \varepsilon^{\text{a}} + \varepsilon_{\text{p}}^{\text{a}} = \frac{\sigma_{\text{vM}}^{\text{a}}}{E} + \left(\frac{\sigma_{\text{vM}}^{\text{a}}}{K'}\right)^{1/m'} \tag{5.23}$$

where $\varepsilon_{\text{vM}}^{\text{a}}$, and $\sigma_{\text{vM}}^{\text{a}}$ are the equivalent von Mises strain and stress amplitudes. The terms ε^{a} and $\varepsilon_{\text{p}}^{\text{a}}$ represent the elastic and plastic strain components, respectively and E is the Young's modulus. K' and m' are empirical functions, based on experimental data from a large variety of structural steels [139], of the uniaxial yield and ultimate tensile strengths and the fractional reduction of area (ε_{RA}):

$$K'/\text{MPa} = \begin{cases} 2.16 \times 10^{-4}\sigma_{\text{UTS}}^{2.1} + 738 & \sigma_{\text{UTS}}/\sigma_{\text{y}} \leq 1.2 \\ 3.63 \times 10^{-4}\sigma_{\text{UTS}}^{2} + 0.68\sigma_{\text{UTS}} + 570 & 1.2 < \sigma_{\text{UTS}}/\sigma_{\text{y}} < 1.4 \\ 1.21\sigma_{\text{UTS}} + 555 & 1.4 < \sigma_{\text{UTS}}/\sigma_{\text{y}} \end{cases}$$

and

$$m' = \left[\ln\{K'\} - \ln\left\{0.089\left(1+\varepsilon_{\text{RA}}\right)^{1.35}\sigma_{\text{UTS}}^{1.35}\left(-\frac{0.002}{\ln\{1-\varepsilon_{\text{RA}}\}}\right)^{0.216} + 120\right\}\right] \Big/ \ln\{500\}$$

These relationships are not dimensionally correct, but are recognised to be empirical and are valid over the ranges $262 \leq \sigma_{\text{y}} \leq 1372\,\text{MPa}$, $621 \leq \sigma_{\text{UTS}} \leq 2027\,\text{MPa}$, $0.14 \leq \varepsilon_{\text{RA}} \leq 0.8$; they can be used to estimate the elastic and plastic strain tensors at the sample surface by implementing them into a finite element model that incorporates specimen geometry [81]. The amplitudes calculated using Equation 5.23 form the boundary conditions in the finite element model so the local strain tensors must be consistent with $\sigma_{\text{min}} \leq \sigma_{\text{a}} \leq \sigma_{\text{max}}$ as must the components of the elastic and plastic strain amplitude tensors ϵ^{a} and $\epsilon_{\text{p}}^{\text{a}}$ respectively.

The assumption is that cracks evolve within slip bands. In the case of a banded $\alpha+\text{P}$ microstructure, Ito et al. [153] assumed that both the ferrite and pearlite (which is a composite microstructure) can be defined as grains with appropriate size- and shape-distributions and different mechanical properties.[12] The Hall-Petch equation is then used to define the yield strength of these two components of the structure as a function of the friction stress σ^{f} and the Hall-Petch coefficient $k_{\text{HP}} \approx 0.74\,\text{MPa}\sqrt{\text{m}}$. The friction stress is expressed as a weighted mean:

$$\frac{1}{2}\sigma^{\text{f}} = (1 - V_{\text{V}}^{\text{P}})\tau_{\alpha}^{\text{f}} + V_{\text{V}}^{\text{P}}\tau_{\text{P}}^{\text{f}} \tag{5.24a}$$

and since the ratio of the shear friction-stresses is set equal to the ratio of their Vickers hardnesses, $\tau_{\alpha}^{\text{f}}/\tau_{\text{P}}^{\text{f}} \approx 198/276$, it follows that

$$\tau_{\alpha}^{\text{f}} = \frac{198}{2(198+78V_{\text{V}}^{\text{P}})}\sigma^{\text{f}} \quad \text{and} \quad \tau_{\text{P}}^{\text{f}} = \frac{276}{2(198+78V_{\text{V}}^{\text{P}})}\sigma^{\text{f}}. \tag{5.24b}$$

Equation 4.8 therefore defines the microscopic phenomena involving the stresses needed to move dislocations within slip bands that form in the proeutectoid-α and pearlite, whereas the cyclic stress-strain relation, Equation 5.23, forms the constitutive equation in the finite element framework to enable the strain amplitude tensors (ε^{a}, $\varepsilon_{\text{p}}^{\text{a}}$) to be estimated within the active zone.

5.6.3 CRACK EVOLUTION IN A DUCTILE MATERIAL

Crack growth during fatigue starts as a short crack forms in a single crystal on a particular slip plane that has the greatest resolved shear stress in the shear direction (i.e., largest Schmid factor). Such cracks then grow by the emission of dislocations, creating a plastic zone at the crack tip. The ratio of the crack length to the size plastic zone size determines the crack tip sliding displacement range which has the form [155, 158–160]

$$\Delta\text{CTSD} = \tau^{\text{f}} \times \frac{8z_{\ell}(1-\nu)}{\pi E_{\text{s}}}\ln\left\{\frac{z_{\ell\text{p}}}{z_{\ell}}\right\} \tag{5.25}$$

where E_{s} is the shear modulus, ν the Poisson's ratio, $2z_{\ell}$ the crack length and $2z_{\ell\text{p}}$ the sum of the crack length and the extent of the plastic zones at the crack tips.

[12]For simplicity, the pearlite is treated as a single phase with slip band growth in the pearlite being treated in an identical manner to that in ferrite grains.

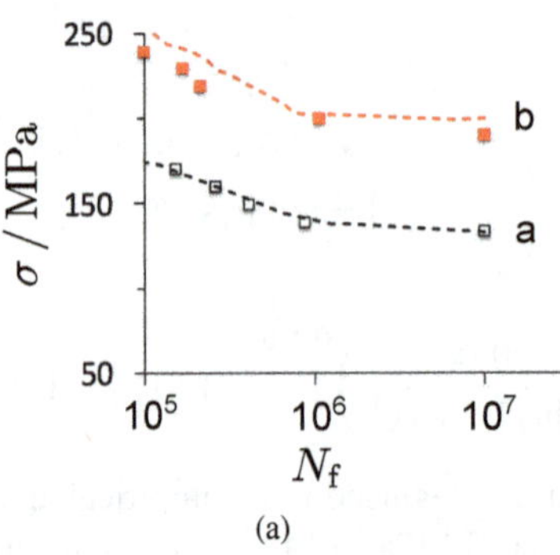

(a)

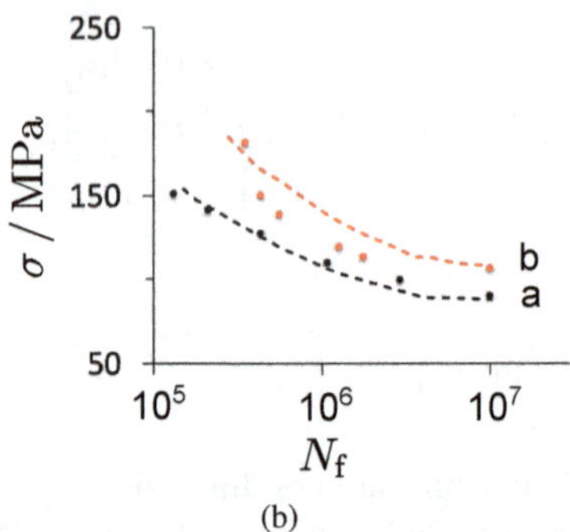

(b)

Figure 5.40 Measured (points) and calculated (dashed lines) plots of nominal stress versus the number of cycles to failure, for two different steels described in Table 5.2 and $R_F = -1$. (a) Smooth fatigue specimens. (b) Three-point bend specimens with sharp notch. Selected data from Ito et al. [153].

The dislocation field that accompanies the crack tip was assumed originally to correspond to the slip line field of macroscopic plasticity, where the lines represent the loci where the shear stress matches the shear-yield stress. To model fatigue in a polycrystalline steel, it is assumed that a linear slip band is interrupted periodically by grain boundaries, but with the effective friction stress calculated according to the orientation of the grain in which the crack tip exists [153]:

$$\Delta\text{CTSD} = \frac{4(1-\nu)}{\pi E_s}\left[2z_\ell \tau_j^f \ln\left\{\frac{z_{\ell p}}{z_\ell}\right\} + \Delta\tau_j \sum_{u=j+1}^{\infty}\left(\frac{\tau_i^f}{\Delta\tau_i} - \frac{\tau_{i-1}^f}{\Delta\tau_{i-1}}\right) \times f\{z_\ell, z_{\ell p}, L_{i-1}\}\right] \quad (5.26)$$

$$f\{z_\ell, z_{\ell p}, L\} = L\ln\left\{\frac{(z_{\ell p}^2 - L^2)^{1/2} + (z_{\ell p}^2 - z_\ell^2)^{1/2}}{(z_{\ell p}^2 - L^2)^{1/2} - (z_{\ell p}^2 - z_\ell^2)^{1/2}}\right\} - z_\ell \ln\left\{\frac{z_\ell(z_{\ell p}^2 - L^2)^{1/2} + (z_{\ell p}^2 - z_\ell^2)^{1/2}}{z_\ell(z_{\ell p}^2 - L^2)^{1/2} - (z_{\ell p}^2 - z_\ell^2)^{1/2}}\right\}$$

where the second term is an addition to Equation 5.25 to account for the role of grain boundaries. The subscript j refers to the grain in which the crack tip is located and i the grain identity in the sequence and L_i the location of the ith grain boundary. $\Delta\tau_i$ and τ_i^f represent the effective resolved shear stress range and the friction strength for the ith grain. The shear stress range $\Delta\tau_i$ for each grain is by $(\hat{\mathbf{n}}_i)\Delta\sigma_{i-1}[\hat{\mathbf{m}}_i]$ where $(\hat{\mathbf{n}})$ is a unit row-vector normal to the slip plane in grain i, and $[\hat{\mathbf{m}}_i]$ the corresponding unit slip direction expressed as a column vector; it is assumed here that of all possible slip systems in the grain, this combination represents the maximally stressed slip system. The $\Delta\sigma_{i-1}$ is the 2nd-rank stress tensor at crack length z_ℓ when $i = j$, otherwise, it is given by $\Delta\tau_{i-1}[\hat{\mathbf{n}}_{i-1}](\hat{\mathbf{m}}_{i-1})$ for $i \geq j+1$.[13]

The crack growth in micrometres per fatigue cycle is then given empirically for α + P steels by [153]

$$\frac{dz}{dN} = \begin{cases} 11.8 \times (\Delta\text{CTSD})^2 & \text{for} \quad z_\ell \leq L_1 \\ 11.8 \times [(\Delta\text{CTSD})^2 - 0.021] & \text{for} \quad z_\ell > L_1. \end{cases}$$

For a ductile steel, the number of cycles to failure N_f is when the crack length has reached the sample thickness. Figure 5.40 shows that for two different specimen shapes and steels, the method gives good estimates of the crack growth rate and fatigue limit, with the input parameters to the models determined from uniaxial tensile tests and some empirical data which are claimed to be

[13] Square brackets imply a column vector, round brackets a row vector.

Table 5.2
Steel composition and material parameters [153] used in the calculations illustrated in Figure 5.40.

Steel	C wt %	Si wt%	Mn wt%	V_V^α	$\overline{L}_\alpha$ µm	τ_P^f MPa	τ_α^f MPa	σ_y/ MPa	σ_{UTS}/ MPa	ε_{RA}
a	0.18	0.15	1.00	0.73	56.6	74	53	216	430	0.72
b	0.14	0.36	1.54	0.79	15.4	116	83	368	538	0.78

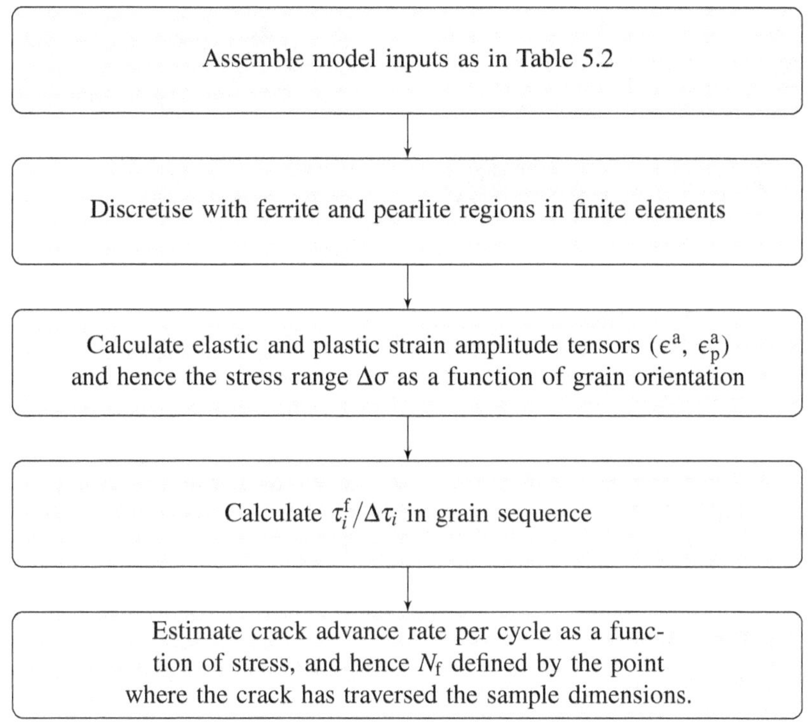

Figure 5.41 Sequence in the calculation of the fatigue properties of ferrite-pearlite steels

generic. The specimen shape, size and manner of loading influences the stress intensity at the crack, calculable using linear elastic fracture mechanics. The process is summarised in Figure 5.41 and if the approximations, assumptions and estimates are considered to be reasonable, then it should no longer be necessary to measure the fatigue properties based on crack growth alone – the model can in principle be used to design better steels that resist such damage. It would be useful to examine cases where the fatigue cracks propagate at orientations different to those of the bands.

5.7 FATIGUE CRACK GROWTH: EFFECT OF BANDING

Fatigue in a banded microstructure is associated with crack branching when the steel is tested in an orientation where the growth occurs normal to the bands, Figure 5.42 [161]. The hard pearlite retards growth by inducing large-angle (almost 90°) deflections in the crack path. As a result, the

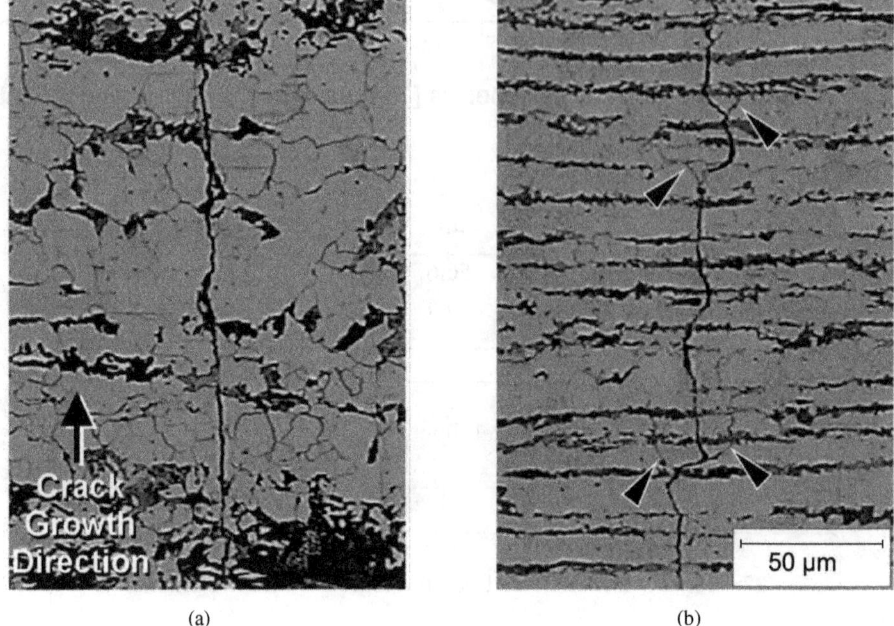

Figure 5.42 Fatigue crack path in $\alpha + \mathrm{P}$ steel. (a) Un-banded steel where the advancing crack is essentially undeflected. (b) Banded steel with considerable crack-branching. Courtesy of Y. Mutoh.

threshold threshold stress intensity range ΔK_{th} is greater for propagation normal to the bands when compared with growth between the bands; the average crack advance per cycle, in the Paris regime, also is reduced. The branching occurs at the α/pearlite interfaces with the resulting roughness in the crack surface diminishing the mode I crack driving force. The root cause of the branching is the difference in the strengths of the ferritic and pearlitic regions. Crack propagation parallel to the bands is similar to that in un-banded structures so it seems that the crack is less hindered in its traverse through the α [162]. The extent of crack branching depends on the distribution and size of the pearlite within the microstructure [147].

Figure 5.43 shows that the dependence of the orientation of the crack path relative to the banding becomes greater when the tests are conducted in a pressurised hydrogen atmosphere.

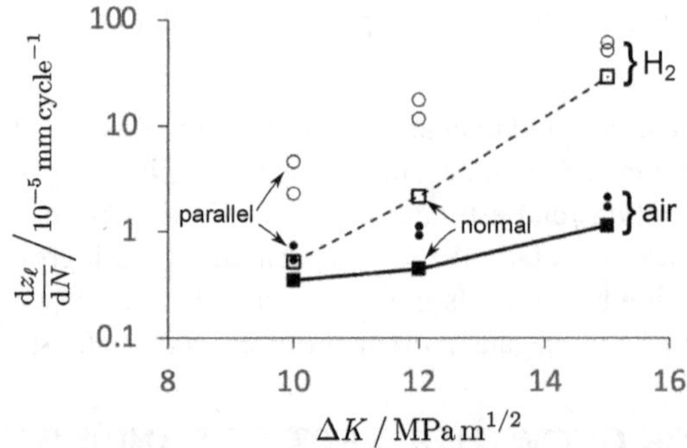

Figure 5.43 Experimental measurements of the crack growth rate as a function of the stress intensity range, whether the crack is conducted in air or in hydrogen at a pressure of 21 MPa. The square and circular points represent crack advance normal or parallel to the bands, respectively. Selected data from Ronevich et al. [162].

REFERENCES

1. E. Dobbs, and G. O. Jones: 'Theory and properties of solid argon', *Reports on Progress in Physics*, 1957, **20**, 516–564.
2. C. Barrett, and L. Meyer: 'X-ray diffraction study of solid argon', *The Journal of Chemical Physics*, 1964, **41**, 1078–1081.
3. F. B. Pickering: Physical Metallurgy and the Design of Steels: Essex, U.K.: Applied Science Publishers, page 104, 1978.
4. H. K. D. H. Bhadeshia: 'Neural networks in materials science', *ISIJ International*, 1999, **39**, 966–979.
5. F. Suárez, F. Sket, J. C. Gálvez, D. A. Cendón, J. M. Atienza, and J. Molina-Aldareguia: 'The evolution of internal damage identified by means of X-ray computed tomography in two steels and the ensuing relation with Gurson's numerical modelling', *Metals*, 2019, **9**, 292.
6. O. Fisher: 'A few notes on Clacton, Essex', *Geological Magazine*, 1868, **47**, 213–215.
7. H. Williams, and O. M. B. Bulman: 'The geology of the Dolwyddelan syncline (North Wales)', *Quarterly Journal of the Geological Society*, 1931, **87**, 425–458.
8. P. J. Noell, J. D. Carroll, and B. L. Boyce: 'The mechanisms of ductile rupture', *Acta Materialia*, 2018, **161**, 83–98.
9. M. Ohata, and M. Toyoda: 'Damage mechanism for controlling ductile cracking of structural steel with heterogeneous microstructure', *Materials Science Forum*, 2006, **512**, 31–36.
10. L. E. Miller, and G. C. Smith: 'Tensile fractures in carbon steels', *Journal of the Iron and Steel Institute*, 1970, **11**, 998–1005.
11. K. Ryttberg, M. K. Wedel, V. Recina, P. Dahlman, and L. Nyborg: 'The effect of cold ring rolling on the evolution of microstructure and texture in 100Cr6 steel', *Materials Science & Engineering A*, 2010, **527**, 2431–2436.
12. H. Shoji, M. Ohata, and F. Minami: 'Simulation-based method for hierarchal material design to improve ductile crack growth resistance of structural component', *International Journal of Fracture*, 2015, **192**, 167–178.
13. P. F. Thomason: 'Tensile plastic instability and ductile fracture criteria in uniaxial compression test', *Journal of Mechanical Sciences*, 1969, **11**, 187–198.
14. B. Berisha, C. Raemy, C. Becker, M. Gorji, and P. Hora: 'Multiscale modeling of failure initiation in a ferritic–pearlitic steel', *Acta Materialia*, 2015, **100**, 191–201.
15. P. W. Anderson: 'Local moments and localized states': Nobel Lecture, Bell Telephone Laboratories, Princeton University, 1977.
16. H. K. D. H. Bhadeshia: 'Mathematical models in materials science', *Materials Science and Technology*, 2008, **24**, 128–135.
17. D. Bancroft, E. L. Peterson, and S. Minshall: 'Polymorphism of iron at high pressure', *Journal of Applied Physics*, 1956, **27**(3), 291–298.
18. R. L. Clendenen, and H. G. Drickamer: 'The effect of pressure on the volume and lattice parameters of ruthenium and iron', *Journal of the Physics and Chemistry of Solids*, 1964, **25**, 865–868.
19. P. M. Giles, M. H. Longenbach, and A. R. Marder: 'High pressure alpha to epsilon martensitic transformation in iron', *Journal of Applied Physics*, 1971, **42**, 4290–4295.
20. B. Crossland: 'The effect of fluid pressure on the shear properties of metals', *Proceedings of the Institution of Mechanical Engineers*, 1954, **168**, 935–946.
21. H. K. D. H. Bhadeshia, and R. W. K. Honeycombe: Steels – structure, properties and design: 5th edition ed., London, U.K.: Elsevier Butterworth-Heinemann, 2024.
22. P. W. Bridgman: Studies in large plastic flow and fracture: New York, USA: McGraw-Hill, 1952.

23. A. Brownrigg, W. A. Spitzig, O. Richmond, D. Teirlinck, and J. D. Embury: 'The influence of hydrostatic pressure on the flow stress and ductility of a spherodized 1045 steel', *Acta Metalllurgica*, 1983, **31**, 1141–1150.
24. T. E. Davidson, and G. S. Ansell: 'The structure sensitivity of the effects of pressure upon the ductility of Fe-C materials', *Transactions of the ASM*, 1968, **61**, 242–254.
25. T. E. Davidson, and G. S. Ansell: 'Some observations on the relationship between the effects of pressure upon the fracture mechanisms and the ductility of Fe-C materials', *Transactions of the Metallurgical Society of the AIME*, 1969, **245**, 2383–2390.
26. E. D. Scott, and J. Blakewell: 'The assessment of ball bearing steels of modified chemical composition', *Wear*, 1978, **46**, 273–279.
27. A. Considère: 'l'emploi du fer et de l'acier dan les constructions (the use of iron and steel in construction)', *Annales des Ponts et Chausées*, 1885, **9**, 574–775.
28. G. Laschet, P. Fayek, T. Henke, H. Quade, and U. Prahl: 'Derivation of anisotropic flow curves of ferrite–pearlite pipeline steel via a two-level homogenisation scheme', *Materials Science & Engineering A*, 2013, **566**, 143–156.
29. E. Corona, J. A. Shaw, and M. A. Iadicola: 'Buckling of steel bars with Lüders bands', *International Journal of Solids and Structures*, 2002, **39**, 3313–3336.
30. M. Dollar, I. M. Bernstein, and A. W. Thompson: 'Influence of deformation substructure on flow and fracture of fully pearlitic steel', *Acta Metallurgica*, 1988, **36**, 311–320.
31. E. Smith, and J. T. Barnby: 'Nucleation of grain-boundary cavities during high-temperature creep', *Metal Science*, 1967, **1**, 1–4.
32. E. Smith: 'The formation of a cleavage crack in a crystalline solid – I', *Acta Metallurgica*, 1966, **14**, 985–989.
33. G. Langford: 'Deformation of pearlite', *Metallurgical Transactions A*, 1977, **8**, 861 – 875.
34. A. J. Perez-Unzueta, and J. H. Beynon: 'Microstructure and wear resistance of pearlitic rail steels', *Wear*, 1993, **162-163**, 173–182.
35. K. Nakase, and I. M. Bernstein: 'The effect of alloying elements and microstructure on the strength and fracture resistance of pearlitic steel', *Metallurgical Transactions A*, 1988, **19**, 2819–2829.
36. J. M. Hyzak, and I. M. Bernstein: 'The role of microstructure on the strength and toughness of fully pearlitic steels', *Metallurgical Transactions A*, 1976, **7**, 1217–1221.
37. J. P. Houin, A. Simon, and G. Beck: 'Relationship between structure and mechanical properties of pearlite between 0.2% and 0.8%C', *Transactions of the Iron and Steel Institute of Japan*, 1981, **21**, 726–731.
38. D. J. Alexander, and I. M. Bernstein: 'Cleavage fracture in pearlitic eutectoid steel', *Metallurgical Transactions A*, 1989, **20**, 2321–2335.
39. C. M. Bae, C. S. Lee, and W. J. Nam: 'Effect of carbon content on mechanical properties of fully pearlitic steels', *Materials Science and Technology*, 2002, **18**, 1317–1321.
40. T. Takahashi, M. Nagumo, and Y. Asano: 'Microstructures dominating ductility of eutectoid pearlitic steels', *Journal of the Japan Institute of Metals*, 1978, **42**, 708–715.
41. E. Ghassemali, J. C. Hernando, D. M. Stefanescu, A. Dioszegi, A. E. W. Jarfors, J. Dluhoš, and M. Petrenec: 'Revisiting the graphite nodule in ductile iron', *Scripta Materialia*, 2019, **161**, 66–69.
42. D. M. Ştefănescu, G. Alonso, P. Larrañaga, E. de La Fuente, and R. Suárez: 'Reassessment of crystal growth theory of graphite in cast iron', *Materials Science Forum*, 2018, **925**, 36–44.
43. F. Iacoviello, O. Di Bartolomeo, V. Di Cocco, and V. Piacente: 'Damaging micromechanisms in ferritic-pearlitic ductile cast irons', *Materials Science & Engineering A*, 2008, **478**, 181–186.
44. G. R. Yoder, and V. Weiss: 'Superplasticity in eutectoid steels', *Metallurgical Transactions*, 1972, **3**, 675–681.

45. T. Furuhara, E. Sato, T. Mizoguchi, S. Furimoto, and T. Maki: 'Grain boundary character and superplasticity of fine-grained ultra-high carbon steel', *Materials Transactions*, 2002, **43**, 2455–2462.
46. A. K. Ghosh, and R. A. Ayres: 'On reported anomalies in relating strain-rate sensitivity (m) to ductility', *Metallurgical Transactions A*, 1976, **7**, 1589–1591.
47. M. Umemoto, Y. Todaka, T. Takahashi, P. Li, R. Tokumiya, and K. Tsuchiya: 'High temperature deformation behavior of bulk cementite produced by mechanical alloying and spark plasma sintering', *Materials Science & Engineering A*, 2004, **375-377**, 894–898.
48. J. F. Knott: Fundamentals of Fracture Mechanics: London, U.K.: Butterworths, 1973.
49. C. O. Mohr: 'Über die darstellung des spannungszustandes und des deformationszustandes eines kŕpcrelemcntes (about the representation of the state of stress and state of deformation of a body element)', *Zivilingenieur*, 1882, **28**, S. 113.
50. D. J. Alexander, and I. M. Bernstein: 'The cleavage plane of pearlite', *Metallurgical Transactions A*, 1982, **13**, 1865–1868.
51. A. Fernández-Vincente, M. Pellizzari, and J. L. Arias: 'Feasibility of laser surface treatment of pearlitic and bainitic ductile irons for hot rolls', *Journal of Materials Processing Technology*, 2012, **212**, 989–1002.
52. J. J. Coronado, and S. A. Rodriguez: 'Cementite characterization with chromium and vanadium contents using indentation technique', *Journal of Iron and Steel Research International*, 2015, **22**, 366–370.
53. M. Umemoto, and H. Ohtsuka: 'Mechanical properties of cementite', *ISIJ International*, 2022, **62**, 1313–1333.
54. R. K. Barik, A. Ghosh, M. Basiruddin Sk, S. Biswal, A. Dutta, and D. Chakrabarti: 'Bridging microstructure and crystallography with the micromechanics of cleavage fracture in a lamellar pearlitic steel', *Acta Materialia*, 2021, **214**, 116988.
55. A. Fernández-Vicente, M. Pellizzari, and J. L. Arias: 'Feasibility of laser surface treatment of pearlitic and bainitic ductile irons for hot rolls', *Journal of Materials Processing Technology*, 2012, **212**, 989–1002.
56. M. Hillert: 'The formation of pearlite', In: V. F. Zackay, and H. I. Aaronson, eds. *Decomposition of Austenite by Diffusional Processes*. New York, USA: Interscience, 1962:197–237.
57. R. J. Dippenaar, and R. W. K. Honeycombe: 'The crystallography and nucleation of pearlite', *Proceedings of the Royal Society A*, 1973, **333**, 455–467.
58. S. Behera, R. K. Barik, M. B. Sk, R. Mitra, and D. Chakrabarti: 'Recipe for improving the impact toughness of high-strength pearlitic steel by controlling the cleavage cracking mechanisms', *Materials Science & Engineering A*, 2019, **764**, 138256.
59. W. Thomson: 'On the division of space with minimum partitional area', *Acta mathematica*, 1887, **11**, 121–134.
60. E. E. Underwood: Quantitative Stereology, chap. 4: Addison-Wesley Publication Company, 1970:90–93.
61. F. Zhang, Y. Zhao, Y. Tan, X. Ji, and S. Xiang: 'Study on the nucleation and growth of pearlite colony and impact toughness of eutectoid steel', *Metals*, 2019, **9**, 1133.
62. O. N. Romaniv, E. A. Shur, A. N. Tkach, T. N. Kiseleva, and V. N. Simin'kovich: 'Crack resistance of pearlitic eutectoid steels. I. fracture of steels in short-term loading', *Soviet materials science: translation of Fiziko-khimicheskaya mekhanika materialov / Academy of Sciences of the Ukrainian SSR*, 1983, **18**, 325–331.
63. S. Zhou, Y. Zuo, Z. Li, X. Wang, and Q. Yong: 'Microstructural analysis on cleavage fracture in pearlitic steels', *Materials Characterization*, 2016, **119**, 110–113.
64. T. Sirithanakorn, M. Tanaka, and K. Higashida: 'Two-step brittle-to-ductile transitions in pearlitic steels', *Advanced Materials Research*, 2014, **922**, 706–710.

65. F. B. Pickering, and B. Garbarz: 'The effect of transformation temperature and prior austenite grain size on the pearlite colony size in vanadium treated pearlitic steels', *Scripta Metallurgica*, 1987, **21**, 249–253.
66. T. Sato, and K. Tomomoto: 'Effects of the structures to impact properties for high carbon steel', *Tetsu-to-Hagané*, 1980, **66**, 231–238.
67. F. P. L. Kavishe, and T. J. Baker: 'Effect of prior austenite grain size and pearlite interlamellar spacing on strength and fracture toughness of a eutectoid rail steel', *Materials Science and Technology*, 1986, **2**, 816–822.
68. J. Toribio: 'Evolution of fracture behaviour in progressively drawn pearlitic steel', *ISIJ International*, 2002, **42**, 656–662.
69. A. B. El-Shabasy, and J. J. Lewandowski: 'Effects of microstructural changes, loading conditions and test temperature on toughness of fully pearlitic eutectoid steel used in transportation industry', *Materials Science and Technology*, 2009, **25**, 369–378.
70. M. Cieśla, J. Herian, and G. Junak: 'Influence of pearlite morphology on creeping characteristic curves of R260 steel', *Key Engineering Materials*, 2014, **598**, 39–44.
71. K. Mishra, and A. Singh: 'Effect of interlamellar spacing on fracture toughness of nano-structured pearlite', *Materials Science & Engineering A*, 2017, **706**, 22–26.
72. F. Yu, P. Y. B. Jar, M. T. Hendry, C. Jar, and K. Nishanth: 'Fracture toughness estimation for high-strength rail steels using indentation test', *Engineering Fracture Mechanics*, 2018, **204**, 469–481.
73. L. B. Godefroid, A. T. Souza, and M. A. Pinto: 'Fracture toughness, fatigue crack resistance and wear resistance of two railroad steels', *Journal of Materials Research and Technology*, 2020, **9**, 9588–9597.
74. M. Srinivas, G. Malakondaiah, and P. R. Rao: 'A first report on fracture toughness of bcc iron alloys as influenced by solutes: opposite effects of silicon and cobalt', *Bulletin of Materials Science*, 1988, **11**, 329–336.
75. Y. Ohmori, and F. Terasaki: 'Effects of nickel and manganese on cleavage fracture of ferritic-pearlitic steels', *Transactions of the Iron and Steel Institute of Japan*, 1976, **16**, 561–568.
76. A. R. Rosenfield, G. T. Hahn, and J. D. Embury: 'Fracture of steels containing pearlite', *Metallurgical Transactions*, 1972, **3**, 2797–2804.
77. K. Shibanuma, Y. Nemoto, T. Hiraide, K. Suzuki, and S. Aihara: 'Investigation of micro-crack initiation as a trigger of cleavage fracture in ferrite pearlite steels,', *ISIJ International*, 2017, **57**, 365–373.
78. M. D. Chapetti, T. Kitano, T. Tagawa, and T. Miyata: 'Fatigue limit of blunt-notched components', *Fatigue and Fracture in Engineering Materials and Structures*, 1998, **21**, 1525–1536.
79. M. F. Garwood, H. H. Zurburg, and M. A. Erickson: Correlation of Laboratory Tests and Service Performance: Interpretation of tests and correlation with service, chap. 1: ASM, 1951:1–77.
80. Y. Murakami, S. Kodama, and S. Konuma: 'Quantitative evaluation of effects of non-metallic inclusions on fatigue strength of high strength steels. I: basic fatigue mechanism and evaluation of correlation between the fatigue fracture stress and the size and location of non-metallic inclusions', *International Journal of Fatigue*, 1989, **11**, 291–298.
81. K. Shibanuma, Y. Nemoto, T. Hiraide, K. Suzuki, S. Sadamatsu, Y. Adachi, and S. Aihara: 'A strategy to predict the fracture toughness of steels with a banded ferrite pearlite structure based on the micromechanics of brittle fracture initiation', *Acta Materialia*, 2018, **144**, 386–399.
82. J. H. Tweed, and J. F. Knott: 'Micromechanisms of failure in C-Mn welds metals', *Acta Metallurgica*, 1987, **35**, 1401–1414.
83. E. J. Song, H. K. D. H. Bhadeshia, and D.-W. Suh: 'Effect of hydrogen on the surface energy of ferrite and austenite', *Corrosion Science*, 2013, **77**, 379–384.

84. H. K. D. H. Bhadeshia: 'Cementite', *International Materials Reviews*, 2020, **65**, 1–27.
85. J. I. S. Martin, and J. M. Rodriguez-Ibabe: 'Determination of energetic parameters controlling cleavage fracture in a Ti-V microalloyed ferrite-pearlite steel.', *Scripta Materialia*, 1999, **40**, 459–464.
86. K. Shibanuma, S. Aihara, and K. Suzuki: 'Prediction model on cleavage fracture initiation in steels having ferrite-cementite microstructures – Part I: Model presentation', *Engineering Fracture Mechanics*, 2016, **151**, 161–180.
87. B. Strnadel, and P. Haušild: 'Statistical scatter in the fracture toughness and Charpy impact energy of pearlitic steel', *Materials Science & Engineering A*, 2008, **486**, 208–214.
88. D. Zeng, L. Lu, Y. Gong, N. Zhang, and Y. Gong: 'Optimization of strength and toughness of railway wheel steel by alloy design', *Materials and Design*, 2016, **92**, 998–1006.
89. E. A. Almond: 'Delamination in banded steels', *Metallurgical Transactions*, 1970, **1**, 2038–2041.
90. J. Embury, N. Petch, A. Wraith, and E. Wright: 'The fracture of mild steel laminates', *Transactions of the Metallurgical Society of AIME*, 1967, **239**, 114–118.
91. X. Arnoult, M. Ružicková, K. Kunzova, and A. Materna: 'Short review: Potential impact of delamination cracks on fracture toughness of structural materials', *Frattura ed Integrità Strutturale*, 2016, **10**, 509–522.
92. D. M. Fegredo: 'Effect of rolling at different temperatures on the fracture toughness anisotropy of a C-Mn structural-steel', *Canadian Metallurgical Quarterly*, 1975, **14**, 243–255.
93. A. Ray, S. K. Paul, and S. Jha: 'Effect of inclusions and microstructural characteristics on the mechanical properties and fracture behavior of a high-strength low-alloy steel', *Journal of Materials Engineering and Performance*, 1995, **4**, 679–688.
94. J. S. Joo, D. W. Suh, and H. K. D. H. Bhadeshia: 'Mechanical anisotropy in steels for pipelines', *ISIJ International*, 2013, **53**, 1305–1314.
95. A. S. Bor: 'Effect of pearlite banding on mechanical properties of hot-rolled steel plates', *ISIJ International*, 1991, **31**, 1445–1446.
96. W. A. Spitzig: 'Effect of sulphide inclusion morphology and pearlite banding on anisotropy of mechanical properties in normalized C-Mn steels', *Metallurgical Transactions A*, 1983, **14**, 271–283.
97. W. S. Owen, M. Cohen, and B. L. Averbach: 'Influence of ferrite banding on the impact properties of mild steel', *Welding Journal, Research Supplement*, 1958, **37**, 368s–374s.
98. P. Shanmugam, and S. D. Pathak: 'Some studies on the impact behaviour of banded microalloyed steel', *Engineering Fracture Mechanics*, 1996, **53**, 991–1005.
99. M. S. Joo, D.-W. Suh, J.-H. Bae, and H. K. D. H. Bhadeshia: 'Toughness anisotropy in X70 and X80 linepipe steels', *Materials Science and Technology*, 2014, **30**, 439–440.
100. H. L. Haskel, E. Pauletti, J. de Paula Martins, and A. L. Moreira de Carvalho: 'Microstructure and microtexture assessment of delamination phenomena in Charpy impact tested specimens', *Materials Research*, 2014, **17**, 1238–1250.
101. A. Wöhler: 'Bericht über die versuche, welche auf der Königl. Niederschlesisch-Märkischen Eisenbahn mit apparaten zum messen der biegung und verdrehung von eisenbahnwagenachsen wḧrend der fahrt angestellt wurden (report on the experiments ... for measuring the bending and torsion of railway carriage axles during travel)', *Zeitschrift für Bauwesen*, 1858, **8**, 641–652.
102. G. Oates, and D. V. Wilson: 'The effects of dislocation locking and strain ageing on the fatigue limit of low-carbon steel', *Acta Metallurgica*, 1964, **12**, 21–33.
103. B. Mintz, and D. V. Wilson: 'Strain ageing during the fatigue of carbon steels', *Acta Metallurgica*, 1965, **13**, 947–956.
104. N. J. Petch: 'Theory of yield point and of strain ageing in steel', In: J. A. Charles, and G. A. Smith, eds. *Advances in Physical Metallurgy*. London, U.K.: Institute of Metals, 1990:11–22.

105. T. Svensson, and J. de Maré: 'Random features of the fatigue limit', *Extremes*, 1999, **2**, 165–176.
106. Y. Mrakami, and M. Endo: 'Effects of hardness and crack geometries on δk_{th} of small cracks emanating from small defects,', In: K. J. Miller, and E. R. de los Rios, eds. *The behaviour of short fatigue cracks*. London, U.K.: Mechanical Engineering Publishers, 1986:275–293.
107. Y. Murakami, and M. Endo: 'Effects of defects, inclusions and inhomogeneities on fatigue strength', *International Journal of Fatigue*, 1994, **16**, 163–182.
108. S. Hamada, D. Sasaki, M. Ueda, and H. Noguchi: 'Fatigue limit evaluation considering crack initiation for lamellar pearlitic steel', *Procedia Engineering*, 2011, **10**, 1467–1472.
109. G. T. Gray III, A. W. Thompson, and J. C. Williams: 'Influence of microstructure on fatigue crack initiation in fully pearlitic steels', *Metallurgical Transactions A*, 1985, **16**, 753–760.
110. W. X. Sun, S. Nashida, N. Hattori, and I. Usui: 'Fatigue properties of cold-rolled notched eutectoid steel', *International Journal of Fatigue*, 2004, **26**, 1139–1145.
111. W. Sun, S. Nishida, and N. Hattori: 'Fatigue properties of pre-strained eutectoid steel', *Journal of the Society of Materials Science, Japan*, 2003, **52**, 210–215.
112. A. Querales, and J. G. Byrne: 'Improvement of fatigue life of 1080 steel by thermomechanical processing', *Fatigue of Engineering Materials and Structures*, 1979, **1**, 371–382.
113. O. H. Basquin: 'The experimental law of endurance', *Proceedings of ASTM*, 1910, **10**, 625–630.
114. R. Yadav, and K. Ajit: 'A new stress-based model to predict the fatigue life of 880-grade pearlitic rail steel under uniaxial ratcheting conditions', *International Journal of Fatigue*, 2023, **171**, 107588.
115. S. Nishiyama: 'Statistical analysis of fatigue test data', *Journal of the Society of Materials Science, Japan*, 1980, **29**, 24–29.
116. G. Chalant, and B. M. Suyitno: 'Effects of microstructure on low and high cycle fatigue behaviour of a micro-alloyed steel', In: M. Jono, and T. Inoue, eds. *Proceedings of the 6th International Conference on Mechical Behaviour of Materials*, vol. VI. Kyoto, Japan: Pergamon Press, Oxford, Elsevier Science, 1991:511–516.
117. R. E. Cohen, D. K. Matlock, and G. Krauss: 'Specimen edge effects on bending fatigue of carburizedsteel', *Journal of Materials Engineering and Performance*, 1992, **1**, 695–704.
118. M. J. Peet, P. Hill, M. Rawson, S. Wood, and H. K. D. H. Bhadeshia: 'Fatigue of extremely fine bainite', *Materials Science and Technology*, 2011, **27**, 119–123.
119. J. Lankford, and F. N. Kusenberger: 'Initiation of fatigue cracks in 4340 steel', *Metallurgical Transactions*, 1973, **4**, 553–559.
120. H. W. Höppel, M. Prell, L. May, and M. Göken: 'Influence of grain size and precipitates on the fatigue lives and deformation mechanisms in the VHCF-regime', *Procedia Engineering*, 2010, **2**, 1025–1034.
121. G. Qian, Y. Hong, and C. Zhou: 'Investigation of high cycle and very-high-cycle fatigue behaviors for a structural steel with smooth and notched specimens', *Engineering Failure Analysis*, 2010, **17**, 1517–1525.
122. V. Gaur, V. Doquet, E. Persent, C. Mareau, E. Roguet, and J. Kittel: 'Surface versus internal fatigue crack initiation in steel: Influence of mean stress', *International Journal of Fatigue*, 2016, **82**, 437–448.
123. H. Gerber: 'Bestimmung der zulässigen spannungen in eisen-constructionen', *eitschrift des Bayerischen Architekten – und Ingenieur-Vereins*, 1874, **6**, 101–110.
124. N. Narasaiah, and K. K. Ray: 'Small crack formation in a low carbon steel with banded ferrite–pearlite structure', *Materials Science & Engineering A*, 2005, **392**, 269–277.
125. D. J. Gooch: 'The effect of cold work on low temperature ($0.35T_m$) creep crack growth in C-Mn steels', *Materials Science & Engineering*, 1984, **64**, 183–196.

126. K. Hussain, and R. R. de Los Rios: 'Microstructural effect on tensile and fatigue behaviour of C-Mn steel', *Journal of Materials Science*, 1997, **32**, 3565–3569.
127. Y. J. Park, and I. M. Bernstein: 'The process of crack initiation and effective grain size for cleavage fracture in pearlitic eutectoid steel', *Metallurgical Transactions A*, 1979, **10**, 1653–1664.
128. S. Nisida, T. Urashima, K. Sugino, and H. Masumoto: 'A study of fatigue crack propagation of rail steels', In: *Strength of Metals and Alloys*. Oxford, U. K.: Pergamon, 1979:1255–1260.
129. H. Sunwoo, M. Fine, M. Meshii, and D. Stone: 'Cyclic deformation of pearlitic eutectoid rail steel', *Metallurgical Transactions A*, 1982, **13**, 2035–2047.
130. S. Sankaran, V. S. Sarma, and K. A. Padmanabhan: 'Low cycle fatigue behavior of a multiphase microalloyed medium carbon steel: comparison between ferrite-pearlite and quenched and tempered microstructures', *Materials Science & Engineering A*, 2003, **345**, 328–335.
131. K. Cvetkovski, J. Ahlström, and B. Karlsson: 'Monotonic and cyclic deformation of a high silicon pearlitic wheel steel', *Wear*, 2011, **271**, 382–387.
132. Y. Wang, Y. Tomota, S. Harjo, W. Gong, and T. Ohmura: 'In-situ neutron diffraction during tension-compression cyclic deformation of a pearlite steel', *Materials Science & Engineering A*, 2016, **676**, 522–530.
133. H. Mayer: 'Ultrasonic torsion and tension-compression fatigue testing: Measuring principles and investigations on 2024-T351 aluminium alloy', *International Journal of Fatigue*, 2006, **28**, 1446–1455.
134. C. Bathias: 'Piezoelectric fatigue testing machines and devices', *International Journal of Fatigue*, 2006, **28**, 1438–1445.
135. Y. J. Chen, B. S. Wei, and W. N. Hsu: 'Ultrasonic fatigue testing of aluminium and magnesium alloys', *Journal of the Chinese Society of Mechanical Engineers*, 2008, **29**, 299–308.
136. Anonymous: 'The technical basis document for API RP 941': Tech. Rep. API TR 941-A, American Petroleum Institute, Washington, D. C., USA, 2019.
137. D. Sornette, T. Magnin, and Y. Brechet: 'The physical origin of the Coffin-Manson law in low-cycle fatigue', *Europhysics Letters*, 1992, **20**, 433–438.
138. W. RAMBERG, and O. WR: 'Description of stress-strain curves by three parameters': Tech. Rep. 902, National Advisory Committee for Aeronautics, Washington, D. C., USA, 1943.
139. J. Li, Z. Zhang, and C. Li: 'An improved method for estimation of Ramberg–Osgood curves of steels from monotonic tensile properties', *Fatigue and Fracture of Engineering Materials and Structures*, 2015, **39**, 412–426.
140. J. Li, Q. Sun, Z. P. Zhang, C. W. Li, and Y. J. Qiao: 'Theoretical estimation to the cyclic yield strength and fatigue limit for alloy steels', *Mechanics Research Communications*, 2009, **36**, 316–321.
141. Y. Gorash, and D. MacKenzie: 'On cyclic yield strength in definition of limits for characterisation of fatigue and creep behaviour', *Open Engineering*, 2017, **7**, 126–140.
142. P. C. Paris, M. P. Gomez, and W. E. Anderson: 'A rational analytic theory of fatigue life', *The Trend in Engineering*, 1961, **13**, 9–14.
143. I. N. Sneddon: 'The stress distribution in the neighborhood of a crack in an elastic solid', *Proceedings of the Royal Society of London A*, 1946, **187**, 229–260.
144. M. L. Williams: 'On the stress distribution at the base of a stationary crack', *Journal of Applied Mechanics*, 1957, **24**, 109–114.
145. G. R. Irwin: 'Analysis of stresses and strains near the end of a crack traversing a plate', *Journal of Applied Mechanics*, 1957, **24**, 361–364.
146. R. O. Ritchie: 'Near threshold fatigue-crack propagation in steels', *International Metals Reviews*, 1979, **4**, 205–230.

147. V. Igwemezie, P. Dirisu, and A. Mehmanparast: 'Critical assessment of the fatigue crack growth rate sensitivity to material microstructure in ferrite-pearlite steels in air and marine environment', *Materials Science & Engineering A*, 2019, **754**, 750–765.
148. Y. Ogawa, and K. Iwata: 'Resistance of pearlite against hydrogen-assisted fatigue crack growth', *International Journal of Hydrogen Energy*, 2022, **47**, 31703–31708.
149. R. O. Ritchie: 'Mechanisms of fatigue-crack propagation in ductile and brittle solids', *International Journal of Fracture*, 1999, **100**, 55–83.
150. Y. Mutoh, A. K. Korda, Y. Miyashita, and T. Sadasue: 'Stress shielding and fatigue crack growth resistance in ferritic–pearlitic steel', *Materials Science & Engineering A*, 2007, **468-470**, 114–119.
151. M. S. Mustapa, and Y. Mutoh: 'Effects of size and spacing of uniformly distributed pearlite particles on fatigue crack growth behavior of ferrite–pearlite steels', *Materials Science & Engineering A*, 2010, **527**, 2592–2597.
152. E. G. Mahin: 'Inclusions and ferrite crystallization in steels', *The Journal of Industrial and Engineering Chemistry*, 1919, **11**, 739–745.
153. H. Ito, Y. Suzuki, H. Nishikawa, M. Kinefuchi, M. Enoki, and K. Shibanuma: 'Multiscale model prediction of ferritic steel fatigue strength based on microstructural information, tensile properties, and loading conditions (no adjustable material constants)', *International Journal of Mechanical Sciences*, 2020, **170**, 105339.
154. K. Shibanuma, K. Ueda, H. Ito, Y. Nemoto, M. Kinefuchi, K. Suzuki, and M. Enoki: 'Model for predicting fatigue life and limit of steels based on micromechanics of small crack growth', *Materials and Design*, 2018, **139**, 269–282.
155. K. Tanaka, Y. Akiniwa, Y. Nakai, and R. P. Wei: 'Modelling of small fatigue crack growth interaction with grain boundary', *Engineering Fracture Mechanics*, 1986, **24**, 803–819.
156. F. M. Burdekin, and D. E. W. Stone: 'The crack opening displacement approach to fracture mechanics in yielding materials', *Journal of Strain Analysis*, 1966, **1**, 145–153.
157. J. C. Neuman Jr: 'A crack opening stress equation for fatigue crack growth', *International Journal of Fracture*, 1984, **24**, R131–R135.
158. B. A. Bilby, A. H. Cottrell, and K. H. Swinden: 'The spread of plastic yield from a notch', *Proceedings of the Royal Society of London A: Mathematical, Physical and Engineering Sciences*, 1963, **272**, 304–314.
159. J. Weertman: 'Explicit calculation of the crack tip stress intensity factor for the double slip plane model crack', *Mechanics of Materials*, 1983, **2**, 331–343.
160. W. Schaef, and M. Marx: 'A numerical description of short fatigue cracks interacting with grain boundaries', *Acta Materialia*, 2012, **60**, 2425–2436.
161. A. A. Korda, Y. Mutoh, Y. Miyashita, T. Sadasue, and S. L. Manan: 'In situ observation of fatigue crack retardation in banded ferrite-pearlite microstructure due to crack branching', *Scripta Materialia*, 2006, **8**, 1835–1840.
162. J. A. Ronevich, B. P. Somerday, and C. W. San Marchi: 'Effects of microstructure banding on hydrogen assisted fatigue crack growth in X65 pipeline steels', *International Journal of Fatigue*, 2016, **82**, 497–504.

6 Banded microstructures

6.1 INTRODUCTION

Liquid-crystalline materials in which a molecular axis can be defined, exhibit a one-dimensionally periodic microstructure when subjected to thin-film shear, fibre drawing, injection moulding or similar treatment [1]. In colloidal suspensions, shear-banding is defined as 'where macroscopically large regions (the 'bands') with differing microstructural properties, coexist under the influence of shear flow' [2]. The solid-state process of friction-stir welding deforms microstructures into onion-rings which are most visible when two different alloys become mixed in the process [3]. Precambrian terrains have banded rocks created by the shear and folding of iron oxide and quartz mixtures, with the band spacing about a millimetre [4]. So, what is to be learnt from this? An initial structure subjected to deformation without relaxation will exhibit banding as long as the starting material has some sort of anisotropy, whether that means a mixture of phases or cigar-shaped molecules.

The development of elongated structures is shown schematically in Figure 6.1, where an image of the cast-dendritic structure (revealed due to chemical segregation) is *digitally* compressed vertically. Although a two-dimensional representation of reality, this actually is how bands develop in practice, with solidification-induced chemical segregation being smeared along the main deformation direction.

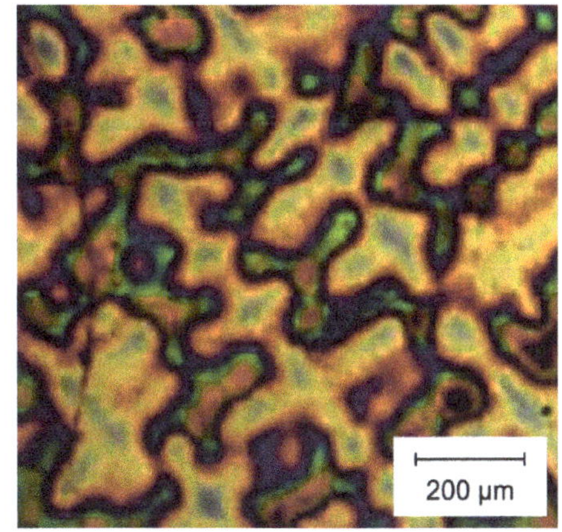

(a) As-cast

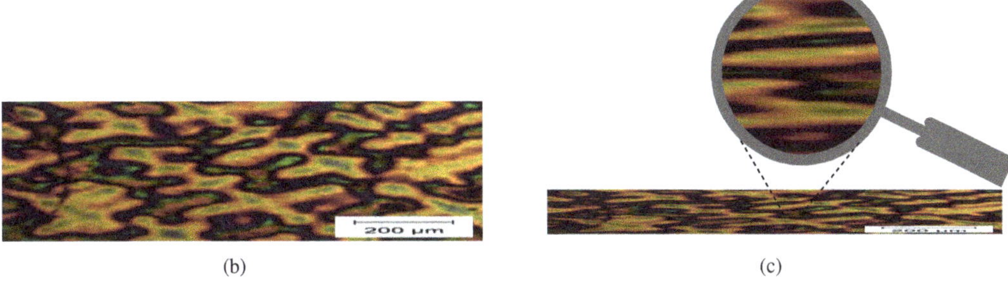

(b) (c)

Figure 6.1 (a) Dendrites in cast Fe-0.8C-2.25Si-1.0Mn-1.1Cr wt% steel, with colours chemical segregation. Reproduced under the CC BY 4.0 (https://creativecommons.org/licenses/by/4.0/) from [5]. (b) As (a) digitally-compressed to 30% of its original height. (c) As (a) but compressed to 10% of its original height.

DOI: 10.1201/9781032631981-6

Low-alloy steels that are cast and then hot-deformed into particular shapes, often exhibit elongated 'bands' in which the microstructure is organised, particularly when the final state is a mixture of proeutectoid-α and pearlite. The bands are elongated along the principal deformation direction, though they lack continuity when the fraction of pearlite is small, as seen on the image of the rolling plane in Figure 6.2a. The three-dimensional form of the banding depends on the strains imposed during hot-deformation. When cast-slabs are hot-rolled, the bands lie parallel to the rolling plane. Hot-rolled round-sectioned bars have elongated ferrite and pearlite in longitudinal sections, but the cross-sections present an apparently uniform distribution of α+P, Figure 6.2b.

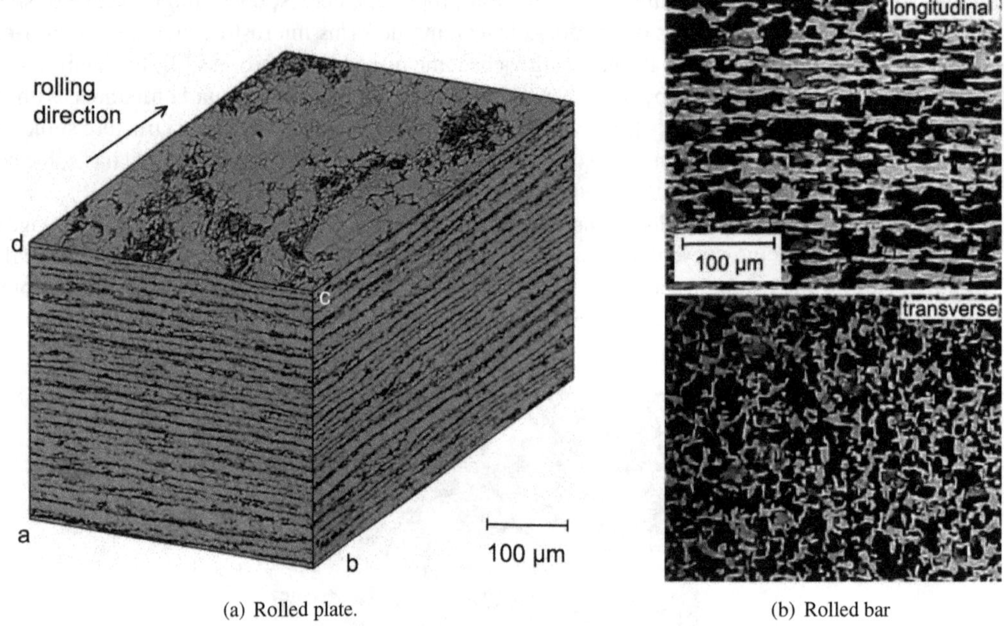

(a) Rolled plate. (b) Rolled bar

Figure 6.2 (a) Banding in hot-rolled ferrite-pearlite steel, Fe-0.15C-0.16Si-1.07Mn wt%. Notice that the patches of pearlite are not continuous when observed normal to the rolling plane Micrograph courtesy of Y. Mutoh [6]. (b) Banding in steel bar, Fe-0.41C-0.21Si-0.72Mn-1.86Ni-0.75Cr-0.21Mo wt%, showing the longitudinal and transverse cross-sections. Reproduced from Penha et al. [7] with permission of Elsevier.

Banding occurs because of the segregation of solutes during solidification at rates inconsistent with the achievement of equilibrium. Dendrite cores and interdendritic regions therefore solidify with different chemical compositions because the solid phases fail to homogenise during solidification.

The last liquid to solidify tends to be the most enriched or depleted in solute, depending on the partition coefficient between the participating phases. Subsequent plastic deformation spreads the segregated-domains into a form consistent with the applied strains. In low-alloy steels, solidification begins with δ-ferrite dendrites which grow while partitioning solutes such as manganese, silicon, phosphorus and sulphur into the interdendritic regions. The distribution of manganese is particularly important given its potent influence on subsequent solid-state transformation (Figure 6.3). Silicon may also segregate but its overall concentration is usually small and it has less of an effect on the transformation of austenite. The dependence of the spatial distribution of carbon on that of substitutional solute in austenite was thought originally to be the cause of banding [8]. Bastien, using metallographic, dilatometric and microanalysis techniques, concluded that microstructural banding can be attributed to the segregation of substitutional solutes during solidification [9], and Kirkaldy et al. [10] using diffusion couples later demonstrated that this is indeed the dominant effect.

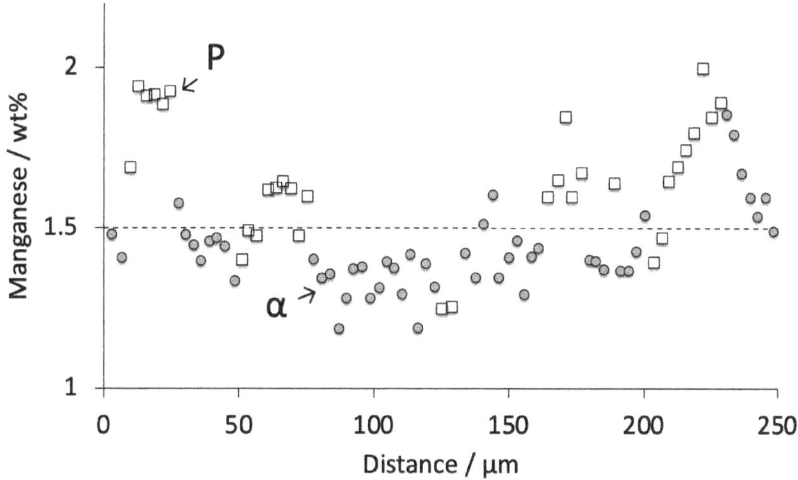

Figure 6.3 The location of pearlite relative to the manganese concentration. Data collected using electron probe microanalyser with a typical depth resolution of 1–2 μm, from Thompson et al. [11].

Any solidification-induced carbon segregation is likely to homogenise during cooling within the austenite phase field, or when the steel is reaustenitised prior to deformation. More precisely, the carbon will achieve a uniform activity in the austenite [10], which means that its *concentration* becomes modulated by any substitutional solute variations. The activity of carbon in dilute austenite is given approximately by [10, 12]

$$a_C^\gamma \approx x_C(1+7.25x_C) \times (1+11.8x_{Si}) \tag{6.1a}$$

$$a_C^\gamma \approx x_C(1+7.25x_C) \times (1-4.0x_{Mn}) \tag{6.1b}$$

$$a_C^\gamma \approx x_C(1+7.25x_C) \times (1+4.9x_{Ni}) \tag{6.1c}$$

$$a_C^\gamma \approx x_C(1+7.25x_C) \times (1-10.5x_{Cr}) \tag{6.1d}$$

with a_C^γ assumed to be independent of temperature. Silicon increases the activity of carbon in austenite so its concentration would be reduced in the silicon-rich regions; manganese reduces a_C^γ so the manganese-rich regions would contain a greater concentration of carbon, giving rise to its exaggerated influence on banding. This is illustrated using diffusion couples, Figure 6.4.

6.1.1 ALTERNATIVE MECHANISM FOR BANDING

An alternative mechanism is found in steels containing large sulphur concentrations [14]. Manganese sulphides then precipitate in the regions that during solidification are enriched in manganese. The region surrounding the sulphide therefore becomes depleted in manganese, thus stimulating the formation there of ferrite. The carbon partitioned during the growth of ferrite enriches the remaining austenite, which then transforms into pearlite during cooling. The position of the ferrite bands is thus shifted into locations where the average Mn concentration is large, but where the Mn is tied up in the form of sulphides (Figure 6.5).

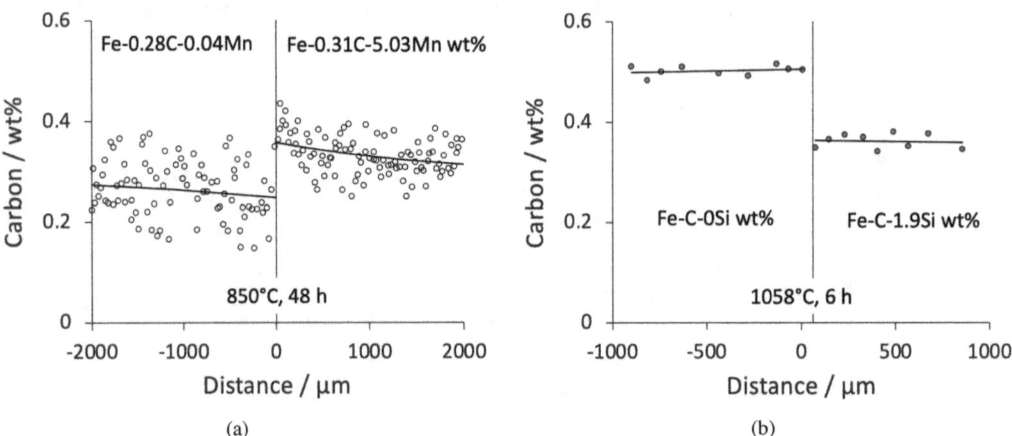

Figure 6.4 Diffusion couples at equilibrium with respect to carbon, but subject to an essentially frozen distribution of the substitutional solute. Selected data from Král et al. [13]. (a) Carbon enriched on the side containing more manganese. (b) Carbon depleted on the side containing more silicon. Selected data from Kirkaldy and Purdy [10].

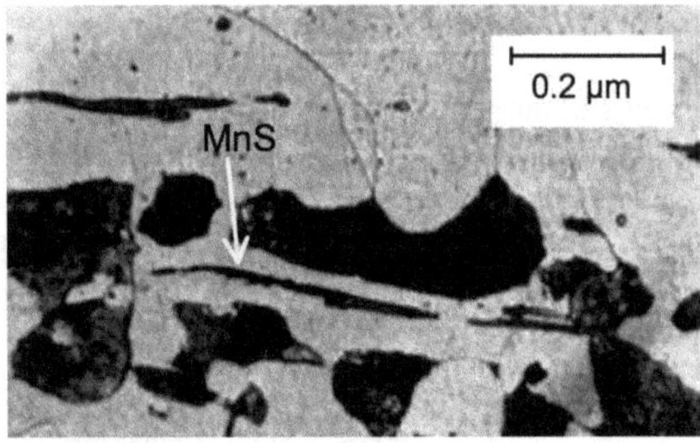

Figure 6.5 Thin, elongated MnS particles surrounded by ferrite within the banded microstructure in Fe-0.36C-0.22Si-0.73Mn-0.02P-0.025S wt%. The larger dark regions are pearlitic. Micrograph from Imai et al. [15], reproduced with the permission of Elsevier.

6.2 ARTIFICIAL BANDING

Banding has been studied using a composite made by roll-bonding high- and low-manganese variants of steel to produce a controlled 'artificially segregated steel' with a step change in the Mn concentration across the bond interfaces [16]. The experiment recognises that manganese is more often than not responsible for the development of proeutectoid-α/pearlite bands. Composite samples with different band-spacings were austenitised and quenched from a variety of temperatures during continuous cooling, in order to follow the progress of transformation. Only faint banding was observed prior to the precipitation of proeutectoid-α because the composite transformed uniformly into martensite, Figure 6.6a. Any contrast in that case could be attributed to more autotempering in the martensite that forms in the low-Mn, high M_S regions.

Proeutectoid-α, when it formed, was confined within the low-Mn regions, Figure 6.6b; the associated partitioning of carbon occurred into the surrounding austenite and the high-Mn region beyond the junctions. Pearlite was able, therefore, to form in the carbon-enriched austenite, particularly at the junctions and grow into the high-Mn regions. Figure 6.6c illustrates the partially-transformed high-Mn steel, the core of which is martensite that forms during quenching from 670 °C. When the temperature reached $\approx$ 600 °C, the entire high-Mn region became pearlitic, with the overall appearance reminiscent of the banding common in low-alloy steels, Figure 6.6d.

A unique observation from these experiments is that wide Mn-rich bands do not transform completely into pearlite because their central regions are not sufficiently enriched in carbon; they transform instead into bainite or martensite during cooling. Pearlite forms only in the vicinity of the carbon-rich bond-interfaces. Narrow high-Mn regions, on the other hand, are able to become fully pearlitic.

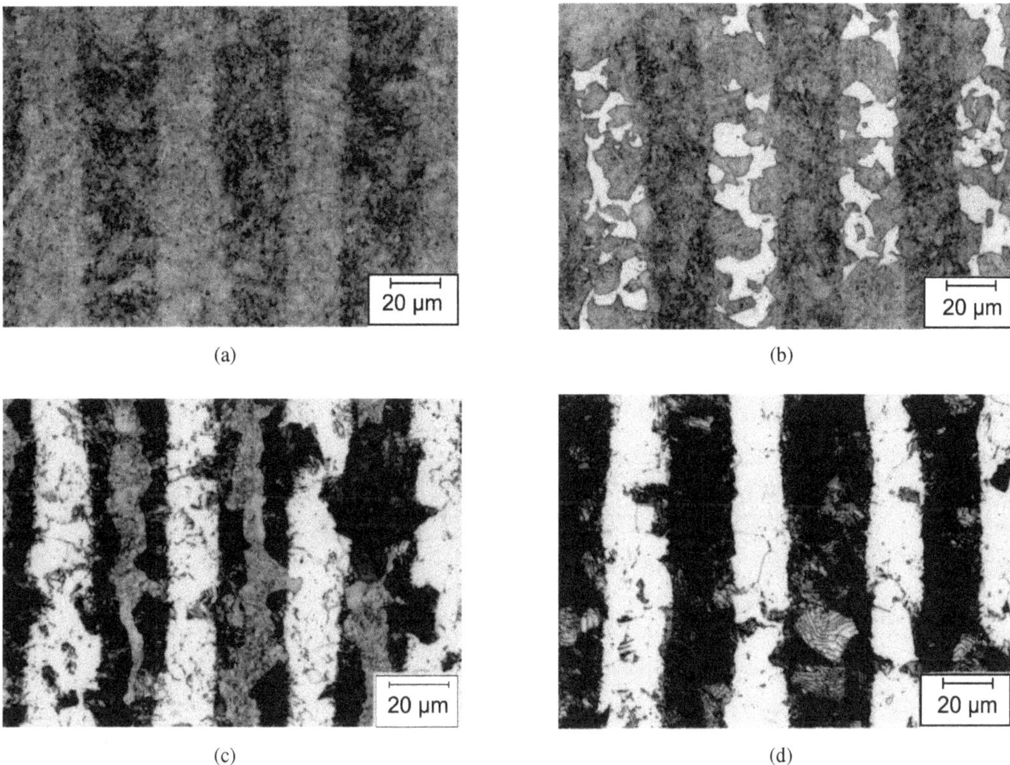

Figure 6.6 Evolution of structure in an artificially banded sample where Mn-rich (Fe-0.41C-1.83Mn-0.81Cr wt%) and Mn-poor (Fe-0.39C-0.82Mn-0.81Cr wt%) steels were bonded together. All the samples were austenitised and then cooled at 1 °C s^{-1} prior to quenching from (a) 750 °C, (b) 700 °C, (c) 670 °C and (d) 600 °C. Reprinted by permission from Springer Nature Customer Service Centre GmbH, Springer Nature, from Majka et al. [16], copyright 2002.

Microstructural banding arises because there are differences in the early stages of ferrite formation as a function of the local substitutional-solute content of the austenite. Nevertheless, aside from the initial transformation events, the overall transformation rates of banded and uniform low-alloy steels are found to be similar [17]. Presumably because the rapid transformation in the low-Mn regions is balanced by slower transformation in Mn-rich domains.

6.3 TENDENCY FOR SEGREGATION DURING SOLIDIFICATION

An indication of the extent to which a particular solute might segregate during non-equilibrium solidification might be gauged from its equilibrium solid-liquid partition coefficient at the point where there is but a small fraction of the liquid left. Although these are equilibrium calculations, they should reflect what happens when the solid and liquid phases cannot homogenise during cooling. In most low-alloy steels, it is the δ-ferrite that forms first from the liquid. In Table 6.1, the fraction of liquid, in a mixture of liquid and δ, is set to 0.1 to represent the enrichment associated with the last liquid to solidify. The average concentration of each solute is set at 1 wt% to be representative of low-alloy steels and all the data are for binary Fe-X mixtures so there may be different behaviours exhibited in multicomponent steels. Chromium has the least tendency to partition between the liquid and ferrite phases and phosphorus is at the other extreme. Nickel, manganese and silicon all partition prominently into the liquid. With nickel and manganese, α-ferrite is expected to form in the solute-depleted regions given that $w^{\alpha}_{\mathrm{Mn,Ni}} < w^{\gamma}_{\mathrm{Mn,Ni}}$ but the case for silicon is opposite, that α should precipitate preferentially in the silicon-rich regions since $w^{\alpha}_{\mathrm{Si}} > w^{\gamma}_{\mathrm{Si}}$. In a multicomponent steel where the manganese concentration tends to be greater than that of silicon, the ferrite would form first in the Mn-depleted regions.

Table 6.1

Thermodynamic equilibrium calculations on binary Fe-1X wt% alloys showing the partitioning of solute. The first two columns represent concentrations when liquid is in equilibrium with δ-ferrite, at a temperature where the weight fraction of liquid is just 0.1. Columns 4 and 5 represent the α/γ equilibrium concentrations when the fraction of α is 0.1. The Fe-1P wt% alloy does not exhibit austenite at low temperatures, rather, $\alpha+\gamma$ can coexist.

Solute **X**	w^{L}	w^{δ}	$w^{\mathrm{L}}/w^{\delta}$	w^{α}	w^{γ}	w^{α}/w^{γ}
Mn	1.29	0.97	**1.33**	0.58	1.05	**0.55**
Si	1.36	0.96	**1.42**	1.17	0.97	**1.21**
Ni	1.21	0.98	**1.24**	0.55	1.01	**0.54**
Cr	1.01	0.93	**1.09**	0.89	1.01	**0.88**
P	1.07	0.35	**3.06**			

6.4 COMPARISON WITH OTHER BANDED MICROSTRUCTURES

Pearlite contains a lot of interfaces between its $\alpha+\theta$ constituents – consequently, it etches readily, appearing dark in a reflection microscope due to the undulating etched-surface that scatters light away from the incident direction. Proeutectoid-α, on the other hand, is more uniform and hence appears white under the same conditions.

Chemically-heterogeneous steels transformed into the bainite also exhibit a high-contrast banding because the bainite has a substructure, and at some stage during its formation, partitions carbon into its surroundings (Figure 6.7a) [18]. When heterogeneous steel is quenched into martensite, only faint banding is observed because of the relative lack of structure within martensite plates (Figure 6.7b) [19]. A technique that enhances the contrast between martensite that forms first from that which follows during further cooling, involves quenching the sample to a temperature $M_{\mathrm{S}} > T > M_{\mathrm{F}}$ [20, 21]. The temperature is then raised to temper the martensite that has formed. Subsequent quenching results in a mixture of tempered (dark-etching) and untempered martensite. Jatczak [8] used this to distinguish the martensite that forms in solute-depleted regions in segregated steel from that which forms subsequently, thus making the banding clearly identifiable.

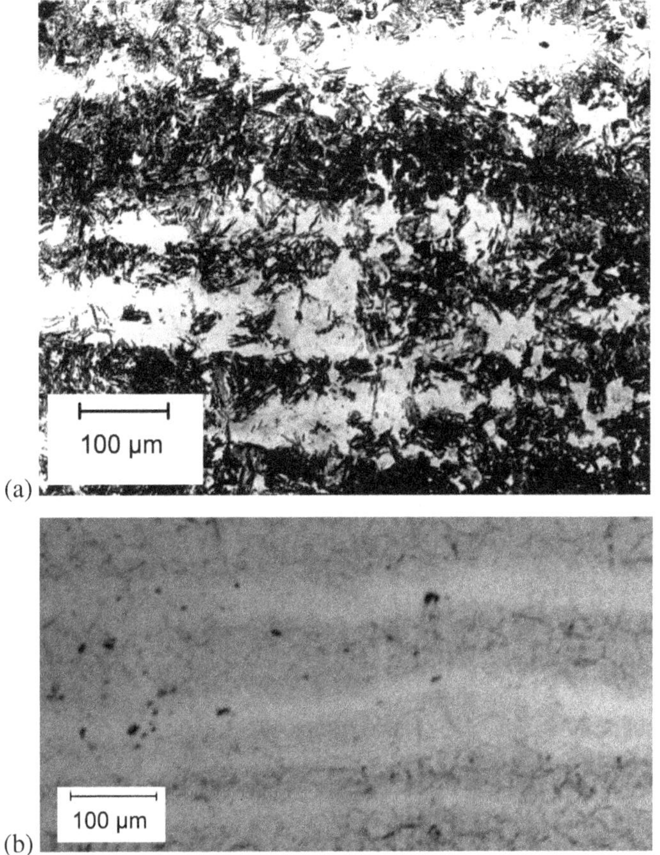

Figure 6.7 (a) Alternating bands of bainite and martensite in Fe-0.44C-1.74Si-0.67Mn-1.85Ni-0.83Cr-0.39Mo-0.09V wt% steel, transformed isothermally at 400 °C for 90 min. The light-etching regions represent untempered martensite. Micrograph courtesy of Shahid A. Khan. (b) Faint martensite banding in Fe-0.98C-1.38Cr-0.28Mn-0.06Mo-0.28Si-0.18Ni wt%, quenched from the austenitisation temperature. Micrograph courtesy of Wilberth Solano-Alvarez.

One important difference between the ferrite-pearlite banding and the bands in fully martensitic steels is that the carbon does not partition when the martensite forms in the solute-depleted regions. The mechanical properties are therefore much more uniform in fully martensitic steels containing chemical segregation. The avoidance of hard, carbon-rich regions can be used to commercial advantage as follows.

Control-rolled steels are cast continuously so they contain pronounced chemical segregation along the mid-thickness of the plate. For example, the manganese concentration at the centre can reach twice the average value. Ferrite naturally forms first in the manganese-depleted regions; the carbon partitioned as the ferrite grows ends up in the manganese-rich regions of austenite. This exaggerates the hardenability of the manganese-rich regions which transform into bands of hard microstructure.

6.5 MITIGATION OF MICROSTRUCTURAL BANDING

6.5.1 HOMOGENISATION

The mitigation of the chemical segregation that causes microstructural banding can be achieved by annealing at a temperature where the steel becomes austenitic, treated approximately as follows.

Consider a binary solution, the composition of which varies sinusoidally as a function of distance z [22]

$$c\{z,t=0\} = \overline{c} + c_{\mathrm{A}} \sin\left\{\frac{2\pi z}{\lambda_{\mathrm{s}}}\right\} \tag{6.2}$$

where c_{A} is the initial amplitude, λ_{s} is the wavelength and t is the annealing time. The decay of the wave as a function of time, temperature and interdiffusion coefficient D is then given by[1]

$$c\{z,t\} = \overline{c} + c_{\mathrm{A}} \sin\left\{\frac{2\pi z}{\lambda_{\mathrm{s}}}\right\} \exp\left\{-\frac{2\pi Dt}{\lambda_{\mathrm{s}}^2}\right\} \tag{6.3}$$

so the initial amplitude decays exponentially $2\pi Dt/\lambda_{\mathrm{s}}^2$ with the time required to reach $1/e$ of the initial amplitude given by

$$t_{\mathrm{e}} = \frac{\lambda_s^2}{4\pi^2 D}. \tag{6.4}$$

In a steel containing an average concentration of only 0.7 wt% manganese (with variations in the range $0.65 \rightarrow 0.9$ wt%, [23]), it was found necessary to anneal at 1250°C for 24 h to eliminate microstructural banding in all subsequent heat-treatments [24]. Assuming a typical wavelength of 200 μm and diffusion data from Fridberg et al. [25], $t_{\mathrm{e}} \approx 26$h, indicating that the amplitude of the segregation would need to be reduced to a third of that inherited during solidification in order to avoid a banded proeutectoid-α+P microstructure. If $\lambda_{\mathrm{s}} = 100$ μm then $t_{\mathrm{e}} = 6.5$ h. Deformation does reduce λ_{s} but in general, the homogenisation heat treatments at high temperatures and time periods are only practical for the slabs in the reheating furnace, where they are kept prior to hot-rolling. Modern steel production does not, however, include reheating – the red-hot solid goes into rolling mills directly from the caster, in order to save energy and facilitate production efficiency.

The concentration profile is unlikely to be sinusoidal, nor one-dimensional, but most actual profiles can be represented by a Fourier series of noninteracting sinusoidal components; Kirkaldy and Purdy point out that since the components with short wavelengths should decay relatively rapidly, the homogenisation time is in practice determined by the longest wavelength $\lambda_{\mathrm{s}}^{\max}$ in the spectrum [22]. Therefore, the substitution of $\lambda_{\mathrm{s}}^{\max}$ into Equation 6.4 can still give a reasonable approximation of the homogenisation time. Order of magnitude estimates are often made assuming a single sine modulation with λ_{s} related to some microstructural scale such as dendrite-arm spacing.

It is possible to extend the analysis for dealing with multicomponent diffusion [22] but this in itself is unlikely to have a significant impact on the homogenisation of low-alloy steels where manganese dominates the development of banding. Furthermore, the composition modulation in general will not be one-dimensional; as seen in Figure 6.2, the bands in reality are not in the form of continuous, alternating layers of ferrite and pearlite. Figure 6.1 illustrates the cored dendrites in a cast-steel structure, together with a simple simulation that vertical compression of that structure will lead to alignment along the horizontal axis but the microscopically segregated layers will be discontinuous. The segregated regions must remain as continuous in three dimensions as they are in the cast microstructure, assuming that the deformation is reasonably consistent with an affine transformation of shape. Notwithstanding these qualifications, the relationship between microstructural banding and the plastic work done on the cored dendrites is well established [26, 27].

An observation from Figure 6.1 is that plastic deformation reduces $\lambda_{\mathrm{s}}^{\max}$, precisely in accordance with the imposed strain.This accelerates homogenisation more than defect-enhanced diffusion. If the plastic strain is anisotropic then $\lambda_{\mathrm{s}}^{\max}$ will vary with orientation, effectively resulting in a change in the dimensionality of the problem (i.e., a one-dimensional diffusion treatment may suffice in the case of rolled samples), with homogenisation occurring via the shortest diffusion path with respect to the sample axes.

[1]This is a solution of Fick's second law, that $\partial c/\partial t = D\nabla^2 c$.

The phase field method applied in three dimensions is well-suited both for the estimation of the segregation pattern following casting and to model the chemical homogenisation process in multi-component alloys, accounting for interactions between the different solutes by using mobility and thermodynamic databases that are available commercially [28]. A simulation of proeutectoid-α+P banding based on idealised manganese segregation bands qualitatively reproduces known phenomena, with additional insights [23]:

- The ferrite is polycrystalline with equiaxed grains featuring different crystallographic orientations when the low-Mn regions are thick relative to the grain size, Figure 6.8a. In contrast, thinner bands tend to transform with just one grain spanning the thickness direction.
- The carbon partitioned by ferrite in Mn-depleted regions is not distributed uniformly in thick high-Mn regions. The centres of these thick bands therefore are able to transform into ferrite as the steel cools.
- The ferrite grains sometimes are elongated along the bands, Figure 6.8b. The mechanism is not understood, but it is possible that growth parallel to the rolling direction is faster than towards the manganese-rich regions that transform ultimately into pearlite. This is consistent with the elongation being more pronounced in high-Mn steels (Alloys [29, 30], Table 5.1, p. 116). Phase field simulations do not predict elongated-α.

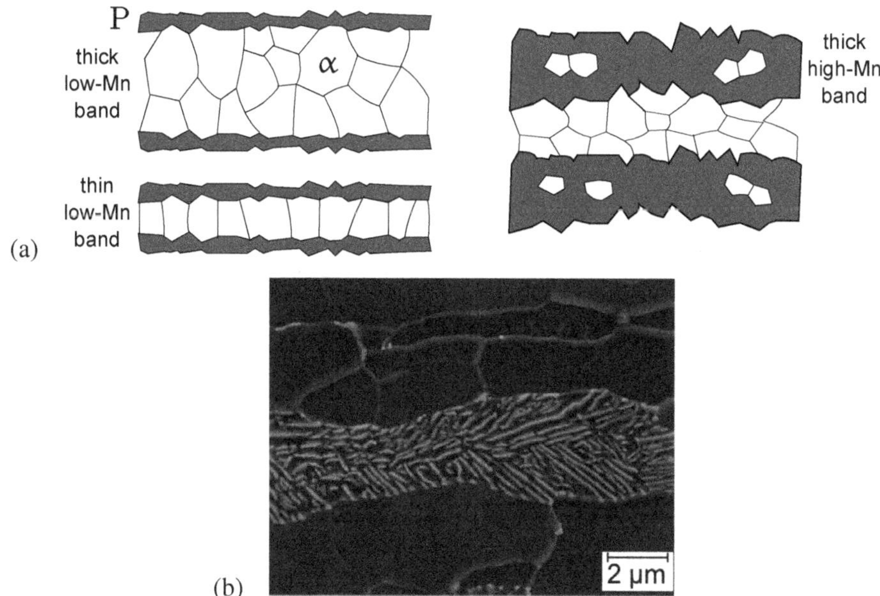

Figure 6.8 (a) The effect of the thickness of segregation bands on the development of microstructure, based on phase field simulations by Maalekian et al. [23]. (b) An illustration of ferrite grains elongated along the direction of the pearlite bands. Reproduced with the permission of Elsevier from Ronevich et al. [29].

6.5.2 ROLE OF AUSTENITE GRAIN SIZE

Austenite grain surfaces are the primary, heterogeneous-nucleation sites for ferrite and the component phases of pearlite. Microstructural banding vanishes when the austenite grain size becomes much larger than the wavelength of the segregation [11, 31]. The nucleation sites then do not correlate with individual solute-depleted regions. There is a decrease in the anisotropy of the microstructure as the austenitisation is conducted at ever greater temperatures [24], even though the time at

T_γ is not sufficient to induce significant homogenisation of substitutional solutes. However, small variations in the austenite grain size (e.g., $L'_\gamma \approx 17 \rightarrow 40\,\mu\text{m}$ [11]) have modest effects on banding.[2]

Another effect comes into play at certain combinations of austenite grain size and the cooling rate. Figure 6.9a shows a banded α+P microstructure for a case where the austenite grain size is small. At the same cooling rate, a large γ-grain size eliminates microstructural banding, Figure 6.9b. The size of the ferrite grains in both samples is identical; once allotriomorphs have decorated the austenite grain boundaries, their penetration into the parent grain becomes the one-dimensional propagation of a broad front, which reduces the growth rate. Therefore, with large L'_γ, there is much austenite left untransformed as the steel cools, triggering eventually the formation of Widmanstätten ferrite (α_W). The plates of α_W have a comparatively large lengthening rate; for the steel concerned, the lengthening rate is $10\,\mu\text{m}\,\text{s}^{-1}$ [32] so they rapidly partition the austenite grains across the segregation bands, thereby eliminating microstructural banding. A Widmanstätten structure is evident in other work where a large L_γ mitigates microstructural banding [Figure 32.6, 31].

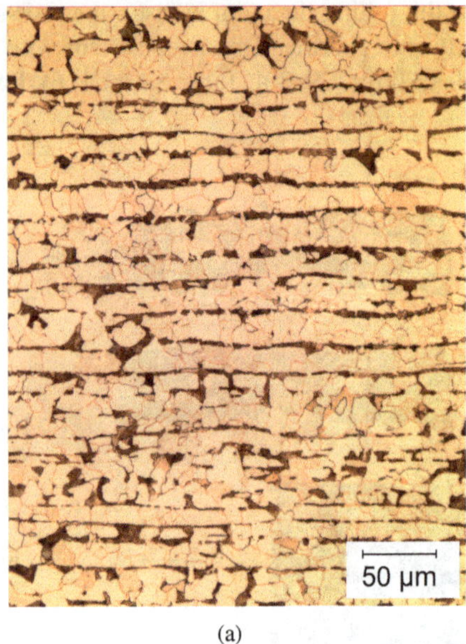

(a)

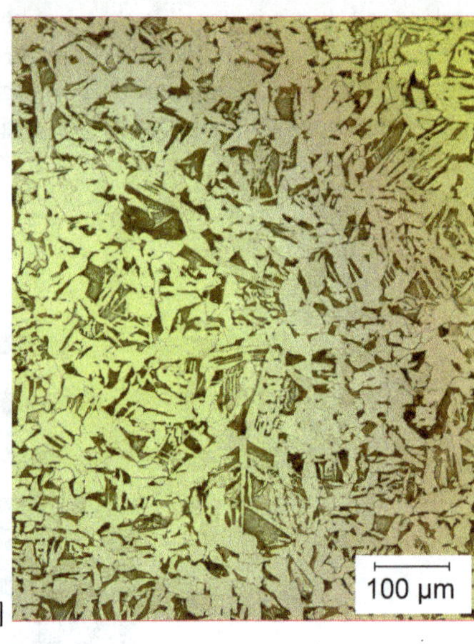

([]

Figure 6.9 Experiments on Fe-0.17C-0.74Mn wt% steel containing manganese concentrations variations between 0.65–0.9 wt%. (a) $L'_\gamma = 15\,\mu\text{m}$, $\dot{T} = 1\,\text{K}\,\text{s}^{-1}$. (b) $L'_\gamma = 100\,\mu\text{m}$, $\dot{T} = 1\,\text{K}\,\text{s}^{-1}$. Micrographs courtesy of Mehran Maalekian [23].

This discussion assumes a uniform austenite grain size. When the size distribution is bimodal and at slow cooling rates, the larger grains lag behind the finer ones in terms of the extent of transformation. They therefore accumulate carbon from the surrounding faster-transforming grains and then decompose into large nodules of pearlite which might span several segregation bands [11].

6.5.3 ACCELERATED COOLING

Banding correlates with the chemical-segregation pattern, particularly when the structure is generated during slow cooling from the parent phase. Rapid cooling suppresses transformation to lower

[2] Austenite grain size data are often reported as 'grain diameters' or other empirical measures. To be useful in stereology, they should be expressed using a mean lineal intercept. The symbol L'_γ is used to indicate that the quantities stated are not lineal intercepts.

temperatures where $|\Delta G^{\gamma\alpha}|$ is large, permitting ferrite to form in all locations, thereby reducing or suppressing microstructural banding.

The variation in $|\Delta G^{\gamma\alpha}|$ as a function of position can be represented by ΔT which is the difference in the Ae_3 temperatures of the solute-enriched and depleted regions. By equating the band spacing to the carbon diffusion distance during the time spent during cooling over the range ΔT, Kirkaldy et al. [10] suggest that the minimum cooling rate ($\dot{T}$) needed to avoid microstructural banding is given by:[3]

$$\dot{T} = \frac{5D_C\Delta T}{\lambda^2} \tag{6.5}$$

where λ is the wavelength of the chemical segregation and D_C is the diffusivity of carbon in the austenite. The factor of five in this equation is an empirical constant. It remains a problem that the cooling rate required cannot be imposed on thick sections of steel; in the case of pipes for fossil-fuel transmission, thick-sectioned pipes (> 20 mm) typically exhibit banding.

The cooling rate can be large enough to entirely eliminate ferrite and pearlite, giving rise instead to bainite but this does not resolve the banding problem (p. 148). Fully martensitic steel exhibits little or no obvious banding.

6.6 QUANTITATIVE CHARACTERISATION OF BANDING

There are different ways in which banding has been characterised historically, either for the correlation of data against steel processing or to enable the estimation of properties. In what follows, it is assumed that the material is *orthotropic*, i.e., the properties vary along three orthogonal directions. However, parameters such as volume fraction of phases and crystallographic texture are independent of the plane of observation, a principle sometimes neglected [33].

6.6.1 LINEAL INTERCEPTS

The anisotropy of the banded microstructure in a rolled plate is sometimes based on the measurement of the ratio of the mean lineal intercepts in directions parallel ($\overline{L}_{||}$) and perpendicular ($\overline{L}_{\perp}$) to the rolling surface, with measurements made on the section transverse to the rolling direction; in Figure 6.2a, this section corresponds to the plane 'abcd' so the measurements of the two intercepts are along ab and ad, respectively [24]. The individual ferrite or pearlite regions are for this purpose assumed to be homogeneous, so for example, α/α grain boundaries are neglected in determining the intercepts. There are two sets of measurements conducted independently for the ferrite and pearlite bands, respectively, Figure 6.10. The somewhat greater anisotropy of the pearlite bands in Figure 6.10 is because of the smaller volume fraction of pearlite in the hypoeutectoid steel, which leads to thinner layers than the intervening ferrite.

The ratio $\overline{L}_{||}/\overline{L}_{\perp}$ helps monitor any changes to the character of banding when, for example, T_γ is changed, or if the nature or magnitude of the rolling deformation is changed. However, neither the lineal-intercept orientations nor the microstructures are random, so they cannot, for example, be used to estimate the boundary area between the pearlite and proeutectoid-α bands. It is not obvious that $(\overline{L}_{||}/\overline{L}_{\perp})_{\mathrm{P}}$ and $(\overline{L}_{||}/\overline{L}_{\perp})_{\alpha}$ independent quantities. In an ideal arrangement of alternating plates of ferrite and pearlite, where the total thickness of the stack of an even number N of plates is z,

$$\overline{L}_{||}^{\alpha} = \overline{L}_{||}^{\mathrm{P}} \quad \text{and} \quad N\overline{L}_{\perp}^{\alpha} = V_{\mathrm{V}}^{\alpha} \times z \quad \text{and} \quad N\overline{L}_{\perp}^{\mathrm{P}} = (1 - V_{\mathrm{V}}^{\alpha}) \times z. \tag{6.6a}$$

It follows that

$$\overline{L}_{\perp}^{\alpha} = \overline{L}_{\perp}^{\mathrm{P}} \times V_{\mathrm{V}}^{\alpha}/(1 - V_{\mathrm{V}}^{\alpha}) \tag{6.6b}$$

[3]If the diffusion distance $\lambda \propto \sqrt{Dt}$, writing $t = \Delta T/\dot{T}$ gives $\dot{T} \propto D\Delta T/\lambda^2$

so for a hypoeutectoid steel

$$\overline{L}_{\perp}^{\alpha} > \overline{L}_{\perp}^{\mathrm{P}} \qquad \text{and} \qquad (\overline{L}_{\|}/\overline{L}_{\perp})_{\alpha} < (\overline{L}_{\|}/\overline{L}_{\perp})_{\mathrm{P}}. \tag{6.6c}$$

This is consistent with the data in Figure 6.10 though the differences between the two ratios are smaller than might be expected based on the estimated value of V_{V}^{α}. And of course, the extent of banding does depend on the austenite grain size. The analysis also shows clearly that the bands of ferrite and pearlite are not to be regarded as continuous (like the ideal stack of plates described). Indeed, the difference between $(\overline{L}_{\|}/\overline{L}_{\perp})_{\alpha}$ and $(\overline{L}_{\|}/\overline{L}_{\perp})_{\mathrm{P}}$ when compared against the ideal scenario defined by Equation 6.6, could represent a measure of the continuity of the banding.

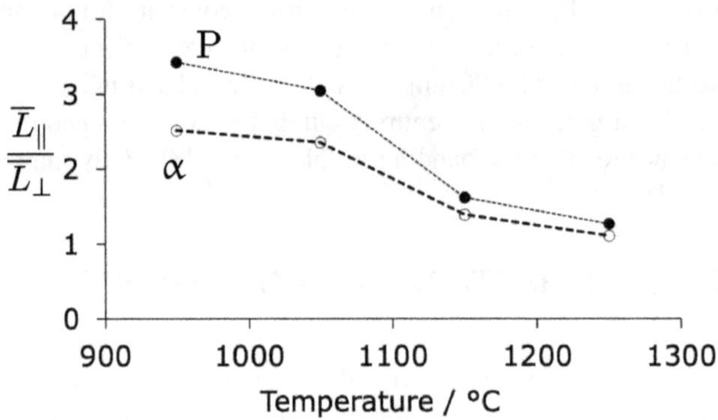

Figure 6.10 Effect of austenitisation temperature (1 h) on the extent of banding in Fe-0.16C-0.69Mn-0.022Si-0.028S wt% steel. The samples were furnace cooled (1.5°C min^{-1}) from the austenitisation temperature. Data for 850°C have been omitted since these were considered unreliable by the original authors [24].

6.6.2 POINTS PER UNIT LENGTH

The general anisotropy and degree of orientation of the microstructure can be measured by counting the number of intersections per unit length made by test lines with proeutectoid-α/pearlite band-interfaces [34, 35],

$$\text{Anisotropy index} = \frac{\overline{P}_{\mathrm{L}_{\perp}}}{\overline{P}_{\mathrm{L}_{\|}}} \tag{6.7}$$

with the subscripts $\perp$ and $\|$ indicate the direction of the test array relative to the orientation axis (in this case, the rolling direction). The anisotropy index and the degree of orientation become 1 and 0 respectively, when the microstructure is random. Saltykov considered a partially oriented structure to consist of superimposed components of oriented and *isometric* structures – if all diameters of a circle superimposed on to a micrograph have the same number of intersections with microstructural features then the structure is said to be isometric. The extent of preferred orientation in a partially-oriented structure was defined as [pp. 219-226, 36]

$$\begin{aligned}\text{Degree of orientation} &= \frac{\Sigma P_{\mathrm{L,oriented}}}{\Sigma P_{\mathrm{L,oriented}} + \Sigma P_{\mathrm{L,isometric}}} \\ &= \frac{\overline{P}_{\mathrm{L}_{\perp}} - \overline{P}_{\mathrm{L}_{\|}}}{\overline{P}_{\mathrm{L}_{\perp}} + (\frac{\pi}{2} - 1)\overline{P}_{\mathrm{L}_{\|}}}\end{aligned} \tag{6.8}$$

where P_{L} is the number of intersections per unit length of test line with the structure; $\overline{P}_{\mathrm{L}_{\perp}}$ is the mean number of intersections per unit length of test lines normal to the bands; and $\overline{P}_{\mathrm{L}_{\|}}$ the corresponding number for test lines parallel to the bands. In deriving this equation, it is assumed that the test array laid parallel to the orientation axis will not intersect oriented line segments in which case the length of the isometric segments per unit area is $\frac{\pi}{2}\overline{P}_{L_{||}}$; the $P_{L_\perp}$ array will intersect both the isometric and oriented line segments, so the intersections per unit length due only to the oriented segments is the difference $P_{L_\perp} - P_{L_{||}}$.

The measure of orientation given by Equation 6.8 changes from 0 for a homogeneous microstructure ($\overline{P}_{L_{||}} = \overline{P}_{L_\perp}$) to 1 for a completely oriented one because $\overline{P}_{L_{||}}$ then becomes zero. Example measurements which illustrate how these parameters change with the plane of observation are given in Table 6.2; they confirm that the anisotropy is most evident in rolled samples when the plane of observation contains the rolling and short-transverse directions (Figure 6.2a).

Table 6.2
Quantitative data on the anisotropy of microstructural banding [37]. 'R, S and T' are the rolling, short transverse and transverse directions.

Plane	$\overline{P}_{L_\perp}/\text{mm}^{-1}$	$\overline{P}_{L_{||}}/\text{mm}^{-1}$	Anisotropy index	Degree of orientation
RT	4.54	1.23	3.69	0.63
TS	40.24	3.85	10.45	0.86
RS	38.48	1.32	29.15	0.95

6.6.3 MICROSTRUCTURAL ANISOTROPY TENSOR

If, in a partially-oriented structure, the mean intercept length is plotted as a radius from the origin, at the angle of measurement, the terminal points all the radii generate the surface of an ellipsoid in three dimensions relative to the sample axes [38]. If the orientation of the measurement line is represented by its direction cosines n_i along the axes of the ellipsoid, then the mean lineal intercept along any direction is given by

$$\frac{1}{(\overline{L})^2} = M_{ij} n_i n_j \tag{6.9}$$

where M_{ij} are the components of a symmetrical 2nd rank tensor that expresses the anisotropy of the structure. The principal axes of the tensor represent the directions of major material orientation and the ratios of their eigenvalues the extent of anisotropy. Figure 6.11 illustrates, in a biological context, how the tensor can be used to visualise the anisotropy.

The anisotropy tensor can be used to calculate property variations [38], for example how the modulus might vary with direction within an aligned structure. The experimental determination of the tensor obviously requires more work than the two-dimensional measures described previously. Measurements, for example of P_L, are made on three orthogonal but otherwise arbitrary faces of the specimen. Errors in the measurements lead to corresponding incompatibilities in M_{ij}, for example,

$$\left(\frac{M_{11}}{M_{22}}\right)_{1,2} \times \left(\frac{M_{22}}{M_{33}}\right)_{2,3} \times \left(\frac{M_{33}}{M_{11}}\right)_{3,1} \neq 1 \tag{6.10}$$

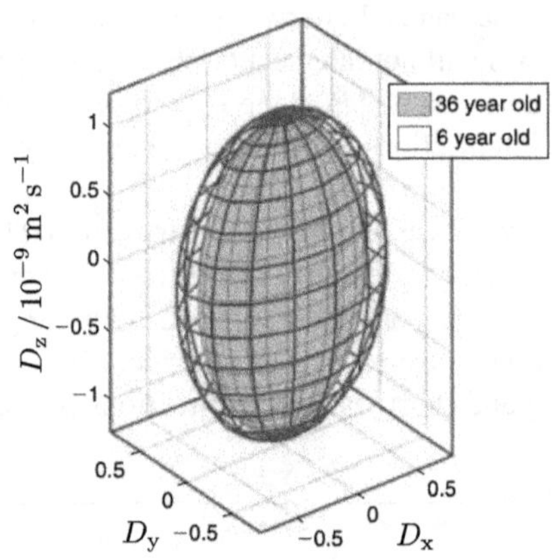

Figure 6.11 An ellipsoid representing the anisotropy of water diffusion in a neural tract. There are two ellipsoids plotted with respect to the principal axes x, y and z as a function of age. Reproduced from Beaulieu [39] with permission of Elsevier.

where the subscripts refer to the face on which the measurement is conducted. To correct for this, the axes z_1 and z_2 defining plane 1,2 are distorted $x' = b_1 x_1$, $x_2' = b_2 x_2$ so that $M_{11}' = b_1^2 M_{11}$, $M_{22}' = b_2^2 M_{22}$ and $M_{12}' = b_1 b_2 M_{12}$, where b_1 are the empirical distortions with similar considerations applying to the data from the other two faces, resulting in six b_i coefficients. To ensure an equality in Equation 6.10,

$$\frac{b_1^2 b_3^2 b_6^2}{b_2^2 b_4^2 b_5^2} = \left(\frac{M_{11}}{M_{22}}\right)_{1,2} \times \left(\frac{M_{22}}{M_{33}}\right)_{2,3} \times \left(\frac{M_{33}}{M_{11}}\right)_{3,1}.$$

These distortions are of the order of a few percent. To minimise the distortions on each face, Harrigan and Mann impose the condition

$$\frac{b_1}{b_2} = \frac{b_3}{b_4} = \frac{b_6}{b_5} = \left[\left(\frac{M_{11}}{M_{22}}\right)_{1,2} \times \left(\frac{M_{22}}{M_{33}}\right)_{2,3} \times \left(\frac{M_{33}}{M_{11}}\right)_{3,1}\right]^{\frac{1}{6}}$$

Using Figure 6.2a, with the transverse, rolling and normal directions labelled [100], [010] and [001] respectively, P_L values can be determined along all the edges and face diagonals. $P_L\{uvw\}$ represents measurements along the direction $[uvw]$, then by placing test lines parallel to the axes, $P_L^2\{100\} = M_{11}$, $P_L^2\{010\} = M_{22}$ and $P_L^2\{001\} = M_{33}$. Measurements along the diagonal to each of the faces gives, for example,

$$P_L^2\left\{\frac{1}{\sqrt{2}}\frac{1}{\sqrt{2}}0\right\} = \frac{M_{11}}{2} + \frac{M_{12}}{2} + \frac{M_{21}}{2} + \frac{M_{22}}{2}$$

which yields the value of the off-diagonal term $M_{12} = M_{21}$. The result of such measurements in Figure 6.2, at a magnification such that 1 cm on the micrograph corresponds to 22 μm in the specimen, are presented below:[4]

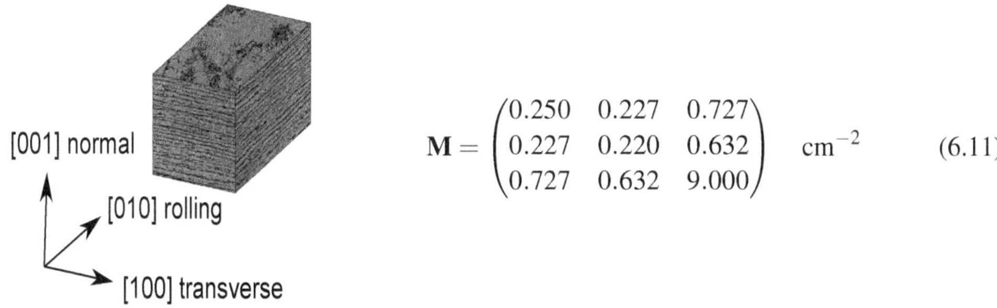

$$\mathbf{M} = \begin{pmatrix} 0.250 & 0.227 & 0.727 \\ 0.227 & 0.220 & 0.632 \\ 0.727 & 0.632 & 9.000 \end{pmatrix} \quad \text{cm}^{-2} \tag{6.11}$$

which has the eigenvectors ($\hat{\mathbf{v}}_i$, unit vectors with length in cm) and eigenvalues as follows:

$$\begin{aligned} \hat{\mathbf{v}}_1 &= [0.0834\ \ 0.0727\ \ 0.9939] & \lambda_1 &= 9.1073 = P_{\rm L}^2\{\mathbf{v}_1\} \\ \hat{\mathbf{v}}_2 &= [\overline{0.7178}\ \ \overline{0.6874}\ \ 0.1106] & \lambda_2 &= 0.3553 = P_{\rm L}^2\{\mathbf{v}_2\} \\ \hat{\mathbf{v}}_3 &= [\overline{0.6912}\ \ 0.7226\ \ 0.0051] & \lambda_3 &= 0.0073 = P_{\rm L}^2\{\mathbf{v}_3\}. \end{aligned} \tag{6.12}$$

Large eigenvalues indicate a small distance between proeutectoid-α/P interfaces and the eigenvectors are directions along which the material exhibits the extremities of $P_{\rm L}$. It is interesting that $\hat{\mathbf{v}}_1$ is almost parallel to [001] which is the direction normal to the banding; as expected, it is associated with the largest eigenvalue indicating closely spaced intercepts; since $L = P_{\rm L}^{-1}$, it follows that $L\{\mathbf{v}_1\} = (9.1073)^{-\frac{1}{2}}\,\text{cm} \equiv 7.3\,\mu\text{m}$, i.e., the spacing of the bands in a direction normal to the rolling plane. Within the rolling plane, $\hat{\mathbf{v}}_2$ is almost parallel to the face-diagonal $[\bar{1}10]$ with the largest distance between intercepts.

Suppose that the measurements had been conducted on a different set of mutually perpendicular faces, none of which are parallel to the rolling plane and may not contain the rolling direction. The tensor **M** would nevertheless reveal the principal directions of the microstructure.

It is a standard result from stereology that the amount of surface per unit volume $S_{\rm V} = 2\overline{P}_{\rm L}$ where the mean of intercepts per unit length is determined from random test lines imposed onto the microstructure [40]. Since along any direction represented by a unit vector $\hat{\mathbf{n}}$, $P_{\rm L}^2\{\hat{\mathbf{n}}\} = M_{ij}n_in_j$, the mean of all intercepts is obtained by integrating over all possible orientations [41],

$$\begin{aligned} \overline{P}_{\rm L}^2 &= \frac{1}{4\pi}\int_A M_{ij}n_in_j\ \text{d}A \\ &= \frac{M_{ij}}{4\pi}\delta_{ij}\frac{4\pi}{3} = \frac{M_{ii}}{3}. \end{aligned} \tag{6.13}$$

It follows that for an isotropic structure, $\overline{P}_{\rm L} = \sqrt{M_{ii}/3} = S_{\rm V}/2$. Using Equation 6.11 and allowing for the magnification, $S_{\rm V}^{\alpha\rm P} \approx 632\,\text{m}^{-1}$. This represents the amount of interface between the ferrite and pearlite bands at the resolution of the measurements; as such, it is quite a small, perhaps two orders of magnitude smaller than if the boundaries within a typical ferritic grain structure are characterised. The interfaces measured here are between macroscopic regions of the two microstructural constituents.

It is important to measure parameters such as $S_{\rm V}$ for rough interfaces such as those between the proeutectoid-α/P bands at a resolution appropriate for the problem. For hydrogen trapping at interfaces, it would be necessary to measure at a much greater resolution than for fracture mechanics

[4]The data for **M** are presented here without correcting using the b_i parameters, nor has any account been taken of experimental error in this illustrative example.

where the plastic zone at the crack tip is likely to be much greater than a grain size. This is because S_V for rough objects is really a function of the resolution of measurement.

The microstructural anisotropy tensor is a good way to characterise banding and does not involve much more work than the empirical indices sometimes used to describe anisotropy or the degree of orientation.

6.6.4 COVARIANCE FUNCTION

Suppose that a phase is identified as a bounded region of space, labelled X. A covariance function $C\{X,\mathbf{h}\}$ represents the probability that both extremities of the vector $\mathbf{h}$ belong to X, whatever the position of $\mathbf{h}$ [42]. If $|\mathbf{h}| = 0$ the probability becomes the volume fraction V_V of X.

If $|\mathbf{h}|$ is very large, it may be assumed that the chance that each extremity lying in X is independent and equal to V_V, so the chance that both extremities lie in X is $C\{X,\mathbf{h}\} = V_V \times V_V = V_V^2$.

Therefore, the covariance function exhibits a monotonic decrease from V_V to an asymptotic value V_V^2 as $|\mathbf{h}|$ is increased. For an anisotropic structure, the direction of $\mathbf{h}$ matters so the curve develops a maximum and a minimum (Figure 6.12); in a banded microstructure, $|\mathbf{h}|_{max}$ corresponds to the band spacing and ΔC to the extent of banding when compared against a homogeneous two-phase microstructure.

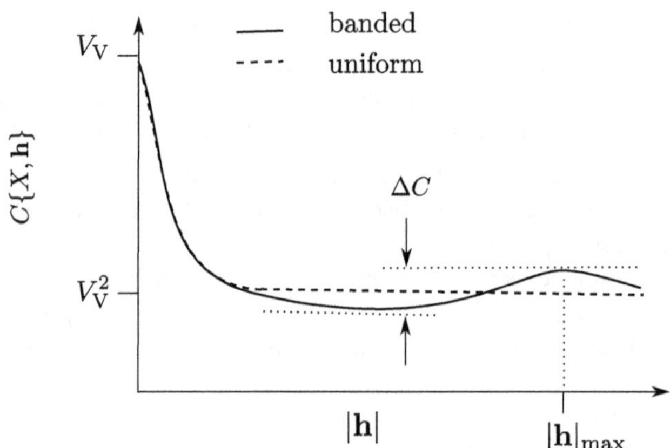

Figure 6.12 Covariance as a function of the magnitude of $\mathbf{h}$. When microstructure is isotropic, $|\mathbf{h}| = 0$, $C\{X,0\} = V_V$ and as $|\mathbf{h}|$ increases the function tends monotonically towards V_V^2 (dashed curve). An anisotropic structure results in a minimum and maximum as illustrated by the continuous curve.

The method is suitable for automatic implementation using image analysis. It is argued that because it is based on global measurements, it avoids border effects, such as those in which linear intercepts are measured [42]. However, deciding where the extremities of $\mathbf{h}$ lie has a similar resolution problem.

6.6.5 CONTINUITY AND CONTIGUITY

If bands of pearlite are perfectly straight along the rolling direction then comparing $L_{||}$ with the sample size would immediately give an idea of continuity along the band direction. However, they are a consequence of the deformation of cored dendrites so definitely cannot be in the form of continuous slabs (Section 6.5.1).

The term *continuity* of a particular component of a microstructure, such as pearlite in α+P mixture, describes whether it is possible to percolate through the sample with the path confined to that component. Note that the term 'component' is used instead of a 'phase' to acknowledge that pearlite is not a phase but a mixture of two phases, but for some purposes it is appropriate to treat pearlite as a whole.

There are degrees of continuity; if all the grains of the component are connected then there is complete continuity but partial continuity implies that although a significant number of grains may be connected, there will be smaller sets of continuous entities for the component concerned. If N represents the number of separate parts of the component in question, $N = 1$ for complete continuity; $1 < N < N_g$ for partial continuity where N_g is the number of grains of the component.

On the other hand, *contiguity* describes the extent of contact between adjacent particles of the same component [43]. Contiguity can be an important parameter when, for example, studying the permeation of hydrogen through a banded microstructure. The parameter is the average fraction of the interfacial area shared by a grain of a particular component with all the neighbouring grains of that component. In a two-component, *isotropic* mixture of ferrite and pearlite, the contiguity C_g would therefore be [43, 44]

$$C_g^\alpha = \frac{S_V^{\alpha\alpha}}{S_V^{\alpha\alpha} + S_V^{\alpha P}} \equiv \frac{2P_L^{\alpha\alpha}}{2P_L^{\alpha\alpha} + P_L^{\alpha P}} \qquad \text{with} \qquad C_g^P = 1 - C_g^\alpha. \tag{6.14}$$

In a perfectly banded structure, both C_g^α and C_g^P can be close to unity because the only α/P interfaces would be at the band junctions.

6.7 CRYSTALLOGRAPHIC TEXTURE

The focus here is on α+P banded-steel that has not been cold-deformed.

The ferrite grains within banded steels are not randomly oriented, presumably inheriting some of the crystallographic texture from the parent austenite because of α/γ orientation relationships. This inheritance is not rigorous because of the reconstructive nature of the transformation and the fact that the ferrite grains are not restricted to grow in the γ-grain that determines the orientation relationship [45–48]. It is worth emphasising that banded steels have served and continue to serve reliably without the implementation of texture control. There does not seem to be a case for engineering the crystallographic texture of the proeutectoid-α, and the methods proposed for such manipulation may not be suitable for mass production.

The way in which thermomechanical processing is applied to produce α+P can influence the texture of the ferrite, Figure 6.13. The textural changes alter the proportion of high-energy grain boundaries;[5] a reduction in those boundaries can improve the resistance to hydrogen embrittlement [50], but the processing is complicated and the short heat-treatment required is not practical for the sizes dealt with in a production environment.

6.8 TENSILE & CHARPY PROPERTIES OF BANDED MICROSTRUCTURES

The strength and ductility of a banded microstructure is identical to that of a corresponding uniform microstructure; tensile tests are insensitive to their orientation relative to the banding. Figure 6.14 shows a low-sulphur steel [51] where the known effect [17] of elongated sulphides on mechanical anisotropy is minimised. The insensitivity to banding is because pearlite forms the minority phase; during tensile testing, the whole of the sample must deform with macroscopic yielding defined by the long-range propagation of yield over distances that are greater than the heterogeneity of structure.

The pearlite bands are not two-dimensional slabs but discontinuous layer-like structures generated by the deformation of three-dimensional interdendritic segregation. They contain variations

[5] That high-energy interfaces can be assumed to be the same as high-misorientation boundaries is not strictly justified given coincidence-site and O- lattices, [e.g., pp.219-227, 49].

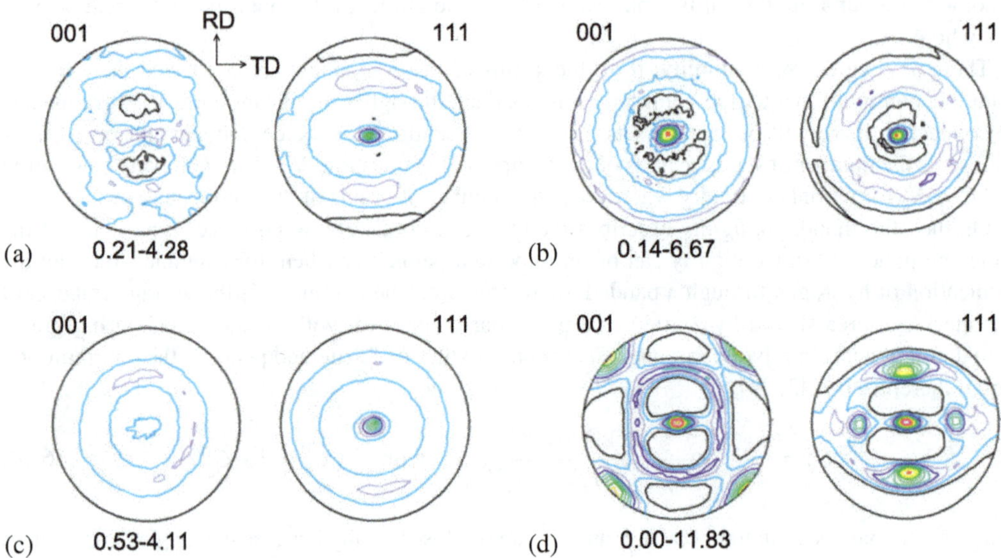

Figure 6.13 Crystallographic texture changes in the proeutectoid-α in a banded α+P steel of composition Fe-0.21C-1.33Mn-0.04Si wt%.The intensity range is indicated below the $\{001\}_\alpha$ pole figures. All samples were austenitised at 1040,°C, thermomechanically processed (details in sub-captions) and then annealed at 850 °C for 2 min. (a) 75% total reduction, in 2 steps, each 50% reduction, finish rolling at 800 °C. (b) As (a) but finish rolling at 600 °C. (c) As (a) but finish rolling at 1040 °C. (d) 50% total reduction, in 15 steps, each 5% reduction, finish rolling at 1000 °C. Images from Vengas et al. [50], reproduced with permission of Elsevier.

in thickness and large discontinuities. The partitioning of strain between the proeutectoid-α and pearlite depends not just on their independent mechanical properties, but also on the band-interface roughness and local thickness. Figure 6.15 shows that the local plastic strain within the harder pearlite can exceed that in ferrite if the band is locally narrow [52]. Ribbons of shear that initiate in the ferrite during the early stages of deformation, focus the strain at the junction with the pearlite so a thin band cannot resist the propagation of the shear. Thicker bands of pearlite can resist such shear more effectively, resulting in much-reduced strain relative to the average value. Such large strain incompatibilities lead eventually to void formation at the pearlite/proeutectoid-α boundaries, a precursor for macroscopic fracture [53].

Attempts have been made, using small test-samples to compare deformation in directions normal and parallel to the bands [54]. The stress-strain (force-displacement) behaviours were found to be almost identical for both orientations, consistent with the insensitivity of the tensile properties described by Spitzig, Figure 6.14. Large deformations such as those associated with tensile tests cannot be achieved without the overall strain being accommodated by both parts of the microstructure.

Spitzig also concludes, based on a study of three steels with different sulphur concentrations, that banding *per se* does not significantly affect the upper shelf energy in a plot of Charpy toughness versus temperature [51], possibly because the plastic zone at the notch tip, in those circumstances, exceeds the length scale of the banding. The main culprit in causing toughness and crack-initiation to be orientation sensitive is undoubtedly sulphide inclusions, whose role becomes pernicious when the concentration of sulphur exceeds about 0.015 wt% [15].

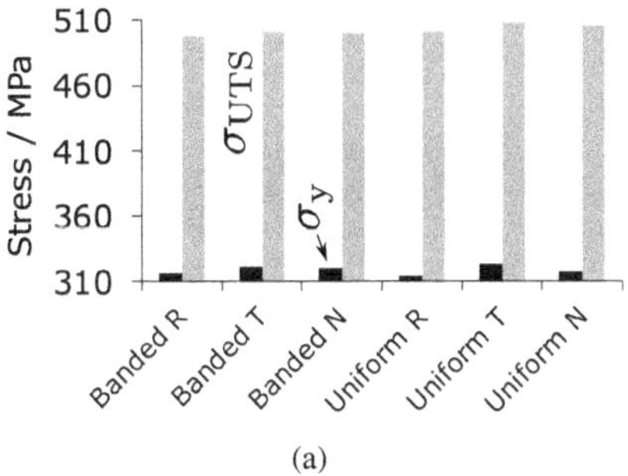

(a)

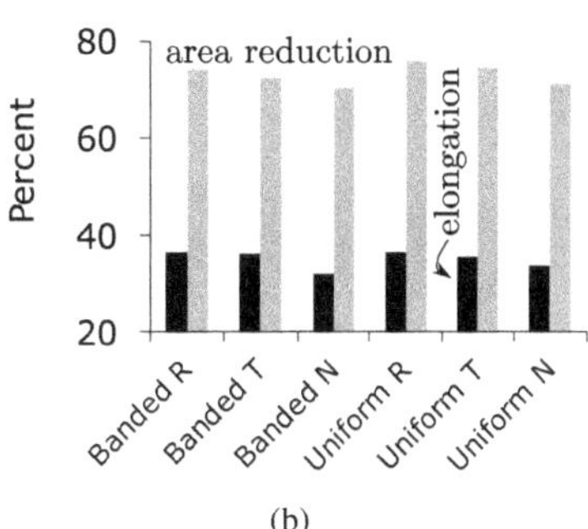

(b)

Figure 6.14 Tensile test data as a function of sample orientation [51]. The steel composition in Fe-0.21C-1.06Mn-0.25Si-0.004S wt% (a) Strength. (b) Ductility. R, T and N are the rolling, transverse and normal orientations (Equation 6.12) of the banded proeutectoid-α+P mixture.

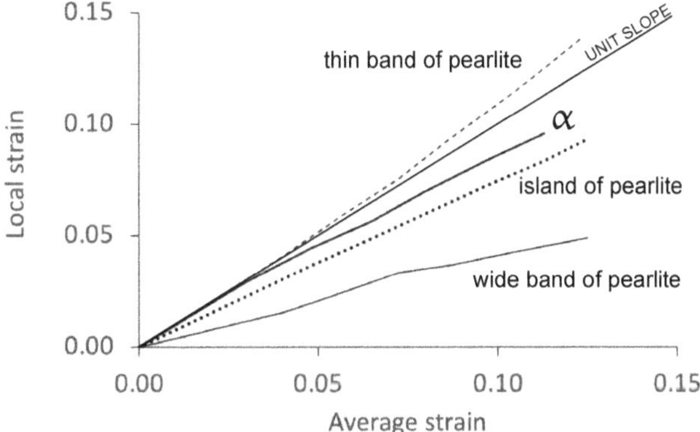

Figure 6.15 Relationship between the overall strain during tensile testing by pulling along the rolling direction, measured using digital image correlation, and local strain measured using a scanning electron microscopy technique, expressed as a von Mises equivalent strain. Selected data from Tasan et al. [52].

6.9 CORROSION OF BANDED MICROSTRUCTURES

The corrosion rate of uniform and banded α+P steel in an aqueous solution at 50 °C with a CO_2 partial pressure of 0.38 MPa mixed with methane to a total pressure of 2.1 MPa, has been studied by monitoring the penetration depth and overall rate of material loss [55]. The penetration of the corrosion into the banded structure was found to be 5.7-6.4 mm y^{-1}, slightly greater than for the uniform structure at 3.3-6.0 mm y^{-1}. The mean rate of corrosion of the banded form was greater at 1.0-2.2 mm y^{-1} compared with 0.8-2.6 mm y^{-1} for the uniform structure. The corroded surface of the banded structure sometimes exhibited ridges because of the preferential attack on the proeutectoid-α bands; the ferrite within the pearlite also corrodes more rapidly than the cementite but iron carbonate forms and hinders the further degradation of that ferrite.[6]

It is known that at low current densities, such as those encountered in practical corrosion in acidic solutions, cementite acts as an easy site for hydrogen evolution, rather like platinum [58, 59]. Assuming that the corrosion of the ferrite within pearlite is blocked by its confinement between cementite plates [55], it is reasonable to expect that in a banded microstructure, the ferrite bands will be anodic relative to the pearlite regions. In weathering steels that rely on a protective oxide

[6]In fully pearlitic steels, a finer S_I is associated with more uniform corrosion with a reduced corrosion current [56, 57].

patina, banding similarly causes a reduced corrosion resistance and exhibits a less uniform corrosion product, Figure 6.16, [57].

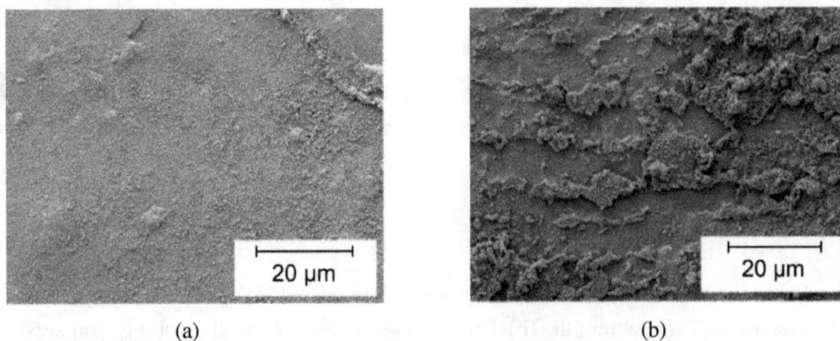

Figure 6.16 Weathering α+P steels subjected to a neutral salt-spray corrosion test. (a) Uniformly distributed pearlite, $V_V^P = 0.1$. (b) Banded steel, $V_V^P = 0.19$. Reproduced from Wang et al [57] with permission of Springer Nature.

REFERENCES

1. C. Viney, and W. S. Putnam: 'The banded microstructure of sheared liquid-crystalline polymers', *Polymer*, 1995, **36**, 1731–1741.
2. J. K. Dhont, M. P. Lettinga, Z. Dogic, T. A. Lenstra, H. Wang, S. Rathgeber, P. Carletto, L. Willner, H. Frielinghaus, and P. Lindner: 'Shear-banding and microstructure of colloids in shear flow', *Faraday discussions*, 2003, **123**, 157–172.
3. W. M. Thomas, P. L. Threadgill, and E. D. Nicholas: 'Feasibility of friction stir welding steel', *Science and Technology of Welding and Joining*, 1999, **4**, 365–372.
4. M. A. Mamtani, A. Mukherji, and A. Chaudhuri: 'Microstructures in a banded iron formation (Gua mine, India)', *Geological Magazine*, 2007, **144**, 271–287.
5. A. J. Basso, I. Toda-Caraballo, A. Eres-Castellanos, D. San-Martin, J. A. Jimenez, and F. G. Caballero: 'Effect of the microsegregation on martensitic and bainitic reactions in a high carbon-high silicon cast steel', *Metals*, 2020, **10**, 574.
6. A. A. Korda, Y. Mutoh, Y. Miyashita, T. Sadasue, and S. L. Manan: 'In situ observation of fatigue crack retardation in banded ferrite-pearlite microstructure due to crack branching', *Scripta Materialia*, 2006, **8**, 1835–1840.
7. R. N. Penha, J. Vatavuk, A. A. Couto, S. A. de Lima Pereira, S. A. de Sousa, and L. C. F. Canale: 'Effect of chemical banding on the local hardenability in AISI 4340 steel bar', *Engineering Failure Analysis*, 2015, **53**, 59–68.
8. C. F. Jatczak, D. J. Girardi, and E. S. Rowland: 'On banding in steel', *Transactions of the ASM*, 1956, **48**, 279–305.
9. P. G. Bastien: 'The mechanism of formation of banded structures', *Journal of the Iron and Steel Institute*, 1957, **187**, 281–291.
10. J. S. Kirkaldy, J. von Destinon-Forstmann, and R. J. Brigham: 'Simulation of banding in steels', *Canadian Metallurgical Quarterly*, 1962, **1**, 59–81.
11. S. W. Thompson, and P. R. Howell: 'Factors influencing ferrite/pearlite banding and origin of large pearlite nodules in a hypoeutectoid plate steel', *Materials Science and Technology*, 1992, **8**, 777–784.
12. J. S. Kirkaldy, and G. R. Purdy: 'The thermodynamics of dilute ternary austenite solutions', *Canadian Journal of Physics*, 1962, **40**, 202–207.

13. L. Král, B. Million, and J. Čermák: 'Diffusion of carbon and manganese in Fe-C-Mn', *Defect and Diffusion Forum*, 2007, **263**, 153–158.
14. J. S. Kirkaldy, R. J. Brigham, H. A. Domian, and R. G. Ward: 'A study of banding in Skelp by electron-probe microanalysis', *Canadian Metallurgical Quarterly*, 1963, **2**, 233–241.
15. T. Imai, Y. Nishida, and S. Kogiso: 'Anisotropy of the Charpy impact value of carbon steels and corrective heat treatment', *Journal of Mechanical Working Technology*, 1982, **7**, 147–161.
16. T. F. Majka, D. K. Matlock, and G. Krauss: 'Development of microstructural banding in low-alloy steel with simulated Mn segregation', *Metallurgical & Materials Transactions A*, 2002, **33**, 1627–1637.
17. R. A. Grange: 'Effect of microstructural banding on steel', *Metallurgical Transactions*, 1971, **2**, 417–426.
18. S. A. Khan, and H. K. D. H. Bhadeshia: 'The bainite transformation in chemically heterogeneous 300M high-strength steel', *Metallurgical Transactions A*, 1990, **21A**, 859–875.
19. W. Solano-Alvarez, and H. K. D. H. Bhadeshia: 'White-etching matter in bearing steel. Part 1: Controlled-cracking of bearing steel', *Metallurgical & Materials Transactions A*, 2014, **45**, 4907–4915.
20. H. Hanemann, and H. J. Wiester: 'Die martensitkristallisation in hochkohlenstoffhaltigen stählen', *Archiv für das Eisenhüttenwesenttenwesen*, 1932, **5**, 377–382.
21. A. B. Greninger, and A. R. Troiano: 'Kinetics of the austenite to martensite transformation in steel', *Transactions of the ASM*, 1940, **28**, 537–563.
22. G. R. Purdy, and J. S. Kirkaldy: 'Homogenisation by diffusion', *Metallurgical Transactions*, 1971, **2**, 371–378.
23. M. Maalekian, H. Azizi-Alizamini, and M. Militzer: 'Phase field modeling of microstructure banding in steels', *Metallurgical & Materials Transactions A*, 2015, **47**, 608–622.
24. W. S. Owen, M. Cohen, and B. L. Averbach: 'Influence of ferrite banding on the impact properties of mild steel', *Welding Journal, Research Supplement*, 1958, **37**, 368s–374s.
25. J. Fridberg, L.-E. Torndähl, and M. Hillert: 'Diffusion in iron', *Jernkontorets Annaler*, 1969, **153**, 263–276.
26. J. D. Verhoeven: 'A review of microsegregation induced banding phenomena in steels', *Journal of Materials Engineering and Performance*, 2000, **9**(3), 286–296.
27. J. D. Verhoeven: 'The role of the divorced eutectoid transformation in the spheroidization of 52100 steel', *Metallurgical & Materials Transactions A*, 2000, **31**, 2431–2438.
28. N. Warnken, D. Ma, A. Drevermann, R. C. Reed, S. G. Fries, and I. Steinbach: 'Phase-field modelling of as-cast microstructure evolution in nickel-based superalloys', *Acta Materialia*, 2009, **57**, 5862–5875.
29. J. A. Ronevich, B. P. Somerday, and C. W. San Marchi: 'Effects of microstructure banding on hydrogen assisted fatigue crack growth in X65 pipeline steels', *International Journal of Fatigue*, 2016, **82**, 497–504.
30. H. L. Haskel, E. Pauletti, J. de Paula Martins, and A. L. Moreira de Carvalho: 'Microstructure and microtexture assessment of delamination phenomena in Charpy impact tested specimens', *Materials Research*, 2014, **17**, 1238–1250.
31. L. E. Samuels: ASM Handbook, chap. Optical microscopy of carbon steels: ASM International, Ohio, USA, 1980:117–167.
32. H. K. D. H. Bhadeshia: 'Critical assessment: Diffusion-controlled growth of ferrite plates in plain carbon steels', *Materials Science and Technology*, 1985, **1**, 497–504.
33. M. A. Beltrán, J. L. Gonzállez, D. Rivas, F. Hernńdez, and H. Dorantes: 'On the role of microstructural properties on mechanical behavior of API-X46 steel', *Procedia Structural Integrity*, 2017, **3**, 57–67.
34. E. E. Underwood: 'Stereology, or the quantitative evaluation of microstructures', *Journal of Microscopy*, 1969, **89**, 161–180.

35. ASTM: Annual Book of ASTM Standards, Designation E, vol. 3.01: American Society of Testing Methods, 1996.
36. S. A. Saltykov: Stereometricheskaya Metallografiya: 2nd ed., Moscow: Metallurgizdat Publishers (English translation US Department of Defence AD267700), 1958.
37. D. Chae, D. A. Koss, A. L. Wilson, and P. R. Howell: 'Effect of microstructural banding on failure initiation of HY-100 steel', *Metallurgical & Materials Transactions A*, 2000, **31**, 995–1005.
38. T. P. Harrigan, and R. W. Mann: 'Characterization of microstructural anisotropy in orthotropic materials using a second rank tensor', *Journal of Material Science*, 1984, **19**, 761–767.
39. C. Beaulieu: 'The biological basis of diffusion anisotropy': In: *Diffusion MRI for quantitative measurement*. Academic Press, 2014:155–183.
40. R. T. DeHoff, and F. N. Rhines: Quantitative Microscopy: New York, USA: McGraw Hill, 1968.
41. C. S. Hartley: 'Quantitative specification of microstructural anisotropy', *Le Journal de Physique Colloques*, 1990, **51 (C1)**, C1–173–C1–177.
42. B. Krebs, A. Hazotte, L. Germain, and M. Gouné: 'Quantitative analysis of banded structures in dual-phase steels', *Image Analysis & Stereology*, 2010, **29**, 85–90.
43. J. Gurland: 'The measurement of grain contiguity in two-phase alloys', *Transactions of the AIME*, 1958, **212**, 452–453.
44. Z. Fan, A. P. Miodownik, and P. Tsakiropoulos: 'Microstructural characterisation of two phase materials', *Materials Science and Technology*, 1993, **9**, 1094–1100.
45. A. D. King, and T. Bell: 'Crystallography of grain boundary proeutectoid ferrite', *Metallurgical Transactions A*, 1975, **6**, 1419–1429.
46. H. K. D. H. Bhadeshia: 'Diffusional formation of ferrite in iron and its alloys', *Progress in Materials Science*, 1985, **29**, 321–386.
47. R. K. Ray, and J. J. Jonas: 'Transformation textures in steels', *International Materials Reviews*, 1990, **35**, 1–37.
48. D. W. Kim, D. W. Suh, R. S. Qin, and H. K. D. H. Bhadeshia: 'Dual orientation and variant selection during diffusional transformation of austenite to allotriomorphic ferrite', *Journal of Material Science*, 2010, **45**, 4126–4132.
49. H. K. D. H. Bhadeshia: Geometry of Crystals, Polycrystals and Phase Transformations: Florida, USA: CRC Press, Taylor and Francis, 2017.
50. V. Venegas, F. Caleyo, T. Baudin, J. H. Espina-Hernandez, and J. Hallen: 'On the role of crystallographic texture in mitigating hydrogen-induced cracking in pipeline steels', *Corrosion Science*, 2011, **53**, 4204–4212.
51. W. A. Spitzig: 'Effect of sulphide inclusion morphology and pearlite banding on anisotropy of mechanical properties in normalized C-Mn steels', *Metallurgical Transactions A*, 1983, **14**, 271–283.
52. C. C. Tasan, J. P. M. Hoefnagels, and M. G. D. Geers: 'Microstructural banding effects clarified through micrographic digital image correlation', *Scripta Materialia*, 2010, **62**, 835–838.
53. L. E. Miller, and G. C. Smith: 'Tensile fractures in carbon steels', *Journal of the Iron and Steel Institute*, 1970, **11**, 998–1005.
54. X. Zhang, Y. Wang, J. Yang, Z. Qjao, C. Ren, and C. Chen: 'Deformation analysis of ferrite/pearlite banded structure under uniaxial tension using digital image correlation', *Optics and Lasers in Engineering*, 2016, **85**, 24–28.
55. D. Clover, B. Kinsella, B. Pejcic, and R. De Marco: 'The influence of microstructure on the corrosion rate of various carbon steels', *Journal of Applied Electrochemistry*, 2005, **35**, 139–149.

56. P. K. Katiyar, S. Misra, and K. Mondal: 'Effect of different cooling rates on the corrosion behavior of high-carbon pearlitic steel', *Journal of Materials Engineering and Performance*, 2018, **27**, 1753–1762.
57. Z. Wang, X. Zhang, H. Yu, J. Liu, L. Cheng, S.-E. Hu, and K. Wu: 'Effects of pearlite on corrosion initiation and propagation in weathering steels in marine environments', *Journal of Materials Science*, 2022, **57**, 6039–6055.
58. D. N. Staicopolus: 'The role of cementite in the acidic corrosion of steel', *Journal of the Electrochemical Society*, 1963, **110**, 1121–1124.
59. J. A. Green, and R. N. Parkins: 'Electrochemical properties of ferrite and cementite in relation to stress corrosion of mild steels in nitrate solutions', *Corrosion*, 1968, **24**, 66–69.

7 Hypereutectoid steels

Steels with a carbon concentration exceeding the eutectoid value stimulate mixed emotions amongst metallurgists. This is because of their susceptibility to the precipitation of proeutectoid cementite, which in particulate form is tolerable or even useful. However, covering the austenite grain boundaries with cementite allotriomorphs is an anathema to most. There is nothing to indicate that such continuous networks can be used to advantage. If the networks can be avoided, for example by using an appropriate heat treatment, the large carbon concentration becomes useful. Most bearing steels, which are critical components in machines of all kinds and sizes, have roughly 1 wt% of carbon [1].

The term *hypereutectoid* refers to the binary Fe-C system when the average carbon concentration exceeds 0.76 wt%, which is the composition at which $\alpha+\gamma+\theta$ can coexist in equilibrium at 727 °C. It is, nonetheless, applied widely to alloy steels where the three phases can coexist in equilibrium at a variety of temperatures, with the reference concentration still taken to be 0.76 wt%.

The locus of the eutectoid temperature in the Fe-C-X system as a function of w_X is given by the intersection of the $\alpha+\gamma/\gamma$ and $\theta+\gamma/\gamma$ equilibrium surfaces in the ternary equilibrium diagram (Figure 7.1), where 'X' is a substitutional solute [2]. Austenite persists below the eutectoid temperature so it is possible for a three phase $\alpha+\gamma+\theta$ equilibrium to continue to exist below T_E^{max} until the two phase $\alpha+\theta$ phase field is reached on cooling.

The chemical composition of the residual austenite changes as the pearlite grows in the three-phase field so a fully pearlitic state is not possible until the steel is cooled into the $\alpha+\theta$ field. In richly-alloyed steels, it may not be possible to access this field. Hadfield steel (Fe-13Mn-1C wt%) never transforms completely so has been used to study the three-phase crystallography of pearlite [3], and to characterise the influence of magnetic fields on its transformation behaviour [4]. The two-phase $\alpha+\theta$ field is reached only at temperatures where diffusion diffusion become imperceptibly slow (Table 7.1) – a fully pearlitic microstructure has never been reported in Hadfield steel.

Table 7.1
Calculated equilibrium phases and their chemical compositions in Fe-13Mn-1C wt%. The mass fractions are rounded-off.

Temperature °C	Stable phases	Fe	Mn	C
		mass fractions		
467	α	0.96	0.04	0.000
467	θ	0.27	0.66	0.067
467	γ	0.76	0.24	0.001
457	α	0.96	0.04	0.000
457	θ	0.26	0.68	0.067

The eutectoid temperature and composition clearly are dependent on the substitutional-solute content. The eutectoid carbon concentration in the austenite can be greater or less than that in the Fe-C system. This is illustrated in Table 7.2 and Figure 7.1a where w_C^γ is seen to be less than 0.76 when the three phases are in equilibrium for both the Mn and Si alloyed steels.[1] In the latter case

[1]An extreme example of a steel said to be of eutectoid composition contains only 0.35C wt% in addition to 0.89Si-0.43Mn-5.5Cr-1.2Mo-0.52V wt% [5]. Surprisingly, the steel transforms completely into pearlite during isothermal holding for 265 min at 715 °C. The data seem to be backed using optical microscopy. It would be useful to verify that the carbide phase within the pearlite is cementite.

DOI: 10.1201/9781032631981-7

the eutectoid temperature increases with silicon concentration whereas manganese has the opposite effect. Trends calculated on the basis of thermodynamics show how T_E and the eutectoid carbon concentration are expected to vary with substitutional solute content in ternary steels are illustrated in Figures 7.1b,c. Aluminium additions have a large effect in raising the eutectoid temperature.[2]

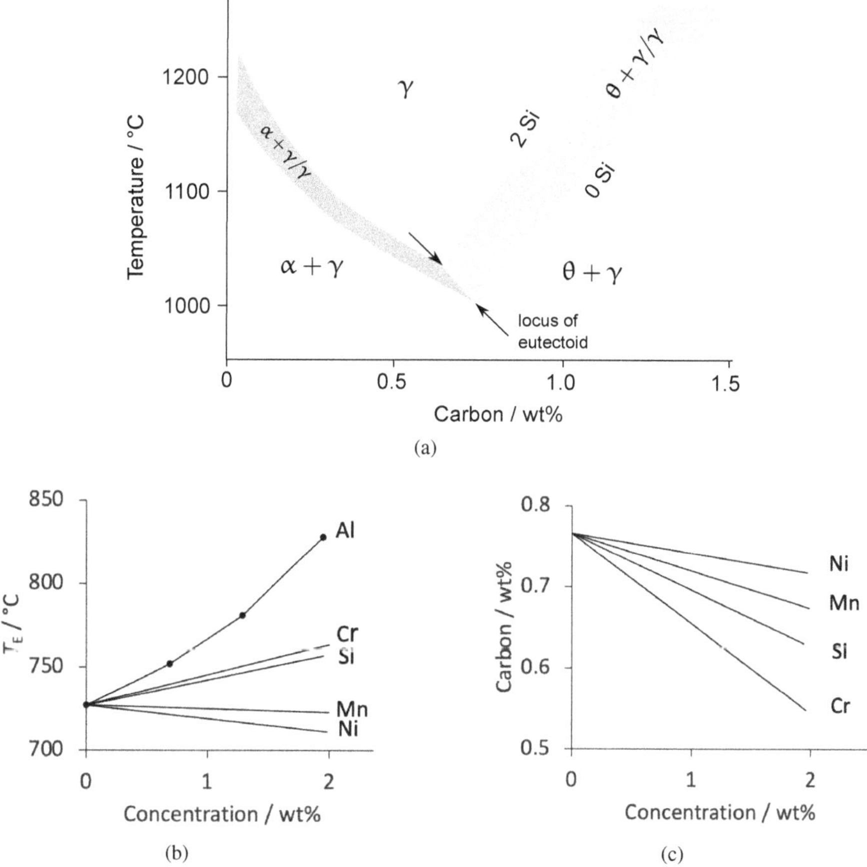

Figure 7.1 (a) The intersection of the $\alpha+\gamma/\gamma$ and $\theta+\gamma/\gamma$ equilibrium surfaces defines T_E in the Fe-Si-C system. The arrows indicate the locus of T_E as a function of Si in wt%. (b) Calculated T_E versus substitutional-solute content. The aluminium data are measured values [6]. (c) Calculated eutectoid carbon-concentration versus substitutional-solute content. Adapted using data from Kirkaldy et al. [2] and Jang et al. [6]. All of these plots should be taken to be approximate indications because for substitutionally-alloyed steels, the eutectoid reaction occurs over a range of temperatures.

With the exception of the measured aluminium data, the calculations represent equilibrium; in practice, the partitioning of solutes between cementite and ferrite may be sluggish, leading to large deviations from equilibrium [e.g., Figure 5, ref. 7], especially when the transformation is suppressed to $T \lessapprox 600\,^{\circ}\mathrm{C}$ where atomic mobility is limited. Such deviations are not currently calculable, which makes it impossible to estimate kinetics [8]. Most kinetic estimates rely on local or constrained equilibria at the transformation front. The complexity of the problem is exaggerated because two phases grow simultaneously from the austenite and their interlamellar spacing is an uncertain variable in its entanglement with capillarity.

[2]The aluminium data refer to hypereutectoid steels containing 0.83-9.92C, 0.44-0.47Mn, 0.29-0.21Si, 0.4Cr, 0.1V and 0-1.95Al wt% [6].

Table 7.2
Dependence of the equilibrium between $\gamma+\alpha+\theta$ on composition. For Fe-C, such equilibrium is at a unique temperature.

Alloy	Temperature / °C	Phase	w_C	w_{Mn}
Fe-0.76C wt%	727	α	0.01	
		θ	6.70	
		γ	0.76	
Fe-2Mn-0.8C wt%	697	α	0.01	0.79
		θ	6.60	8.58
		γ	0.67	3.19
Fe-2Mn-0.8C wt%	707	α	0.01	0.54
		θ	6.70	5.76
		γ	0.70	2.14

7.1 MICROSTRUCTURE

It is inevitable that cementite allotriomorphs precipitate at the austenite grain boundaries in hypereutectoid steels, either during cooling or during isothermal transformation at a temperature where pearlite can form. In the latter case, the growth of allotriomorphs competes with that of pearlite.

The allotriomorphs are thin when the austenite grain size is small (Figure 7.2), because the increased number density of nucleation sites allows pearlite to form more rapidly, which then stifles the growth of the allotriomorphs. Consistent with this interpretation, the θ-allotriomorphs also become thinner as the isothermal transformation temperature is reduced [9]. The continuity of θ-allotriomorphs at the austenite grain boundaries can be disrupted by the prior precipitation there of vanadium carbides [10] which consume the easy nucleation sites and physically hinder the cementite. The vanadium carbides can even lead to the precipitation of allotriomorphic ferrite in hypereutectoid steel, by locally depleting the carbon concentration, the α then interfering with the precipitation of θ-allotriomorphs [11].

There remain aspects of pearlite in hypereutectoid steels that are not understood. Figure 7.3 shows that the interlamellar spacing of pearlite transformed isothermally from a fully austenitic state, has a $\overline{L}_\gamma$ dependence. The grain size was controlled by varying the austenitisation temperature T_γ; a possible interpretation is that a higher T_γ requires a greater dissipation of heat when the sample is quenched to the intended transformation temperature. This may allow the pearlite to form before the intended transformation temperature is reached. This could be investigated by manipulating $\overline{L}_\gamma$ using different time periods at a fixed T_γ.

7.1.1 CARBIDE-NETWORK CONNECTIVITY

Networks of proeutectoid-cementite allotriomorphs in hypereutectoid steels compromise the ductility and toughness of the steel; Figure 7.4a shows cracks that propagate along the cementite. Indeed, it is possible during deformation at ambient temperature, to detect bursts of acoustic emissions as the cementite fractures [13].

The extent to which toughness deteriorates depends on the geometrical characteristics of the cementite network [14]. The network can be described as slabs that offer continuous pathways for

(a)

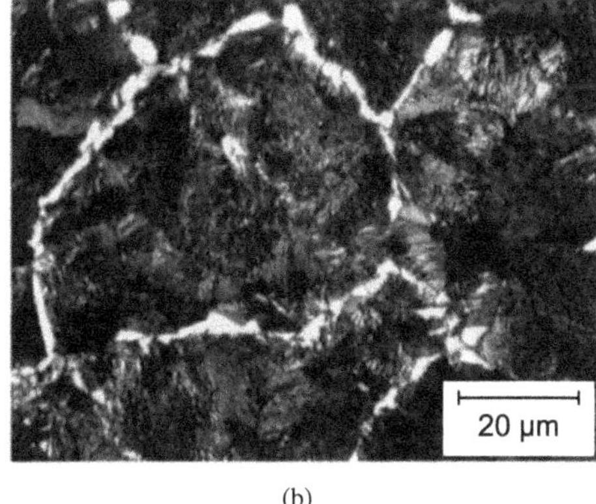

(b)

Figure 7.2 Fe-0.91C-0.49Mn-0.21Si wt% transformed isothermally at 620°C showing layers of cementite decorating the austenite grain surfaces, in an otherwise pearlitic microstructure. The heat treatment in each case ensured a fully austenitic state prior to transformation. (a) Austenitised at 900 °C. (b) Austenitised at 1200 °C. Reproduced with the permission of Elsevier from Elwazri et al. [12], © 2015.

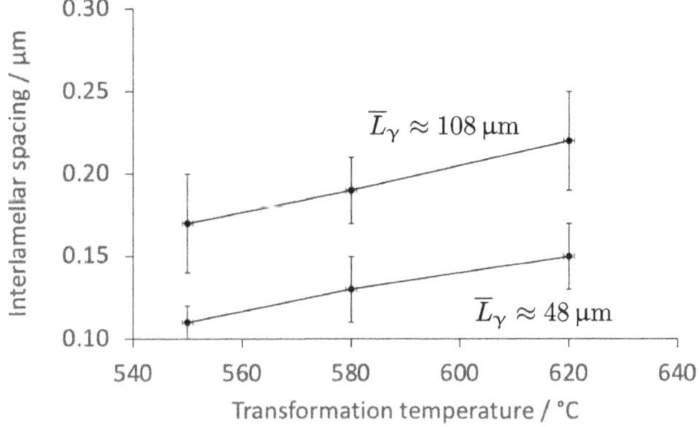

Figure 7.3 Pearlite interlamellar-spacing during isothermal transformation from a fully austenitic condition, in Fe-0.91C-0.49Mn-0.21Si wt%. Selected data from Elwazri et al. [12].

crack propagation. Junctions between differently oriented slabs would deflect cracks, and connectivity between cracks would be lost at the end-points of allotriomorphs (Figure 7.4b). The connectivity index is defined in terms of the numbers of the three different types of pixels in the image representing the cementite network, i.e., $N_{\text{slab}}/(N_{\text{node}}+N_{\text{end}})$. Because junctions deflect cracks and ends disrupt continuity, larger numbers of N_{node} and N_{end} should benefit toughness, although strictly, the correlation should be against number *densities*. Figure 7.4c shows this to be the case, with the caveat that the toughness at high connectivity improves if the average spacing between the nodes is reduced, because this would necessitate more crack deflections.

The representation of allotriomorph network in terms of a connectivity index is based on the analysis of two-dimensional sections of the microstructure. Fracture topology, on the other hand, is three-dimensional. In three dimensions, the nodes and end-points become edges, slabs no longer have a well-defined length and additional points are generated where more edges meet. The idea of a three-dimensional fracture network is well developed in geology [15] where the exploitation of underground resources requires a good understanding of flow paths. That level of sophistication

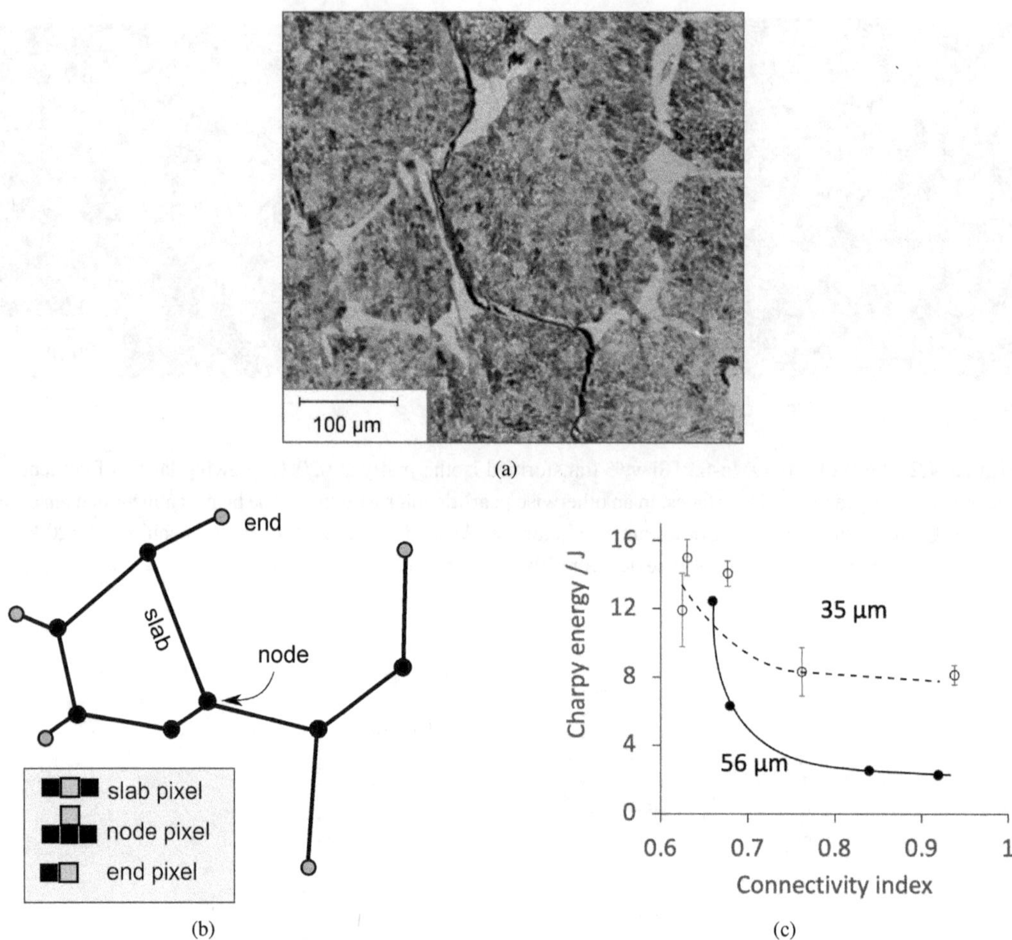

Figure 7.4 (a) Crack propagating through cementite network in a predominantly pearlitic hypereutectoid steel. Micrograph reproduced with the permission of Elsevier from Hecht et al. [14]. (b) Interpretation of cementite network. The light-coloured pixels in the inset define the characteristics of slabs, nodes and ends. (c) Correlation of Charpy impact energy against the connectivity of the cementite network and the average distance between junctions. Adapted using data from [14].

may not be justified in the present context – the methods available for controlling the structure of allotriomorphic cementite are limited, although admittedly, a better understanding of the topology may progress the issue.

7.1.2 ALLOTRIOMORPHIC FERRITE IN HYPEREUTECTOID STEELS

Pearlite forms if the austenite is supercooled such that both ferrite *and* cementite can precipitate. The transformation temperature must simultaneously fall below the $\gamma/(\alpha+\gamma)$ and $\gamma/(\theta+\gamma)$ phase boundaries that are extrapolated to $T < T_E$; this condition, described often as the Hultgren extrapolation was first described by Belaiew [16]. It can alternatively be expressed in terms of concentrations, that $c^{\gamma\theta} \leq \overline{c}^{\gamma} \leq c^{\gamma\alpha}$.

The greatest reduction in free energy occurs when the ferrite and cementite grow together in such a way that the overall chemical composition of the remaining austenite is not altered in the process. The reduction is smaller when only one of the constituent phases of pearlite precipitates before the

$\alpha_P + \theta$ cooperative growth is established.

Whatever the reason, there is no doubt that against intuition, ferrite allotriomorphs can be found in hypereutectoid steel. They have been observed in both eutectoid (Fe-0.74C-1Mn wt%) and hypereutectoid (Fe-1.05C-1Mn wt%) steels that were transformed at temperatures where pearlite is expected [17]. Figure 7.5a shows an example where the cementite first forms on the austenite grain boundaries, locally depletes the carbon concentration, thus stimulating allotriomorphic ferrite. The thermodynamic data in Figure 7.6 reveal that the reduction in ΔG for cementite precipitation from austenite exceeds that for ferrite in Fe-1.05C-1Mn wt% so the α is unlikely to precede θ precipitation.

Similarly, in Fe-1.25C-1.45Cu wt% cooled slowly from the austenitic state to 710 °C, allotriomorphic cementite uniformly decorated the γ-grain surfaces [18].[3] This stimulated the growth of ferrite in association with the proeutectoid cementite; pearlite followed with the cooperative growth of $\alpha + \theta$. The allotriomorphic ferrite is stimulated by carbon depletion in the vicinity of the proeutectoid cementite.

It is only when the driving force for ferrite precipitation becomes significant at large undercoolings that allotriomorphic ferrite that is not associated with solute depletion seems to form independently of cementite allotriomorphs, as illustrated in Figure 7.5b. It is clear from Figure 7.6 that in the Fe-0.74C-1Mn wt% steel transformed at 600 °C, the reduction of free energy accompanying the formation of ferrite in isolation is much greater than the corresponding term for cementite to precipitate in isolation.

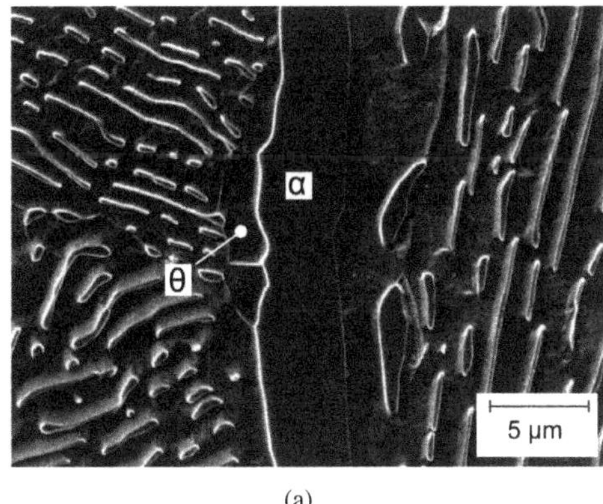

(a)

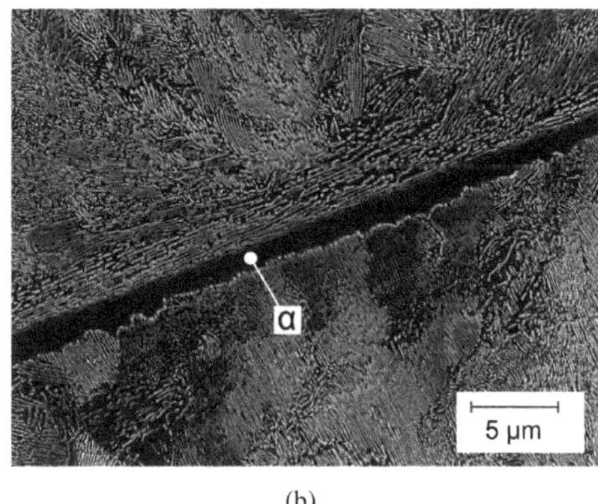

(b)

Figure 7.5 Fe-1.05C-1Mn steel. (a) Transformed at 700 °C, showing parallel cementite and ferrite allotriomorphs at a prior austenite-grain boundary. (b) Transformed at 600 °C, showing a ferrite allotriomorph at a prior austenite grain boundary, in contact only with lamellar cementite that is a part of the pearlite. Images courtesy of Professors G. Miyamoto and T. Furuhara.

When the carbon concentration of the steel is reduced to approximately the eutectoid composition, the driving force for the independent precipitation of ferrite becomes greater than for $\gamma \rightarrow \theta + \gamma'$, at all the transformation temperatures studied by Miyamoto et al. [17], Figure 7.6. Allotriomorphic ferrite is able to form independently at the austenite grain surfaces. Even Widmanstätten ferrite plates (α_W) can be induced in a hypereutectoid steel transformed at a large undercooling, Figure 7.7.

It is interesting that the primary α_W plates seem to have formed without association with allotriomorphic ferrite, implying direct nucleation at the austenite grain surfaces. It appears that these

[3]The samples were austenitised while enclosed in a quartz tube, which then was transferred intact into a furnace at the isothermal transformation temperature [19].

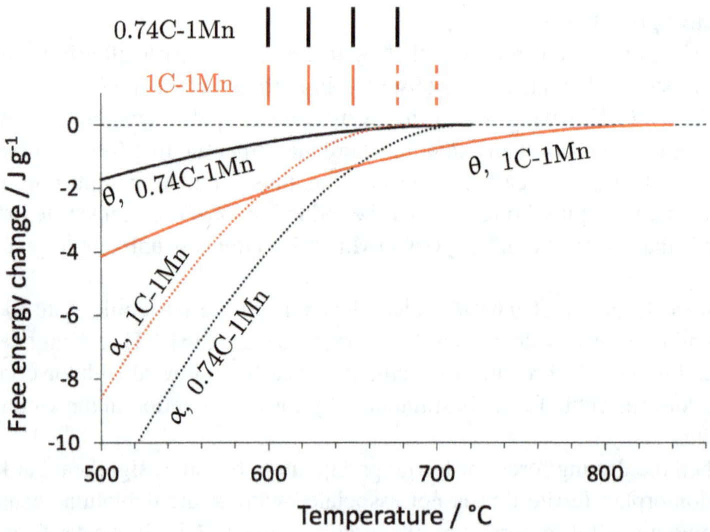

Figure 7.6 Calculated free energy changes accompanying the $\gamma \rightarrow \alpha + \gamma'$ and $\gamma \rightarrow \theta + \gamma'$ reactions in both Fe-0.74C-1Mn and Fe-1.05C-1Mn wt% steels. The calculations used MTDATA [20] with the SGTE database, permitting only the phases defined in the reactions to exist. γ' refers to residual austenite whose composition differs from the fully austenitic state. The bar codes at the top indicate the temperatures at which allotriomorphic ferrite was observed [17], with the dashed bars reserved for the case where the ferrite was observed in association with allotriomorphic cementite.

microstructures, which occupy only a fraction 0.015 of the whole, have no significant effect on the strength or fatigue resistance of the structure, because the fine interlamellar spacing generated during patenting at 500 °C has the predominant influence on the properties [21].

It is interesting that the primary α_w plates seem to have formed without association with allotriomorphic ferrite, implying direct nucleation at the austenite grain surfaces. Given the length of the α_w plates, there can be no doubt that they formed first, before the surrounding pearlite. The transformation temperature was 500 °C, far below T_E, making it possible to sustain Widmanstätten ferrite formation. It appears that these microstructures, which occupy only a fraction 0.015 of the whole, have no significant effect on the strength or fatigue resistance of the structure, because the fine interlamellar spacing generated during patenting at 500 °C has the predominant influence on the properties [21].

7.2 MECHANICAL PROPERTIES

Hypereutectoid steels tend to be strong but are limited to low levels of plasticity at ambient temperature. As a result, mechanical properties are characterised in compression or in experiments such as the shear punch test [22], where monotonic properties are derived indirectly with some empirical fitting.

The ultimate tensile strength measured using the shear punch test has a constant offset relative to the yield strength: $\sigma_{\text{UTS}} = \sigma_y + 344 \pm 18\,\text{MPa}$ [12]. Normal tensile tests conducted on steels with 0.88-1.04C wt%, that contain proeutectoid cementite, indicate a much larger offset between $\sigma_{\text{UTS}} - \sigma_y = 514 \pm 29\,\text{MPa}$ (Figure 7.8). The punch test implements the deformation of a clamped sheet by a cylindrical punch until failure, while monitoring the load and displacement. The deformation involved is complicated and it is not clear that the σ_{UTS} for hypereutectoid steels with their limited plasticity, corresponds to the maximum load noted in a punch test.

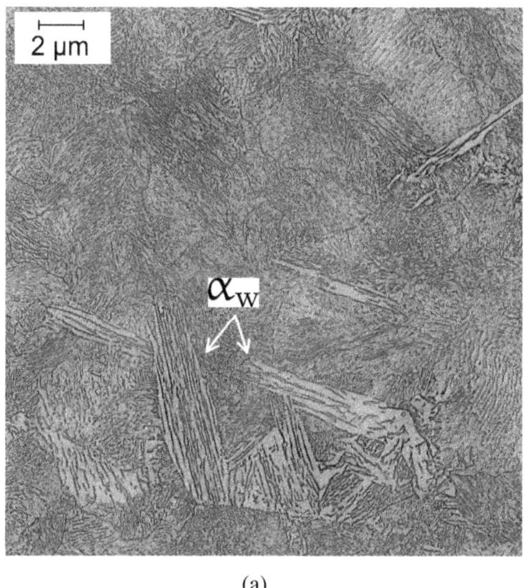

(a)

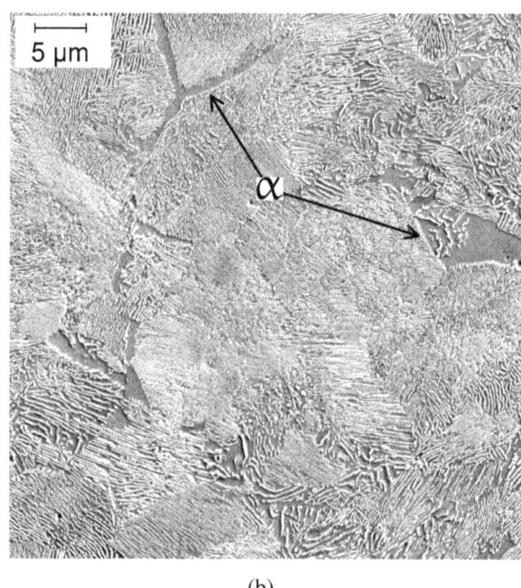

(b)

Figure 7.7 Fe-0.85C-0.58Mn-0.22Si-0.06Ni-0.08Cu wt% hypereutectoid steel transformed isothermally at 500 °C. (a) Illustrating the presence of Widmanstätten ferrite (image with inverted contrast for clarity). (b) Allotriomorphic ferrite with proeutectoid cementite absent at the austenite grain boundaries. Note that the amount of α_W and α is only a fraction 0.015 of the total microstructure when larger areas are examined. Images kindly provided by Steffen Scherbring and Javad Mola [21] with permission of Elsevier.

These observations do not tally with many data on hypoeutectoid steels [23] where $\sigma_{UTS} \approx 1.31\sigma_y + 19\,\text{MPa}$. The work-hardening ability of steels with α+P microstructures is definitely less than of hypereutectoid alloys.

Hypereutectoid steels lack ductility, in part because of any proeutectoid-θ present at the γ/γ boundaries, but also because they are relatively strong. The reduction in area during tensile testing is only about 6% with much scatter [24]. The stress-strain curve in tension does not exhibit a yield point, nor is there a maximum because fracture occurs after limited plasticity.

The fracture toughness K_{IC} decreases as the fraction of cementite increases, dropping sharply when the carbon concentration exceeds about 0.9 wt% because avoiding cementite networks then becomes difficult, Figure 7.8b. Much greater ambient-temperature toughness has been reported for pearlite with a eutectoid, as opposed to a hypereutectoid composition [25].

Large carbon concentrations in hypereutectoid steels are not always detrimental to toughness. Bearing steels and even steels containing more than 1.5 wt%C are often austenitised in the $\gamma + \theta$ phase field in such a way that they contain spheroidal cementite; on cooling, the steel undergoes a divorced eutectoid reaction in which the spheroids of θ simply coarsen to absorb excess carbon at the advancing γ/α boundaries [p.299, 1]. Re-heating into the $\gamma + \theta$ phase field followed by cooling, leaves a microstructure containing spheroidal cementite and pearlite, Figure 7.9a. Networks of cementite are therefore eliminated, leading to dramatic improvements in toughness and ductility, especially when the austenite grain size is small, Figure 7.9b [27–29].

7.3 ALLOYING WITH ALUMINIUM

Cementite at the austenite grain surfaces can be disrupted using a cyclic heat-treatment, that in a 1.24C wt% alloy has been shown to boost the tensile ductility from 6% to 13% without reducing strength [31]. The treatment consisted of repeated, short-duration (6 min) excursions into the $\theta + \gamma$ phase field, followed in each case by forced air cooling to ambient temperature. Technology in

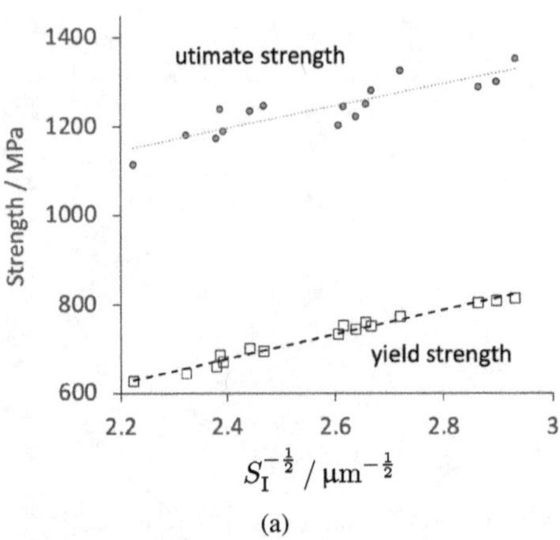

(a)

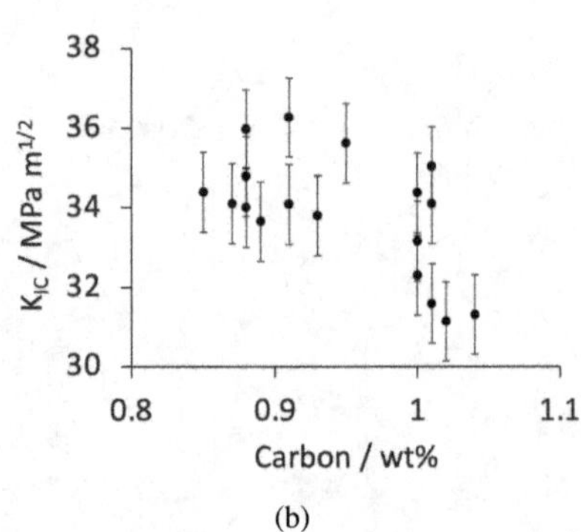

(b)

Figure 7.8 Yield and ultimate tensile strengths of a number of hypereutectoid steels what also contained chromium or microalloying additions such as vanadium and niobium. Selected data from Qiao et al. [24]. These data were not included in the derivation of Equation 4.17 because the steels were not free from proeutectoid cementite. (b) Selected ambient temperature fracture toughness data for the same steels; the error bars are consistent with the standard error reported in Qiao et al. [26].

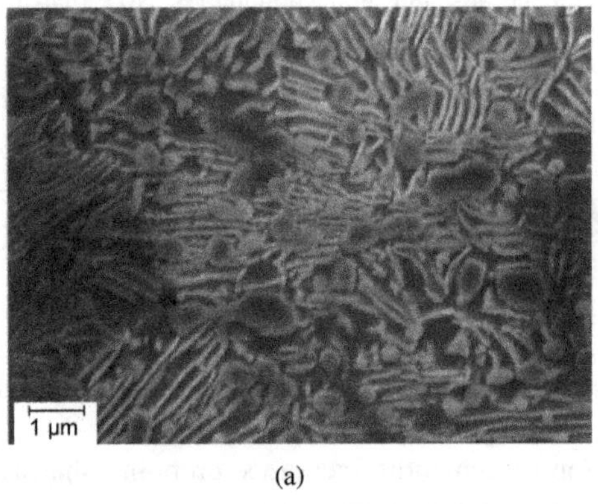

(a)

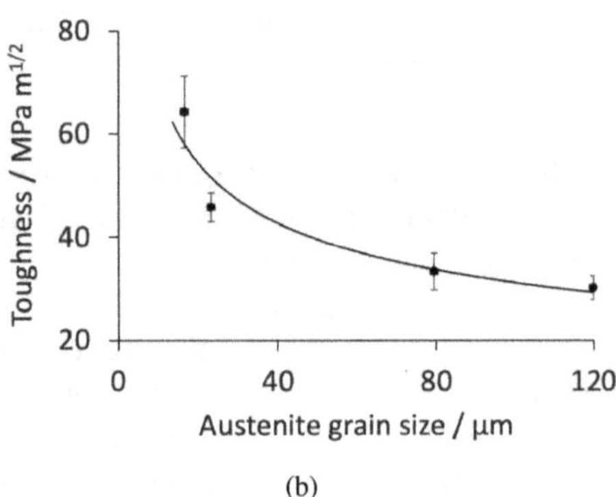

(b)

Figure 7.9 (a) Microstructure generated in Fe-1.8C-0.5Mn-0.51Si-0.22Cr wt% by transforming first to a divorced eutectoid, heating into the $\gamma + \theta$ phase field at 870 °C and then cooling to generate pearlite. Image reproduced from Taleff et al. [30] with permission of Springer Nature. (b) The fracture toughness K_{IC} of Fe-1.3C-0.5Mn-0.51Si-0.22Cr wt% as a function of the austenite grain size. The microstructure consisted of spheroidal cementite and pearlite. Data from Fernandez-Vicente [27].

principle could be developed to make this into a commercial process, for example, using induction heating.

Aluminium in solid solution retards the precipitation of cementite during the tempering of martensite [32, 33] or of supersaturated ferrite [34], and even precipitation from austenite [35]. The solubility of aluminium in cementite which is in equilibrium with α or γ [36, 37] is rather small, leading to a reduction in its thermodynamic stability. Observations indicate that allotriomorphic cementite is prevented when > 1 wt% of aluminium is added to hypereutectoid steel [6].

The mechanism by which aluminium suppresses or retards the precipitation of cementite is not clear. If it is assumed that local equilibrium is maintained at the θ/γ interface, then simulations show that the growth rate of cementite from austenite is reduced at 450 °C [37]. There is, however, no experimental evidence to suggest that equilibrium is maintained at the θ/γ interface.

Aluminium also reduces the thermodynamic stability of the austenite relative to ferrite so $|\Delta G^{\gamma\rightarrow\alpha}|$ increases, which is why T_E also increases (Figure 7.1b). Since ferrite is the predominant phase in pearlite, $|\Delta G^{\gamma\rightarrow\alpha+\theta}|$ increases, leading to a reduction in the interlamellar spacing and pearlite colony size. This in turn makes the steel stronger [6], although the full effect cannot be explained. The influence of interlamellar spacing as calculated using the penultimate term in Equation 4.13 is not sufficient to explain the enhancement in σ_y, even when solid-solution strengthening due to Al is accounted for (Table 7.3).

Table 7.3

The strength of fully pearlitic hypereutectoid steels (0.83-0.93C wt%) as a function of aluminium concentration. The solid solution strengthening of ferrite, $\sigma^{\alpha}_{SS,Al}$, is taken to be 70 MPa per wt% of aluminium [38] with the assumption that cementite is similarly strengthened.

Al concentration / wt%	σ_y / MPa	$\Delta\sigma_{S_I}$ / MPa	$\sigma^{\alpha}_{SS,Al}$	$\sigma_y - \Delta\sigma_{S_I} - \sigma^{\alpha}_{SS,Al}$
0.00	452	281	0	171
0.69	674	305	48	321
1.29	691	370	90	230
1.95	767	433	137	197

A peculiar effect of adding aluminium to fully pearlitic steels is that the engineering stress-strain curve shows a sharp deflection at the onset of plasticity when compared with the aluminium-free steel which yields continuously, Figure 7.10. Even more pronounced discontinuous yielding has been observed in pearlitic Fe-(1.3-1.8)C-1.6Al-1.5Cr-0.5Mn wt% steels [30]. Experiments on Fe-Cr-Al interstitial-free ferritic steel have demonstrated that aluminium atoms pin dislocations [39], which may explain the discontinuous yielding apparent in Figure 7.10.

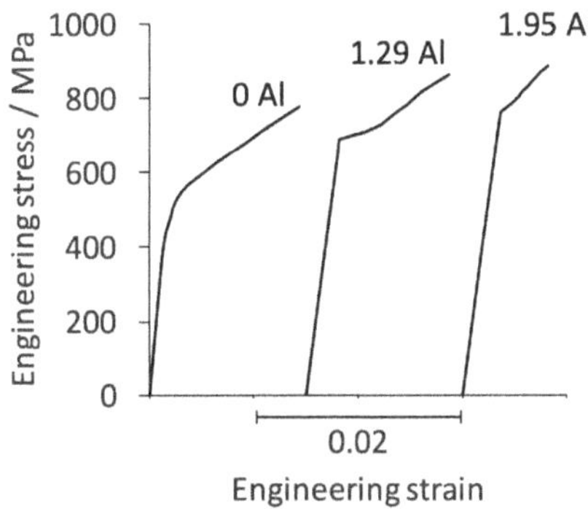

Figure 7.10 $\sigma - \varepsilon$ curves of hypereutectoid steels (Fe- 0.85C-0.4Mn-0.3Si-0.4Cr-0.1V wt%) as a function of the aluminium concentration (wt%). Selected data from Jang et al. [6].

7.4 ALLOYING WITH SILICON

It has been shown both experimentally [40] and through calculations [41] that silicon and cementite simply do not mix. Silicon-rich hypereutectoid steels are therefore slow to spheroidise from the pearlitic state. At the same time, silicon in high-carbon (1.2-1.8 wt%) steels is said to be more effective in disrupting the continuity of proeutectoid-θ networks at the austenite grain surfaces, although silicon-containing steels is more susceptible to graphite formation, Figure 7.11 [42].

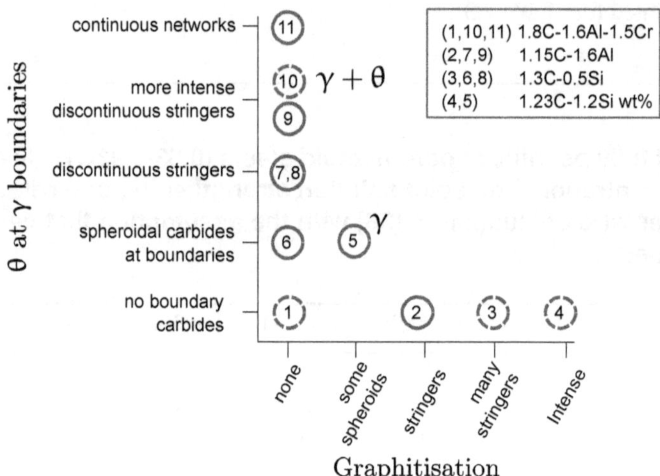

Figure 7.11 Tendency to form cementite at the γ/γ boundaries, and to form graphite. Continuous circles imply that the steel was extruded at a temperature where it was fully austenitic, whereas dashed circles represent extrusion of a mixture of austenite and cementite. Adapted using data from Lesuer et al. [42].

REFERENCES

1. H. K. D. H. Bhadeshia: 'Steels for bearings', *Progress in Materials Science*, 2012, **57**, 268–435.
2. J. S. Kirkaldy, B. A. Thomson, and E. Baganis: 'Prediction of multicomponent equilibrium and transformation diagrams for low alloy steels', In: D. V. Doane, and J. S. Kirkaldy, eds. *Hardenability concepts with applications to steels*. Materials Park, Ohio, USA: TMS-AIME, 1978:82–125.
3. R. J. Dippenaar, and R. W. K. Honeycombe: 'The crystallography and nucleation of pearlite', *Proceedings of the Royal Society A*, 1973, **333**, 455–467.
4. Y. Xu, H. Ohtsuka, and H. Wada: 'Effects of high magnetic fie1d on pearlite transformation behavior and structure', *Transactions of the Materials Research Society of Japan*, 2000, **25**, 509–512.
5. M. Umemoto, N. Komatsubara, and I. Tamura: 'Effect of austenite grain size on the hardenability of eutectoid steel', *Tetsu-to-Hagané*, 1980, **66**, 400–409.
6. Y. S. Jang, M. P. Phaniraj, D. I. Kim, J. H. Shim, and M. Y. Huh: 'Effect of aluminum content on the microstructure and mechanical properties of hypereutectoid steels', *Metallurgical & Materials Transactions A*, 2010, **41**, 2078–2084.
7. J. Chance, and N. Ridley: 'Chromium partitioning during isothermal transformation of a eutectoid steel', *Metallurgical Transactions A*, 1981, **12A**, 1205–1213.
8. H. K. D. H. Bhadeshia: Theory of Transformations in Steels: London, U.K.: CRC Press, Taylor and Francis Group, 2021.

9. A. M. Elwazri, P. Wanjara, and S. Yue: 'Effect of prior-austenite grain size and transformation temperature on nodule size of microalloyed hypereutectoid steels', *Metallurgical & Materials Transactions A*, 2005, **36**, 2297–2305.
10. K. Han, T. D. Mottishaw, G. D. W. Smith, D. V. Edmonds, and A. G. Stacey: 'Effects of vanadium additions on microstructure and hardness of hypereutectoid pearlitic steels', *Materials Science & Engineering A*, 1995, **190**, 207–214.
11. F. A. Khalid, and D. V. Edmonds: 'Effect of vanadium on the grain boundary carbide nucleation of pearlite in high-carbon steels.', *Scripta Metallurgica et Materialia*, 1994, **30**, 1251–1255.
12. A. M. Elwazri, P. Wanjara, and S. Yue: 'The effect of microstructural characteristics of pearlite on the mechanical properties of hypereutectoid steel', *Materials Science & Engineering A*, 2005, **404**, 91–98.
13. T. J. C. Webborn, and R. D. Rawlings: 'Acoustic emission from structural steels and Fe-C alloys', *Metal Science*, 1981, **15**, 533–540.
14. M. D. Hecht, B. A. Webler, and Y. N. Picard: 'Digital image analysis to quantify carbide networks in ultrahigh carbon steels', *Materials Characterization*, 2016, **117**, 134–143.
15. D. J. Sanderson, and C. W. Nixon: 'The use of topology in fracture network characterization', *Journal of Structural Geology*, 2015, **72**, 55–66.
16. N. T. Belaiew: 'The inner structure of the pearlite grain', *Journal of the Iron and Steel Institute*, 1922, **105**, 201–227.
17. G. Miyamoto, Y. Karube, and T. Furuhara: 'Formation of grain boundary ferrite in eutectoid and hypereutectoid pearlitic steels', *Acta Materialia*, 2016, **103**, 370–381.
18. T. Chairuangsri, and D. V. Edmonds: 'The precipitation of copper in abnormal ferrite and pearlite in hyper-eutectoid steels', *Acta Materialia*, 2000, **48**, 3931–3949.
19. T. Chairuangsri: 'Private communication to H. K. D. H. Bhadeshia', 2021: Heat treatment of samples referred to in Acta Materialia 48 (2000) 3931-3949.
20. NPL: 'MTDATA': Software, National Physical Laboratory, Teddington, U.K., 2006.
21. S. Scherbring, B. Adams, and J. Mola: 'Impact of interlamellar spacing and non-pearlitic features on mechanical properties and cyclic damage initiation in near-eutectoid pearlitic steels', *Materials Science and Engineering: A*, 2024, **889**, 145846.
22. G. Lucas: 'The development of small specimen mechanical test techniques', *Journal of Nuclear Materials*, 1983, **117**, 327–339.
23. E. J. Pavlina, and C. J. V. Tyne: 'Correlation of yield strength and tensile strength with hardness for steels', *Journal of Materials Engineering and Performance*, 2008, **17**, 888–893.
24. L. Qiao, Z. Wang, and J. Zhu: 'Application of improved GRNN model to predict interlamellar spacing and mechanical properties of hypereutectoid steel', *Materials Science & Engineering A*, 2020, **792**, 139845.
25. A. B. El-Shabasy, and J. J. Lewandowski: 'Effects of microstructural changes, loading conditions and test temperature on toughness of fully pearlitic eutectoid steel used in transportation industry', *Materials Science and Technology*, 2009, **25**, 369–378.
26. L. Qiao, Y. Liu, and J. Zhu: 'Application of generalized regression neural network optimized by fruit fly optimization algorithm for fracture toughness in a pearlitic steel', *Engineering Fracture Mechanics*, 2020, **235**, 107105.
27. A. Fernandez-Vicente, J. Chao, F. Penalba, M. Carsi, and O. Ruano: 'Fracture behavior of two ultrahigh carbon steels; comportamiento a fractura de dos aceros ultraalto contenido en carbono', *Revista de Metalurgia (Madrid)*, 2001, **37**, 381–385.
28. E. M. Taleff, J. J. Lewandowski, and B. Pouladian: 'Microstructure-property relationships in pearlitic eutectoid and hypereutectoid carbon steels', *Journal of Metals*, 2002, **54**, 25–30.
29. A. Fernandez-Vicente, M. Carsí, F. Penalba, and O. A. Ruano: 'The fracture toughness of a ultrahigh carbon steel containing 1.5 wt% C', *Fatigue & Fracture of Engineering Materials & Structures*, 2006, **29**, 817–828.

30. E. M. Taleff, C. K. Syn, D. R. Lesuer, and O. D. Sherby: 'Pearlite in ultrahigh carbon steels: Heat treatments and mechanical properties', *Metallurgical & Materials Transactions A*, 1996, **27A**, 111–118.
31. D. C. Saha, S. Han, K. G. Chin, I. Choi, and Y. D. Park: 'Weldability evaluation and microstructure analysis of resistance-spot-welded high-Mn steel in automotive application', *Steel Research International*, 2012, **83**, 352–357.
32. A. G. Allten: 'Discussion to "the effect of silicon on the kinetics of tempering"', *Transactions of the ASM*, 1954, **46**.
33. M. S. Bhat: 'Microstructure and mechanical properties of AISI 4340 steel modified with Al and Si': Ph.D. thesis, Lawrence Berkley Laboratories, California, USA, 1977.
34. W. C. Leslie, and G. C. Rauch: 'Precipitation of carbides in low-carbon Fe-AI-C alloys', *Metallurgical Transactions A*, 1978, **9**, 343–349.
35. J. Mahieu, J. Maki, B. C. de Cooman, and S. Claessens: 'Phase transformation and mechanical properties of Si-free CMnAl TRIP-aided steel', *Metallurgical & Materials Transactions A*, 2002, **33**, 2573–2580.
36. K. Kuo, and A. Hultgren: 'La repartition des elements d'alliage dans les aciers faiblement allies transformes', *Revue de Métallurgie*, 1953, **50**, 847–856.
37. S. Khare, and P. Mahajan: 'Effect of Al on growth kinetics of cementite in Fe–C–Al alloys', *National Academy of Sciences Letters*, 2016, **39**, 441–444.
38. B. Mintz, W. D. Gunawardana, and H. Su: 'Aluminium as solid solution hardener in steels', *Materials Science and Technology*, 2008, **24**, 596–600.
39. X. Zhou, L. Guo, Y. Wei, H. Wang, C. Chen, Y. Chen, W. Zhang, S. Liu, R. Liu, and S. Mo: 'Effect of aluminum content on dislocation loops in model FeCrAl alloys', *Nuclear Materials and Energy*, 2019, **21**, 100718.
40. S. J. Barnard, G. D. W. Smith, A. J. Garratt-Reed, and J. V. Sande: 'Atom probe studies: Role of Si in tempering of martensite, Cr diffusion in bainite', In: H. I. Aaronson, D. E. Laughlin, M. Sekerka, and C. M. Wayman, eds. *Solid-Solid Phase Transformations*. Materials Park, Ohio, USA: TMS-AIME, 1981:881–885.
41. J. H. Jang, I. G. Kim, and H. K. D. H. Bhadeshia: 'Substitutional solution of silicon in cementite: a first-principles study', *Computational Materials Science*, 2009, **44**, 1319–1326.
42. D. Lesuer, C. Syn, J. Whittenberger, and O. Sherby: 'Microstructure-property relations in as-extruded ultrahigh-carbon steels', *Metallurgical and Materials Transactions A*, 1999, **30**, 1559–1568.

8 Kinetics

There are two important kinds of polymorphic transformations from a structural point of view [1],

1. Those involving only a slight displacement of parts of the structure, of the order of heat motion. This kind of transformation may be termed *displacive transformation*....

2. Others requiring a bit-by-bit tearing down of one structure and a rebuilding of the bits to form a differently linked structure. This may be called a *reconstructive transformation*. It is of the sluggish variety because it requires a distillation of one structure into the other ...

Buerger went on to discuss these mechanisms in terms of twinning in minerals, where mechanical twins are of the displacive variety whereas the growth [annealing] twins form when a crystal completely disintegrates into bits and then is reconstructed into the twin structure. Mechanical twins cause a deformation, a change in shape, which annealing twins do not.

Pearlite falls in Buerger's second category of transformation mechanisms, whereby the austenite is reconstructed into a new pair of crystal structures. It is 'sluggish' because matter must flow by the diffusion of all atoms, whether they exist in interstices or form the host lattice. It can only form at temperatures where atomic mobility is comparable to the length scale of the 'micro'-structure. An advantage is that the flow of matter dramatically reduces the strain energy associated with the transformation when compared with the mechanism that causes deformation. This is reflected in a simple experiment, Figure 8.1. A sample of austenite that is polished organically flat and then allowed to transform into pearlite does not exhibit the shape deformation involving shear, that personifies *plates* of martensite, bainite or Widmanstätten ferrite [2–4]. The reconstruction of the lattice accommodates changes in crystal structure without causing macroscopic strains other than small and ill-defined rumples related to differences in density. The diffusion also facilitates changes in the chemical compositions of the product phases relative to that of austenite. Pearlite never is found to form without the partitioning of substitutional solutes.

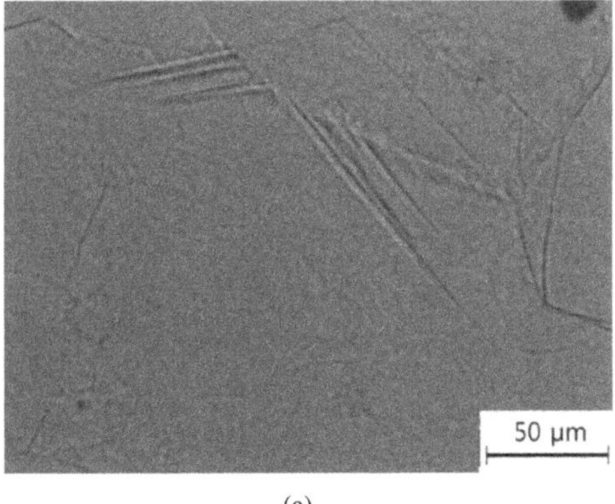

(a)

(b)

Figure 8.1 (a) Austenite in a eutectoid steel polished flat and transformed isothermally at 823 K for 3 s followed by quenching; the image shows systematic surface-displacements caused by the growth of bainite. (b) Image corresponding to (a), after etching to expose the microstructure of pearlite which does not cause surface upheavals [4]. Similar observations were first reported by Watson [2].

DOI: 10.1201/9781032631981-8

The haphazard movement of atoms during transformation means that the growth of pearlite is immune to the presence of crystallographic discontinuities such as austenite grain boundaries. Pearlite nodules can therefore exceed the size of the austenite grains [5]. Neither is it hindered by twin boundaries present in the austenite [6].

8.1 DRIVING FORCES

Pearlite exists as a two-phase mixture but during its earliest stages, before the complimentary growth sets in, one or the other of its component phases is likely to form first. The motivation for a new phase to form depends on whether there eventually is a reduction in the thermodynamic quantity, the free energy. Consider first, a binary steel of eutectoid composition. Figure 8.2a shows that the reduction in free energy is greatest when austenite decomposes into a mixture of $\alpha + \theta$, as opposed to being permitted to transform into just one phase, α or θ.

It is interesting that the difference between $|\Delta G^{\gamma \rightarrow \alpha+\theta}|$ and $|\Delta G^{\gamma \rightarrow \alpha+\gamma'}|$ decreases at large undercoolings below T_E. The motivation for the cooperative growth of $\alpha + \theta$ thus diminishes, explaining why spiky pearlite (p. 237) is generated at $T \ll T_E$, with the ferrrite leading at the transformation front, while maintaining a semblance of cooperation with the cementite.

The magnitude of the driving force for transformation decreases in the hypereutectoid 1C wt% steel (Figure 8.2b) for all scenarios when compared against the lower carbon version in Figure 8.2a. As a consequence, the free energy change accompanying the precipitation of just cementite becomes more significant, especially at temperatures in the vicinity of T_E, so it is possible the cementite can become the leading phase at the transformation front, but suitable observations of partially transformed samples are absent.

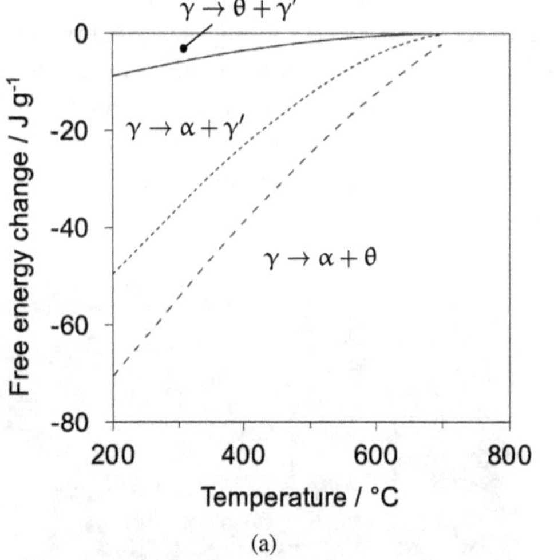

(a)

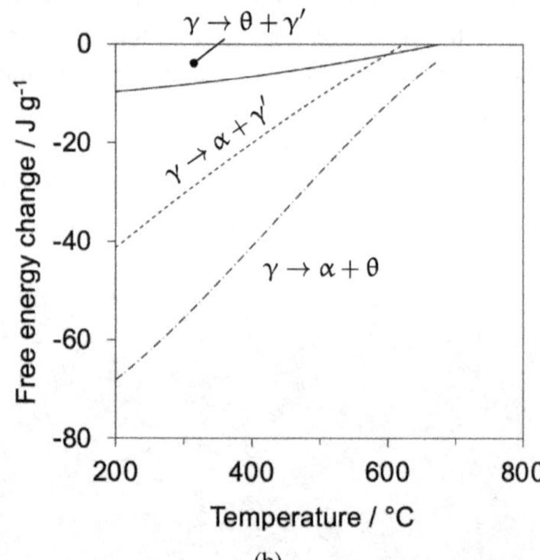

(b)

Figure 8.2 Gibbs free energy changes associated with the decomposition of austenite, as a function of temperature, in a binary Fe-C alloy. γ' refers to austenite with composition altered due to the formation of α or θ in isolation. (a) Eutectoid composition, 0.76 wt% carbon. (b) Hypereutectoid steel containing 1 wt%.

With some caveats, the evolution of pearlite should on thermodynamic grounds begin with the nucleation of ferrite in hypoeutectoid steels; indeed, it sometimes is assumed pragmatically that the nucleation rate is identical to that of allotriomorphic ferrite at γ/γ [7].[1] We shall see later that it is necessary to define the nucleation process in terms of both α and θ.

[1]Pearlite can nucleate intragranularly on niobium carbide or on a cerium-based rare earth compound [8] but the mechanisms are not clear and rates of nucleation have not been measured. It is known in other contexts that intragranular nucleation is more difficult than at γ/γ interfaces [9, 10].

When transformation is not from the fully austenitic state, the presence of prior phases must influence the evolution of the pearlite. Figure 8.3a shows cementite, as a component of what eventually becomes pearlite, evolving from an allotriomorph of cementite. The bulbous cementite particle has the same crystallographic orientation as the allotriomorph of cementite, which is related crystallographically to γ_1 on the other side [11]. It is presumed that the cementite can grow more readily into γ_2, with which it has a less-coherent and hence mobile interface.

Figure 8.3b,c illustrates two further situations developing in the same sample during isothermal transformation, with pearlite nucleating from the α/γ interface, from an isolated cementite particle that is in contact with the austenite.

There is an interesting dependence of what would be recognisable as pearlite, on the size of the austenite available. Carbon-enriched retained austenite can occur as films or as somewhat more substantial blocks in low-alloy steels containing mixed microstructures. When tempered, it is only the blocks that transform into pearlite, Figure 8.3d. The films decompose into discrete θ-particles accompanied by the epitaxial growth of bainitic-ferrite, Figure 8.3e. In the limited confines of films, it is not possible to establish the cooperative growth of α_P and θ before the austenite is consumed [p.95, 10].

8.2 NUCLEATION

8.2.1 THE NUCLEATION EVENT

Both cementite and ferrite are necessary to establish the growth process that defines pearlite. Either α or θ can nucleate first, but the event must then stimulate the creation of the other phase in close proximity. If α forms first then the austenite in its vicinity would be enriched in carbon, thus increasing $|\Delta G^{\gamma\rightarrow\theta}|$, and *vice versa*. The nucleation events are most likely heterogeneous, at γ/γ boundaries, but pearlite can nucleate at free surfaces and cracks within the steel. Cracks are favoured over free surfaces, which may be contaminated with oxide etc. [14]. The evolution of a pearlite nodule occurs in stages:

1. Single phase (α or θ) nucleates (length scale $\approx 30\,$nm), grows sufficiently to alter the chemical composition of the austenite in its vicinity, length scale $\approx 0.3\,\mu$m.

2. The complementary phase nucleates and grows in the solute-affected region alongside the first phase.

3. Cooperative growth initiates, length scale $\approx 1\,\mu$m.

4. The growing phases permeate to establish a pearlite colony containing many apparently parallel lamellae within a bicrystal, length scale $\approx 5\,\mu$m. The word *permeate* implies that the lamellae of each phase are connected in three dimensions, which is not possible if each lamella is a continuous plate with no connections to any other.

5. Branching to generate a pearlite nodule containing several colonies growing in different directions but essentially maintaining the original crystallography throughout, length scale $\approx$ 10-15 μm (p. 18).

Classical theory suggests that the nucleation rate should increase dramatically with the undercooling below the equilibrium temperature [p.431, 15]. This is not the case with pearlite, where it increases much more gently with supersaturation, with the greatest increase at undercoolings in excess of $T_E - T \approx 100\,^\circ$C, Figure 8.4. Therefore the nucleation of pearlite is not surprisingly, different from that of a single-phase.

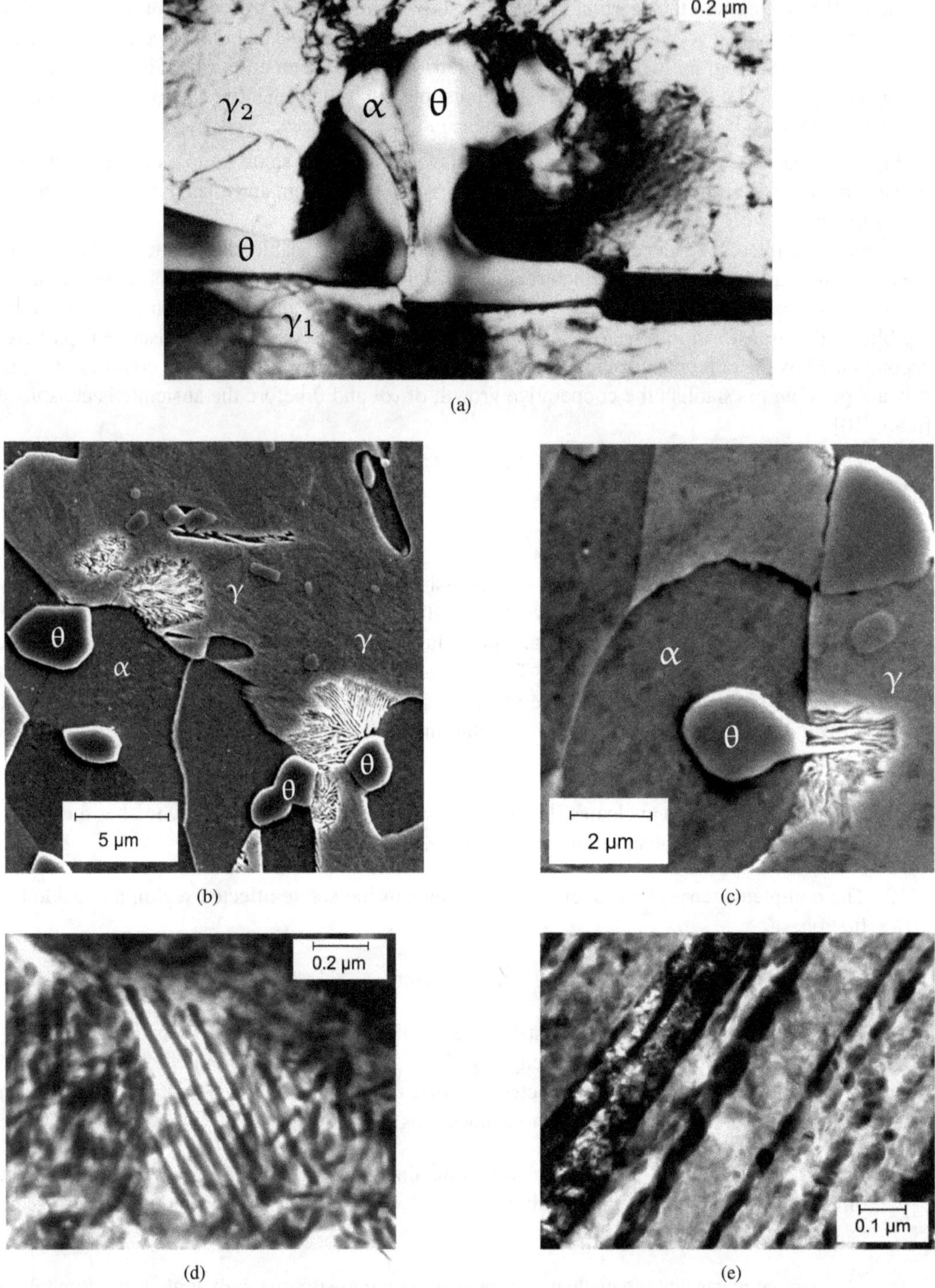

Figure 8.3 Evolution of a pearlite colony. (a) From proeutectoid cementite layer present at the austenite grain boundary, prior to the onset of pearlite (micrograph courtesy of R. Dippenaar) [12]. (b) From pre-existing ferrite/austenite interface, and in the lower half, probably from pre-existing cementite. (c) From pre-existing cementite/austenite interface. Micrographs b,c, courtesy of Hala Salman Hasan. (d) Blocks of carbon-enriched γ able to establish cooperative growth to pearlite during tempering. (e) Discrete cementite precipitated from films of γ during tempering of $\alpha_b + \gamma$. After [13].

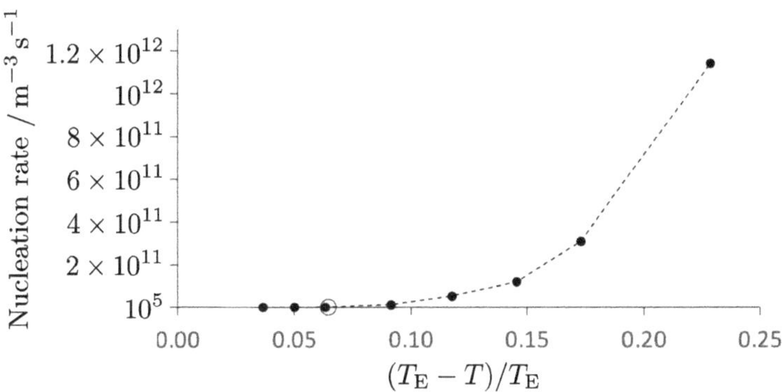

Figure 8.4 Measured nucleation rate of pearlite in Fe-0.78C-0.63Mn wt% steel. In plotting the data, it is assumed here that $T_E = 727\,°C$. An undercooling of 100 °C corresponds to 0.14 on the abscissa. Selected data from Mehl and Dubé [5]. The large open point refers Fe-0.78C wt% steel austenitised at 870 °C and transformed at 680 °C [16].

Fisher dealt with the nucleation of pearlite as a two-stage process involving both α and θ [17], assuming that the cementite nucleates first at the austenite grain boundaries. The θ/γ interfacial area thus created and the depletion of carbon in its vicinity stimulates ferrite and it is this combination that is regarded as a pearlite nucleus.

If the cementite particle, following nucleation, grows under diffusion control, then its radius would increase with the square root of time, $r = \eta t^{\frac{1}{2}}$, where η is an appropriate parabolic thickening rate constant [e.g., p.187, 18]. The θ/γ area assuming a spherical particle is therefore $4\pi\eta^2 t$ so

$$I^{\alpha} = I^{\alpha}_{A} \times 4\pi\eta^2 t \tag{8.1a}$$

where I^{α} is the number of ferrite nuclei per second on the individual cementite particle, and I^{α}_{A} the nucleation rate per unit area of θ/γ interface. It may be that more than one ferrite nucleus emerges, but if p is the probability of the first ferrite particle nucleating on the θ/γ interface, representing the genesis of a pearlite nucleation event, then in a time interval $\mathrm{d}t$,

$$\mathrm{d}p = (1-p)I^{\alpha}\,\mathrm{d}t = (1-p)I^{\alpha}_{A}\,4\pi\eta^2 t\,\mathrm{d}t \tag{8.1b}$$

$$\text{so} \quad p = 1 - \exp\{-\underbrace{2\pi I^{\alpha}_{A}\eta^2}_{m_o} t^2\} = 1 - \exp\{-m_o t^2\}. \tag{8.1c}$$

The rate of nucleation of pearlite *per cementite particle* becomes

$$I^{P} = \mathrm{d}p/\mathrm{d}t = 2m_o t \exp\{-m_o t^2\}. \tag{8.1d}$$

where the units of I^{P} are in number per unit time. If I^{θ}_{A} is the nucleation rate of cementite per unit area of γ/γ boundary, the number of cementite particles that form in the time interval $\mathrm{d}\tau$ in the unit area is $I^{\theta}_{A}\mathrm{d}\tau$, so the nucleation rate of pearlite at these particles becomes, at time t,

$$\mathrm{d}I^{P}_{A} = [2m_o(t-\tau)\exp\{-m_o(t-\tau)^2\}]\,I^{\theta}_{A}\,\mathrm{d}\tau \tag{8.1e}$$

where the term in square brackets corresponds to the nucleation of ferrite on a single θ-particle (Equation 8.1d). Therefore,

$$I^{P}_{A} = \int_{\tau=0}^{t} \mathrm{d}I^{P}_{A} = I^{\theta}_{A}[1 - \exp\{-m_o t^2\}]. \tag{8.1f}$$

Equation 8.1f is derived on the assumption that cementite is the precursor to pearlite nucleation, but a similar form would be obtained if ferrite nucleated first, followed by cementite on α/γ interfaces – all that is needed is to substitute the superscripts in $I^{\theta}, \eta^{\theta}$ by the corresponding terms for α. Note also that to convert the nucleation rate per unit area to the rate per unit volume, $I_V^P = I_A^P \times S_V$ where S_V is the γ/γ interfacial area per unit volume, related to the lineal intercept size of the austenite grains, $2/\overline{L}_{\gamma}$.

The fit of Equation 8.1f to published experimental data [19] is illustrated in Figure 8.5, using $m_o = 0.001\,\mathrm{s}^{-2}$ and $I_A^{\theta} = 3960\,\mathrm{mm}^{-3}\,\mathrm{s}^{-1}$; the latter value indicates that the first phase to nucleate (whether θ or α) has a much greater nucleation rate than pearlite (I_A^P). This should not surprise because the nucleus of pearlite is a two-staged event, consistent with the fact that the nucleation rate does not rise as sharply as would be expected for a single phase with undercooling (Figure 8.4).

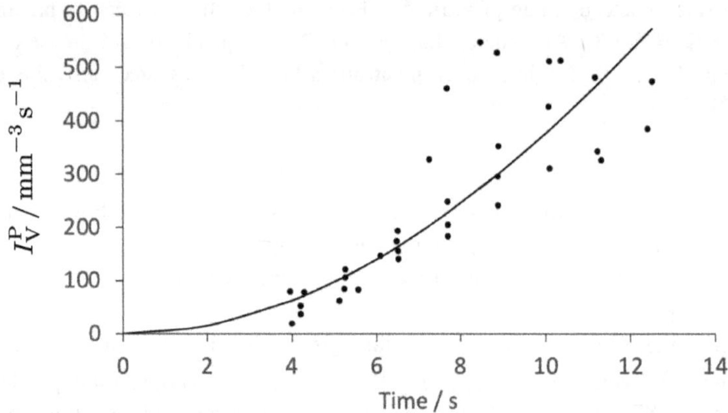

Figure 8.5 The data on the nucleation rate of pearlite are measured [19], the curve fitted to the data using Equation 8.1f, as described in the text. The steel has the composition Fe-1.02C-0.31Si-0.2Mn wt%, transformed isothermally at 680 °C, $\overline{L}_{\gamma} \approx 95\,\mu\mathrm{m}$ following austenitisation at 1100 °C for 30 min. Equilibrium thermodynamic phase diagram calculations indicate that there is no proeutectoid cementite present in the sample at the austenitisation temperature [20].

The number of pearlite nodules has been measured *in situ* using a three-dimensional neutron depolarisation technique during the isothermal transformation in Fe-0.72C-0.61Mn-0.35Si-0.27Cr-0.09Ni wt%, Figure 8.6 [21].

Table 8.1 shows that at the same value of time (12 s) at 680 °C, the lower carbon steel has a much smaller nucleation rate than the hypereutectoid steel, even though the former has a much smaller austenite grain size. The same trend applies to the pearlite growth rate. The explanation must lie in the reduction of the free energy change available for transformation, which has a smaller magnitude in the lower carbon steel, because of its greater substitutional solute content.

8.3 FERRITE-CEMENTITE INTERFACIAL ENERGY

The cooperative growth of cementite and ferrite creates α/θ interfaces that are defects in the sense that their excess energy is stored within the pearlite. This stored energy increases as the interlamellar spacing decreases and must, during transformation, be provided by the free energy of transformation $\Delta G^{\gamma\rightarrow\alpha+\theta}$, so the spacing becomes a function of the driving force. The interfacial energy per unit area, $\sigma^{\alpha\theta}$ determines the magnitude of the stored energy, and although the crystallography and interface orientation are important, $\sigma^{\alpha\theta}$ often is taken to be single-valued, based on macroscopic measurements.

Table 8.1

Measured data on alloys transformed isothermally at 680 °C. Using the steel designations in [19], Steels G (nucleation rate data) and F (growth rate data) have almost identical chemical composition; growth data from F are assumed here to apply to G. The nucleation rates are after 12 s at 680 °C, measured [19] and by extrapolation of the curve in Figure 8.6 to the required time and taking its slope. The driving forces are calculated using MTDATA [22] with the SGTE database.

Composition	$\overline{L}_\gamma$	I_V^P	v_P	$\Delta G^{\gamma \rightarrow \alpha+\theta}$	Ref.
wt%	µm	$\mathrm{mm^{-3}\,s^{-1}}$	$\mathrm{\mu m\,s^{-1}}$	$\mathrm{J\,kg^{-1}}$	
Fe-1.02C-0.31Si-0.2Mn	95	360	4.7	−4940	[19]
Fe-0.72C-0.61Mn-0.35Si-0.27Cr-0.09Ni	27	28	0.12	−4030	[21]

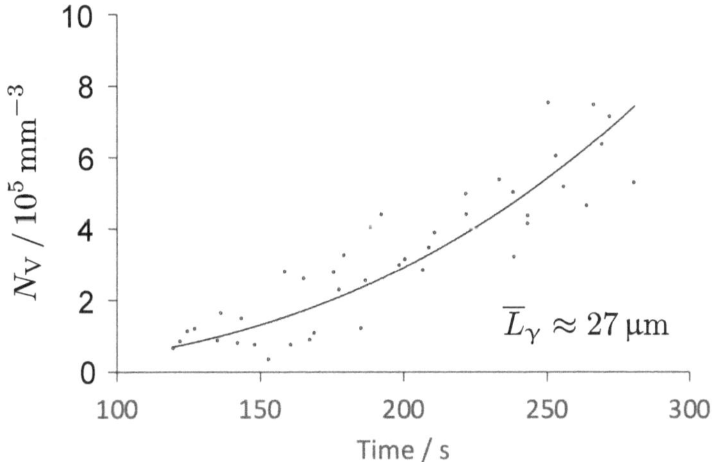

Figure 8.6 Number of pearlite nodules as a function of time at 680 °C in Fe-0.72C-0.61Mn-0.35Si-0.27Cr-0.09Ni wt%. The data are represented by an empirical equation $N_V \approx 0.123t^{2.77}$ with the same units as the plot. The austenite grain size was not reported but is calculated using Equation 2.5. Selected data from Offerman et al. [21].

There are approximate determinations of the $\alpha/\theta_{\mathrm{pearlite}}$ orientation relationship [11, 23, 24], but these have not had practical consequences other than indicate scatter. Studies that use conventional electron diffraction or electron backscattered diffraction are not of sufficient accuracy to establish crystallography. The popular α/θ Bagaryatskii orientation relationship (Chapter 3) is an imprecise representation of the actual Isaichev orientation [25–27]. When nucleation begins with cementite, the ferritic component forms later in contact with γ and θ, while adopting the Isaichev α/θ orientation [28]. Other orientations develop in hypoeutectoid steels where ferrite nucleates first.

The activation energy of nucleation of a single phase scales with $(\sigma^{\alpha\theta})^3$. But given the orthorhombic symmetry of cementite, $\sigma^{\alpha\theta}$ will vary with the orientation of the interface plane. Data on this subject are sparse. For an isolated cementite particle at equilibrium with zero contamination of the surface, density functional calculations suggest the surface energies listed in Table 8.2; surprisingly, the energy does not vary much with the crystallographic indices of the surface. This gives an indication of anisotropy but not the actual behaviour of cementite in contact with either γ or α. We shall see that the calculated surface-energy values in Table 8.2 are much greater than $\sigma^{\alpha\theta}$ estimated experimentally (p. 187).

Whereas termination at a free surface leaves incomplete Fe-Fe bonds, an interface between cementite and α or γ ensures continuity of the metallic bond, so it is not surprising that $\sigma^{\alpha\theta}$ represents less of a 'defect' than a free surface. For a wide range of metals, the solid-liquid interfacial energy is far smaller than the corresponding liquid-vapour value [29]. This conclusion persists when a comparison is made between grain-boundary energies and solid-gas surface energies [30].

Table 8.2

Calculated surface energy per unit area of an isolated cementite particle as a function of the crystallographic indices of the surface. After Li et al. [31].

{hkl}	$\mathrm{J\,m^{-2}}$	{hkl}	$\mathrm{J\,m^{-2}}$
(011)	2.294	{111}	2.520
(001)	2.357	{101}	2.556
{110}	2.432	(100)	2.681
(010)	2.494		

Interfaces can determine mechanical properties, the cohesion of a polycrystalline aggregate and kinetics by introducing a barrier to nucleation. The interface is characterised by its fine structure, which is one of the features that determines its mobility and its energy per unit area.

When considering a bicrystal, there are five degrees of freedom associated with an interface. Two are used in defining the orientation of the interface (i.e., the unit-normal to the interfacial plane) and a further three to deal with the orientation relationship between the two crystals. Each of these degrees of freedom can be varied independently to generate different combinations of interfacial structure and energy. The defect structure of an interface cannot be unambiguously calculated because for identical degrees of freedom, the Burgers vector content crossing an arbitrary vector in the interface then has to be distributed across assumed defect structures. All practical steels will contain impurity atoms that tend to segregate to the interface and consequently alter its energy.

Interfacial energies usually are derived by fitting theory to data on the coarsening of cementite precipitates, from pearlite growth rate measurements, or using calorimetric measurements of the enthalpy of transformation. As such, the data are susceptible to imperfections in the theory available and to the fact that singular derived-values necessarily neglect the difficulties listed above. Published data showing wide variations, are plotted in Figure 8.7, emphasising that $\sigma^{\alpha\theta}$ is strictly a fitted parameter.

Atomistic calculations or interatomic potentials have been used to estimate $\sigma^{\alpha\theta}$. One such simulation of 4442 block of atoms assumed the Bagaryatskii orientation with the interface at $\{112\}_\alpha \parallel (001)_\theta$. However, the wrong cementite space group *Pnma* was assumed, so the $\{112\}_\alpha$ should have been set parallel to $(010)_\theta$ [p.383, 18]. The interface at the centre of the block was in forced coherency, so the excess energy due to the interface is a linear function of the block size [32], giving $\sigma^{\alpha\theta} = 0.615\,\mathrm{J\,m^{-2}}$ when extrapolated to a block size to zero. This means that the derived value refers only to the chemical component of $\sigma^{\alpha\theta}$ since the coherency strains are neglected. One difficulty is that the interfacial area in the simulation is small ($4.12\,\mathrm{nm^2}$); in reality, coherency stresses become unbearable as the area increases so dislocations are introduced periodically to mitigate long-range strains by localising them approximately to a distance comparable to the dislocation spacing.

Larger simulations that generate dislocation-like defects within the α/θ interface while accounting for microscopic degrees of freedom[2] have estimated interfacial energies using interatomic potentials and assuming the Bagaryatskii orientation relationship [33]. For the optimum *Pnma*-cementite terminations at the $(010)_\theta$ plane parallel to $\{211\}_\alpha$, the range of $\sigma^{\alpha\theta}$ was from $\approx 0.52 \rightarrow 1.23\,\mathrm{J\,m^{-2}}$ depending on the interatomic potential selected.

It is difficult to see how the atomistic simulations help identify an interfacial energy that is of use in kinetic theory. The Bagaryatskii relationship, although much quoted, does not exist [25–27] and the α-θ orientation relationship in pearlite is far from unique [34]. The interfacial planes considered in the simulations are themselves are simply assumed along with the orientation relationship.

There are some particularly low values of interfacial energy evident in Figure 8.7, derived by fitting cementite coarsening to simplistic theory [35, 36], but there are difficulties in doing this. The growing particles tend to locate at grain boundaries, with evidence that boundary diffusion has a prominent role. Figure 8.8 which shows the bimodal θ-size distribution, with coarse allotriomorphs at α/α boundaries and fine precipitates at intragranular locations, emphasising the role of boundary diffusion in the former case. If $\sigma^{\alpha\theta}$ values based on coarsening data are neglected, the interfacial energy is in the range $0.8 \rightarrow 1.4\,\mathrm{J\,m^{-2}}$, depending on temperature for reasons that are not known.

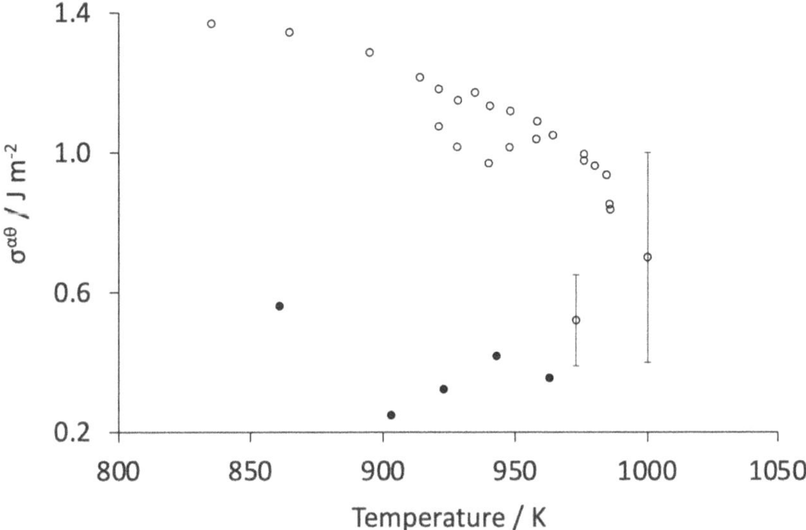

Figure 8.7 A compilation of α/θ interfacial energies per unit area, data from [35–41]. The techniques used for the derivation $\sigma^{\alpha\theta}$ are from coarsening data (filled points) [35, 36], enthalpy measurements [37], dihedral angle determinations [38], and pearlite growth measurements [40, 41]. Thermodynamic calculations for an unspecified temperature give $\sigma^{\alpha\theta} = 0.24\,\mathrm{J\,m^{-2}}$ but for austenite, a much lower value of $\sigma^{\gamma\theta} = 0.07\,\mathrm{J\,m^{-2}}$ [42], smaller than an assumed $\sigma^{\gamma\theta} = 0.4\,\mathrm{J\,m^{-2}}$ for an incoherent interface [43].

[2]Because cementite has an ordered arrangement of carbon and iron atoms, the atomic structure that the cementite presents to the ferrite at the interface can be different depending on where the plane is located within the unit cell. For example, the $(100)_\theta$ plane has six possible terminations.

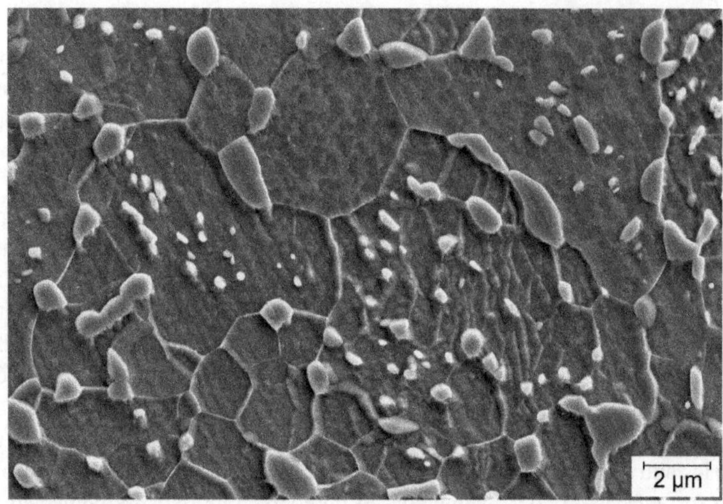

Figure 8.8 Eutectoid steel spheroidised at 700 °C for 11 h from an original microstructure that was pearlitic. The large cementite particles all are located at the α/α grain boundaries. Micrograph courtesy of Apparao Chintha.

8.4 INTERLAMELLAR SPACING

There is no fundamental method for predicting the spacing of pearlite. Common assumptions include that of Zener, that the spacing adopted would maximise the rate of growth. Alternatively, that the spacing leads to a maximum entropy production rate, which during isothermal transformation, is proportional to the maximum free energy dissipation rate.

Experimental data on S_I can of course be expressed empirically as a function of chemical composition and the transformation temperature. In one case, a neural network was applied but then used to develop a linear relationship with the input variables treated as if they have independent effects (with T expressed in °C) [44]:

$$\begin{aligned}\log\{S_I/\mu\mathrm{m}\} \approx\; & -2.212+0.0514w_{\mathrm{Mn}}-0.1171w_{\mathrm{Cr}}+0.0967w_{\mathrm{Ni}}-0.002w_{\mathrm{Si}} \\ & -0.4812w_{\mathrm{Mo}}-\log\left\{\frac{T_E-T}{T_E}\right\}. \qquad (8.2)\end{aligned}$$

The term for chromium has been recalculated here to exclude concentrations greater than the maximum in the dataset:

	w_{Mn}	w_{Ni}	w_{Cr}	w_{Si}	w_{Mo}	$T/$°C
Minimum	0.00	0.00	0.00	0.0	0.0	567
Maximum	1.80	9.00	1.94	0.48	3.00	728
Mean	0.42	1.01	1.87	0.25	0.48	661
Standard deviation	0.64	0.61	0.52	0.63	0.25	41

The effect of silicon was perceived to be minimal, but this is inconsistent with careful experiments using model alloys, Figure 8.9, which shows that the addition of 1 wt% leads to a large reduction in S_I but adding more than that has little effect. This is similar to the effect of manganese, emphasising that the real relationship must be non-linear. There will exist interdependencies between the solutes, as evident for the Fe-C-Mn-Si alloy where the addition of Mn has virtually no effect in the presence of 1Si wt%. A neural network model is ideal for capturing these complexities

– however, the practice of reducing its output to a linear equation for ease of use [44] must lead to a loss of information. Other work in multicomponent steels confirms the role of silicon [45].

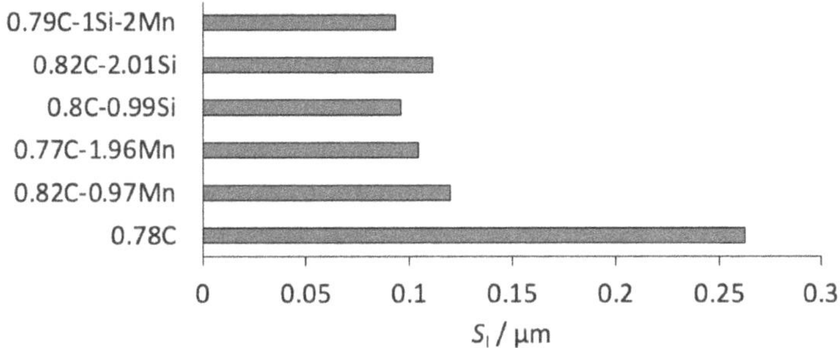

Figure 8.9 Interlamellar spacings from pearlite generated by isothermal transformation at 600 °C, using a series of model alloys. Selected data from Fu et al. [46].

Equation 8.2 has a eutectoid temperature that has been set constant at 728 °C [47] but must be different for substitutionally-alloyed steels and indeed, the eutectoid reaction then occurs over a range of temperatures so T_E is difficult to define. An alternative based on a collection of data listed in [48] therefore avoids T_E (with T expressed in °C):

$$S_I/\mu m = (127.351 - 0.17368T - 4.9195w_{Mn} + 1.7868w_{Cr})^{-1} \quad (8.3)$$

	w_{Mn}	w_{Cr}	$T/°C$
Minimum	0.00	0.00	600
Maximum	2.00	2.00	790

8.4.1 EFFECT OF CARBON ON INTERLAMELLAR SPACING

Equations such as 8.2 and 8.3 assume that carbon itself does not influence the interlamellar spacing. However, fully pearlitic steels can form over a range of carbon concentrations if the steel is transformed at a temperature below T_E where both α and θ can form simultaneously from austenite (p. 193). Figure 8.10a shows that there is a greater reduction in free energy when pearlite forms from carbon-rich austenite. Since the critical interlamellar spacing at which all of $\Delta G = G^{\alpha+\theta} - G^{\gamma}$ is consumed in creating α/θ interfaces decreases as $|\Delta G|$ increases (Equation 8.8a), the spacing adopted by pearlite is expected to be finer at greater carbon concentrations. This indeed is found to be the case, Figure 8.10b. This effect is independent of the accompanying increase in the volume fraction of cementite within the pearlite. The greater dependence of the interlamellar spacing with temperature at the higher carbon concentration, is consistent with the corresponding variation of ΔG with T and the carbon concentration.

8.4.2 DEPENDENCE OF INTERLAMELLAR SPACING ON DRIVING FORCE

None of the admittedly useful, empirical equations or machine learning models described have a physical basis. It will be shown in Section 8.5 that the minimum possible interlamellar spacing, S_I^C is a function of the chemical driving force available because at that spacing, all of the available free energy is consumed in creating α/θ interfaces, i.e.,

$$S_I^C = \frac{2\sigma^{\alpha\theta}}{\Delta G^{\alpha+\theta\rightarrow\gamma}} \quad \text{with} \quad S_I = f\{S_I^C\}$$

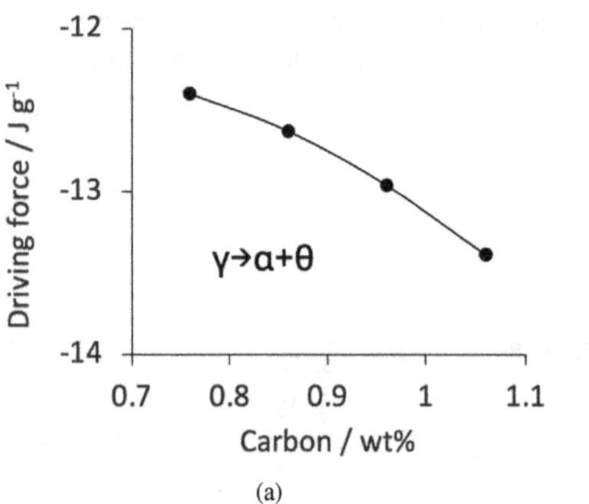

(a)

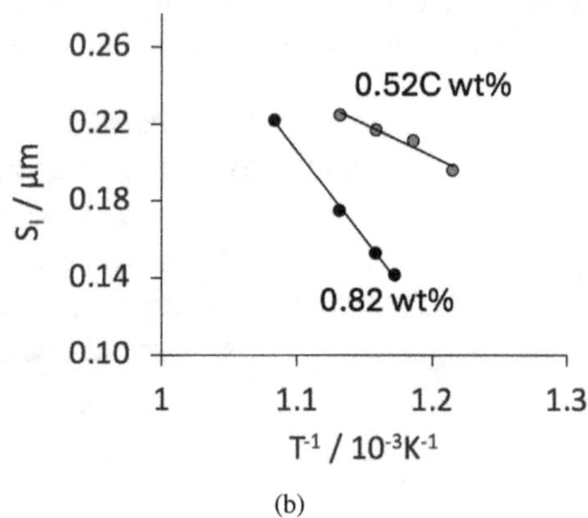

(b)

Figure 8.10 (a) Calculated free energy change (driving force $\Delta G = G^{\alpha+\theta} - G^{\gamma}$) when austenite transforms into pearlite, as a function of the carbon concentration of that austenite. The calculations were implemented using the MTDATA with the SGTE database [22]. (b) The lower carbon steel has a coarser interlamellar spacing; both alloys contained 0.5Mn-0.2Si wt% and were transformed into fully pearlitic microstructures. Selected data from Bae et al. [49].

where $\sigma^{\alpha\theta}$ is the interfacial energy per unit area, $\Delta G^{\alpha+\theta\rightarrow\gamma}$ is written to be positive by reversing the reaction, $S_{\rm I} \geq S_{\rm I}^{\rm C}$ and the functional relationship between these two spacings turns out to be simple. It is logical on the basis of Equations 8.8a that there is an inverse relationship between $S_{\rm I}$ and $\Delta G^{\alpha+\theta\rightarrow\gamma}$. Such a relationship would remove the need to consider individual alloying elements or solutes for which pearlite has not been characterised, since the entire effect of the solute would be embedded in the driving force. A knowledge of the eutectoid temperature is not necessary, a clear advantage given that it is not uniquely defined for steels containing substitutional solutes. Thermodynamic methods that enable the calculation of $\Delta G^{\gamma\rightarrow\alpha+\theta}$ as a function of alloying elements in multicomponent steels are now routinely available. Figure 8.11 shows, for a large variety of steels [49–60], the relationship between the spacing and driving force. The fitted line has a zero intercept to express the fact that the spacing should be vanishingly small in the hypothetical scenario where the driving force approaches infinity, expressed by the equation:

$$S_{\rm I}\,/\,{\rm m} = -\frac{8.6}{\Delta G^{\gamma\rightarrow\alpha+\theta}\,/\,{\rm J\,m^{-3}}} \qquad \text{correlation coefficient 0.91.} \tag{8.4}$$

The coefficient 8.6 would indicate an interfacial energy per unit area $\sigma^{\alpha\theta} \approx 4.3\,{\rm J\,m^{-2}}$, which is too large. However, it should be borne in mind that the driving force represents equilibrium, with the equilibrium partitioning of solutes. This does not happen in practice, with considerable solute trapping accompanying the growth of pearlite [61]; pearlite never has been found to grow with the equilibrium partitioning of solutes between the ferrite and cementite. Mehl and Dubé [5] pointed out during 1951, that the initial composition of the pearlitic-cementite in substitutionally-alloyed steels is not as expected from equilibrium; prolonged holding at the reaction temperature after the pearlite had stopped growing, caused the composition to approach equilibrium. Therefore, the equilibrium driving force $\Delta G^{\gamma\rightarrow\alpha+\theta}$ is exaggerated in magnitude, which would inflate the interfacial energy to account for the observed spacings. None of the other empirical equations, when applied to the same dataset, are able to produce greater correlation than illustrated in Figure 8.11, and are limited in dealing with a few solutes whereas a generic calculation of the free energy is able to cope with any solute concentration or type, if the basis of Equation 8.4 is justified.

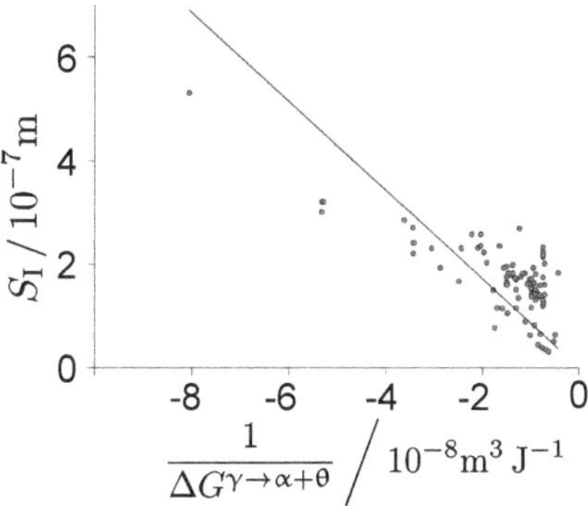

Figure 8.11 Plot of experimentally measured interlamellar spacings against the inverse of the calculated driving force available at the isothermal transformation temperature. The line has a zero intercept.

8.4.3 CAST IRON

In the presence of graphite, the equilibrium carbon concentration of the austenite is determined by the temperature within the graphite $+\gamma$ phase field. There are limited data on the interlamellar spacing of pearlite that may form during cooling from that field, but one study is revealing [62]. During continuous cooling transformation, S_I is found to be proportional to $\Delta T = T_E - T$, Figure 8.12 with the caveat that T_E is ill-defined. The cast irons studied had $\overline{w}_C = 2.80 \rightarrow 3.62$ with $\approx$1.8Si, 0.55Mn, 0.14Cr and 0.4Cu wt%. The equilibrium 'eutectoid temperature' was estimated using thermodynamic calculations and expressed empirically as

$$T_E/^\circ\text{C} \approx 727 + 21.6w_{\text{Si}} + 0.023w_{\text{Si}}^2 - 21w_{\text{Cu}} - 25w_{\text{Mn}} + 8w_{\text{Mo}} + 13w_{\text{Cr}} - 33_{\text{Ni}} + 21w_{\text{Sn}} \qquad (8.5)$$

covering $w_{\text{Si}} \leq 3$, $w_{\text{Mn,Cu,Cr,Ni}} \leq 1$ and $w_{\text{Mo}} \leq 0.5$. Because graphite does not absorb substitutional solutes or iron, the concentrations w_i should be set 1.05 times greater than the average in the cast iron. Tin often is an impurity in cast irons but its effect on the pearlite reaction is established. It raises the eutectoid temperature and eutectoid carbon-concentration. In low-carbon steels this results in a reduction in the fraction of pearlite [63].

In Figure 8.12, the line has been drawn so that the spacing becomes zero when $\Delta T = \infty$; while the method overestimates the interlamellar spacing at large undercoolings, the agreement is reasonable given the complexities of defining a eutectoid temperature in a multicomponent alloy (p. 166) and the fact that the pearlite isn't generated isothermally.

Pearlite, particularly when present in grey cast iron with its typically 2 wt% silicon, can decompose into a mixture of flake-graphite and cementite during heat treatment for several hours at temperatures in excess of about 600 °C [64], Figure 8.13. Tempering in the presence of flake-graphite accelerates the process; there is good metallography to show ferrite allotriomorphs in contact with graphite where pearlite originally existed [65]. During the decomposition of pearlitic cementite, nearby-graphite is a sink for carbon, leaving carbide-free ferrite in its vicinity. In regions somewhat remote from the flakes, it would be necessary for graphite to nucleate in order to progress the decomposition of cementite. For the same reason, it is common in as-cast spheroidal cast iron to see nodules of graphite enclosed by thick layers of ferrite in the otherwise pearlitic matrix [66, 67], Figure 8.13c.

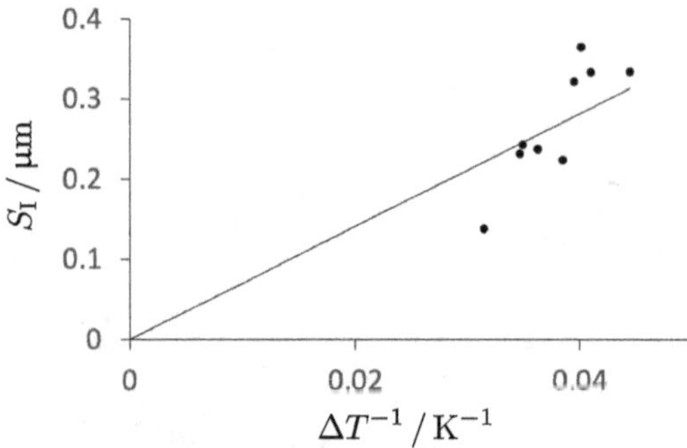

Figure 8.12 The measured interlamellar spacing of pearlite in cast irons as a function of the reciprocal of the undercooling below the eutectoid temperature. $S_I\,/\,\mu\text{m} \approx 7.05/\Delta T$. Data from Fourlakidis et al. [62].

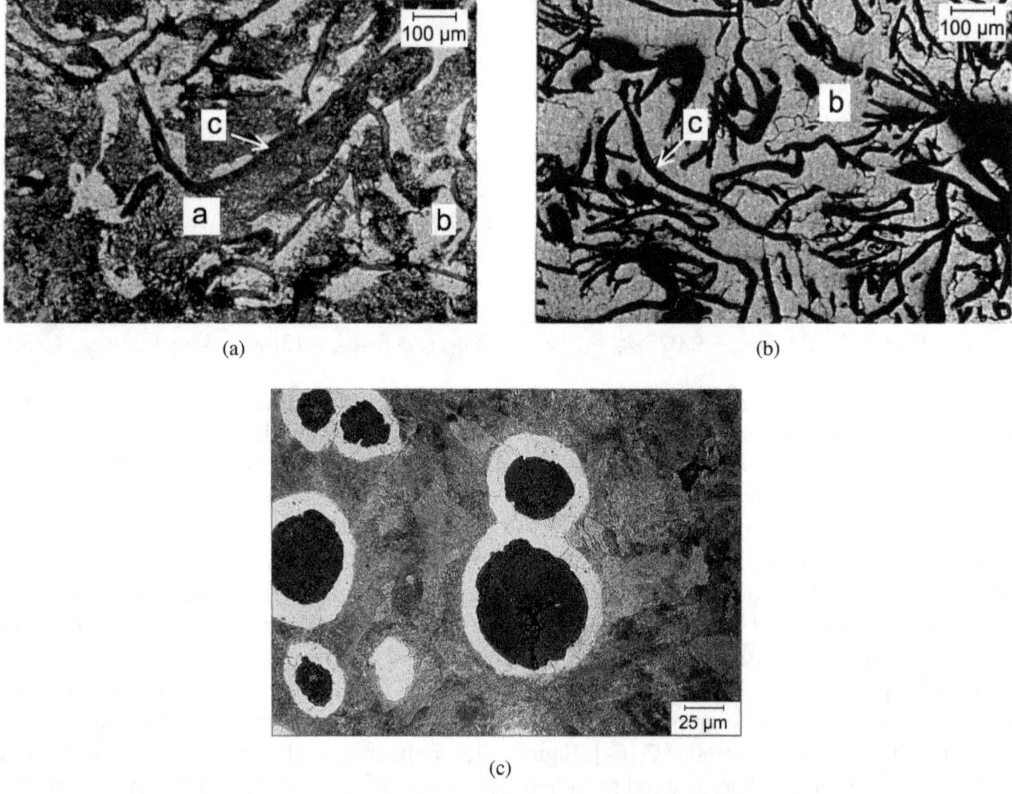

Figure 8.13 (a) Grey cast iron from a seventeenth century cannon. The letters a, b, c represent pearlite, ferrite and graphite, respectively. (b) Following heat treatment at 1000 °C for 6 h. Reproduced with the permission of Taylor and Francis from North et al. [64], with minor adaptation. (c) Spheroidal graphite cast-iron showing nodules of graphite encased by ferrite, in an otherwise pearlitic matrix. Image reproduced from Lumley [67] with permission of Springer Nature.

8.5 GROWTH

The pearlite transformation in steels is unique in having both a source and a sink for solute that is partitioned into the austenite. Cementite is rich in carbon whereas ferrite accommodates very little when it is in equilibrium with either cementite or austenite. It therefore is necessary for carbon to be redistributed at the common transformation front that the α, θ have with the parent austenite. In a binary Fe-C system, the cementite would absorb all the carbon that is partitioned into the austenite by the growth of ferrite. This can happen by diffusion through the adjacent austenite, in a direction parallel to the transformation front, and also by a flux of carbon through the transformation interface. There is no net accumulation of carbon in the austenite, leading to a constant growth rate [19, 68–70] and the ability of the pearlite to consume all of the austenite, Figure 8.14.

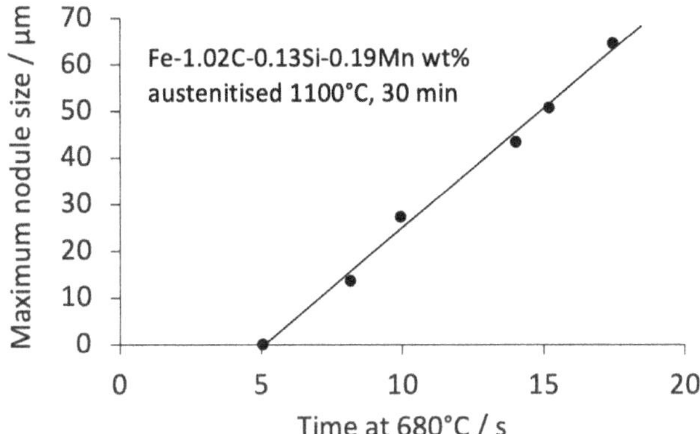

Figure 8.14 Each point is measured on a different nodule, isolated for its maximum size in a metallographic section, after the given amount of transformation time. Selected data from Hull et al. [19].

The growth models to be described here are necessarily simplistic given the reality of the elaborate three-dimensional structure of pearlite (Chapter 2) [11, 71–73]. Growth occurs in three-dimensions whereas the models treat the problem essentially as the one-dimensional advance of the transformation front. The cementite must branch to accommodate changes in growth-orientation consistent with the shape of the pearlite nodule, resulting in irregularities in growth rate. The cementite and ferrite within pearlite are known to be feature-rich, with curvature, striations, discontinuities and holes [72, 73]. Some of the complexity is illustrated in Figure 8.15; the ferrite bulges out at the transformation front, presumably because $|\Delta G^{\gamma \rightarrow \gamma' + \alpha}|$ is greater than that for cementite precipitation (Figure 8.2).

The simplified process is illustrated in Figure 8.16, where all the fluxes essentially are parallel to the transformation front, J_V through the austenite in the vicinity of the transformation front, and J_B through the averaged interface between the austenite and pearlite. The proportion of ferrite and cementite within the colony is determined by the average carbon concentration of the austenite $\overline{c}^{\gamma}$ together with the equilibrium concentrations $c^{\alpha\theta}$ and $c^{\theta\alpha}$. Using the lever rule, it follows that

$$\frac{S_I^{\alpha}}{S_I} = \frac{c^{\theta\alpha} - \overline{c}^{\gamma}}{c^{\theta\alpha} - c^{\alpha\theta}}, \qquad \frac{S_I^{\theta}}{S_I} = \frac{\overline{c}^{\gamma} - c^{\alpha\theta}}{c^{\theta\alpha} - c^{\alpha\theta}}, \qquad V_V^{\theta} = \frac{S_I^{\theta}}{S_I} \tag{8.6}$$

where V_V^{θ} is the fraction of cementite within the pearlite. Both α and θ are necessary to form pearlite, so the austenite must be supersaturated with respect to them both [74] with $c^{\gamma\theta} \leq \overline{c}^{\gamma} \leq c^{\gamma\alpha}$, (Figure 8.17a) – this condition defines the Hultgren extrapolation region.[3] In that region, it is possible for the austenite to transform completely into pearlite, even though the austenite does not have the eutectoid composition. In a time-temperature-transformation diagram for a hypoeutectoid steel, ferrite would not have to precede pearlite at an undercooling where $c^{\gamma\theta} \leq \overline{c}^{\gamma} \leq c^{\gamma\alpha}$ (Figure 8.17b).

[3] Although attributed often to Hultgren, the extrapolation was described and illustrated first by Belaiew in his 1922 paper on the inner structure of pearlite [74].

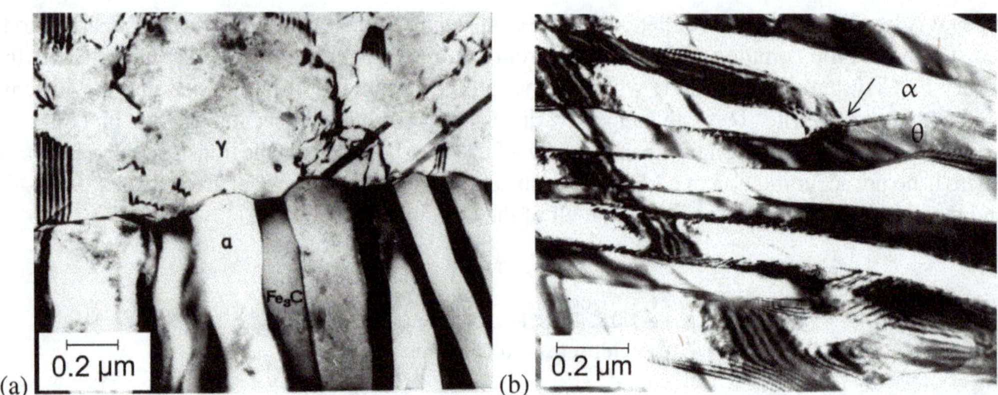

Figure 8.15 Transmission electron micrographs of pearlite in a Fe-0.79C-11.9Mn wt% steel transformed partially at 630 °C. (a) The shape of the transformation front with austenite, and an illustration of the fact that the local apparent-spacing at the interface with austenite is not uniform. (b) The arrow indicates a cementite branching event. The α/θ interface-plane can be identified by both depth fringes and the interfacial dislocation structure. The interface clearly is not flat along the length of a single θ-lamella. Micrographs courtesy of Professor R. Dippenaar [12].

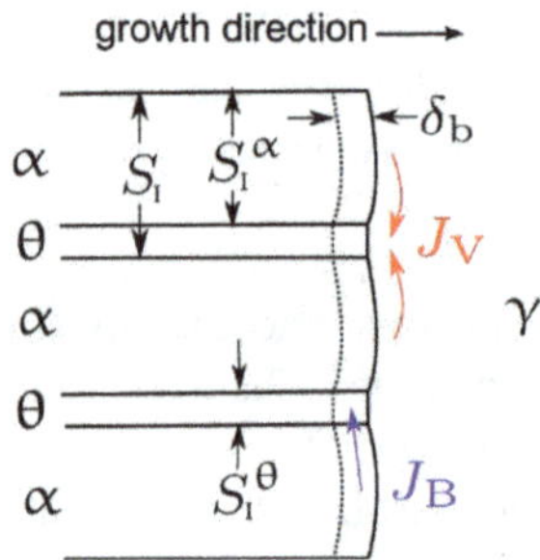

Figure 8.16 Schematic representation of a growing pearlite colony. The fluxes indicate the diffusion paths for the carbon, either through the austenite ahead of and parallel to the interface (J_V), or through the γ/θ and γ/α interfaces (J_B). The interfaces are assumed to be of uniform thickness δ_b. S_I is the interlamellar spacing. This diagram represents a simplified view of the three-dimensional structure of pearlite.

The diffusion distance parallel to the interface can be approximated as $b_{19}S_I$ where b_{19} is a fraction usually set to $\frac{1}{2}$, on the basis that diffusion occurs in both directions parallel to the transformation front and normal to the assumed lamellar planes. Assuming that the diffusion flux is entirely through the volume of the austenite (i.e., J_V), the rate at which solute is partitioned per unit area of the transformation front (taking account of the fractional areas presented by each phase at the transformation front) from α and absorbed into θ must be equal,

$$\underbrace{v\frac{S_I^\alpha(\overline{c}^\gamma - c^{\alpha\gamma})}{S_I}}_{\text{partitioning of C into austenite}} = \underbrace{v\frac{S_I^\theta(c^{\theta\gamma} - \overline{c}^\gamma)}{S_I}}_{\text{absorption of C from austenite}}$$

where v is the speed of the growth front. Using Equations 8.6 each of these terms becomes:

$$\frac{vS_I^\theta S_I^\alpha}{S_I^2}(c^{\theta\alpha} - c^{\alpha\theta}).$$

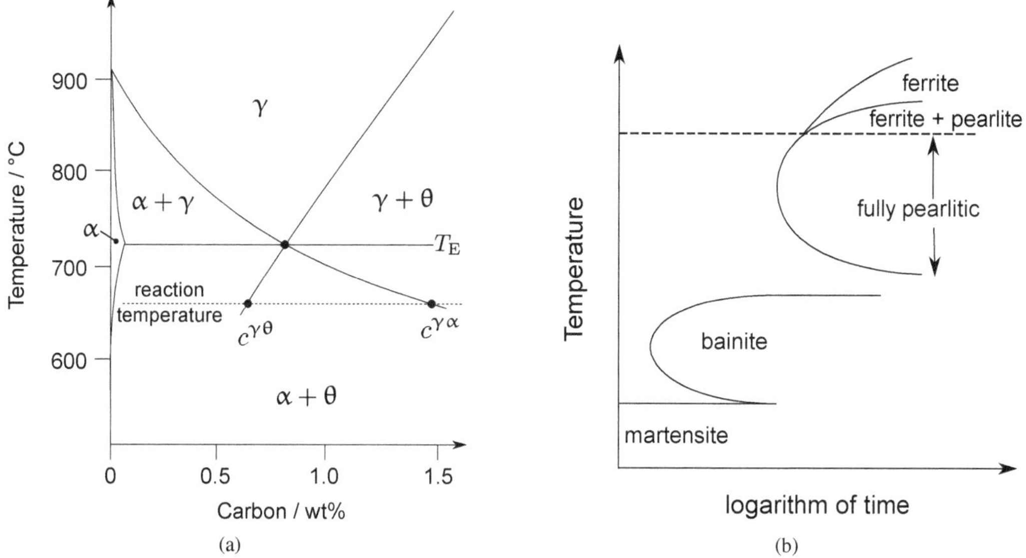

Figure 8.17 (a) Phase diagram with extrapolated phase boundaries to identify the concentrations in the austenite that are in equilibrium with cementite or ferrite. The region below T_E, where θ and α can precipitate simultaneously from γ, represents the Belaiew-Hultgren extrapolation region. (b) Schematic time-temperature-transformation diagram for a hypoeutectoid steel, showing the undercooling required to generate a fully pearlitic microstructure even when the γ is not of eutectoid composition.

Since the rate of solute partitioning must equal the flux of solute at the interface, it follows that

$$\frac{vS_I^\theta S_I^\alpha}{S_I^2}(c^{\theta\alpha} - c^{\alpha\theta}) = D_C^\gamma \frac{(c^{\gamma\alpha} - c^{\gamma\theta})}{b_{19}S_I}$$

$$\therefore \quad v = D_C^\gamma \frac{S_I}{b_{19}S_I^\theta S_I^\alpha} \frac{c^{\gamma\alpha} - c^{\gamma\theta}}{c^{\theta\alpha} - c^{\alpha\theta}} \tag{8.7}$$

where D_C^γ is the diffusivity of carbon in austenite.This equation does not uniquely define v, which is a function of the unknown interlamellar spacing; it increases indefinitely as the spacing decreases. After all, the diffusion distance decreases with S_I.

However, S_I cannot decrease without constraint, because the creation of α/θ interfaces consumes energy. The amount of α/θ interfacial area per unit volume is, from stereology, $2/S_I$ which when multiplied by $\sigma^{\alpha\theta}$ gives the free energy consumed as interfaces. Therefore, the minimum value of interlamellar spacing possible is a critical spacing

$$S_I^C = \frac{2\sigma^{\alpha\theta}}{|\Delta G^{\gamma\rightarrow\alpha+\theta}|} \tag{8.8a}$$

where $\Delta G^{\gamma\rightarrow\alpha+\theta}$ is here defined in units of J m^{-3}. The effective driving force is therefore given by:

$$\Delta G_{net} = |\Delta G^{\gamma\rightarrow\alpha+\theta}| - \frac{2\sigma^{\alpha\theta}}{S_I} \tag{8.8b}$$

and since $\Delta G_{net} = 0$ when $S_I = S_I^C$, it follows that

$$\Delta G_{net} = \frac{2\sigma^{\alpha\theta}}{S_I^C} - \frac{2\sigma^{\alpha\theta}}{S_I} \qquad \text{so that} \qquad \frac{\Delta G_{net}}{\Delta G} = 1 - \frac{S_I^C}{S_I}. \tag{8.8c}$$

To allow for the energy consumed in the process of interface creation, Equation 8.7 is scaled by a term $1-[S_I^C/S_I]$:

$$v = D_C^\gamma \frac{S_I}{b_{19} S_I^\theta S_I^\alpha} \frac{c^{\gamma\alpha} - c^{\gamma\theta}}{c^{\theta\alpha} - c^{\alpha\theta}} \left(1 - \frac{S_I^C}{S_I}\right). \tag{8.8d}$$

Figure 8.18 plots the growth rates using Equations 8.7 and 8.8d for two values of α/θ interfacial energy and reasonable values of all the other terms. The curve corresponding to Equation 8.7 has $\sigma^{\alpha\theta} = 0\,\mathrm{J\,m^{-2}}$, whereas the one with a maximum in the growth rate has $\sigma^{\alpha\theta} = 0.3\,\mathrm{J\,m^{-2}}$. The physical reason for the maximum in the growth rate expressed as a function of interlamellar spacing, is that a large spacing S_I increases the diffusion distance, whereas a small S_I means that a greater amount of the driving force is consumed in creating α/θ interfaces, making growth impossible when $S_I = S_I^C$.

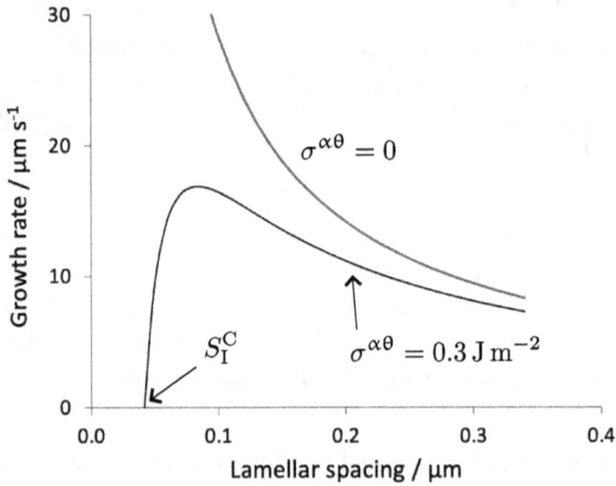

Figure 8.18 Calculated growth rate of pearlite, controlled by the diffusion of carbon in the austenite ahead of the transformation front. The rate increases indefinitely as the interlamellar distance is reduced, if the α/θ interfacial energy is zero. For the case where $\sigma^{\alpha\theta} = 0.3\,\mathrm{J\,m^{-2}}$, the critical interlamellar spacing when the growth rate becomes zero is $0.042\,\mu\mathrm{m}$.

The growth rate is a function of the interlamellar spacing so something else is needed to fix it. One assumption is that the spacing will correspond to that consistent with the maximum growth rate, i.e., when $S_I = 2S_I^C$ [69]. Alternatively, during isothermal transformation, the pearlite may adopt a spacing consistent with the maximum entropy production rate [75]; for isothermal transformation, this is proportional to the maximum free-energy dissipation rate. The rate of entropy production is [75, 76]

$$\dot{S} = \frac{v \Delta G_{\mathrm{net}}}{T}. \tag{8.9}$$

At small undercoolings, the free energy change $\Delta G \approx \Delta H \Delta T / T_E$, where ΔH is the enthalpy change assumed to be independent of temperature, and $\Delta T = T_E - T$. Using Equation 8.8c, the rate of entropy production is

$$\begin{aligned} \dot{S} &= v \frac{\Delta H \Delta T}{T T_E} \left(1 - \frac{S_I^C}{S_I}\right) \\ &= D_C^\gamma \frac{S_I}{b_{19} S_I^\theta S_I^\alpha} \left(\frac{c^{\gamma\alpha} - c^{\gamma\theta}}{c^{\theta\alpha} - c^{\alpha\theta}}\right) \frac{\Delta H \Delta T}{T T_E} \left(1 - \frac{S_I^C}{S_I}\right)^2. \end{aligned} \tag{8.10}$$

The maximum in entropy production $\dot{S}$ occurs at $S_I = 3S_I^C$. The two criteria, maximum growth rate and maximum entropy production rate, yield different interlamellar spacings that might persist during growth, so the relevant criterion must be selected by experiment.

Supposing now that in addition to the flux through the austenite, there is solute transport through the averaged interface of thickness δ^b between the pearlite and austenite, defined by an overall boundary diffusion coefficient D_C^b. An additional flux must therefore be considered in Equation 8.7, leading to [41]

$$v = \left(\frac{D_C^\gamma}{b_{19}} + \frac{\delta^b D_C^b}{b_{20} S_I}\right) \frac{S_I}{S_I^\theta S_I^\alpha} \left(\frac{c^{\gamma\alpha} - c^{\gamma\theta}}{c^{\theta\alpha} - c^{\alpha\theta}}\right) \left(1 - \frac{S_I^C}{S_I}\right). \tag{8.11}$$

The concentration terms for the boundary are not likely to be identical to those determined by local equilibrium between the product phases and austenite. One assumption is that the concentration in the boundary is still represented by the phase diagram but scaled by an empirical *distribution coefficient*; since this coefficient is assumed to apply to all of the concentration terms, it can simply be incorporated into b_{20}. Figure 8.19 shows how the flux through the boundary becomes the dominant mechanism of carbon transport during pearlite growth at temperatures not far below T_E. For the calculations illustrated, because there now are two diffusion paths, the interlamellar spacing based on a maximum $\dot{S}$ is between 2.01-2.17S_I^C, and the corresponding values for the maximum growth rate criterion are 1.36-1.53S_I^C [41].

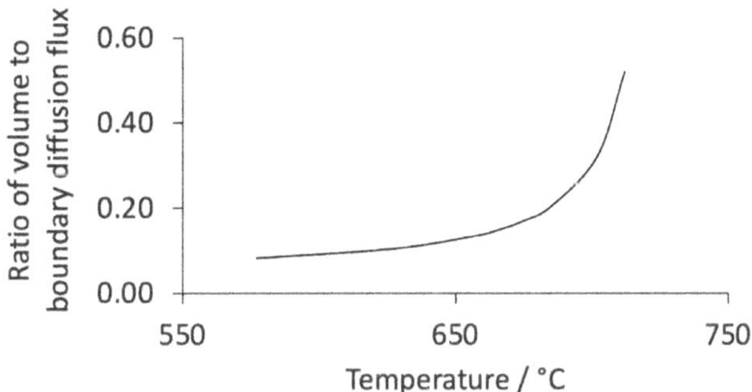

Figure 8.19 Relative contributions of volume diffusion through austenite, and boundary diffusion fluxes, during the growth of pearlite in a Fe-0.8C wt% steel. The details of the parameters used in the calculations can be found in [41].

8.6 Fe-C-X: GROWTH WITH LOCAL EQUILIBRIUM

The well-known complication is that the diffusivities of the substitutional and interstitial solutes are vastly different, so unlike the case for binary Fe-C, it becomes necessary to discover conditions where the two or more solute fluxes can keep pace whilst maintaining equilibrium locally at the interface [75, 77–80]. And there is no doubt that substitutional solutes are partitioned between the phases at all temperatures where pearlite is observed [61, 81–83]. Para-equilibrium pearlite or 'para-pearlite' [84–86] where both of its phases have the same X/Fe ratio as the parent austenite has never been observed experimentally. This is because the interface with austenite is sufficiently incoherent with no shape deformation associated with its translation, other than irregular minor-effects associated with volume change. All atoms, including iron, must diffuse during the growth of pearlite. In fact, there are fundamental difficulties in applying either the para-equilibrium or the negligible partitioning local equilibrium growth conditions to reconstructive transformations [87].

Fe-C-X steels contain a substitutional solute (X) in addition to interstitial carbon. Local equilibrium requires the compositions at the interface to be maintained at levels that are consistent with a tie-line of the Fe-C-X phase diagram. This is not in general possible at a fixed temperature with the tie line passing through $\overline{c}_X, \overline{c}_C$, because the rate at which each solute is partitioned must then equal that at which it is carried away from the interface by diffusion. It is necessary therefore that

$$\text{at } \alpha/\gamma \text{ interface: } \begin{cases} v(c_C^{\gamma\alpha} - c_C^{\alpha\gamma}) = -D_C^{\gamma}\dfrac{\partial c_C}{\partial z} \\ v(c_X^{\gamma\alpha} - c_X^{\alpha\gamma}) = -D_X^{\gamma}\dfrac{\partial c_X}{\partial z} \end{cases} \tag{8.12}$$

$$\text{at } \theta/\gamma \text{ interface: } \begin{cases} v(c_C^{\gamma\theta} - c_C^{\theta\gamma}) = -D_C^{\gamma}\dfrac{\partial c_C}{\partial z} \\ v(c_X^{\gamma\theta} - c_X^{\theta\gamma}) = -D_X^{\gamma}\dfrac{\partial c_X}{\partial z} \end{cases} \tag{8.13}$$

where the subscripts identify the solute. Given that $D_X^{\gamma} \ll D_C^{\gamma}$, it becomes impossible to simultaneously satisfy either Equation sets 8.12 or 8.13 if the tie-line passing through $\overline{c}_X, \overline{c}_c$ is selected. The solution of the conditions appropriate for the interface requires the following two equations [4]:

$$v_C = \left(2\frac{D_C^{\gamma}}{b_{19}} + 12\frac{D_C^{b}\delta^{b}}{b_{20}S}\right)\frac{S_I}{S^{\alpha}S^{\theta}}\left(\frac{c_C^{\gamma\alpha} - c_C^{\gamma\theta}}{c_C^{\theta\gamma} - c_C^{\alpha\gamma}}\right)\left(1 - \frac{S_I^C}{S_I}\right) \tag{8.14a}$$

$$v_X = \left(2\frac{D_X^{\gamma}}{b_{19}} + 12\frac{D_X^{b}\delta^{b}}{b_{20}S_I}\right)\frac{S_I}{S^{\alpha}S^{\theta}}\left(\frac{c_X^{\gamma\alpha} - c_X^{\gamma\theta}}{c_X^{\theta\gamma} - c_X^{\alpha\gamma}}\right)\left(1 - \frac{S_I^C}{S_I}\right) \tag{8.14b}$$

where the velocities v_C and v_X are calculated on the basis of the diffusion of only carbon or only manganese, respectively. Since there is only one transformation front, the equations must be solved such that $v_C = v_X$. The interlamellar spacing is also identical in these equations, so a further condition is that:

$$\frac{D_C}{D_X} = \frac{R_X}{R_C} \quad \text{with} \quad \begin{cases} D_i \equiv \frac{D_i^{\gamma}}{b_{19}} + \frac{6D_i^{b}\delta^{b}}{b_{20}S_I} \\ R_i \equiv \frac{c_i^{\gamma\alpha} - c_i^{\gamma\theta}}{c_i^{\theta\gamma} - c_i^{\alpha\gamma}}. \end{cases} \tag{8.15}$$

This ensures that the weighted average of the ferrite and cementite yields the mean composition of the steel, assuming that there is no long-range diffusion of manganese into the bulk of the parent phase. With these two constraints and the local equilibrium condition, it becomes possible to find unique interface compositions at the growth front by coupling the conditions and the velocity equations to thermodynamic data using the following procedures:

- a trial θ/γ interface composition is set, consistent with a selected tie-line for the transformation temperature.
- The α/γ interface composition tie-line is selected such that $\overleftrightarrow{c_{C,X}^{\alpha\gamma},\, c_{C,X}^{\theta\gamma}} \ni \overline{c}_{C,X}$, where the double-headed arrow implies a tie-line connecting the equilibrium compositions of ferrite and cementite. $\overline{c}_{C,X}$ is the average composition.
- If Equation 8.15 is not satisfied by these choices then the process is repeated until a solution is found.
- This solution provides the interface compositions which when substituted into Equation 8.14b permits the estimation $v = v_C = v_X$ of the transformation interface.

One of the difficulties in applying the type of theory described here is data on boundary-diffusion coefficients. For substitutional solutes, a good source is an assessment by Fridberg et al. [88], where the following applies approximately to iron and X in α and γ boundaries:

$$D_{\mathrm{b}}^{\mathrm{Fe\ or\ X}} \times \delta_{\mathrm{b}} \approx 5.4 \times 10^{-12} \exp\left(-\frac{155000\,\mathrm{J\,mol^{-1}}}{RT}\right) \quad \mathrm{m^2\,s^{-1}}. \tag{8.16}$$

and the boundary diffusion coefficient for carbon has been estimated to be [4]

$$D_{\mathrm{b}}^{\mathrm{C}} \approx 1.84 \times 10^{-3} \exp\left(-\frac{124995\,\mathrm{J\,mol^{-1}}}{RT}\right) \quad \mathrm{m^2\,s^{-1}}. \tag{8.17}$$

Figure 8.20 shows that an order of magnitude agreement can be achieved between the calculated [4] and measured [89] pearlite growth rates in Fe-Mn-C steels, assuming combined volume and boundary diffusion. The chemical compositions of the cementite and ferrite are also well predicted [4]. This level of consistency between theory and experiment might be disconcerting, but approximations abound: unspecified errors in the experimental data, the oversimplification of shape, lack of confidence in interfacial energies, uncertain relation between $S_{\mathrm{I}}^{\mathrm{C}}$ and S_{I}, and the assumption of local equilibrium at all interfaces. This makes it difficult to apply the theory – there never has been a demonstration of its practical utility.

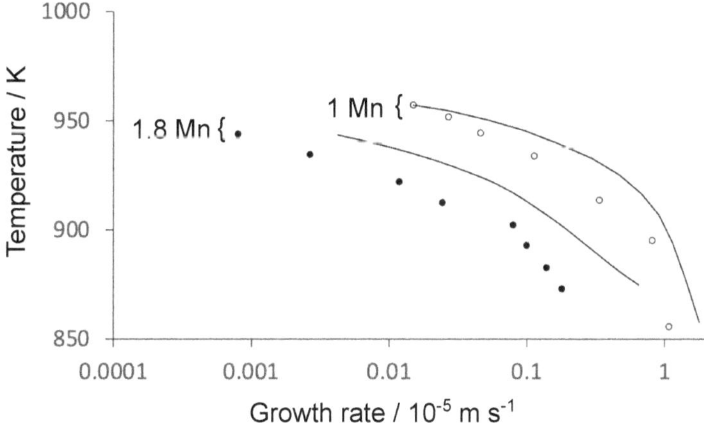

Figure 8.20 Curves showing the calculated diffusion-controlled growth rate of pearlite, in 1.0 and 1.8 wt% manganese eutectoid steels as a function of the transformation temperature [4]. The solute flux is both through the austenite ahead of the interface and through the interface itself. The corresponding data are from [89].

It is possible in a ternary system for the austenite to be supercooled into the $\alpha+\gamma+\theta$ phase field where they can all co-exist in equilibrium. The overall composition of the pearlite no longer matches that of the austenite. Solutes partition between the parent and products so the driving force decreases as pearlite grows, causing its interlamellar spacing to progressively increase with the size of the colony, Figure 8.21; such pearlite is said to be *divergent* [90]. Transformation stops when the $\alpha+\gamma+\theta$ reach equilibrium.

8.6.1 LOCAL EQUILIBRIUM?

Both the θ and α inherit the X/Fe atomic ratio of the austenite during the bainite transformation; this is most definitely not the case with pearlite, even when it grows at the same temperature as bainite.

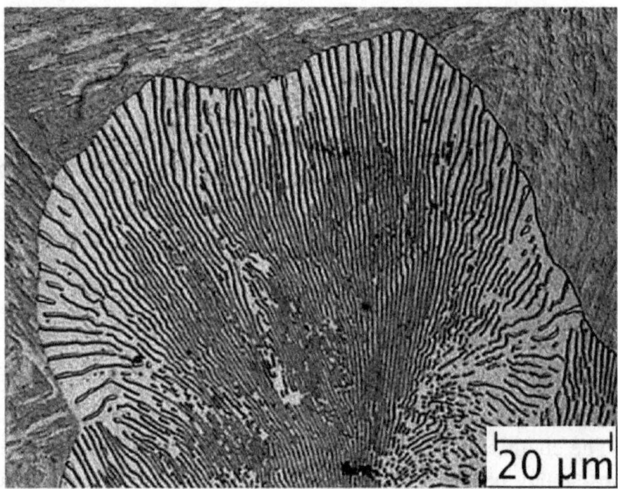

Figure 8.21 Fe-2.5C-5.4Mn at.% steel transformed from austenite supercooled into the $\alpha+\gamma+\theta$ phase field. The interlamellar spacing increases as the colony enlarges. After Hutchinson et al. [83], reproduced with the permission of Elsevier.

Tsivinsky et al. [91] first reported that chromium and tungsten partitioned from $\gamma \rightarrow \theta$ during pearlite growth, but not during that of bainite. Chance and Ridley [61] found that for upper bainite in a Fe-0.81C-1.41Cr wt% alloy, the partition coefficient $k_{\mathrm{Cr}} = w_{\mathrm{Cr}}^{\theta}/w_{\mathrm{Cr}}^{\alpha}$ could not be distinguished from unity (Figure 8.22). The different behaviour is because pearlite and bainite grow by distinct atomic mechanisms, reconstructive and displacive, respectively. The transformation front of pearlite is relatively incoherent P/γ interface, so diffusion must accompany its growth, hence the inevitable partitioning of substitutional solutes. Nevertheless, k_{Cr} for pearlite is quite different from that for equilibrium, Figure 8.22. The local equilibrium condition at the α/γ and θ/γ interface fails at small undercoolings below T_{E}. The growth process therefore requires more than one interface response function rather than focussing on diffusion-controlled growth in isolation [18].

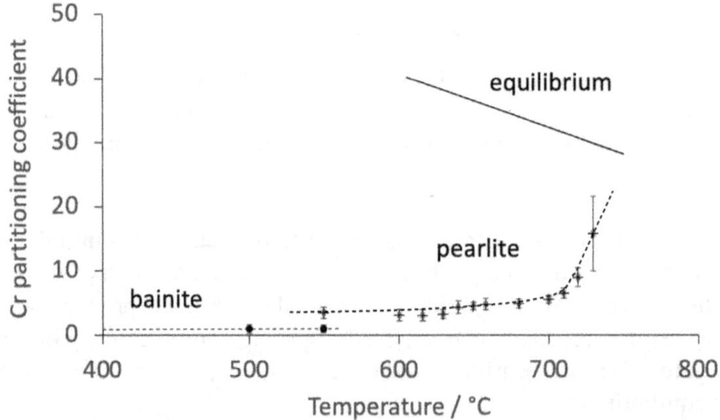

Figure 8.22 Measured ratio of chromium in cementite to that in ferrite (i.e., partition coefficient $k_{\mathrm{Cr}} = w_{\mathrm{Cr}}^{\theta}/w_{\mathrm{Cr}}^{\alpha}$ based on wt%), when the cementite is a part of bainite or pearlite. Adapted using data from [61].

Morooka S An interesting experiment has been conducted [92] in which a mixture of bainite and pearlite was obtained at the same temperature, during isothermal transformation of austenite at 450 °C, Figure 8.23. There is no partitioning observed of Mn or Al during the bainite transformation – the X/Fe ratio remained identical to that of the parent austenite. In contrast, the α and θ in pearlite changed dramatically, with $w^{\alpha}_{\mathrm{Mn}} < \overline{w}_{\mathrm{Mn}} < w^{\theta}_{\mathrm{Mn}}$. The measurements are particularly pertinent given that the microanalysis scan is across the junction between the bainitic and pearlitic regions. The results are entirely expected from the displacive and reconstructive transformation mechanisms of bainite and pearlite, respectively, and all the more important since the microstructures were generated at the same temperature.

We note in passing that atom-probe data are sometimes interpreted to show manganese segregation at the $\alpha_{\mathrm{P}}/\gamma$ interface though not at the less coherent $\theta_{\mathrm{P}}/\gamma$ boundary [93]. However, the measured manganese concentration profile extends over a distance of 5 nm which is far greater than the thickness of the interface, so is likely to reflect a degree of partitioning into γ during the pearlite transformation.

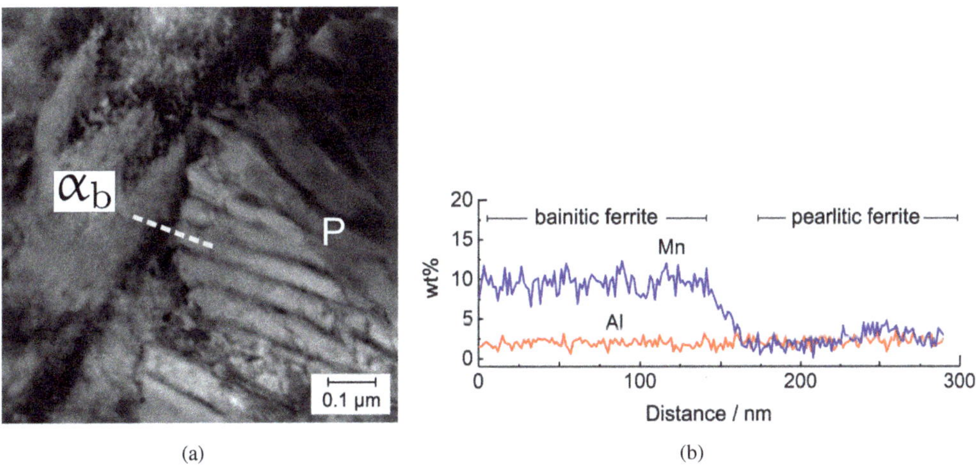

(a) (b)

Figure 8.23 A mixture of bainite and pearlite, in Fe-0.37C-10.2Mn-2.06Al-0.27V wt% obtained by transformation at 450 °C for 24 h [92]. (a) The dashed line shows the path along which microanalytical data were obtained, from bainitic ferrite to the ferrite within pearlite. (b) Plots show the variation in the manganese and aluminium concentrations in the ferrite along the dashed line in (a). The pearlitic ferrite is depleted in manganese relative to the average in the alloy, due to partitioning into the cementite within the pearlite [Fig. 10d, 92]. Images courtesy of M. X. Huang.

8.7 FLUX OF CARBON THROUGH FERRITE

It has been assumed thus far that during diffusion-controlled growth, transport occurs either in the volume of the austenite ahead of the interface with pearlite, which may be supplemented significantly by a flux through that interface itself, the exact contribution to the total flux being dependent on the transformation temperature (Figure 8.19). Onsager suggested an additional possibility, that carbon diffusion in the ferrite behind the transformation front may control the growth of pearlite [94], Figure 8.24.[4] The diffusion of carbon in the ferrite would lead to its precipitation on existing

[4]Onsager is quoted frequently, to have suggested the diffusion path through ferrite at a 1948 conference held at Cornell University. There is no record to this effect in the published proceedings [95], though he was present at the meeting. After Tisza's talk on the theory of phase transitions, he outlined a famous, exact formula for spontaneous magnetisation that he had derived with Kaufman [96]. That too is not recorded formally in the discussion section at the end of Tisza's paper.

cementite behind the transformation front, so the cementite lamellae would become tapered with greater thickness away from the front. This is not observed experimentally.

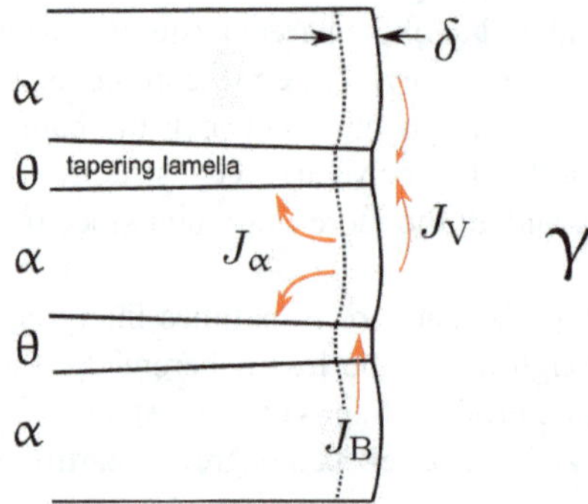

Figure 8.24 Three fluxes, J_V that occurs through the austenite, J_B that occurs through the interface of thickness δ_b and J_α via diffusion through the ferrite behind the transformation front.

There are several reasons why diffusion through ferrite behind the transformation front does not play a role in the growth of pearlite. The diffusion of carbon through the boundary is much faster than through ferrite, as illustrated in Figure 8.25, where the boundary diffusivity is calculated using Equation 8.17 and that for ferrite according to [97]. Note that it is not appropriate to use the diffusivity of carbon in martensite because its solubility in ferrite is very small; diffusion through tetragonal martensite is even slower than through ferrite [98], [p. 113, 18].

The concentration gradient that drives diffusion[5] will be much gentler in ferrite $\propto (x^{\alpha\gamma} - x^{\alpha\theta})$ than that within the boundary, which should scale with $(x^{\gamma\alpha} - x^{\gamma\theta})$. Furthermore, the unidirectional diffusion through the boundary is along the shortest distance between the source and sink; this is in contrast to the need for the carbon in ferrite to meander towards the cementite lamellae. For all these reasons, the predominant fluxes must in practice be a combination of diffusion through the boundary and through the volume of the austenite in the vicinity of the transformation front..

The phase field method has been applied to simulate pearlite growth that includes carbon migration within the ferrite [99]. The method has not been applied rigorously because the flux through the γ/P boundary with the austenite is not included, and S_I was assumed based on model behaviour. Given the neglect of J_B, it is not surprising that the growth rate is predicted to be faster than if only diffusion in the volume of the austenite is permitted. The shape of the cementite lamellae was found to be 'conical', which is inconsistent with experimental observations. It was assumed that D_C^γ is concentration-independent (thus significantly underestimating D_C^γ) and there are other parameters, the origin of which was not stated.The concentrations at the interfaces were assumed to be given by a 'linearisation of the binary Fe-C diagram' whereas in fact, the $\alpha/\alpha+\gamma$ phase boundary is retrograde in shape (Figure 8.26).

[5]Strictly, the free energy gradient, but for the case in hand, the concentration and chemical potential gradients have the same sign.

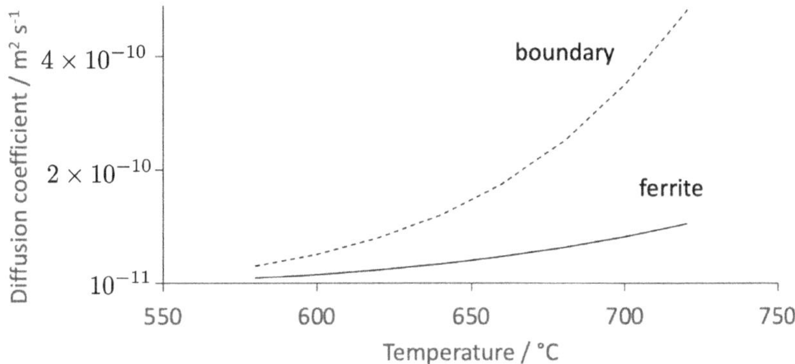

Figure 8.25 Diffusion coefficient for carbon through ferrite (D^{C}_{α}) and for carbon diffusion in a boundary ($D^{\mathrm{C}}_{\mathrm{b}}$).

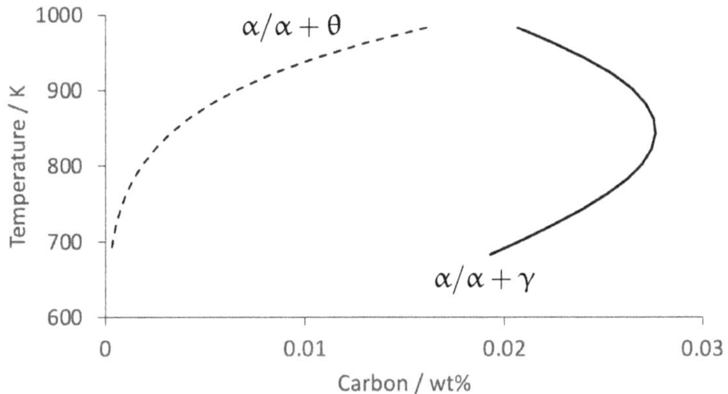

Figure 8.26 The solubility in weight percent, of carbon in ferrite when it is in equilibrium with austenite or with cementite, in a Fe-C alloy. The curves are calculated using MTDATA [22].

8.8 OVERALL TRANSFORMATION KINETICS

8.8.1 ISOTHERMAL TRANSFORMATION

The dependence of pearlite fraction as a function of parameters such austenite grain size, temperature, free energy, chemical composition is important in the design of steels. Experimental data of this kind are expressed as time-temperature-transformation diagrams, such as Figure 8.27 which is for a Fe-Mn-C-Si steel transformed isothermally into pearlite. Temperatures less than about 600 °C induce some bainite to form because the mobility of iron atoms then is at a premium, rendering reconstructive transformations sluggish.

To generate Figure 8.27, each set of isothermal data for the initiation of pearlite was fitted to an equation of the form $V^{\mathrm{P}}_{\mathrm{V}} = 1 - \exp\{-b_{27}t^{b_{28}}\}$, where b_i are constants, enabling the estimation of the curves for 1%, 50% and 99% transformation. Although applied empirically, this equation has a basis that is discussed in Section 8.8.2.

The alloy on which Figure 8.27 is based, is quaternary; similar measurements are difficult in the binary Fe-C of eutectoid composition because of the speed of transformation. One experiment using a hypereutectoid Fe-0.93C wt% steel with a large austenite grain size that limits the nucleation rate

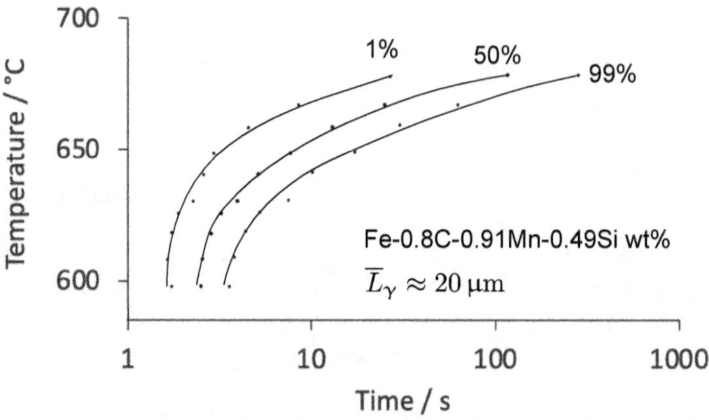

Figure 8.27 Time-temperature transformation diagram for a near-eutectoid steel illustrating the amount of pearlite obtained isothermally from a fully austenitic state as a function of temperature. Selected data from Hawbolt et al. [100]. Notice that fully pearlitic microstructures can be obtained even though the alloy is strictly hypereutectoid, because of the undercooling that allows both α and θ to grow from austenite (Figure 8.10a).

per unit volume, is illustrated in Figure 8.28, though the large carbon concentration increases the driving force for pearlite formation, Figure 8.10a, which would accelerate transformation.

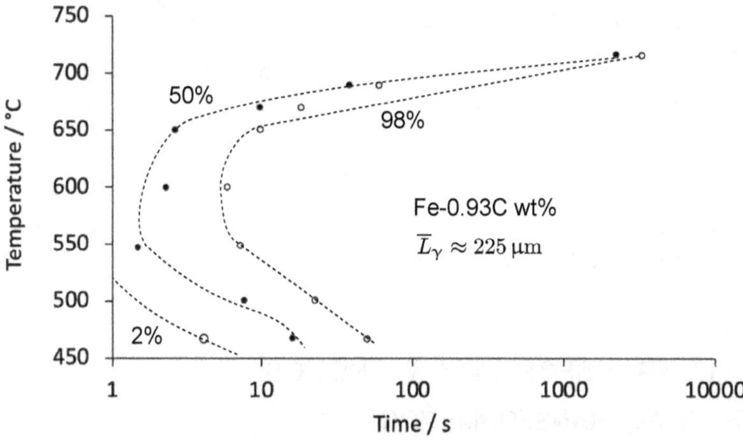

Figure 8.28 TTT diagram for a high-purity Fe-0.93C wt% steel showing the amount of pearlite obtained as a function of transformation temperature. Data from Hull et al. [19].

Methods have evolved that are semi-empirical but based on thermodynamic information, which enable the estimation of TTT diagrams. A comparison of Fe-0.76C wt% curve in Figure 8.29 with the hypereutectoid Fe-0.93C wt% case in Figure 8.27 reveals that the exact eutectoid steel with its much small austenite grain size transforms slower, presumably because of its smaller $|\Delta G^{\gamma \rightarrow \alpha+\theta}|$. The possibility of some transformation during cooling to the isothermal temperature may not be ruled out for the experimental diagram for the hypereutectoid steel.

Figure 8.29 shows also that the addition of just 1 wt% of manganese causes a large retardation in pearlite formation. Manganese not only stabilises austenite but it must diffuse during transformation, which prolongs the transformation relative to the binary Fe-C at all transformation temperatures.

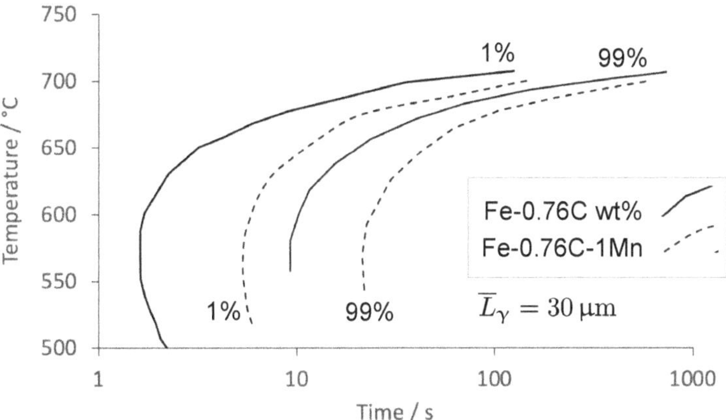

Figure 8.29 Calculated time-temperature transformation diagrams for an exactly eutectoid steel and another containing 1Mn wt% beginning with a fully austenitic state. The calculations, truncated below a temperature where bainite begins, use JMatPro® which originated in the work of Kirkaldy and Venugopalan [101]. Calculations courtesy of Apparao Chintha.

8.8.2 JOHNSON-MEHL-AVRAMI-KOLMOGOROV APPROACH

The nucleation and growth rates of individual particles can be studied in isolation, but the estimation of the fraction of transformation requires a treatment of *impingement*, i.e., the interaction between particles originating from different spatial locations. When precipitates touch, they cannot grow through each other, nor can nucleation occur in already transformed regions of the parent austenite.

To deal with impingement, an imaginary, *extended volume* V_e is defined in which particles can grow through each other and nucleation is permitted in all regions, including those that already have transformed [102–106]. Figure 8.30 shows two pearlite colonies existing at time t; a small interval Δt later, the original particles grow and overlap in the region labelled c, and new particles a, b form. The net increase in the extended volume dV_e^P therefore comes $a+b+c$. Only those newly transformed regions that lie in previously untransformed matrix can contribute to a change in the real volume dV^P of the product phase:

$$dV^P = \underbrace{\left(1-\frac{V^P}{V}\right)}_{\text{probability of unreacted }\gamma} \times dV_e^P \tag{8.18}$$

where it is assumed that the microstructure develops randomly, and V is the total volume. Multiplying the change in extended volume by the probability of finding untransformed γ has the effect of excluding regions such as b, that do not contribute to the real change in volume of the product. For a random distribution of precipitated particles, this equation is integrated to obtain the real volume fraction:

$$\frac{V^P}{V} = 1-\exp\left\{-\frac{V_e^P}{V}\right\}. \tag{8.19}$$

The extended volume V_e^P is straightforward to estimate using nucleation and growth models while neglecting any impingement effects. Suppose the pearlite grows isotropically at a constant rate v and where the constant nucleation-rate per unit volume is I_V; the volume of a particle nucleated at time $t=\tau$ is given by

$$v_\tau = \frac{4}{3}\pi v^3 (t-\tau)^3. \tag{8.20a}$$

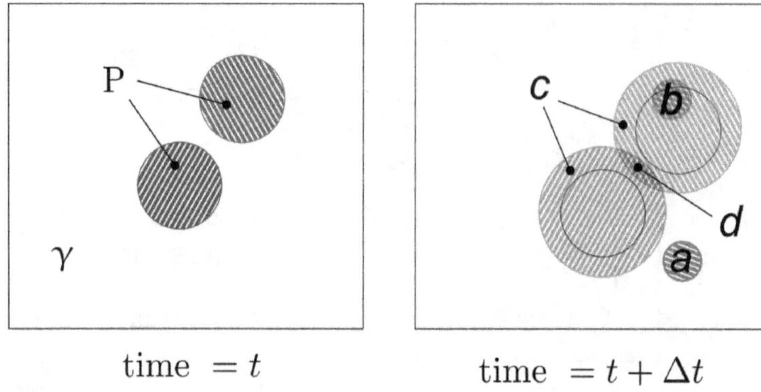

Figure 8.30 The concept of extended volume. Two colonies of pearlite exist at time t. With an increment Δt, particles a and b form, of which b lies within previously transformed γ. The original particles have grown during Δt by the annular rings marked c, with overlap in the region d.

The change in extended volume over the interval τ and $\tau + \mathrm{d}\tau$ is

$$\mathrm{d}V_\mathrm{e}^\mathrm{P} = \frac{4}{3}\pi v^3 (t-\tau)^3 \times I_\mathrm{V} \times \mathrm{V} \times \mathrm{d}\tau. \tag{8.20b}$$

Substituting into Equation 8.19 and writing $\xi = V^\mathrm{P}/\mathrm{V}$ gives

$$\mathrm{d}V^\mathrm{P} = \left(1 - \frac{V^\mathrm{P}}{\mathrm{V}}\right)\frac{4}{3}\pi v^3 (t-\tau)^3 I_\mathrm{V}\mathrm{V}\ \mathrm{d}\tau$$

$$\therefore \qquad -\ln\left\{1 - \frac{V_\mathrm{P}}{\mathrm{V}}\right\} = \frac{4}{3}\pi v^3 I_\mathrm{V}\int_0^t (t-\tau)^3\ \mathrm{d}\tau \tag{8.20c}$$

$$\text{and} \qquad \frac{V^\mathrm{P}}{\mathrm{V}} = 1 - \exp\{-\pi v^3 I_\mathrm{V} t^4/3\}. \tag{8.20d}$$

This equation has been derived for the specific assumptions of random nucleation, a constant nucleation rate and a constant growth rate. There are different possibilities but they often reduce to the general form:

$$\xi = 1 - \exp\{-k_A t^n\} \tag{8.21}$$

where ξ is the volume fraction of transformed product,[6] k_A and n characterise the reaction as a function of time, temperature and other variables. This equation often is used empirically, but only as an economic way of representing experimental data rather than having a significant predictive capability. The temptation to deduce mechanistic information from an empirical application of the Avrami equation should be avoided even if it accurately fits the data, since the fitting parameters can be ambiguous. Equation 8.21 often is applied empirically but this is not a reflection quality of the theory, because the application can be rigorous as evident for the specific case described by Equation 8.20d.

8.8.3 ADDITIVITY RULE

It may be possible to adapt TTT diagrams, for dealing with the many variables that determine continuous cooling transformation. In the Scheil *additive reaction rule* [107], a cooling curve is treated

[6] When a transformation reaches a limiting equilibrium fraction < 1, ξ represents its volume fraction normalised by the equilibrium fraction.

as a combination of i isothermal reaction steps. In Figure 8.31, a fraction $\xi = 0.05$ of transformation is achieved during continuous cooling when

$$\sum_i \frac{\Delta t_i}{t_i} = 1 \tag{8.22a}$$

with the summation beginning as soon as the parent phase cools below the equilibrium temperature. If the $t_{\mathrm{TTT}}\{\xi', T\}$ is now defined as the time taken at temperature T to achieve a specific fraction ξ' of isothermal transformation, and $t_{\mathrm{CCT}}\{\xi', \dot{T}\}$ the time taken in continuous cooling at a constant rate $\dot{T}$, then Equation 8.22a can be expressed as an integral [108]

$$\int_0^{t_{\mathrm{CCT}}\{\xi', \dot{T}\}} \frac{\mathrm{d}t}{t_{\mathrm{TTT}}\{\xi', T\}} = 1 \tag{8.22b}$$

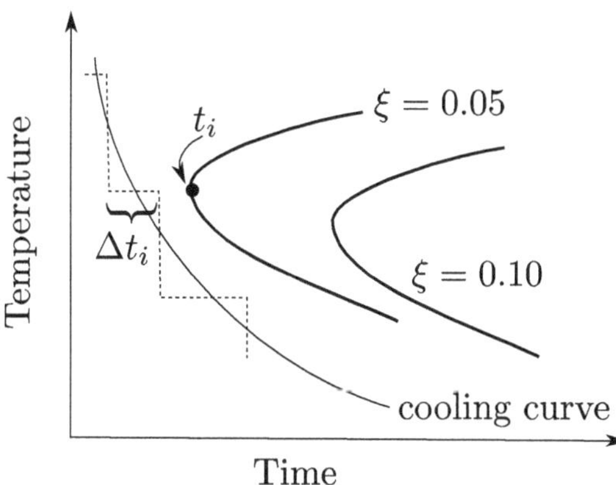

Figure 8.31 The Scheil method for converting between isothermal and anisothermal transformation data. The C-curves plotted represent isothermal transformation to two different volume fractions. The continuous cooling curve is divided into isothermal steps (dashed line).

The rule is appropriate if the reaction rate depends solely on ξ and T. An analysis of the isothermal and continuous cooling transformation data on pearlite (Fe-0.81C-0.4Mn-0.2Si wt%) indicates that the use of the additivity rule is not always accurate, but may be so when pearlite forms at large undercoolings [108].

Figure 8.32 shows that continuous cooling at a high rate suppresses pearlite to greater undercoolings below T_{E} and extends the temperature range over which the pearlite is generated. This must cause a dispersion of true S_{I} in the microstructure, perhaps reflected in a more gradual transition to plasticity in the stress-strain curve, with the coarser S_{I} colonies yielding first. Using the curve illustrated in Figure 8.12 and the Hall-Petch relationship (Equation 4.17), the variation in σ_{y} at $1\,^{\circ}\mathrm{C\,s^{-1}}$ would be 683-787 MPa, but 806-917 MPa when the structure is generated at $10\,^{\circ}\mathrm{C\,s^{-1}}$.

8.8.4 SITE SATURATION

Nucleation sites can all be consumed during the early stages of transformation; the sites are said to be saturated with subsequent transformation involving only growth. If the austenite grain surfaces where pearlite initiates become saturated, then the pearlite growth can be treated as the one-dimensional thickening of layers that completely decorate the austenite grain surfaces and advance

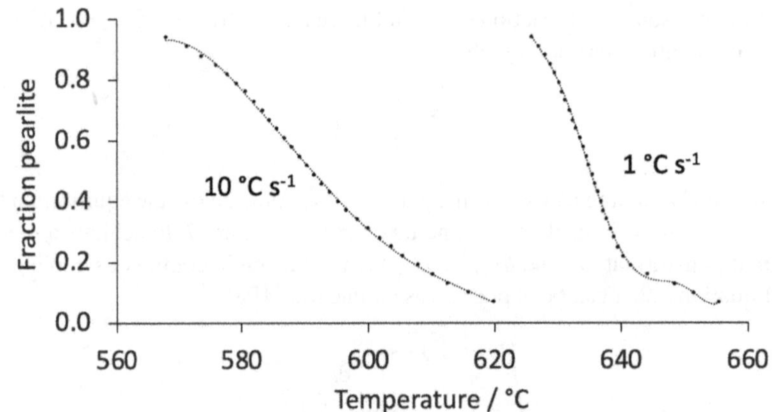

Figure 8.32 Continuous cooling transformation data for Fe-0.81C-0.4Mn-0.2Si wt% as it transforms into pearlite, as a function of two cooling rates from the austenitic condition. Selected data from Kim et al. [108].

at a constant growth rate v_P in both directions normal to the γ/γ boundaries. τ is then zero so the change in extended volume in an interval dt' is

$$\frac{dV_e^P}{V} = 2S_V\, v_P\, dt' \tag{8.23a}$$

where S_V is the γ/γ boundary area per unit volume. On substituting into Equation 8.18, the true change in volume is related to that of the extended change as follows:

$$\int_0^{V_P/V} \frac{dV^P/V}{1-V^P/V} = 2S_V\, v_P \int_0^t dt' \tag{8.23b}$$

so the volume fraction of pearlite evolves as

$$\frac{V_P}{V} = 1 - \exp\{-2S_V\, v_P t\}. \tag{8.23c}$$

Following this logic, if L_V represents the total length of grain edges per unit volume and assuming they are rapidly site-saturated and grow as cylinders on the edges,

$$\frac{V_P}{V} = 1 - \exp\{-\pi L_V\, v_P^2 t^2\}. \tag{8.23d}$$

For nucleation at corners that are site-saturated, the pearlite should grow as spheres at each corner. If the number of corners per unit volume is C_V then

$$\frac{V_P}{V} = 1 - \exp\left\{-\frac{4\pi}{3} C_V\, v_P^3 t^3\right\}. \tag{8.23e}$$

If all three types of site-saturated events happen simultaneously, then Equation 8.23a becomes

$$dV_e^P = \left(2S_V\, v_P + \pi L_V\, v_P^2\, 2t' + \frac{4\pi}{3} C_V\, v_P^3\, 3t'^2\right) dt' \tag{8.23f}$$

and hence,

$$\frac{V_P}{V} = 1 - \exp\left\{-2S_V\, v_P t - \pi L_V\, v_P^2 t^2 - \frac{4\pi}{3} C_V\, v_P^3 t^3\right\}. \tag{8.23g}$$

An equiaxed grain structure can be approximated by truncated octahedra that are stacked in a body-centred cubic arrangement that enables space to be filled [109]. These octahedra consist of 8 hexagonal and 6 square faces, with 36 edges, each of length L_e and 24 corners. If the mean lineal intercept defining the austenite grain size is $\overline{L}_\gamma$, then $S_V = 2/\overline{L}_\gamma$. Each of the 36 edges is shared between 3 grains and since $\overline{L}_\gamma = 1.69 L_e$ and the volume per grain is $11.314\overline{L}_e = 2.344\overline{L}_\gamma$ [110], it follows that $L_V = 3.02/\overline{L}_\gamma^2$. The number of corners per grain is 24, but each is shared with four grains so the number of corners per unit volume becomes $C_V = 2.56/\overline{L}_\gamma^3$. Substituting these quantities into Equation 8.23g gives

$$\frac{V_P}{V} = 1 - \exp\left\{ -\frac{4}{\overline{L}_\gamma} v_P t - \frac{9.49}{(\overline{L}_\gamma)^2} v_P^2 t^2 - \frac{12.65}{(\overline{L}_\gamma)^3} v_P^3 t^3 \right\}. \tag{8.23h}$$

An important approximation in deriving Equation 8.23h is that all three sites are saturated at the same time, effectively at $t = 0$, so for example, the austenite grain surfaces would be uniformly covered with a thin layer of pearlite. Figure 8.33 shows calculations using this equation for two cases. When the rate of growth is small, the corners contribute little to the overall transformation for the given austenite grain size. At a larger growth rate, all sites make a significant contribution although as expected, that from the grain surfaces make the dominant fraction of the total pearlite content.

It is worth emphasising that heterogeneous nucleation at corners should be easier than at edges, which in turn would be more effective than grain surfaces. So the assumptions associated with Equation 8.23h would be most valid at large undercoolings below T_E where the nucleation rates at all sites would be large, consistent with the greater driving force for transformation.

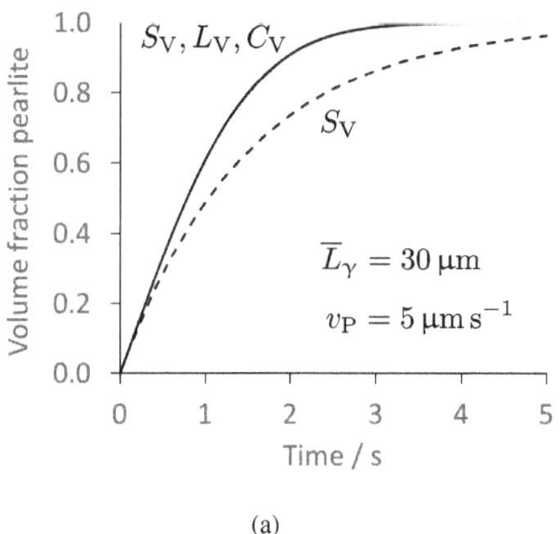

(a)

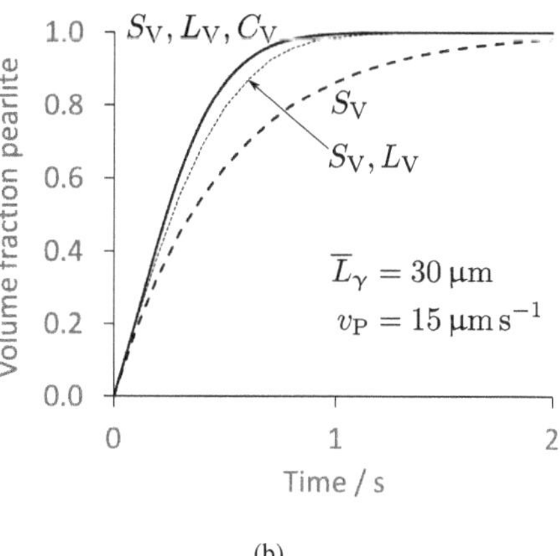

(b)

Figure 8.33 Illustration of Equation 8.23h, which assumes that all nucleation sites are saturated at $t = 0$ and transformation then occurs simultaneously from all three types of sites. The terms $(S_V + L_V + C_V)$, $(S_V + L_V)$ and S_V refer to surface, edge and corner; surface and edge; and surface only nucleation, respectively. (a) Small undercooling below T_E that leads to a low growth rate. The curves for $(S_V + L_V + C_V)$, $(S_V + L_V)$ are virtually identical. (b) Undercooling below T_E consistent with a large growth rate.

8.8.5 TOWARDS SITE SATURATION

Cahn [111] has derived rate laws based on non-random nucleation that are more representative of events at grain boundary features. This section is based on that work which is relevant particularly to pearlite though it can be adapted for phases that do not have a constant growth rate.

Beginning with a plane γ/γ boundary of 'infinite extent', and an arbitrary test-plane labelled Z parallel to that but located a distance z from it, a spherical nodule that forms at a time τ will have a circular intersection with Z, of radius

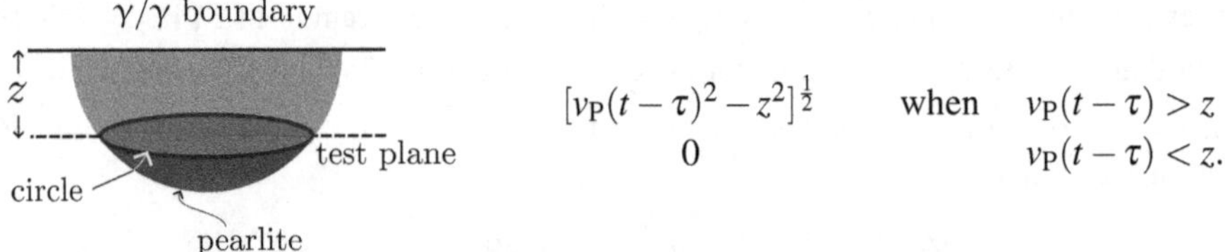

$$\begin{array}{lll} [v_P(t-\tau)^2 - z^2]^{\frac{1}{2}} & \text{when} & v_P(t-\tau) > z \\ 0 & & v_P(t-\tau) < z. \end{array}$$

The extended area *fraction* A_e for nodules initiating in the interval τ to $\tau + d\tau$ is

$$\begin{aligned} dA_e &= I_A d\tau \times \pi[v_P(t-\tau)^2 - z^2] & v_P(t-\tau) > z \\ &= 0 & v_P(t-\tau) < z \\ A_e &= \pi I_A \int_0^{t-z/v_P} [v_P(t-\tau)^2 - z^2] d\tau. \end{aligned}$$

If the variable z is redefined $z_* = z/v_P t$ then

$$\begin{aligned} A_e &= \pi I_A v_P^2 t^3 \left[\tfrac{1}{3}(1 - z_*^3) - z_*^2(1 - z_*) \right] & z_* < 1 \\ &= 0 & z_* > 1. \end{aligned}$$

On the test plane Z therefore, the pearlite nodules originating randomly at the boundary plane have the true total area fraction of intersection $A = 1 - \exp\{-A_e\}$ (*cf.* Equation 8.19). The total volume V_* of material originating per *unit area* of the boundary is given by integrating over all z, taking account of both sides of the boundary,

$$\begin{aligned} V_* &= 2 \int_0^\infty A \, dz_* \\ &= 2v_P t \int_0^1 1 - \exp\{-\pi I_A v_P^2 t^3 [\tfrac{1}{3}(1 - z_*^3) - z_*^2(1 - z_*)]\} \, dz_* \\ &\approx 2v_P t \left[\tfrac{1}{6}(4(1 - \exp\{-0.166375a\}) - \exp\{-0.333a\} + 1)\right] \quad \text{with} \quad a = \pi I_A v_P^2 t^3 \end{aligned}$$

where the last line uses Simpson's rule for the preceding integral. The extended volume fraction of pearlite is then calculated by taking account of the area of γ/γ boundary per unit volume (S_V)

$$\frac{V_e^{P,f}}{\mathrm{V}} = \frac{S_V V_*}{\mathrm{V}} \quad \text{so} \quad \frac{V^{P,f}}{\mathrm{V}} = 1 - \exp\left\{-\frac{S_V V_*}{\mathrm{V}}\right\} \tag{8.24}$$

where V is the total volume, the superscript 'f' is to indicate that the pearlite is that nucleated at grain faces. Equation 8.19 is used to convert the extended volume fraction to the real fraction V/V.

Figure 8.34 illustrates calculations using Equation 8.24 based on measured nucleation and growth rates. The effect of a large but experimentally measured nucleation rate is to induce site saturation at the γ/γ boundaries well before the pearlite has consumed all of the austenite (dashed curve). There is then a significant change in slope with slower reaction rate associated simply with the growth of pearlite nodules; the fraction $V_V^{P,f}$ thus increases noticeably gradually beyond saturation with an extended time required to complete transformation. If I_V is reduced by a factor of 100, site saturation is delayed to a point where transformation is almost complete, with the V_V^P versus t (continuous) curve exhibiting the classic smooth, 'S-shape'. The experimental data are not well represented by the measured nucleation rates, and marginally consistent with the curve based on the much-reduced I_V indicating that site saturation is not activated.

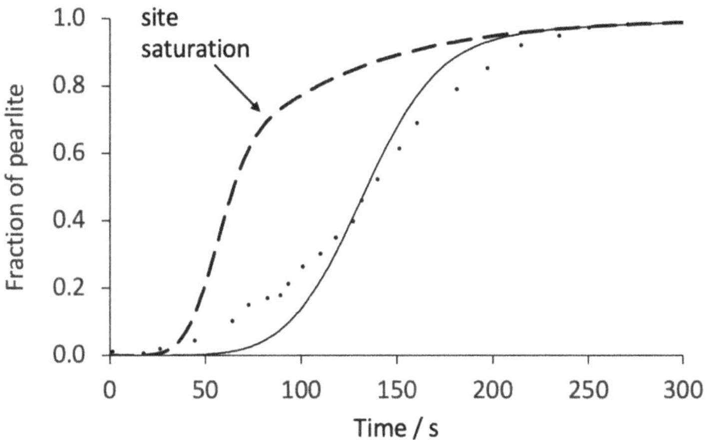

Figure 8.34 Calculations using Equation 8.24, assuming that all the pearlite originates from γ/γ grain faces, and data from the second row [21] of Table 8.1. The dashed curve uses the measured growth rate and the nucleation rate from Figure 8.6, where $N_V \approx 0.123t^{2.77}$ so $I_V \approx 0.341t^{1.77}$ with $S_V = 2/\overline{L}_\gamma$; the arrow indicates where site-saturation sets in. Referring to the continuous curve, the nucleation rate has had to be divided by 100 in order to delay site saturation to $V_V^{P,b} > 0.97$ and get a semblance of agreement with the points that represent experimental data for isothermal transformation at 680 °C [21].

Cahn derived, using similar extended space considerations in which a straight edge is parallel to an arbitrary line, and the intersection of a nodule with that line is assessed over a distance $0 \le z \le \infty$. in the first instance. Setting the edge length per unit volume is L_V and I_L is the nucleation rate per unit length, $a_e = \sqrt{I_L v_P} t$ and $z_* = z/v_P t$ gives

$$\frac{V^{P,e}}{V} = 1 - \exp\left\{ -\frac{2\pi v_P L_V}{I_L} a_e^2 \int_0^1 z_* \left[1 - \exp\left\{ -a_e^2 \left((1-z_*^2)^{1/2} - z_*^2 \ln \frac{1+\sqrt{1-z_*^2}}{z^*} \right) \right\} \right] \right\} \quad dz_*$$

$$\approx 1 - \exp\left\{ -\frac{2\pi v_P L_V}{I_L} a_e^2 [0.163333(0.01(1 - \exp\{-0.99942 a_e^2\}) \right.$$

$$\left. + 2(1 - \exp\{-0.536786 a_e^2\}) + 0.99(1 - \exp\{-0.001879 a_e^2\})] \right\}$$

where the superscript 'e' defines pearlite originating from grain edges.

As Cahn pointed out, each grain-corner may be assumed to be associated with just one nucleation event so the problem is similar to Avrami's model for nucleation from a limited number per unit volume N_V° of randomly distributed point sites [104, 112]. If the nucleation rate at a single corner is I_C then the change in the number of corners in a small time interval dt is $dN_V = -N_V I_C dt$ where N_V is the number per unit volume remaining after a time t, so $N_V = N_V^\circ \exp\{-I_C t\}$. The nucleation rate per unit volume is therefore $I_V = -dN_V/dt = N_V^\circ I_C \exp\{-I_C t\}$. This can be substituted into a modification of Equation 8.20c to account for the fact that I_V is not constant, to obtain the fraction $V_{P,c}/V$ that originates from grain corners

$$-\ln\left\{ 1 - \frac{V^{P,c}}{V} \right\} = \frac{4}{3} \pi v_P^3 \int_0^t I_V (t-\tau)^3 \, d\tau,$$

so integrating by parts gives [112],

$$\frac{V^{\mathrm{P,c}}}{\mathrm{V}} = 1 - \exp\left\{ -\left(\frac{3}{4\pi N_{\mathrm{V}}^{\circ}}\right)^{\frac{1}{3}} \frac{I_{\mathrm{C}}}{v_{\mathrm{P}}} \left(a_{\mathrm{c}}^3 - 3a_{\mathrm{c}}^2 + 6a_{\mathrm{c}} - 6 + 6\exp\{-a_{\mathrm{c}}\}\right) \right\} \quad \text{with} \quad a_{\mathrm{c}} = I_{\mathrm{C}}t. \tag{8.25}$$

In practice, nucleation will occur at all three locations, faces, edges and corners of grains, albeit at different rates. It is possible to deal with this by generalising the extended space concept and implementing the following equations in a numerical scheme that can be progressed in a stepwise manner with small increments of transformation: [7, 113, 114]:

$$\mathrm{d}V^{\mathrm{P,f}} = \left(1 - \frac{V^{\mathrm{P,f}} + V^{\mathrm{P,e}} + V^{\mathrm{P,c}}}{\mathrm{V}}\right) \mathrm{d}V_{\mathrm{e}}^{\mathrm{P,f}}$$

$$\mathrm{d}V^{\mathrm{P,e}} = \left(1 - \frac{V^{\mathrm{P,f}} + V^{\mathrm{P,e}} + V^{\mathrm{P,c}}}{\mathrm{V}}\right) \mathrm{d}V_{\mathrm{e}}^{\mathrm{P,e}}$$

$$\mathrm{d}V^{\mathrm{P,c}} = \left(1 - \frac{V^{\mathrm{P,f}} + V^{\mathrm{P,e}} + V^{\mathrm{P,c}}}{\mathrm{V}}\right) \mathrm{d}V_{\mathrm{e}}^{\mathrm{P,c}}.$$

8.9 STRESS- AND STRAIN-AFFECTED KINETICS

The pearlite transformation is accelerated when the austenite is stressed in tension. The time to reach a detectable rate of transformation[7] is reduced, as is the interval $t_{\mathrm{f}} - t_{\mathrm{s}}$ between the start and finish of the reaction, Figure 8.35 [115, 116]. The pearlite growth rate is not affected when $\sigma_{\mathrm{a}} \leq \sigma_{\mathrm{y}}^{\gamma}$, but the nucleation rate increases [117]. Even at small stresses, a degree of creep occurs that creates defects within the austenite [115], that may enhance nucleation rate. The maximum creep-strain recorded was $\lesssim 5 \times 10^{-4}$ over the duration of the experiments, which may be large enough to roughen the austenite grains boundaries, rendering them more susceptible to nucleation events. The same effect of stress is observed during the continuous cooling transformation of eutectoid steels [116].

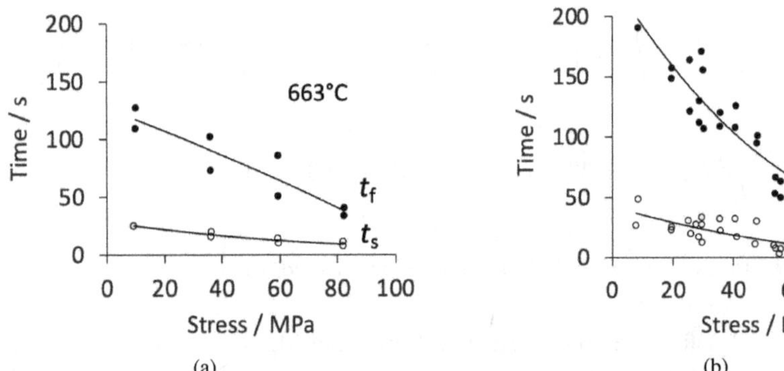

Figure 8.35 The transformation-start time (t_{s}) and that at which the transformation finishes (t_{f}) for a Fe-0.82C-0.28Si-0.80Mn wt%, following austenitisation at 850 °C for 15 min as a function of applied tensile stress. (a) Isothermal transformation at 663 °C, (b) at 673 °C. Adapted from Gautier et al. [115] with data from both dilatometry and electrical resistance measurements.

Pearlite is less dense than austenite. The volume change accompanying the $\gamma \rightarrow \mathrm{P}$ transformation in polycrystalline austenite is isotropic but not so when pearlite forms under the influence of an applied stress. The volume expansion induces plastic strain that accommodates the applied stress.

[7]Often referred to as the incubation period, but dependent on the sensitivity of the detection method.

This *transformation plasticity* was expressed first by Greenwood and Johnson in the context of the deformation of uranium during its repeated $\alpha \leftrightarrow \beta$ phase changes [118, 119]. Assuming ideal plasticity, i.e., zero work hardening beyond yield, and in the absence of a time-dependence, the excess strain due to transformation, along the tensile axis, is (Section 8.11)

$$\varepsilon_{\text{TRIP}} = \frac{5\sigma \Delta V/\text{V}}{3\sigma_y^\gamma} \tag{8.26}$$

where $\Delta V/\text{V}$ is the volume strain due to transformation, which becomes greater at lower transformation temperatures because the thermal expansion coefficient of austenite is much greater than that of ferrite.

Experimental data shown in Figure 8.36 do not show much of a dependence on temperature presumably because the range covered is small. The maximum values of $\varepsilon_{\text{TRIP}}$ corresponding to about $\sigma = 80\,\text{MPa}$ are larger than estimated using Equation 8.26, although uniaxial compression seems to be associated with smaller transformation strain along the stress axis [120, 121].

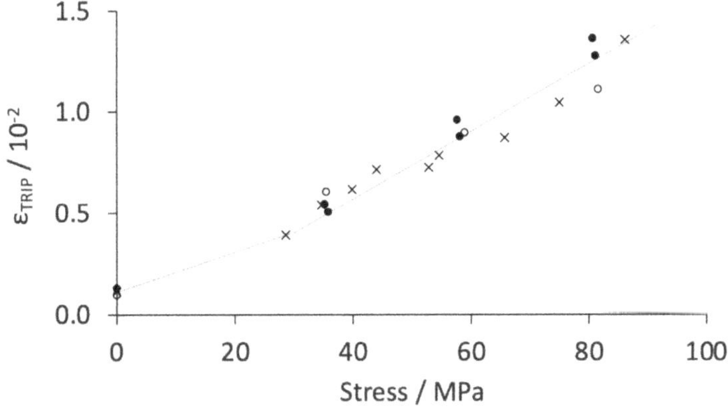

Figure 8.36 Observed transformation plasticity as a function of tensile stress, during the formation of pearlite at 663 °C (•), 673 °C (×) and 683 °C (∘). Aggregated data from Gautier et al. [115], who corrected them to remove any creep strain, except for stresses less than 10 MPa where creep was not detectable.

An experiment in which the strain was monitored along two orthogonal sample-axes while the eutectoid steel was transformed at 550 °C whilst subjected to a uniaxial compressive stress of 80 MPa ($\sigma_y^\gamma \approx 110\,\text{MPa}$), revealed the anisotropy of transformation strain, Figure 8.37.

The strain along the stress axis is smaller than in the radial directions of the cylindrical samples studied. The volume strain is isotropic during stress-free transformation. As might be expected from Le Chatelier's principle, during stress-affected transformation there is a smaller expansion along the compression axis than in the radial directions, Figure 8.37.

The strain anisotropy for pearlite is much smaller than that observed for bainite. In the latter case, the total transformation strain is a combination of large shear of ≈ 0.26 parallel to the habit plane and a volume expansion normal to the habit plane [122, 123]. There are 24 crystallographic-variants possible within a single austenite grain. However, these variants become biased under the influence of stress because those with transformation strains that comply with the imposed stress are favoured [124]. This *variant selection* is absent with pearlite, but in bainite the effect is to exaggerate the strain along the stress axis. The difference becomes striking when it is realised that in Figure 8.37, the volume fraction of bainite is just 0.23 whereas the pearlite has completely consumed the austenite.

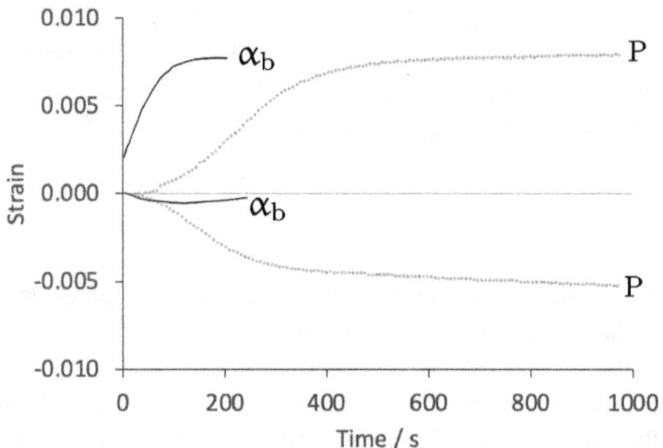

Figure 8.37 Transformation strains measured along the axis of the applied stress and in a direction normal to that. Isothermal transformation at 550 °C, −80 MPa stress, into a fully pearlitic microstructure, Fe-0.5C-1.48Si-0.7Mn-0.78Cr wt%. Selected data from Ueji and Inoue, [120]. Also included, partial isothermal transformation into a volume fraction 0.23 of bainite, at 460 °C, +96 MPa stress, Fe-0.12C-0.27Si-0.84Mn-2.86Cr-1.48Mo wt%. After Bhadeshia et al. [125].

8.9.1 EFFECT OF HYDROSTATIC PRESSURE ON KINETICS

From le Chatelier's [126] principle, a positive hydrostatic pressure will oppose the transformation of austenite into lower-density products such as ferrite. The expansion associated with transformation would have to push against the pressure. This is why in Figure 8.38, the $\alpha+\gamma$ phase field is reduced in extent and suppressed to large undercoolings when the pressure is increased from 0.1 MPa to 3 GPa (there may be minor non-hydrostatic components of stress included [14]). The precipitation of cementite from austenite leads to a decrease in the lattice parameters of austenite, and although the volume per iron atom in cementite is greater than in austenite, there is a net contraction during $\gamma \rightarrow \gamma' + \theta$ [127]. Therefore, the $\gamma+\theta$ phase field expands as the hydrostatic pressure is increased. The estimation of the high-pressure phase diagram is as follows. The change in eutectoid temperature $T_{\rm E}$ is expressed as [128]

$$\frac{\mathrm{d}T_{\rm E}}{\mathrm{d}P} \approx \frac{\Delta V}{\Delta S} = \frac{\Delta V T_{\rm E}}{\Delta H} \tag{8.27}$$

where $\Delta V, \Delta H, \Delta S$ are the total changes in volume, enthalpy and entropy at the temperature $T_{\rm E}$ for $\gamma \rightarrow \alpha+\theta$. The compositional displacement of a point on a phase boundary say γ and $\gamma+\alpha$ field as a function of pressure at a particular temperature T' is similarly

$$RT'\left(\frac{\partial c^{\alpha\gamma}}{\partial P}\right)_T \left[1+\frac{\partial \ln\{\Gamma_{\rm C}^{\alpha}\}}{\partial \ln\{c^{\alpha}\}}\right]_{P,T} \tag{8.28}$$

where $\Gamma_{\rm C}^{\alpha}$ is the activity coefficient of carbon in ferrite. The same rationale can be applied to the $\gamma/\gamma+\theta$ phase boundary.

These effects can lead to a dramatic reduction in the eutectoid composition ($0.76 \rightarrow 0.37$ C wt%) with the eutectoid temperature suppressed to 655 °C. To study the consequences of pressure, Hilliard examined the kinetics of transformation in steels of approximate eutectoid carbon concentration, corresponding to the ambient and high-pressure phase diagrams: Fe-0.73C-0.97Mn and Fe-0.33C-1Mn wt%.

Figure 8.39a shows that at a given transformation temperature, the interlamellar spacing of pearlite is greater – this is to be expected because the driving force for transformation becomes smaller at greater pressures, i.e., the undercooling below the suppressed eutectoid temperature is

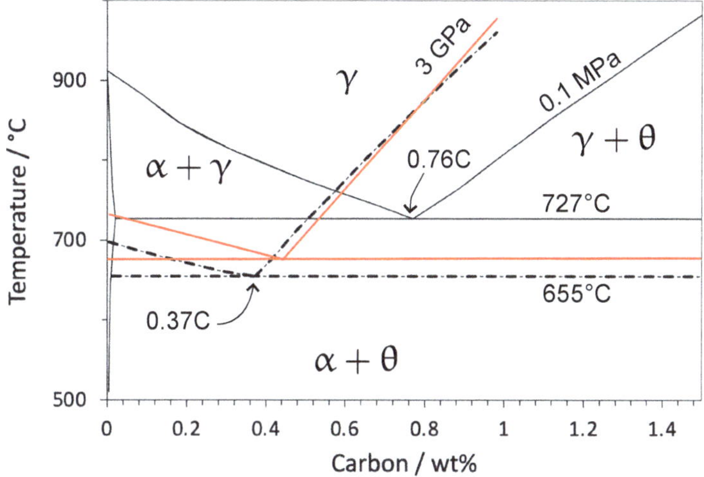

Figure 8.38 The effect of 3 GPa of hydrostatic pressure on the iron-carbon phase diagram – the high pressure version is plotted using calculations from Hilliard [14, 128] and the ambient pressure diagram is from calculations using modern thermodynamic data. The red lines represent the 3 GPa phase diagram also calculated using modern data, giving a eutectoid composition and temperature of 0.44C wt% and 676,°C (courtesy of Apparao Chintha, 2024.)

reduced. The cementite fraction within the pearlite in the lower carbon steel (Fe-0.33C-1Mn wt%) will also be smaller.

The growth rate is dramatically reduced under high-pressure, Figure 8.39b, consistent with the larger S_I but that on its own may not be sufficient to explain the retardation of growth. The diffusion of carbon in pressurised austenite is retarded because the activation volume for diffusion, V^*, which has been reported to be in the range $6 \times 10^{-8} \rightarrow 7 \times 10^{-6}\,\mathrm{m^{-3}}$ mole [129]. The activation volume is the change in the volume of the crystal during a diffusional jump, defined as

$$V^* = \left(\frac{\partial G^*}{\partial P}\right)_T \tag{8.29}$$

and if $D^\gamma = D_o \exp\{-G^*/kT\}$ then if follows that

$$V^* = -kT\left(\frac{\partial \ln\{D^\gamma/D_o\}}{\partial P}\right)_T \tag{8.30}$$

where $G*$ is the activation-free energy for diffusion. This means that the slope of the plot of diffusion coefficient versus pressure will be negative, i.e., diffusion is retarded with increasing pressure. The ratio $D^\gamma\{3\mathrm{GPa}\}/D^\gamma\{0.1\mathrm{MPa}\}$ at 600 °C was estimated from growth rate data to be ≈ 0.36 [128].

8.10 ANNEALING OF COLD-WORKED PEARLITE

Cold-worked pearlitic steels rarely require annealing heat treatments because any deformation has the purpose of enhancing strength, such as in wire drawing which forms the basis of all of the strongest ropes used in engineering (Chapter 13). There is no generic model for estimating the changes expected during annealing. For a given type and level of deformation, and fixed chemical composition, changes in hardness due to annealing at a variety of temperatures below Ae_1 can be rationalised in terms of the kinetic strength of the heat treatment, $t\exp\{-Q/RT\}$, where $Q = 190\,\mathrm{kJ}$

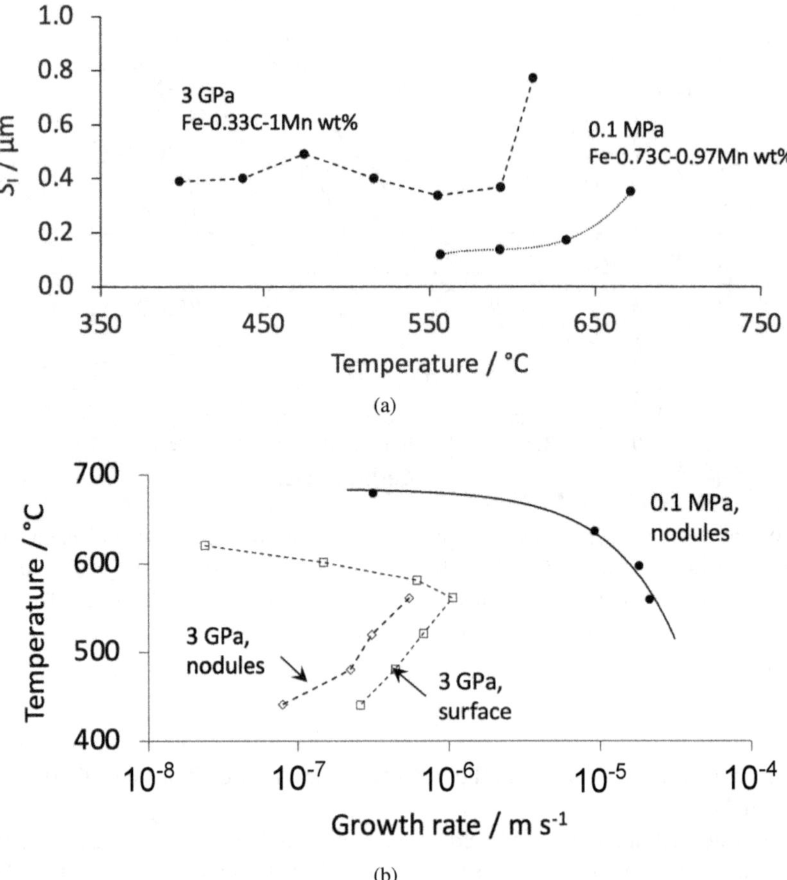

Figure 8.39 (a) Change in the pearlite interlamellar spacing as a function of hydrostatic pressure. (b) The growth rate of pearlite as a function of pressure. Nodules nucleate within the steel; pearlite that nucleates at free surfaces has its maximum rate recorded. Data from Hilliard [14].

as determined in grain growth theory (p. 17). The idea is to combine the influence of time and temperature on the kinetics of the process.

Figure 8.40 shows how the normalised-hardness of deformed pearlite (75% cold reduction) changes with annealing at three different temperatures (400, 500, 600 °C) and a variety of time periods. The combined data are rationalised in terms of the kinetic strength. The normalised hardness is used (Equation 8.31) to broaden the applicability to other scenarios where the maximum and minimum HV values are known, i.e. in the deformed state and after prolonged annealing. The equation corresponding to the line in Figure 8.40 , with $HV_{max} = 420$ and $HV_{min} = 200$ is

$$\frac{HV - HV_{min}}{HV_{max} - HV_{min}} \approx 0.052\left(\frac{Q}{RT} - \ln t\right) - 0.7725 \pm 0.13, \tag{8.31}$$

where the time is expressed in seconds, the temperature in Kelvin.

It is interesting to apply the method to a different degree of cold deformation prior to annealing. A pearlitic sample subjected to $\varepsilon_{vM} = 16.1$ using high-pressure torsion achieved Vickers hardness of $HV_{max} = 750$ [131]; its reported hardness prior to deformation is $HV_{max} = 277$ [132]. The samples were then annealed over the temperature range 420-600 °C, for two-hour intervals in each case. Figure 8.41 shows how Equation 8.31 is able to reproduce the normalised hardness – the agreement

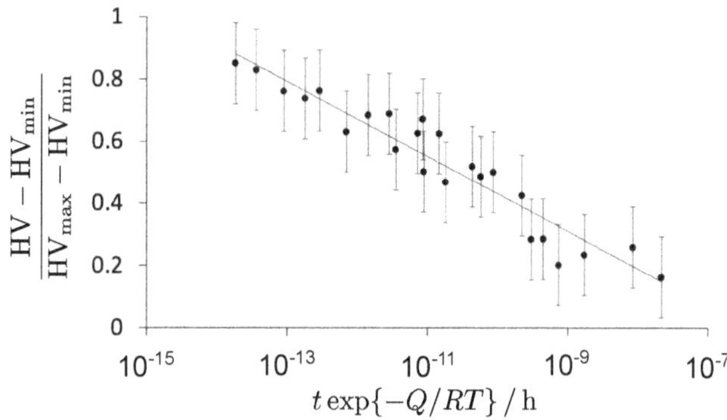

Figure 8.40 Pearlitic steel (Fe-0.85C-0.17Si-0.77Mn wt%) transformed at 690 °C, cold-rolled to a 75% reduction, followed by annealing heat treatments at a variety of temperatures. The normalised Vickers-hardness (0.2 kg) is plotted as a function of the kinetic strength of the heat treatment, with $Q = 190\,\text{kJ mole}^{-1}$, the time is in seconds and the units of temperature are in Kelvin. Data from Chang and Byrne [130].

obtained is remarkable given the extreme plastic strains associated with high-pressure torsion.

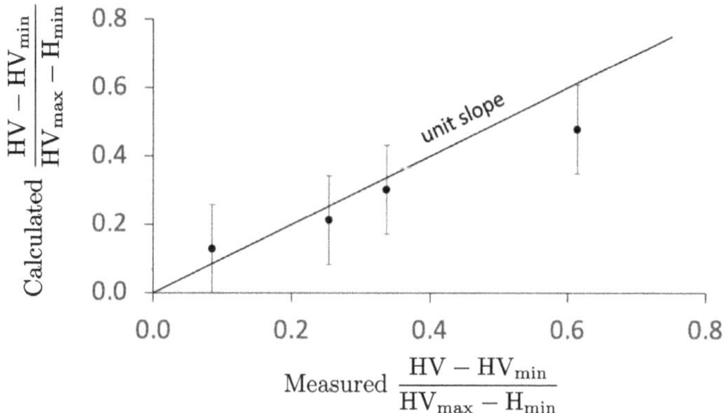

Figure 8.41 The calculated versus measured normalised-hardness determined on annealed samples of pearlitic steel that have been subjected to extreme plastic strains. Selected data from Kammerhofer et al. [131].

The maximum hardness ever achieved in a fully pearlitic, undeformed pearlite, was generated in Fe-0.78C-1.6Si-2.02Mn-0.24Mo-1.01Cr-3.87Co-1.37Al wt% transformed while under the influence of a 30 T magnetic field, with an interlamellar spacing of ≈ 50 nm seems to be 484 ± 13 HV. In the deformed state, the maximum hardness relates to the high-pressure torsion experiment, at 750 HV [131]. The minimum for a fully pearlitic microstructure has been reported as 220 HV in a Fe-0.68C-0.19Si-0.85Mn wt% steel with an interlamellar spacing of 0.24 μm [133].

8.10.1 ANNEALING OF P+α MIXTURES

Railway wheels usually are hypoeutectoid with 0.5-0.6 wt% of carbon, with a structure that is a mixture of α+P. In such a steel (Fe-0.56C-0.8Mn-0.4Si-0.3Ni wt%) with $V_V^\alpha = 0.08 \rightarrow 0.12$, annealing experiments were conducted on the undeformed (initial hardness $HV_{min} = 260$) and cyclically deformed ($HV_{max} = 274$) states [134]. The annealing time periods ranged from 240–14280 s and annealing temperatures between 246-650 °C, Figure 8.42a. The data do rationalise across the different treatments but obviously do not fall on a single straight line with a steeper decrease in hardness when the kinetic strength exceeds about 10^{-9} /h where microstructural studies indicate a significant spheroidisation of the pearlite [134].

There are points in Figure 8.42a, for the deformed condition, where the normalised hardness exceeds unity, i.e., the hardness actually increases on low-temperature annealing at about 300 °C, because of a strain-ageing effect where dislocations in the allotriomorphic ferrite become pinned by carbon atoms. However, cyclic deformation leads to many free dislocations so the hardening effect is not as large as when the sample is monotonically strained in tension to a strain of 0.065 prior to annealing, Figure 8.42b, with the peak hardness reaching 299 HV from a starting level of 274 HV.

The effect of strain-ageing has been attributed here to the presence of allotriomorphic ferrite, because it is known that pearlite does not exhibit this phenomenon. This is because of a level of coherency between the pearlitic cementite and ferrite, which allows dislocations to be emitted from the interface, (p. 52).

8.11 APPENDIX

We deal here with the derivation of the Johnson and Greenwood transformation plasticity [118] Equation 8.26. Given the small strains involved, the material is treated as if it does not work to harden during plastic deformation so the elastic strain-increments are zero. It also is assumed that the principal stress and principal strain directions are identical, in which case, the relationship between the plastic-strain increment and combined applied stress is, according to the Lévy-von Mises equations [135, 136]

$$\dot{\varepsilon}_{ij} = \sigma'_{ij} b_{31}, \qquad \varepsilon_{ij} = \sigma'_{ij} b_{31} t \tag{8.32a}$$

where $\dot{\varepsilon}_{ij}$ represents the strain rate and σ'_{ij} has only the deviatoric (shear) components of stress, b_{31} is a constant and t is the time over which the total strain is achieved. From the von Mises yield criterion,

$$\sigma'^2_{11} + \sigma'^2_{22} + \sigma'^2_{33} + 2\sigma'^2_{12} + 2\sigma'^2_{23} + \sigma'^2_{31} = \frac{2}{3}\sigma^2_{\text{y}} \tag{8.32b}$$

where the yield strength is for the weaker phase. If the stress is applied along the '3' axis, and assuming that the transformation causes a volume dilatation $\Delta = \Delta V/V$ which gives rise to strain components Δ_{ij} along different directions, substituting in Equation 8.32b for the strains according to Equation 8.32a gives

$$[-\frac{1}{2}\varepsilon - \Delta_{11}]^2 + [-\frac{1}{2}\varepsilon - \Delta_{22}]^2 + [\varepsilon - \Delta_{33}]^2 - 2[\Delta_{12}]^2 - 2[\Delta_{23}]^2 - 2[\Delta_{31}]^2 = \frac{2}{3}\sigma^2_{\text{y}}(b_{31}t)^2 \tag{8.32c}$$

$$\frac{3}{2}\varepsilon^2 + \underbrace{\Delta^2_{11} + \Delta^2_{22} + \Delta^2_{33} + 2(\Delta^2_{12} + \Delta^2_{23} + \Delta^2_{31})}_{\frac{2}{3}\Delta^2} + \varepsilon(\Delta_{11} + \Delta_{22} - 2\Delta_{33}) = \frac{2}{3}\sigma^2_{\text{y}}(b_{31}t)^2. \tag{8.32d}$$

Along the '3' axis, $\varepsilon - \Delta_{33} = \sigma'_{33} b_{31} t$, which can be used to eliminate the term $b_{31}t$ so that

$$\left[\frac{3}{2}\varepsilon^2 + \frac{2}{3}\Delta^2 + \varepsilon(\Delta_{11} + \Delta_{22} - 2\Delta_{33})\right]^{\frac{1}{2}} = \frac{\sqrt{2}}{\sqrt{3}}\sigma_{\text{y}}[\varepsilon - \Delta_{33}]/\sigma'_{33}. \tag{8.32e}$$

If Δ_{ii} $(i = 1,2,3)$ can be identified with the principal axes then $\Sigma^3_{i=1}\Delta_{ii} = 0$; also, the deviatoric stress $\sigma'_{33} \equiv \frac{2}{3}\sigma_3$ which is the externally applied stress. It follows that

$$\varepsilon \approx \frac{(2\sigma_3\Delta/3\sigma_{\text{y}}) + \overline{\Delta_{33}}}{(1 - 9(\overline{\Delta_{33}})^2)/4\Delta^2} \approx \frac{5\sigma_3\Delta}{3\sigma_{\text{y}}} \tag{8.32f}$$

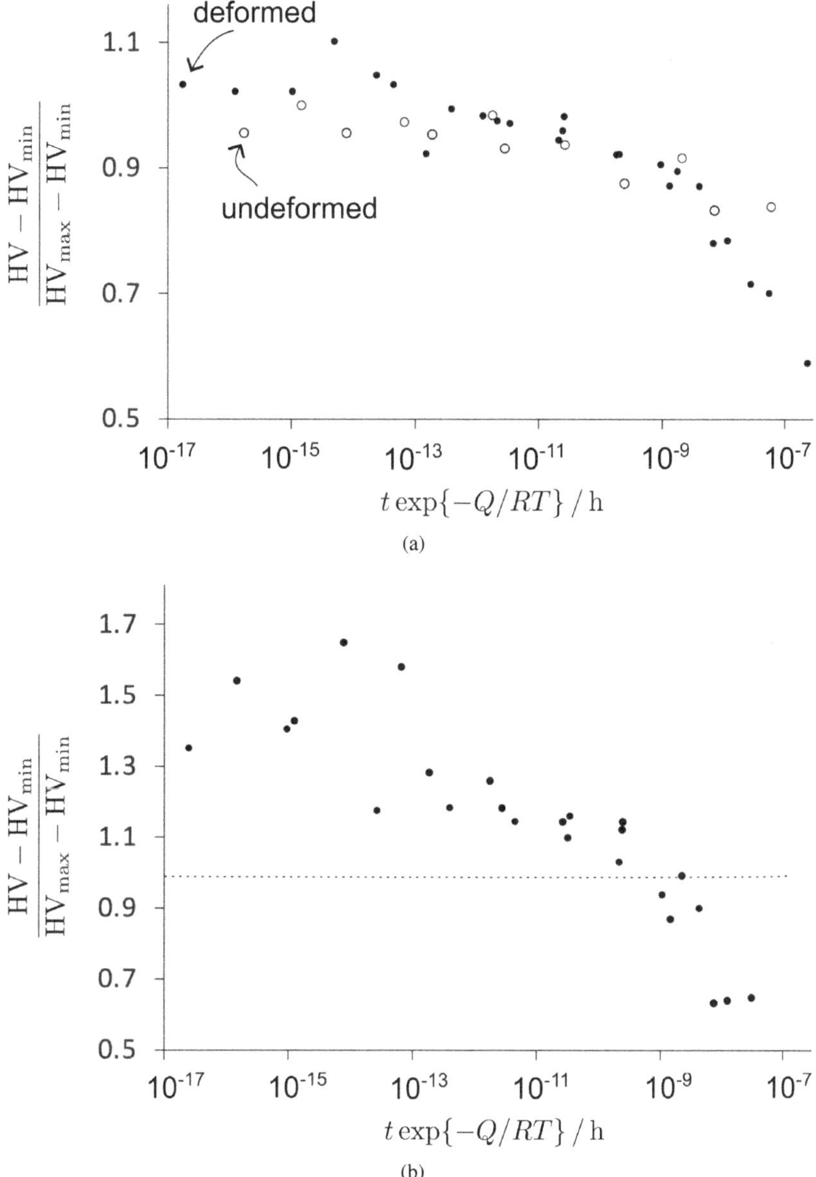

Figure 8.42 (a) Hardness of railway-wheel steel interpreted in terms of kinetic strength, with time in seconds and $Q = 190\,\mathrm{kJ\,mol^{-1}}$. The unfilled points represent annealing of the undeformed samples whereas the filled points are for samples annealed after cyclic deformation. The cyclic straining involved 1300 cycles of fatigue loading to a strain amplitude of ± 0.01. (b) The steel strained in tension to $\varepsilon_p = 0.065$ prior to annealing. Data from Nikas [134].

where $\overline{\Delta_{33}}$ implies an averaged value throughout the phase deforming and with further small simplifications and approximations, Greenwood and Johnson arrive at Equation 8.26 expressed in the term on the right in the above equation.

REFERENCES

1. M. J. Buerger: 'The genesis of twin crystals', *American Mineralogist: Journal of Earth and Planetary Materials*, 1945, **30**, 469–482.
2. J. D. Waston: 'The crystallography and morphology of Widmanstätten precipitates': Ph.D. thesis, The University of New South Wales, Australia, http://hdl.handle.net/1959.4/56295, 1970.
3. J. D. Watson, and P. G. McDougall: 'The crystallography of Widmanstätten ferrite', *Acta Metallurgica*, 1973, **21**, 961–973.
4. S. W. Seo, H. K. D. H. Bhadeshia, and D. W. Suh: 'Pearlite growth rate in Fe-C and Fe-Mn-C steels', *Materials Science and Technology*, 2015, **31**, 487–493.
5. R. F. Mehl, and C. A. Dubé: 'The eutectoid reaction', In: J. E. Mayer, R. Smouluschowski, and W. A. Weyl, eds. *Phase Transformations in Solids*. New York, USA: John Wiley & Sons, Inc., 1951:545–582.
6. G. W. Rathenau, and G. Baas: 'Electron-optical observations of transformations in eutectoid steel', *Acta Metallurgica*, 1954, **2**, 875–883.
7. S. Jones, and H. K. D. H. Bhadeshia: 'Kinetics of the simultaneous decomposition of austenite into several transformation products', *Acta Materialia*, 1997, **45**, 2911–2920.
8. Y. Liang, Z. Shi, and Y. Liang: 'Effects of Ce or Nb addition on nucleation of pearlite in high carbon steel', *Advanced Materials Research*, 2013, **734-737**, 1513–1531.
9. R. A. Ricks, G. S. Barritte, and P. R. Howell: 'The nature of acicular ferrite in HSLA steel weld metals', *Journal of Materials Science*, 1982, **17**, 732–740.
10. H. K. D. H. Bhadeshia: Bainite in Steels: Theory and Practice: 3rd ed., Leeds, U.K.: Maney Publishing, 2015.
11. R. J. Dippenaar, and R. W. K. Honeycombe: 'The crystallography and nucleation of pearlite', *Proceedings of the Royal Society A*, 1973, **333**, 455–467.
12. R. Dippenaar: 'Decomposition of austenite in alloy steels': Ph.D. thesis, University of Cambridge, Cambridge, U. K., 1970.
13. C. Garcia-Mateo, M. Peet, F. G. Caballero, and H. K. D. H. Bhadeshia: 'Tempering of a hard mixture of bainitic ferrite and austenite', *Materials Science and Technology*, 2004, **20**, 814–818.
14. J. E. Hilliard: 'Research and development on the effect of high pressure on various elements and binary alloys. Part 1: effect of pressures on the austenite to pearlite reaction': Tech. Rep. ASD-TDR-62-479, Part 1, General Electric Corporation, New York, USA, 1963.
15. J. W. Christian: Theory of Transformations in Metals and Alloys, Part I: 2 ed., Oxford, U. K.: Pergamon Press, 1975.
16. M. Richman, D. Thomas, and M. Cohen: 'The role of austenitizing temperature in the nucleation of pearlite', *Acta Metallurgica*, 1959, **7**, 814–816.
17. J. Fisher: 'Eutectoid decompositions', In: *Thermodynamics in Physical Metallurgy*. Ohio, U. S. A.: ASM, 1950:201–241.
18. H. K. D. H. Bhadeshia: Theory of Transformations in Steels: London, U.K.: CRC Press, Taylor and Francis Group, 2021.
19. F. C. Hull, R. A. Colton, and R. F. Mehl: 'Rate of nucleation and rate of growth of pearlite', *Transactions of the AIMME*, 1942, **150**, 185–207.
20. H. K. D. H. Bhadeshia: 'Equilibrium in Fe-1.02C-0.13Si-0.2Mn wt%': https://www.phase-trans.msm.cam.ac.uk/2023/Hull_1942.txt, 2023.
21. S. Offerman, L. Van Wilderen, N. Van Dijk, J. Sietsma, M. T. Rekveldt, and S. Van der Zwaag: 'In-situ study of pearlite nucleation and growth during isothermal austenite decomposition in nearly eutectoid steel', *Acta Materialia*, 2003, **51**, 3927–3938.
22. NPL: 'MTDATA': Software, National Physical Laboratory, Teddington, U.K., 2006.

23. S. A. Hackney, and G. J. Shiflet: 'The pearlite-austenite growth interface in an Fe-0.8C-12Mn alloy', *Acta Materialia*, 1987, **35**, 1007–1017.
24. A. Durgaprasad, S. Giri, S. Lenka, S. Kundu, S. Mishra, S. Chandra, R. D. Doherty, and I. Samajdar: 'Defining a relationship between pearlite morphology and ferrite crystallographic orientation', *Acta Materialia*, 2017, **129**, 278–289.
25. M.-X. Zhang, and P. M. Kelly: 'Determination of carbon content in bainitic ferrite and carbon distribution in austenite by using CBKLDP', *Materials Characterization*, 1998, **40**, 159–168.
26. M.-X. Zhang, and P. M. Kelly: 'Crystallography of spheroidite and tempered martensite', *Acta Materialia*, 1998, **46**, 4081–4091.
27. H. K. D. H. Bhadeshia: 'Solution to the Bagaryatskii and Isaichev ferrite-cementite orientation relationship problem', *Materials Science and Technology*, 2018, **34**, 1666–1668.
28. S. Kante, and A. Leineweber: 'Two-phase and three-phase crystallographic relationships in white-solidified and nitrided Fe-C-Si cast iron', *Acta Materialia*, 2019, **170**, 240–252.
29. B. Vinet, L. Magnusson, H. Fredriksson, and P. J. Desré: 'Correlations between surface and interface energies with respect to crystal nucleation', *Journal of colloid and interface science*, 2002, **255**, 363–374.
30. G. Kaptay: 'A coherent set of model equations for various surface and interface energies in systems with liquid and solid metals and alloys', *Advances in colloid and interface science*, 2020, **283**, 102212.
31. Y. Li, X. Yu, C. Zhao, L. Jin, and Q. Yu: 'Surface morphology of Fe_3C catalyst under different CO coverage from DFT and thermodynamics', *Computational and Theoretical Chemistry*, 2023, , 114220.
32. M. Ruda, D. Farkas, and G. Garcia: 'Atomistic simulations in the Fe-C system', *Computational Materials Science*, 2009, **45**, 550–560.
33. M. Guziewski, S. P. Coleman, and C. R. Weinberger: 'Atomistic investigation into the atomic structure and energetics of the ferrite-cementite interface: The Bagaryatskii orientation', *Acta Materialia*, 2016, **119**, 184–192.
34. M.-X. Zhang, and P. M. Kelly: 'Accurate orientation relationships between ferrite and cementite in pearlite', *Scripta Materialia*, 1997, **37**, 2009–2015.
35. P. Deb, and M. C. Chaturvedi: 'Coarsening behavior of cementite particles in a ferrite matrix in 10B30 steel', *Metallography*, 1982, **15**, 341–354.
36. S. K. Das, A. Biswas, and R. N. Ghosh: 'Volume fraction dependent particle coarsening in plain carbon steel', *Acta Metallurgica and Materialia*, 1993, **41**, 777–781.
37. J. J. Kramer, G. M. Pound, and R. F. Mehl: 'The free energy of formation and the interfacial enthalpy in pearlite', *Acta Metallurgica*, 1958, **6**, 763–771.
38. A. G. Martin, and C. M. Sellars: 'Measurement of interfacial energy from extraction replicas of particles on grain boundaries', *Metallography*, 1970, **3**, 259–273.
39. H. O. K. Kirchner, B. G. Mellor, and G. A. Chadwick: 'A calorimetric determination of the interfacial enthalpy of Cu-In and Cu-Al lamellar eutectoids', *Acta Metallurgica*, 1978, **26**, 1023–1031.
40. K. Hashiguchi, and J. S. Kirkaldy: 'Pearlite growth by combined volume and phase boundary diffusion', *Scandinavian Journal of Metallurgy*, 1984, **13**, 240–248.
41. A. S. Pandit, and H. K. D. H. Bhadeshia: 'Mixed diffusion-controlled growth of pearlite in binary steel', *Proceedings of the Royal Society A*, 2011, **467**, 508–521.
42. J. A. Mathews, H. Farahani, J. Sietsma, R. Petrov, M. Mecozzi, and M. Santofimia: 'Microstructures in a carburized steel after isothermal pearlitic treatment', *Journal of Materials Science & Technology*, 2023, **160**, 66–75.
43. A. Schneider, and G. Inden: 'Simulation of the kinetics of precipitation reactions in ferritic steels', *Acta Materialia*, 2005, **53**, 519–531.

44. C. Capdevila, F. G. Caballero, and C. García de Andrés: 'Neural network model for isothermal pearlite transformation. part I: Interlamellar spacing', *ISIJ International*, 2005, **45**, 229–237.
45. J. H. Park, S. W. Jeong, and W. J. Nam: 'The effects of alloying elements, Cr, Si, and transformation temperature on the microstructures and tensile properties of hyper-eutectoid steels', *Korean Journal of Metals and Materials*, 2017, **55**, 460–469.
46. W. Fu, T. Furuhara, and T. Maki: 'Effect of Mn and Si addition on microstructure and tensile properties of cold-rolled and annealed pearlite in eutectoid Fe–C alloys', *ISIJ International*, 2004, **44**, 171–178.
47. H. K. D. H. Bhadeshia: 'Private communication', 2021: Queries about *Neural network model for isothermal pearlite transformation*, ISIJ International 45 (2002) 229.
48. M. Takahashi: 'Reaustenitisation from bainite in steels': Ph.D. thesis, University of Cambridge, http://www.phase-trans.msm.cam.ac.uk/2000/phd.html#Takahashi, 1992.
49. C. M. Bae, C. S. Lee, and W. J. Nam: 'Effect of carbon content on mechanical properties of fully pearlitic steels', *Materials Science and Technology*, 2002, **18**, 1317–1321.
50. T. Gladman, I. D. McIvor, and F. B. Pickering: 'Some aspects of the structure-property relationships in high-C ferrite-pearlite steels', *Journal of the Iron and Steel Institute*, 1972, **210**, 916–930.
51. J. M. Hyzak, and I. M. Bernstein: 'The role of microstructure on the strength and toughness of fully pearlitic steels', *Metallurgical Transactions A*, 1976, **7**, 1217–1221.
52. J. P. Houin, A. Simon, and G. Beck: 'Relationship between structure and mechanical properties of pearlite between 0.2% and 0.8%C', *Transactions of the Iron and Steel Institute of Japan*, 1981, **21**, 726–731.
53. M. Dollar, I. M. Bernstein, and A. W. Thompson: 'Influence of deformation substructure on flow and fracture of fully pearlitic steel', *Acta Metallurgica*, 1988, **36**, 311–320.
54. K. Nakase, and I. M. Bernstein: 'The effect of alloying elements and microstructure on the strength and fracture resistance of pearlitic steel', *Metallurgical Transactions A*, 1988, **19**, 2819–2829.
55. D. J. Alexander, and I. M. Bernstein: 'Cleavage fracture in pearlitic eutectoid steel', *Metallurgical Transactions A*, 1989, **20**, 2321–2335.
56. J. Toribio: 'Relationship between microstructure and strength in eutectoid steels', *Materials Science & Engineering A*, 2004, **387-389**, 227–230.
57. A. M. Elwazri, P. Wanjara, and S. Yue: 'The effect of microstructural characteristics of pearlite on the mechanical properties of hypereutectoid steel', *Materials Science & Engineering A*, 2005, **404**, 91–98.
58. Y. S. Jang, M. P. Phaniraj, D. I. Kim, J. H. Shim, and M. Y. Huh: 'Effect of aluminum content on the microstructure and mechanical properties of hypereutectoid steels', *Metallurgical & Materials Transactions A*, 2010, **41**, 2078–2084.
59. S. Behera, R. K. Barik, M. B. Sk, R. Mitra, and D. Chakrabarti: 'Recipe for improving the impact toughness of high-strength pearlitic steel by controlling the cleavage cracking mechanisms', *Materials Science & Engineering A*, 2019, **764**, 138256.
60. J. A. Mohandesi, and M. Saadatmand: 'The optimization of interlamellar spacing in a nanopearlitic lead-patented hypoeutectoid steel wire', *Journal of Materials Engineering and Performance*, 2011, **20**, 1467–1473.
61. J. Chance, and N. Ridley: 'Chromium partitioning during isothermal transformation of a eutectoid steel', *Metallurgical Transactions A*, 1981, **12A**, 1205–1213.
62. V. Fourlakidis, I. Belov, and A. Diózegi: 'Experimental model of the pearlite interlamellar spacing in lamellar graphite iron', *Tecnologia em Metalurgia, Materials e Mineração*, 2022, **19**, e2634.

63. X. Xu, T.-F. Chung, S. Hu, Q. Zhu, J. Fu, J.-R. Yang, and Q. Tian: 'Effect of tin microalloying on the microstructure of low-carbon free-machining steels', *Journal of Materials Research and Technology*, 2022, **20**, 1172–1185.
64. N. North, M. Owens, and C. Pearson: 'Thermal stability of cast and wrought marine iron', *Studies in Conservation*, 1976, **21**, 192–197.
65. S. Ghodrat, M. Janssen, L. A. Kestens, and J. Sietsma: 'Volume expansion of compacted graphite iron induced by pearlite decomposition and the effect of oxidation at elevated temperature', *Oxidation of metals*, 2013, **80**, 161–176.
66. M. A. Yescas-Gonzalez, and H. K. D. H. Bhadeshia: 'Cast irons': https://www.phase-trans.msm.cam.ac.uk/2001/adi/cast.iron.html, 2023.
67. R. Lumley: 'Inhomogeneous pearlite and ferrite in ductile iron sand castings', *International Journal of Metalcasting*, 2023, **17**, 1467–1492.
68. R. F. Mehl: 'The structure and rate of formation of pearlite', *Transactions of the American Society for Metals*, 1941, **29**, 813–862.
69. C. Zener: 'Kinetics of the decomposition of austenite', *Transactions of the American Institute of Mining and Metallurgical Engineers*, 1946, **167**, 550–595.
70. M. Hillert: 'Role of interfacial energy during solid-state phase transformations', *Jernkontorets Annaler*, 1957, **141**, 757–789.
71. M. Hillert: 'The formation of pearlite', In: V. F. Zackay, and H. I. Aaronson, eds. *Decomposition of Austenite by Diffusional Processes*. New York, USA: Interscience, 1962:197–237.
72. F. C. Frank, and K. E. Puttick: 'Cementite morphology in pearlite', *Acta Metallurgica*, 1956, **4**, 206–210.
73. E. Werner: 'The growth of holes in plates of cementite', *Materials Science & Engineering A*, 1991, **132**, 213–223.
74. N. T. Belaiew: 'The inner structure of the pearlite grain', *Journal of the Iron and Steel Institute*, 1922, **105**, 201–227.
75. J. S. Kirkaldy, and R. C. Sharma: 'Stability principles for lamellar eutectoid(ic) reactions', *Acta Metallurgica*, 1980, **28**, 1009–1021.
76. J. W. Cahn, and W. C. Hagel: 'Theory of pearlite reaction', In: V. F. Zackay, and H. I. Aaronson, eds. *Decomposition of Austenite by Diffusional Processes*. New York: Intersciences, 1962:131–192.
77. M. Hillert: 'Paraequilibrium': Tech. Rep., Swedish Institute for Metals Research, Stockholm, Sweden, 1953.
78. G. R. Purdy, D. H. Weichert, and J. S. Kirkaldy: 'The growth of proeutectoid ferrite in ternary Fe-C-Mn austenites', *Transactions of the Metallurgical Society of the AIME*, 1964, **230**, 1025–1034.
79. D. E. Coates: 'Diffusional growth limitation and hardenability', *Metallurgical Transactions*, 1973, **4**, 2313–2325.
80. H. K. D. H. Bhadeshia: 'Diffusional formation of ferrite in iron and its alloys', *Progress in Materials Science*, 1985, **29**, 321–386.
81. S. A. Al-Salman, G. W. Lorimer, and N. Ridley: 'Pearlite growth kinetics and partitioning in a Cr-Mn eutectoid steel', *Metallurgical Transactions A*, 1979, **10A**, 1703–1709.
82. P. R. Williams, M. K. Miller, and G. D. W. Smith: 'The partitioning of alloy elements during the pearlite transformation: at atom probe study', In: H. I. Aaronson, D. E. Laughlin, M. Sekerka, and C. M. Wayman, eds. *Solid-Solid Phase Transformations*. Warrendale, Pennsylvania, USA: TMS-AIME, 1981:813–817.
83. C. R. Hutchinson, R. E. Hackenberg, and G. J. Shiflet: 'The growth of partitioned pearlite in Fe-C-Mn steels', *Acta Materialia*, 2004, **52**, 3565–3585.

84. M. Hillert: 'An analysis of the effect of alloying elements on the pearlite reaction', In: H. I. Aaronson, D. E. Laughlin, M. Sekerka, and C. M. Wayman, eds. *Solid-Solid Phase Transformations*. Materials Park, Ohio, USA: TMS-AIME, 1981:789–806.
85. J. Lacaze: 'Pearlite growth in cast irons: a review of literature data', *International Journal of Cast Metals Research*, 1999, **11**(5), 431–436.
86. J. Lacaze: ASM Handbook: cast iron science and technology, chap. The Austenite-to-Pearlite/Ferrite Transformation: Ohio, U. S. A., ed. D. M. Stefanescu: ASM International, 2017:106–113.
87. H. K. D. H. Bhadeshia: 'Some difficulties in the theory of diffusion-controlled growth in substitutionally alloyed steels', *Current Opinion in Solid State and Materials Science*, 2016, **20**, 396–400.
88. J. Fridberg, L.-E. Torndähl, and M. Hillert: 'Diffusion in iron', *Jernkontorets Annaler*, 1969, **153**, 263–276.
89. N. A. Razik, G. W. Lorimer, and N. Ridley: 'An investigation of manganese partitioning during the austenite-pearlite transformation using analytical electron microscopy', *Acta Metallurgica*, 1974, **22**, 1249–1258.
90. J. W. Cahn, and W. C. Hagel: 'Divergent pearlite in a manganese eutectoid steel', *Acta Metallurgica*, 1963, **11**, 561–574.
91. S. V. Tsivinsky, L. I. Kogan, and R. I. Entin: 'Investigation of the distribution of chromium and tungsten during the decomposition of austenite using the radioactive tracer method', In: *Problems of Metallography and the Physics of Metals*. Moscow, Russia: State Scientific Press, ed. B. Ya Lybubov, 1955:185–199.
92. L. Huang, F. Liu, and M. Huang: 'Accelerating bainite transformation by concurrent pearlite formation in a medium Mn steel: Experiments and modelling', *Journal of Materials Science & Technology*, 2024, **176**, 211–223.
93. Y.-J. Zhang, T. Umeda, S. Morooka, S. Harjo, G. Miyamoto, and T. Furuhara: 'Pearlite growth kinetics in Fe-C-Mn eutectoid steels: Quantitative evaluation of energy dissipation at pearlite growth front via experimental approaches', *Metallurgical and Materials Transactions A*, 2024, **?**, https://doi.org/10.1007/s11661–024–07518–1.
94. R. F. Mehl, and W. C. Hagel: 'The austenite:pearlite reaction', *Progress in Metal Physics*, 1956, **6**, 74–134.
95. R. Smoluchowski, J. E. Mayer, and W. A. Weyl, eds.: Phase Transformations in Solids: Symposium held at Cornell University, August 23-26, 1948. New York, USA: John Wiley and Sons Inc., 1951.
96. H. C. Longuet-Higgins, and M. E. Fisher: 'Lars Onsager', *Biographical Memoirs of Fellows of the Royal Society*, 1978, **24**, 443–471.
97. R. B. McLellan, M. L. Rudee, and T. Ishibachi: 'The thermodynamics of dilute interstitial solid solutions with dual-site occupancy and its application to the diffusion of carbon in α-iron', *Transactions of the AIME*, 1965, **233**, 1939–1943.
98. B. S. Lement, and M. Cohen: 'A dislocation-attraction model for the first stage of tempering', *Acta Metallurgica*, 1956, **4**, 469–476.
99. K. Nakajima, Y. Tanaka, Y. Hosoya, M. Apel, and I. Steinbach: 'Phase-field simulation of cooperative growth of pearlite', *Materials Science Forum*, 2007, **558-559**, 1013–1020.
100. E. B. Hawbolt, B. Chau, and J. K. Brimacombe: 'Kinetics of austenite-pearlite transformation in eutectoid carbon steel', *Metallurgical Transactions A*, 1983, **14**, 1803–1815.
101. J. S. Kirkaldy, and D. Venugopalan: 'Prediction of microstructure and hardenability in low alloy steels', In: A. R. Marder, and J. I. Goldstein, eds. *Phase Transformations in Ferrous Alloys*. Philadelphia, USA: TMS-AIME, 1984:125–148.
102. W. A. Johnson, and R. F. Mehl: 'Reaction kinetics in processes of nucleation and growth', *TMS-AIMME*, 1939, **135**, 416–458.

103. M. Avrami: 'Kinetics of phase change 1', *Journal of Chemical Physics*, 1939, **7**, 1103–1112.
104. M. Avrami: 'Kinetics of phase change 2', *Journal of Chemical Physics*, 1940, **8**, 212–224.
105. M. Avrami: 'Kinetics of phase change 3', *Journal of Chemical Physics*, 1941, **9**, 177–184.
106. A. N. Kolmogorov: 'On statistical theory of metal crystallisation', *Izvestiya Akad. Nauk SSSR (Izvestia Academy of Science, USSR)*, 1937, **Ser. Math. 3**, 335–360.
107. E. Scheil: 'Anlaufzeit der austenitumwandlung', *Archiv für das Eisenhüttenwesen*, 1935, **12**, 565–567.
108. J. M. Kim, S. Hong, and K. J. Lee: 'Improved method to verify the additivity rule for pearlite transformation in eutectoid steel', *Metals and Materials International*, 2023, **29**, 259–268.
109. W. Thomson: 'On the division of space with minimum partitional area', *Acta mathematica*, 1887, **11**, 121–134.
110. E. E. Underwood: Quantitative Stereology, chap. 4: Addison-Wesley Publication Company, 1970:90–93.
111. J. W. Cahn: 'The kinetics of grain boundary nucleated reactions', *Acta Metallurgica*, 1956, **4**, 449–459.
112. J. W. Christian: Theory of Transformations in Metals and Alloys, Part I: 3 ed., Oxford, U. K.: Pergamon Press, 2003.
113. S. Jones, and H. K. D. H. Bhadeshia: 'Competitive formation of inter- and intragranularly nucleated ferrite', *Metallurgical & Materials Transactions A*, 1997, **28A**, 2005–2103.
114. J. D. Robson, and H. K. D. H. Bhadeshia: 'Modelling precipitation sequences in power plant steels: Part I, kinetic theory', *Materials Science and Technology*, 1997, **13**, 631–639.
115. E. Gautier, A. Simon, and G. Beck: 'Deformation of eutectoid steel during pearlitic transformation under tensile stress', In: P. Haasen, V. Gerold, and G. Kostorz, eds. *Proceedings of the 5th International Conference on the Strength of Metals and Alloys, ICMSA 5*. Oxford, U. K.: Pergamon Press, 1979:867–873.
116. S. Denis, E. Gautier, S. Sjöström, and A. Simon: 'Influence of stresses on the kinetics of pearlitic transformation during continuous cooling', *Acta Metallurgica*, 1987, **35**, 1621–1632.
117. E. Abey-Gautier: 'Transformations perlitique et martensitique sous constrainte de trasnction dans les aciers': Ph.D. thesis, L'Institut National Polytechnique de Lorraine, Lorraine, France, 1985.
118. G. W. Greenwood, and R. H. Johnson: 'Deformation of metals under small stresses during phase transformation', *Proceedings of the Royal Society of London A*, 1965, **238**, 403–422.
119. H. K. D. H. Bhadeshia: 'Geoffrey Wilson Greenwood. 3 February 1929 - 13 December 2022', *Biographical Memoirs of Fellows of the Royal Society*, 2024, **76**, 221–240.
120. R. Ueji, and T. Inoue: 'Acceleration of pearlite transformation in a high-carbon steel by uniaxial compressive stress confirmed by volume measurements', *Materials Letters*, 2019, **256**, 126637.
121. R. Ueji, Y. Kimura, and T. Inoue: 'Preferable resistance against hydrogen embrittlement of pearlitic steel deformed by caliber rolling', *ISIJ International*, 2022, **62**, 368–376.
122. E. Swallow, and H. K. D. H. Bhadeshia: 'High resolution observations of displacements caused by bainitic transformation', *Materials Science and Technology*, 1996, **12**, 121–125.
123. H. K. D. H. Bhadeshia: 'Atomic mechanism of the bainite transformation', *Journal of Heat Treatment and Materials*, 2017, **72**, 340–345.
124. J. R. Patel, and M. Cohen: 'Criterion for the action of applied stress in the martensitic transformation', *Acta Metallurgica*, 1953, **1**, 531–538.
125. H. K. D. H. Bhadeshia, S. A. David, J. M. Vitek, and R. W. Reed: 'Stress induced transformation to bainite in a Fe-Cr-Mo-C pressure vessel steel', *Materials Science and Technology*, 1991, **7**, 686–698.
126. H. L. Le Chatelier: 'On a general statement of the laws of chemical equilibrium (sur un énoncé générale des lois des équilibres chimiques)', *Comptes rendus*, 1884, **99**, 786–789.

127. J. Mola, G. Luan, Q. Huang, C. Schimpf, and D. Rafaja: 'Cementite evolution in medium manganese twinning-induced plasticity steels', *Materialia*, 2018, **2**, 138–147.
128. E. W. Goliber, K. H. McKee, J. S. Kasper, J. E. Hilliard, J. W. Cahn, and V. A. Phillips: 'Research and development on the effects of high pressure and temperature on various elements and binary alloys': Tech. Rep. 59-747, Wright Patterson Air Force Base, Ohio, U. S. A., 1960.
129. J. F. Cox, and C. G. Homan: 'Pressure effect on the diffusion of carbon in α-iron', *Physical Review B*, 1972, **5**, 4755–4761.
130. M. H. Chang, and J. G. Byrne: 'Subcritical thermomechanical treatments of eutectoid steel', *Metal Science*, 1983, **17**, 475–480.
131. C. Kammerhofer, A. Hohenwarter, S. Scherlau, H. P. Brantner, and R. Pippan: 'Influence of morphology and structural size on the fracture behavior of a nanostructured pearlitic steel', *Materials Science & Engineering A*, 2013, **585**, 190–196.
132. O. Yazici, S. Yilmaz, and S. Yildirim: 'Microstructural and mechanical properties examination of high-power diode laser-treated R260 grade rail steels under different processing temperatures', *Metallurgical & Materials Transactions A*, 2019, **50**, 1061–1075.
133. H. Yahyaoui, H. Sidhom, C. Braham, and A. Baczmanski: 'Effect of interlamellar spacing on the elastoplastic behavior of C70 pearlitic steel: Experimental results and self-consistent modeling', *Materials and Design*, 2014, **55**, 888–897.
134. D. Nikas: 'Licenciate thesis: Effect of temperature on mechanical properties of railway wheel steels': Master's thesis, Chalmers University of Technology, Gothenburg, Sweden, 2016.
135. M. Lévy: 'Extrait du mémoire sur les équations générales des mouvements intérieurs des corps solides ductiles au delà des limites où l'élasticité pourrait les ramener à leur premier état (extract from the memoir on the general equations of motion interiors of ductile solid bodies beyond the limits where elasticity could bring them back to their first state)', *Journal de Mathématiques Pures et Appliquées 2e série*, 1871, **16**, 369–371.
136. R. v. Mises: 'Mechanik der festen körper im plastisch-deformablen zustand (mechanics of solid bodies in the plastic-deformable state)', *Nachrichten von der Gesellschaft der Wissenschaften zu Göttingen, Mathematisch-Physikalische Klasse*, 1913, **1913**, 582–592.

9 Spheroidisation

Wilhelm Ostwald is acknowledged as the founder of physical chemistry. In a particular study, he measured the solubility of crystals in contact with a stagnant saturated solution; when he later ground the slurry for a few days, the solubility increased [1]. A logical deduction is that when the solution contains a mixture of crystal sizes, the larger ones must grow at the expense of the smaller crystals. This is the phenomenon we now refer to as *Ostwald ripening* [2, 3].

Cue Geoffrey Greenwood [4], who over half a century later, was investigating how nuclear reactors might be fuelled by a uranium slurry contained in liquid bismuth (+Pb,Sn). The idea was to circulate this fluid by natural convection between the hot reactor-core and the exchanger where the heat could be usefully extracted. The idea failed, because of particle-aggregation and blockages across the temperature gradients, but a shrewd observation was that the suspended particles coarsened over time. Greenwood felt that theory for this could be developed on the basis of Ostwald's work on the size-dependence of solubility. A pattern of diffusional fluxes was conceived, leading to a relationship between particle size and its rate of change in terms of solubility, solute diffusion coefficient, atomic volume, temperature and interfacial energy. The increase in average particle size with time could be deduced as expressed in the now classical equation [5]:

$$r_{\mathrm{F}}^3 - r_{\mathrm{I}}^3 = 3DS\frac{2M\sigma}{RT\rho^2}t \tag{9.1}$$

where r_{F} and r_{I} are the final and initial particle radii respectively of particles growing at the fastest rate throughout, ρ is the density of UPb_3 of molecular weight M, σ the interfacial energy per unit area, S the solubility of uranium in lead. This equation and its variants continue to be the basis of many attempts to design better materials. We shall return to an equation of this form when describing the annealing of pearlite.

Why anneal pearlite? Steels sometimes need to be soft before manipulation into a desired shape, whether that involves machining or forming operations. This is achieved by annealing that induces the lamellar cementite to spheroidise as the system reduces the θ/α interfacial area per unit volume ($S_{\mathrm{V}}^{\theta\alpha}$). Incidentally, we shall also describe an alternative way in which spheroidisation can be achieved in hypereutectoid steels directly during the decomposition of austenite.

9.1 MECHANISM OF SPHEROIDISATION

Atoms added at a curved interface cause an increase in its interfacial area, which requires energy, so the equilibrium there will be different from that at a flat interface.

Consider equilibrium between a cementite particle of radius r, enclosed by ferrite. The addition of a small number n new atoms to the particle chances the interfacial area by $\mathrm{d}O$ with a corresponding increment in energy of $\sigma^{\theta\alpha}\mathrm{d}O$, where $\sigma^{\theta\alpha}$ is the interfacial energy per unit area. The increment of free energy per atom added is $\sigma^{\theta\alpha}\mathrm{d}O/\mathrm{d}n$ as shown in Figure 9.1, so the equilibrium concentration in the ferrite changes from $x^{\alpha\theta}$ to a quantity $x_r^{\alpha\theta}$. The composition of the cementite is assumed not to change because its free energy is sensitive to deviations from the stoichiometric ratio Fe_3C.

For a sphere, $\mathrm{d}O/\mathrm{d}n = 2\mathrm{V_a}/r$ where $\mathrm{V_a}$ is the volume per atom. From the geometry illustrated in Figure 9.1, using similar triangles,

$$\frac{\mu_r^{\alpha\theta} - \mu^{\alpha\theta}}{1 - x^{\alpha\theta}} = \frac{\sigma^{\theta\alpha}}{x^{\theta\alpha} - x^{\alpha\theta}}\frac{\mathrm{d}O}{\mathrm{d}n} \equiv \frac{2\mathrm{V_a}\sigma^{\theta\alpha}}{r(x^{\theta\alpha} - x^{\alpha\theta})}. \tag{9.2a}$$

DOI: 10.1201/9781032631981-9

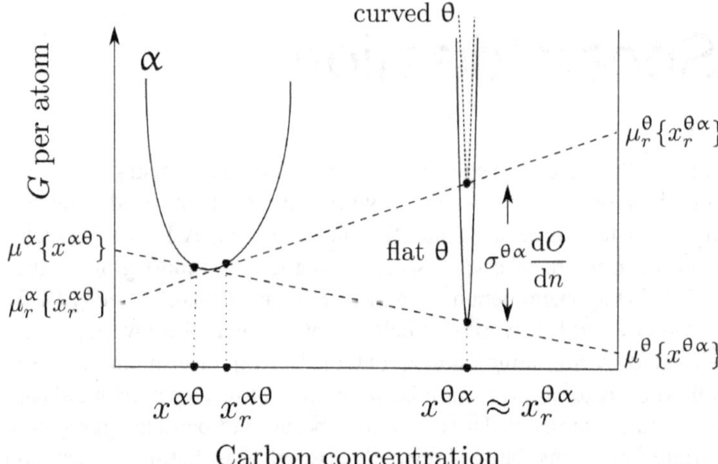

Figure 9.1 An illustration of the Gibbs-Thomson effect due to the curvature of the cementite-ferrite interface. G is the free energy. $x^{\alpha\theta}$ and $x_r^{\alpha\theta}$ are the equilibrium compositions of the ferrite when the two phases are connected by a planar interface or a curved interface with r representing the radius of curvature of the interface, respectively. Note that $\mu\{x\}$ means the chemical potential for the concentration x.

The difference in chemical potential per atom is given by standard thermodynamics[1] as

$$\mu_r^{\alpha\theta} - \mu^{\alpha\theta} = kT\ln\left\{\frac{\Gamma\{x_r^{\alpha\theta}\}x_r^{\alpha\theta}}{\Gamma\{x^{\alpha\theta}\}x^{\alpha\theta}}\right\} \tag{9.2b}$$

where Γ is an activity coefficient. Since ferrite is a dilute solution and the difference in solubility due to interface curvature is small, it may be assumed that the two activity coefficients are equal to unity

$$\mu_r^{\alpha\theta} - \mu^{\alpha\theta} = kT\ln\left\{\frac{x_r^{\alpha\theta}}{x^{\alpha\theta}}\right\} \approx kT\frac{\Delta x_r}{x^{\alpha\theta}} \quad \text{where} \quad \Delta x_r = x_r^{\alpha\theta} - x^{\alpha\theta}. \tag{9.2c}$$

Combining Equations 9.2a and 9.2c gives

$$x_r^{\alpha\theta} = x^{\alpha\theta}\left[1 + \left(\frac{1 - x^{\alpha\theta}}{x^{\theta\alpha} - x^{\alpha\theta}}\frac{2V_a\sigma^{\theta\alpha}}{rkT}\right)\right]. \tag{9.2d}$$

The effect of particle size on $x_r^{\alpha\theta}$ is illustrated in Figure 9.2. If follows that curved particles should dissolve relative to those that are less curved, driven by the diffusion fluxes created within the ferrite because $x_r^{\alpha\theta} \geq x^{\alpha\theta}$.

The structure of a pearlite colony is not represented accurately as a stack of alternating cementite and ferrite plates, Chapter 2. The cementite within a colony is three-dimensionally connected, does not have a habit plane with the ferrite, twists as it grows and contains holes, Figure 2.8. The holes are filled with ferrite that meanders through the cementite. The holes have curved edges – three-dimensional studies have revealed their role in the initiation of spheroidisation [6]. During annealing, some adjacent holes can coalesce. The number density of holes therefore decreases with the annealing time, Figure 9.3, with a simultaneous decrease in $S_V^{\theta\alpha}$. The cementite lamellae also break up into ribbons which then split into particles, resulting in a net decrease in $S_V^{\theta\alpha}$ [6, 7]. Defects in the cementite, such as θ/θ boundaries, terminations, holes and fissures additionally contribute to the spheroidisation process [8].

[1] $\mu = \mu_\circ + RT\ln\{\text{activity}\}$, activity $= \Gamma x$, Γ is the activity coefficient.

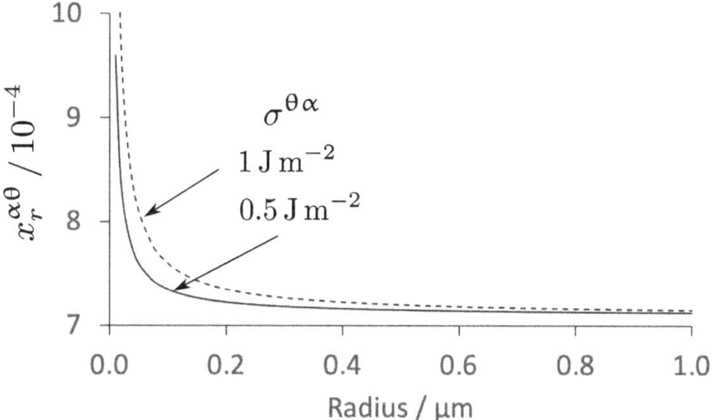

Figure 9.2 Calculations using Equation 9.2d, of the effect of cementite particle-radius on the carbon concentration in ferrite with which it is in equilibrium, at 700 °C.

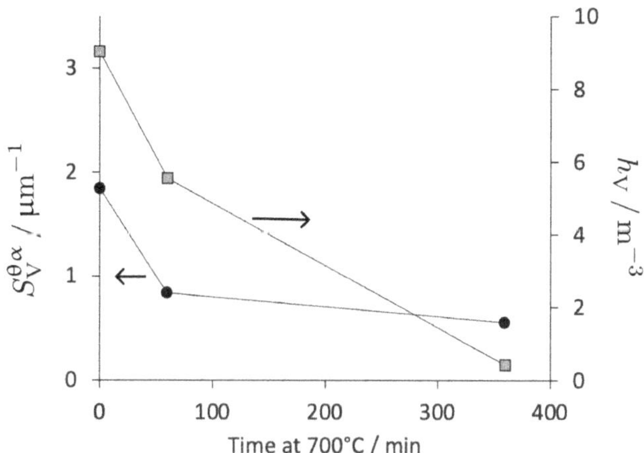

Figure 9.3 Changes in the amount of α/θ interfacial area per unit volume ($S_V^{\theta\alpha}$) and the number of holes per unit volume (h_V) in cementite lamellae, during the annealing of pearlite in Fe-0.8C wt% steel. Adapted using data from Wang et al. [6].

9.2 SPHEROIDISATION BY ANNEALING

9.2.1 DRIVING FORCE

The quantity $S_V^{\theta\alpha}$ can be measured using random lines drawn across the structure of pearlite. If the number of cementite plates intersected by a unit length of line is N, then the number of α/θ interfaces intersected per unit length is $\overline{P}_L = 2N$ and the mean lineal intercept is $\overline{L} = 1/n$. Stereological methods show that [p. 100, 9]

$$S_V^{\theta\alpha} = 2\overline{P}_L = 4/\overline{L}. \tag{9.3}$$

In terms of the true interlamellar spacing, $S_V^{\theta\alpha} \approx 2/S_I$ for an idealised pearlite microstructure [10]. The energy per unit volume of interfaces is therefore the product $\sigma^{\theta\alpha} \times S_V^{\theta\alpha}$. Figure 8.7 shows a compilation of fitted values of $\sigma^{\theta\alpha}$ [11]. The range observed may reflect difficulties in the design and interpretation of experiments; given the large difference in the crystal structures and chemical compositions of ferrite and cementite, it is likely that $\sigma^{\theta\alpha} > 0.5\,\mathrm{J\,m^{-2}}$.

For typical microstructures, the energy locked in interfaces amounts to just 1-2 $\mathrm{J\,mol^{-1}}$, so the driving force $|\Delta G|$ for spheroidisation is much smaller than the typical free energy reduction on phase transformation from austenite. A change in the shape of lamellar cementite into spheroids necessarily requires both the advance and retreat of α/θ interfaces. Since the rate of interfacial motion depends directly on the driving force, spheroidisation is necessarily slow. There are no experimental data on the mobility of the α/θ interface. The Q for the self-diffusion of iron in paramagnetic ferrite is $Q = 240\,\mathrm{kJ\,mol^{-1}}$ [12], so that for spheroidisation should lie in the range 120-240 $\mathrm{kJ\,mol^{-1}}$ depending on the level of coherency in the α/θ interface. The lower limit to Q is taken to be half the value for self-diffusion, for transport across an incoherent boundary [p.458, 13].

The speed of the interface is at small driving forces given by[2]

$$v = |\Delta G| \times M_o \exp\{-Q/RT\} \tag{9.4}$$

where the mobility is assumed here to take on a value $M_o = 60\,\mathrm{mm\,mol\,J^{-1}\,s^{-1}}$, $Q = 120\,\mathrm{kJ\,mol^{-1}}$; the time required to move $10\,\mu\mathrm{m}$ is 0.12 h, which is far too small when compared with typical spheroidisation times of some 50 h or more at temperatures close to T_E. Fitting to 50 h at 700 °C gives a revised lower limit of $Q \approx 170\,\mathrm{kJ\,mol^{-1}}$. This value does depend on the coherency of the θ/α interface. The interfaces are known to be sufficiently coherent to emit dislocations when stressed, so as-transformed pearlite does not exhibit a yield point in the stress-strain curve, even after a strain-ageing heat treatment (p. 52).

The maximum temperature at which a spheroidisation anneal can be implemented is Ae_1 to avoid reversion to austenite.

9.3 KINETICS OF SPHEROIDISATION

The break-up of cementite brings to mind the surface-tension driven instability of a stream of fluid into droplets [15], with an accompanying reduction in surface area. In contrast, the theory for the transition of cementite in the solid state, from a lamellar to spheroidal morphology is less clear, though driven also by interfacial area minimisation. Conventional coarsening theory is applied neglecting complications such as shape change. There may be secondary changes unrelated to coarsening, e.g., the partitioning of substitutional solutes between the θ and α because during the $\gamma \rightarrow$ P transformation in Fe-C-X leaves the products with compositions that deviate from equilibrium [14, 16].

Venugopalan and Kirkaldy dealt with cementite coarsening in multicomponent steels during the tempering of martensite [17]. The site fraction of a substitutional solute i is $y_i = x_i/(1-x_C)$ and the partition coefficient is $k_i^{\theta\alpha} = y_i^{\theta}/y_i^{\alpha}$. Assuming local equilibrium at the α/θ boundary, and neglecting diffusion in the cementite, the interface speed will depend to the diffusivity of solutes in the ferrite, k_i and $\overline{x}_i^{\alpha\theta}$. By analogy to the addition of parallel conductances in electricity, the effective diffusion coefficient is:

$$\frac{1}{D_{\mathrm{eff}}} = \sum_i \frac{(1-k_i^{\theta\alpha})^2}{D_i} \times \frac{y_i}{[1+(k_i^{\theta\alpha}-1)V_V^{\theta}]} \tag{9.5}$$

where the second term is simply the average alloy content in the ferrite, an approximation to the concentration in the ferrite at its interface with cementite.

---

[2]Irreversible thermodynamics assumes that a force and flux are proportional if the rate of entropy production can be written as their product. Here the velocity is equivalent to the flux and the driving force for interface motion is the 'force'. Therefore, the velocity would be proportional to the driving force as long as the latter is small, [e.g., p.88, 14].

Figure 9.4 represents experiments on a series of model alloys that were cold-rolled prior to annealing at 650 °C [18]. The original measurements characterised size by equivalent circle diameters, here represented by $\bar{r}$ assumed to be a mean radius, for consistentency with coarsening theory based on the transfer of matter between particles by diffusion through the matrix [4, 5, 19]. This leads to an equation of the form (*cf.* Equation 9.1)

$$\bar{r}^3 = \frac{8\sigma^{\theta\alpha} D c_\circ V_\mathrm{m}^2}{9RT} t \tag{9.6}$$

where $c_\circ$ is the equilibrium solubility for a flat interface, V_m is the molar volume, and the other terms have their usual meaning. This equation applies strictly to a set of spherical particles, not quite the shapes associated with pearlite. The latter problem has been treated by Tian and Kraft [20] but its application would require a knowledge of the shape and particle number-density changes as they happen during the course of spheroidisation.

Figure 9.4 shows quite large differences in the rate of coarsening, that are unlikely to be explained by interfacial energy considerations alone.

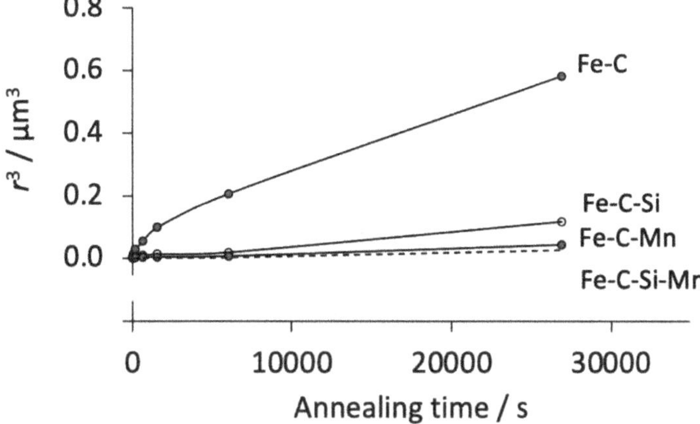

Figure 9.4 A representation of the size of cementite particles as a function of annealing time for pearlite produced by isothermal transformation at 600 °C and cold-rolled to a 90% reduction. The annealing temperature was 650 °C; the steel compositions (wt%) are Fe-0.78C, Fe-0.80C-0.99Si, Fe-0.82C-0.97Mn, Fe-0.79C-1Mn-2Si. Selected data from Fu et al. [18].

Table 9.1 shows calculated values of D_eff for the alloys illustrated in Figure 9.4; the values of D_eff are in the same order as the observed rates of coarsening. And a comparison between D_i and D_eff shows clearly the role of solute partitioning in determining the coarsening rate. Venugopalan and Kirkaldy [17], included the role of dislocations in their analysis; since the experiments analysed here begin with a severely cold-rolled state, dislocations should contribute to the total flux, but their role is not considered because of the lack of information on the dislocation density and how it evolves during annealing.

Figure 9.5 compares the evolution of cementite size during annealing for cases where the starting microstructure is 90% cold-reduced pearlite [18] and for martensite [22]. The two alloys have similar chemical compositions and annealing temperatures. The coarsening rate of cementite is comparable for the two cases, though a little slower for the martensite, possibly because the pearlite was in a cold-deformed state prior to annealing. There is no doubt that starting with a cold-worked steel accelerates spheroidisation, with a correlation against the stored energy due to the deformation – this has been demonstrated quantitatively by measuring the stored energy using differential scanning calorimetry [23].

Table 9.1
Data for the calculation of the effective diffusion coefficient, Equation 9.5. The partition coefficients volume fractions were determined using MTDATA [21] and SGTE database. The diffusion coefficients for 650 °C are from [12] for ferrite in its ferromagnetic state.

D_i / 10^{-18} m^2 s^{-1}		Alloy / wt%	$k_{Fe}^{\theta\alpha}$	$k_{Mn}^{\theta\alpha}$	$k_{Si}^{\theta\alpha}$	V_V^{θ}	D_{eff} / 10^{-18} m^2 s^{-1}
D_{Fe}	1.8321	Fe-0.78C	0.93			0.1155	308.5
D_{Mn}	3.3191	Fe-0.80C-0.99Si	0.93		0	0.1156	182.2
D_{Si}	12.880	Fe-0.82C-0.97Mn	0.93	15.96		0.1215	3.2
		Fe-0.79C-1Mn-2Si	0.93	11.87	0	0.1172	2.4

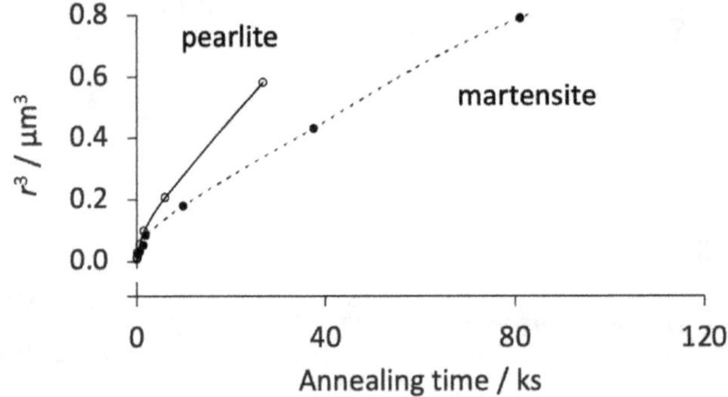

Figure 9.5 Cementite particle-radius cubed versus annealing time. The pearlite (Fe-0.78C wt%) was cold deformed and annealed at 650 °C, while the martensite (Fe-0.8C wt%) was quenched from 825 °C and annealed at 690 °C. Selected data from [18, 22].

9.4 DIVORCED EUTECTOID

When particles of proeutectoid cementite exist in austenite, cooling to a temperature where eutectoid transformation become possible can lead to *divorced pearlite.* In this, the pre-existing cementite particles simply grow to absorb the excess carbon partitioned at the γ/α transformation front. There is, therefore, no cooperative growth of the ferrite and cementite as occurs in conventional pearlite. This kind of transformation is important as a rapid mechanism of obtaining a spheroidised instead of lamellar cementite.

The transformation to divorced pearlite is of considerable technological significance, in both beneficial and detrimental scenarios. High-carbon steels supplied to make bearing raceways are in the form of tubes or forgings. These need to be soft at the machining or forming stages, achieved by generating divorced pearlite [24]. On the other hand, pearlitic rails that are flash butt-welded have a domain within the heat-affected zone that undergoes undesirable softening (360 HV to 260 HV) because of spheroidisation via the divorced-pearlite transformation [25].

The mechanism [26] is illustrated schematically in Figure 9.6. The mixture of austenite and proeutectoid cementite when cooled below Ae_1 leads to the flux of carbon towards pre-existing cementite particles in both the austenite and ferrite, Figure 9.6. To maintain equilibrium at the α/γ

interface as it advances, the carbon partitioned must equal that absorbed in the cementite:

$$(c^{\gamma\alpha} - c^{\alpha\gamma})v = D_\gamma \frac{c^{\gamma\alpha} - c^{\gamma\theta}}{\lambda_\gamma} + D_\alpha \frac{c^{\alpha\gamma} - c^{\alpha\theta}}{\lambda_\alpha} \tag{9.7}$$

where the distances $\lambda_{\gamma,\alpha}$ are defined in Figure 9.6. If ΔT is the undercooling below the temperature at which ferrite may first form, an approximate equation for the velocity of the α/γ interface is given by:[3]

$$v \approx \frac{2D_\alpha}{\lambda_\gamma + \lambda_\alpha} \frac{\frac{\Delta T}{27}\left[\frac{0.28}{D_\alpha/D_\gamma} + 0.009\right]}{0.75 + \frac{\Delta T}{27} \times 0.225} \tag{9.8}$$

where v is the velocity of the α/γ interface. By comparing this velocity against that for the growth of lamellar pearlite, it is possible to identify the domains in which divorced pearlite can be favoured over the lamellar form, Figure 9.6.

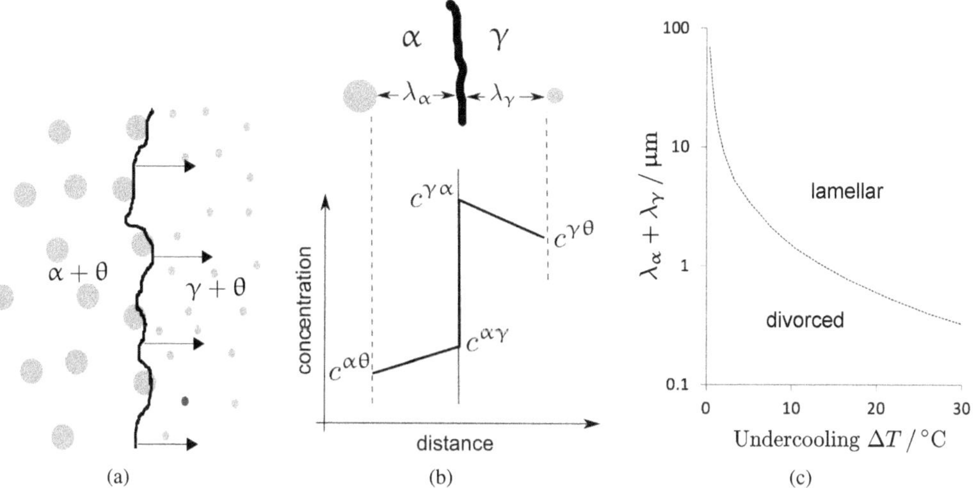

Figure 9.6 (a) Mechanism of the divorced eutectoid transformation of a mixture of austenite and fine cementite [27]. (b) Carbon fluxes in this model. (c) Low undercoolings below the eutectoid temperature and fine spacings between cementite particles favour the formation of divorced pearlite. The calculations [27] are for a plain-carbon eutectoid steel.

The phase field simulation illustrated in Figure 9.7 [28] confirms the basic circumstances by which divorced pearlite can be induced, but reveals also that it is possible for cementite particles both at the γ/α interface and those ahead of it to absorb partitioned carbon.

[3] Some of the terms here are approximated from the phase diagram, for 700°C, are $c^{\gamma\alpha} - c^{\gamma\theta} \approx \Delta T(0.28/0.27)$, $c^{\alpha\gamma} - c^{\alpha\theta} \approx \Delta T(0.009/27)$, $c^{\gamma\alpha} - c^{\alpha\gamma} \approx 0.75 + \Delta T(0.225/27)$.

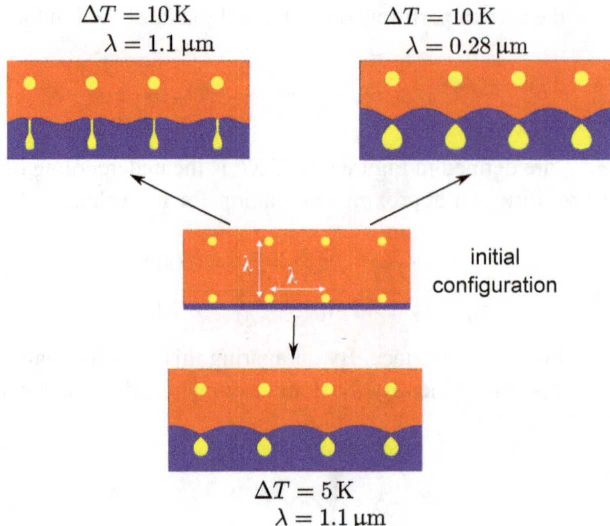

Figure 9.7 Phase field simulation showing the transition from lamellar to divorced eutectoid as a function of spacing and undercooling below the eutectoid temperature. Blue represents the ferrite and red the austenite. The diagram at the top left shows a lamellar eutectoid whereas that on the top right is the divorced form. Reproduced with permission of Elsevier, from Kumar et al. [28].

9.5 SPHEROIDISATION BY INTERCRITICAL ANNEALING

The advantage of heating pearlite to a temperature not much greater than Ae_1, i.e., an excursion into the $\alpha + \gamma$ phase field, is that access to the higher temperatures accelerates the spheroidisation kinetics, but some austenite will inevitably form. One way to avoid difficulties with the subsequent transformation of the austenite is to follow the inter-critical heat-treatment by cooling the sample below Ae_1 for the usual subcritical anneal;[4] any austenite then transforms into ferrite and cementite, with the latter becoming spheroidal during the subcritical anneal [29]. Whether or not the inter-critical anneal is an advantage depends on the relative scales of the pearlite in the original sample and that which forms from the austenite. If the latter is coarse then the kinetics of spheroidisation are faster with just a subcritical anneal [30].

9.6 SPHEROIDISED PEARLITE IN THE HOT ROLLED STATE

9.7 COIL COLLAPSE

It is found that when coiling certain hot-rolled steels, the coil collapses into an unacceptable oval shape, the phenomenon is associated with the spheroidisation of pearlite in the microstructure (Fig. 9.8). This austenite, because of its low volume fraction will contain a large carbon concentration so on cooling, will undergo a divorced eutectoid reaction, leading to coarse, spheroidal cementite. Typical microstructures obtained in this way, then exhibit a bimodal distribution of cementite particles, the coarsest originating from the decomposition of the austenite and the finer particles from the conventional annealing of pearlitic cementite [29].

The hypothesis is that in this steel, the fraction of cementite V_V^θ in the pearlite is too small because of its hypoeutectoid composition and the supercooling at which the pearlite formed, which

---

[4]Inter-critcal means in the $\alpha + \gamma + \theta$ phase field, whereas sub-critical is in the $\alpha + \theta$ field.

(a)

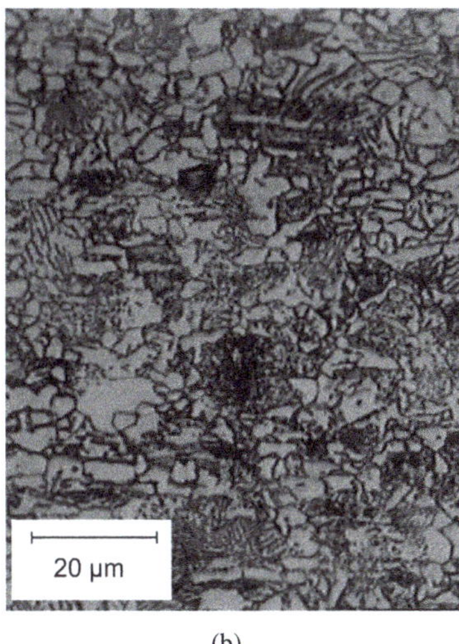

(b)

Figure 9.8 (a) A coil which has collapsed under the influence of gravity, here illustrated after turning through approximately 90°. (b) Optical microstructure showing the partial spheroidisation of the pearlite. The steel has a composition Fe-0.31C-0.18Si-1.37Mn wt% and was coiled at 625°C after rolling to a thickness of 2.25 mm. The outer and inner diameters are 1200 and 1090 mm respectively.

would promote rapid spheroidisation. The resulting weakening of the coil would cause the observed distortion under its own weight.

The solution lies in adjusting the conditions in the hot-strip mill to ensure a greater fraction of α, and hence a larger carbon concentration in the residual austenite which transforms into a pearlite with a larger average carbon concentration (larger V_V^θ).

REFERENCES

1. W. L. Noorduin, E. Vlieg, R. M. Kellogg, and B. Kaptein: 'From Ostwald ripening to single chirality', *Angewandte Chemie International Edition*, 2009, **48**, 9600–9606.
2. W. Ostwald: 'Studien über die bildung und umwandlung fester körper. 1. abhandlung: übersättigung und überkaltung (studies on the formation and transformation of solid bodies. 1st treatise: Supersaturation and overcooling)', *Zeitschrift für Physikalische Chemie*, 1897, **22**, 289–330.
3. W. Ostwald: 'Über die vermeintliche isomerie des roten und gelben quecksilberoxyds und die oberflächenspannung fester körper (about the supposed isomerism of red and yellow mercury oxide and the surface tension of solid bodies)', *Zeitschrift für Physikalische Chemie, Stoeechiometrie und Verwandschaftslehre*, 1900, **34U**, 495–503.
4. H. K. D. H. Bhadeshia: 'Geoffrey Wilson Greenwood. 3 February 1929 - 13 December 2022', *Biographical Memoirs of Fellows of the Royal Society*, 2024, **76**, 221–240.
5. G. W. Greenwood: 'The growth of dispersed precipitates in solutions', *Acta Metallurgica*, 1956, **4**, 243–248.
6. Y. T. Wang, Y. Adachi, K. Nakajima, and Y. Sugimoto: 'Topology and differential geometry-based three-dimensional characterization of pearlite spheroidization', *ISIJ International*, 2012, **52**, 697–703.
7. S. Chattopadhyay, and C. M. Sellars: 'Quantitative measurements of pearlite spheroidization', *Metallography*, 1977, **10**, 89–105.
8. Y. L. Tian, and R. W. Kraft: 'Mechanisms of pearlite spheroidization', *Metallurgical Transactions A*, 1987, **18**, 1403–1414.

9. E. E. Underwood: Quantitative Microscopy, eds R. T. DeHoff and F. N. Rhines, chap. 4: New York, USA: McGraw Hill, 1968:77–125.
10. S. A. Saltykov: Stereometricheskaya Metallografiya: 2nd ed., Moscow: Metallurgizdat Publishers (English translation US Department of Defence AD267700), 1958.
11. A. S. Pandit, and H. K. D. H. Bhadeshia: 'Mixed diffusion-controlled growth of pearlite in binary steel', *Proceedings of the Royal Society A*, 2011, **467**, 508–521.
12. J. Fridberg, L.-E. Torndähl, and M. Hillert: 'Diffusion in iron', *Jernkontorets Annaler*, 1969, **153**, 263–276.
13. J. W. Christian: Theory of Transformations in Metals and Alloys, Part I: 3 ed., Oxford, U. K.: Pergamon Press, 2003.
14. H. K. D. H. Bhadeshia: Theory of Transformations in Steels: London, U.K.: CRC Press, Taylor and Francis Group, 2021.
15. J. W. Strutt (Rayleigh, Lord): 'On the instability of jets', *Proceedings of the London mathematical society*, 1878, **1**, 4–13.
16. J. Chance, and N. Ridley: 'Chromium partitioning during isothermal transformation of a eutectoid steel', *Metallurgical Transactions A*, 1981, **12A**, 1205–1213.
17. D. Venugopalan, and J. S. Kirkaldy: 'New relations for predicting the mechanical properties of quenched and tempered low alloy steels', In: D. V. Doane, and J. S. Kirkaldy, eds. *Hardenability Concepts with Applications to Steels*. Warrendale, PA: The Metallurgical Society of AIME, 1977:249–268.
18. W. Fu, T. Furuhara, and T. Maki: 'Effect of Mn and Si addition on microstructure and tensile properties of cold-rolled and annealed pearlite in eutectoid Fe–C alloys', *ISIJ International*, 2004, **44**, 171–178.
19. R. A. Oriani: 'Ostwald ripening of precipitates in solid matrices', *Acta Metalllurgica*, 1964, **12**, 1399–1409.
20. Y. L. Tian, and R. W. Kraft: 'Kinetics of pearlite spheroidizations', *Metallurgical Transactions A*, 1987, **18**, 1359–1369.
21. NPL: 'MTDATA': Software, National Physical Laboratory, Teddington, U.K., 2006.
22. B. A. Lindsley, and A. R. Marder: 'The morphology and coarsening kinetics of spheroidized Fe–C binary alloys', *Acta Materialia*, 1998, **46**, 341–351.
23. K.-S. Kim, D.-Y. Wi, J.-S. Lee, and K.-A. Lee: 'Adjusting the thermomechanical condition to change the microstructure of C-Mn cold heading quality steel for rapid cementite spheroidization at subcritical temperature: Effect of stored energy', *Metallurgical and Materials Transactions A*, 2022, **53**, 1099–1109.
24. H. K. D. H. Bhadeshia: 'Steels for bearings', *Progress in Materials Science*, 2012, **57**, 268–435.
25. L. P. Nishikawa, and H. Goldenstein: 'Divorced eutectoid on heat-affected zone of welded pearlitic rails', *JOM*, 2019, **71**, 815–823.
26. J. D. Verhoeven: 'The role of the divorced eutectoid transformation in the spheroidization of 52100 steel', *Metallurgical & Materials Transactions A*, 2000, **31**, 2431–2438.
27. J. D. Verhoeven, and E. D. Gibson: 'The divorced eutectoid transformation in steel', *Metallurgical & Materials Transactions A*, 1998, **29**, 1181–1189.
28. A. Kumar, R. Mukherjee, T. Mittnacht, and B. Nestler: 'Deviations from cooperative growth mode during eutectoid transformation: Insights from a phase-field approach', *Acta Materialia*, 2014, **81**, 204–210.
29. D. Henández-Silva, R. D. Morales, and J. G. Cabanas-Moreno: 'Spheroidization of cementite in a medium carbon steel by means of subcritical and intercritical annealing', *ISIJ International*, 1992, **32**, 1297–1305.
30. J. O'Brien, and W. Hosford: 'Spheroidization of medium-carbon steels', *Journal of Materials Engineering and Performance*, 1997, **6**, 69–72.

10 Alloy pearlite

The tremendous success of ordinary pearlite has meant that other varieties where the carbide phase is not cementite, have yet to attract profitable application. The term *alloy pearlite* is used to describe the structure resulting from the simultaneous growth of ferrite and a substitutional-solute rich carbide at a common transformation front with the austenite. The carbide, for example, can be $M_{23}C_6$, M_7C_3, M_6C and even M_2C ('M' stands for metal atoms).

Although the product phases grow together at a common front with the γ, the comparison with ordinary pearlite becomes tenuous because the process *requires* the partitioning of substitutional solutes. Furthermore, the diffusion necessary must occur on a scale comparable to the colony size.[1] When transformed in the $\gamma + \alpha +$ carbide phase-field, the composition of the austenite changes to a point where the reaction must eventually cease before all the austenite is consumed.

The alloy pearlite grows as spherical nodules but can become 'spiky' when the structure is generated at large undercoolings, Figure 10.1a which is from a chromium-rich steel. The same phenomenon is observed with conventional $\alpha + M_3C$ pearlite, as seen in Fig. 10.1b where the steel does not contain strong carbide formers. The spiky morphology occurs because the extent of co-operative growth between the ferrite and carbide diminishes as the driving force for transformation becomes large [p. 299, 1]. The alloy pearlite is sometimes mistaken for bainite but it does not result in a shape change that is characteristic of a displacive transformation [2].

(a)

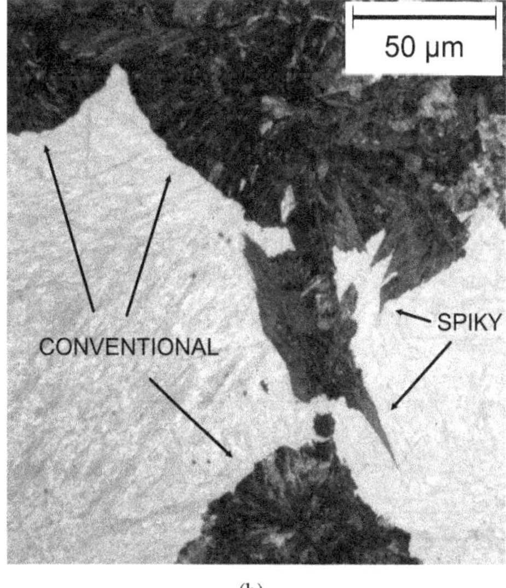

(b)

Figure 10.1 (a) Fe-0.22C-10.6Cr wt% steel transformed isothermally at 550 °C. Reproduced from [3] with permission of Springer Nature. (b) Fe-0.8C-1Mn wt% steel transformed isothermally at 550 °C. Image courtesy of S.-W. Seo with details in [2].

[1] The colony size, rather than the spacing between α and carbide, is emphasised here because there can be a long-range partitioning of solutes ahead of the interface between the pearlite and austenite. This is not just diffusion parallel to the interface as in ordinary pearlite.

DOI: 10.1201/9781032631981-10

Alloy pearlite requires the diffusion of Fe and X atoms over the length scale of the transformation product. A better way to express this condition is that the kinetic strength (p. 17) of the heat treatment must be sufficient because both time and temperature matter – the quantity is defined as the product $\Phi = t\exp\{-190{,}000\,\mathrm{J\,mol^{-1}}/RT\}$, where the units of time t are in hours and the temperature is absolute. Table 10.1 lists experimental observations of alloy pearlite together with values of the kinetic strength. If the case where the carbide phase has not been identified is neglected, then the minimum value required to generate alloy pearlite is $\Phi_{\min} \approx 5.9 \times 10^{-12}\,\mathrm{h}$. Figure 10.2 shows the time-temperature locus generated using $\Phi_{\min}$, indicating that heat treatments below about 500 °C may prove impractical in terms of the time required to generate alloy pearlite.

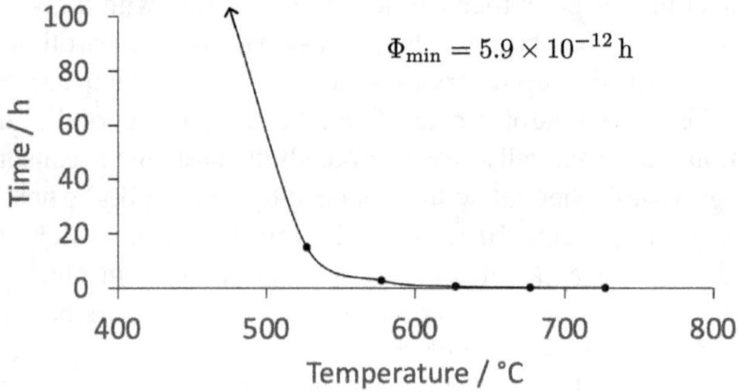

Figure 10.2 The time required to achieve alloy pearlite as a function of temperature, calculated using $\Phi_{\min}$ from Table 10.1 for the cases where time and temperature data were explicitly available.

Table 10.1

Chemical compositions of steels containing alloy carbides that grow cooperatively with ferrite to form pearlite-like colonies. Φ represents the kinetic strength.

The time periods of the heat treatments were not given in [3] so were taken from TTT diagrams corresponding to the initiation of reaction.

Alloy	C	Si	Mn	Ni	Mo	Cr	T/K, t/h	Φ/h	Ref
1. $\alpha + M_{23}C_6$	0.16	0.4	0.55	3.3	0.2	1.40	873, 20	8.6×10^{-11}	[4]
2. $\alpha + M_{23}C_6$	0.45					10.4	998, 0.0833	9.5×10^{-12}	[5]
3. $\alpha + M_{23}C_6$	0.45					10.4	948, 0.833	2.8×10^{-11}	[5]
4. $\alpha + M_{23}C_6$	0.22					10.6	1023, 0.072	1.4×10^{-11}	[3]
5. $\alpha + M_{23}C_6$	0.22					10.6	823, 5.785	5.1×10^{-12}	[3]
6. $\alpha + M_{23}C_6$	0.23					11.8	948, 0.25	8.5×10^{-12}	[6]
7. $\alpha + M_7C_3$	0.30					4.08	751, 1032	6.3×10^{-11}	[p. 512, 7]
8. $\alpha + M_7C_3$	0.96					8.24	993, 0.2	2.0×10^{-11}	[8]
9. $\alpha + M_7C_3$	0.43					3.63	993, 0.25	1.6×10^{-11}	[9]
10. $\alpha + M_7C_3$	0.43					3.63	923, 0.33	5.9×10^{-12}	[9]
11. $\alpha + M_6C$	0.21				3.73		1098, 0.5	4.6×10^{-10}	[5]
12. α+?	0.12		0.44	0.12		12.8	1098, 0.5	9.7×10^{-14}	[10]

Like ordinary pearlite (p. 13), the $\alpha + M_{23}C_6$ in a colony of alloy pearlite is a bicrystal [5, 11]. Both the ferrite and carbide must therefore be interconnected in some way. The three-dimensional morphology has not been characterised and it may be that the $M_{23}C_6$ has a 'fibrous' shape. Neither the ferrite nor the $M_{23}C_6$ exhibit a reproducible orientation relationship with the austenite into which the pearlite grows though the carbide and ferrite are related by an orientation that is close to [6, 11]

$$\{111\}_{M_{23}C_6} \parallel \{011\}_\alpha \qquad \langle 10\bar{1}\rangle_{M_{23}C_6} \parallel \langle 11\bar{1}\rangle_\alpha$$

If the pearlite nucleates at austenite grain boundaries, the orientation relationship may exist with just one of the adjacent austenite grains, with growth proceeding into that with which the transformation products have a less coherent and hence more mobile interface.

Figure 10.3 shows a transmission electron micrograph where the $M_{23}C_6$ lamellae and ferrite are seen to grow at common front with the austenite – notice, however, that there are other, smaller $M_{23}C_6$ precipitates that also form but not involving cooperative growth. It is probable that these small particles nucleate at the α/γ interface and hence, in some regions, appear to be in rows. But the consequence is that the interlamellar spacing of the alloy pearlite is not constant as the fine precipitates alter the growth conditions. It is as if the lamellar growth is suspended when the fine precipitates form.

Figure 10.3 Fe-11.8Cr-0.23C wt% steel transformed isothermally at 675 °C for 15 min. Image from Campbell and Honeycombe [6], with permission of Taylor and Francis.

The approximate and non-unique crystallography has been established in the case of $\alpha + M_7C_3$ pearlite [8]:

$$\{\bar{1}13\}_\alpha \parallel (\bar{1}100)_{M_7C_3} \qquad \{\bar{3}1\bar{1}\}_\alpha \approx\parallel (0001)_{M_7C_3}$$

$$\text{or} \qquad \{\bar{1}13\}_\alpha \parallel (\bar{1}1\bar{2}0)_{M_7C_3} \qquad \{\bar{3}1\bar{1}\}_\alpha \approx\parallel (0001)_{M_7C_3}.$$

We note that there are two representations of the crystal structure of M_7C_3 depending on the exact lattice parameters. The orthorhombic crystal structure, space group *Pmcn* has lattice parameters

$a = 0.701$ nm, $b = 1.2142$ nm and $c = 0.4526$ nm [12]. If the ratio b/a is equal to $\sqrt{3}$, the structure can also be represented as a hexagonal cell with space group $P6_3mc$, $a = b = 1.402$ nm and $c = 0.4526$ nm. There is no information available on which phase nucleates first from the austenite. Given that the transformation is reconstructive and γ/γ- boundary nucleated (can also initiate from α/γ interfaces [13]), the orientation relationships are not expected to be unique. The penetration of the pearlite into the adjacent austenite grains is not uniform [13], presumably because of crystallographic differences influencing interface mobilities.

The structure of a $\alpha + M_{23}C_6$ alloy-pearlite colony is illustrated in Figure 10.4. The interface with the austenite is smooth with a regular interlamellar spacing in some regions, but elsewhere, there are bits of the colony that seem to break away to form protrusions. In these latter regions, there are indications that the interlamellar spacing is large and that the advance of the ferrite in the central positions between lamellae is slower than that of the carbide. It is likely that these cusps in the α/γ interfaces are caused by the partitioning of carbon from α into the austenite in its vicinity. Figure 10.5 illustrates the free energy changes accompanying the transformation of austenite into an equilibrium mixture of products, with γ' represented austenite that is changed in chemical composition from the average composition of the steel. The calculations in each case allow only the designated transformation products to form. As observed in the original work [4], ΔG is similar for all cases where ferrite and carbides form. In assessing nucleation, $|\Delta G^{\gamma \rightarrow \gamma' + \alpha}| \gg |\Delta G^{\gamma \rightarrow \gamma' + \theta}|$ or $|\Delta G^{\gamma \rightarrow \gamma' + M_{23}C_6}|$ implying that ferrite may be the first phase to form. Similarly, the free energy change is greater for the alloy carbide than for cementite, perhaps explaining why alloy pearlite is initiated rather than the form based on cementite.

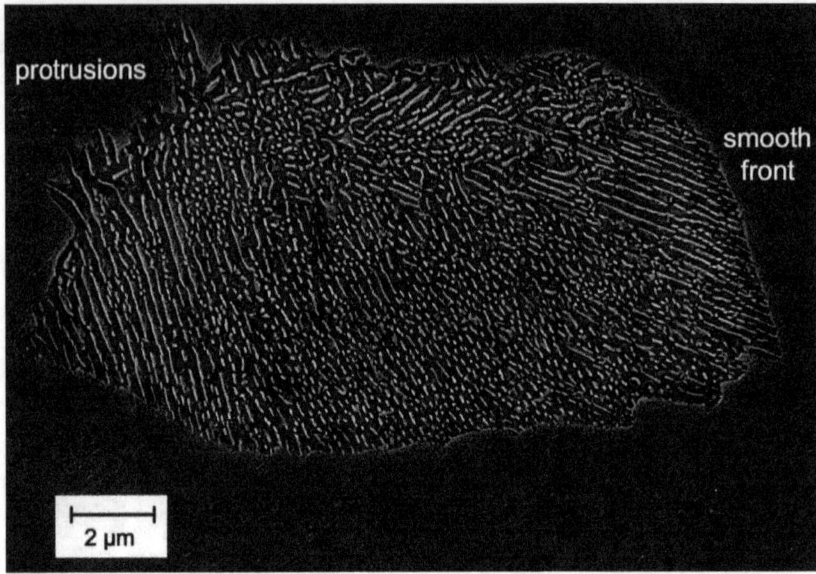

Figure 10.4 An image showing a pearlite-like colony in which the carbide phase is $M_{23}C_6$ chromium-rich carbide [4]. The structure was generated in Alloy 1 (Table 10.1) during isothermal transformation at 600 °C for 20 h. Micrograph courtesy of J. A. Mathews.

The ease of nucleation depends also on the interfacial energy between the embryo and austenite, but these energies are so uncertain that it is difficult to distinguish between the two kinds of carbides.

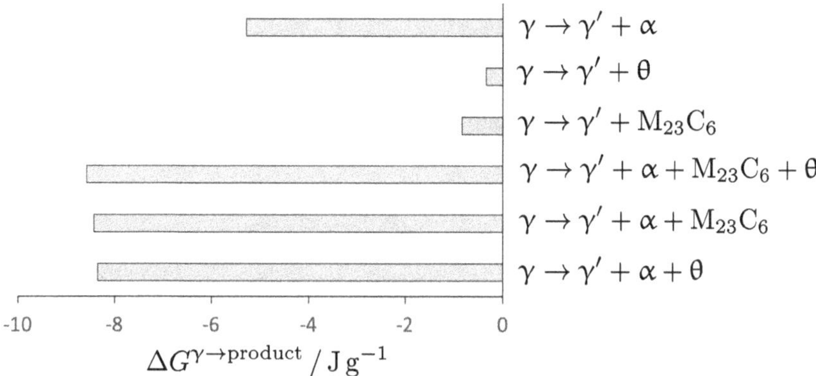

Figure 10.5 Free energy changes accompanying a variety of transformation products of austenite in Alloy 1 (Table 10.1) when transformed isothermally at 600 °C. The calculations utilised MTDATA [14].

10.1 FIBROUS ALLOY-PEARLITE

Pearlite is more often than not, described as lamellar. However, this involves the creation of a significant amount of interfacial area within the pearlite, and there is an energy cost associated with that process (p. 195), a cost that scales linearly with the α/carbide interfacial energy per unit area. To obtain a planar interface requires good fit along two directions within the plane; Mo_2C is not able to achieve this, the fit being reasonable only along one direction [p. 477, 7], thus explaining the needle or fibre shape that is observed experimentally.

Many of the studies of α | Mo_2C pearlite have been conducted on stoichiometric ternary-alloys, where the amount of molybdenum is equal to that needed to combine completely with the carbon. This means that it is possible for the alloy pearlite reaction to go to completion [13, 15] if other phases such as ferrite do not precede the pearlite.

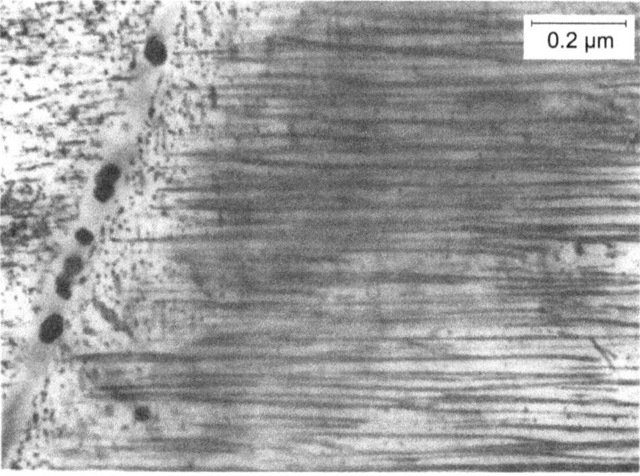

Figure 10.6 Fe-3.9Mo-0.23C wt% transformed isothermally at 700 °C for 30 min. Fibrous Mo_2C in a region of alloy pearlite. Reproduced from Berry and Honeycombe [13], with permission of Springer Nature.

REFERENCES

1. H. K. D. H. Bhadeshia: Bainite in Steels: Theory and Practice: 3rd ed., Leeds, U.K.: Maney Publishing, 2015.
2. S. W. Seo, H. K. D. H. Bhadeshia, and D. W. Suh: 'Pearlite growth rate in Fe-C and Fe-Mn-C steels', *Materials Science and Technology*, 2015, **31**, 487–493.
3. J. Bee, P. Howell, and R. Honeycombe: 'Isothermal transformations in iron-chromium-carbon alloys', *Metallurgical Transactions A*, 1979, **10**, 1207–1212.
4. J. A. Mathews, H. Farahani, J. Sietsma, R. Petrov, M. Mecozzi, and M. Santofimia: 'Microstructures in a carburized steel after isothermal pearlitic treatment', *Journal of Materials Science & Technology*, 2023, **160**, 66–75.
5. J. V. Bee, and D. V. Edmonds: 'Metallographic study of the high-temperature decomposition of austenite in alloy steels containing Mo and Cr', *Metallography*, 1979, **12**, 3–21.
6. K. Campbell, and R. W. K. Honeycombe: 'The isothermal decomposition of austenite in simple chromium steels', *Metal Science*, 1974, **8**, 197–203.
7. H. K. D. H. Bhadeshia: Theory of Transformations in Steels: London, U.K.: CRC Press, Taylor and Francis Group, 2021.
8. D. V. Shtansky, K. Nakai, and Y. Ohmori: 'Crystallography and interface boundary structure of pearlite with M_7C_3 carbide lamellae', *Acta Materialia*, 1999, **47**, 1105–1115.
9. A. Kaya, and D. Edmonds: 'Nonclassical decomposition products of austenite in Fe-C-Cr alloys', *Metallurgical and Materials Transactions A*, 1998, **29**, 2913–2924.
10. M. C. Tsai, C. S. Chiou, J. S. Du, and J. R. Yang: 'Phase transformations in AISI 410 stainless steel', *Materials Science & Engineering A*, 2002, **A332**, 1–10.
11. P. R. Howell, J. V. Bee, and R. W. K. Honeycombe: 'The crystallography of the austenite-ferrite/carbide transformation in Fe-Cr-C alloys', *Metallurgical Transactions A*, 1979, **10**, 1213–1222.
12. F. Z. Kayser: 'A re-examination of Westbrooks's X-ray diffraction pattern for C_7C_3', *Materials Research Bulletin*, 1996, **31**, 635–638.
13. F. G. Berry, and R. W. K. Honeycombe: 'The isothermal decomposition of austenite in Fe-Mo-C alloys', *Metallurgical Transactions*, 1970, **1**, 3279–3286.
14. NPL: 'MTDATA': Software, National Physical Laboratory, Teddington, U.K., 2006.
15. Y. Ohmori: 'The precipitation of molybdenum carbides during the isothermal decomposition of an Fe-4.10% Mo-0.23% C austenite', *Transactions of the Iron Steel Institute of Japan*, 1972, **12**, 350–357.

11 Hydrogen effects

Henry Cavendish in his experiments on 'factitious air' discovered a gas that is flammable and less dense than air [1]; he later showed that when burnt, it produces water.[1] This contradicted Aristotle's proposal of the elements, Earth, fire, air, water and aether [2], by demonstrating that water is not elementary. Lavoisier later confirmed the experiments and named the element *hydrogen* – Cavendish had hugely advanced the subject by showing it as an element and therefore introducing the concept of a chemical reaction. In all these experiments, metals including iron, were reacted to produce the hydrogen. Here we talk of molecular H_2, not H or H^+. At atmospheric pressure and 300 K, the reaction $H_2 \rightleftharpoons 2H$ would at equilibrium result in a fraction of H/H_2 of only $\approx 2\times10^{-36}$; at 5000 K, this fraction increases to ≈ 0.96 [3]. The more likely absorption reaction into iron is $H_2 + 2e \rightarrow 2H^- \rightarrow [2H]_{Fe}$, for example, during a corrosion reaction. This is important because hydrogen can only enter iron in a dissociated state, and when it does so, it does harm. The hydrogen in α iron is not strictly $[H]_{Fe}$, rather, it is slightly positive due to charge transfer to the neighbouring iron atoms [4]. The irony (!) is that the lightest element in the Universe debilitates the most stable element, Fe^{56}, in the same universe [5].

Hydrogen embrittles pearlite, as it does all steels. The blame lies on the atoms of hydrogen inside iron that are able to diffuse and accumulate at a susceptible location where the quantity becomes large enough to induce damage [6]. Hydrostatic stress at a crack tip or similar locations can drive this diffusion, but the strain-rate must be slow ($\approx 10^{-6}\,s^{-1}$) to allow time for the flux to enrich the concentration in the critical region. The emphasis therefore is on diffusible hydrogen within the steel, not any that might be trapped by defects.

It no longer is useful to demonstrate that hydrogen embrittles iron – that is written in stone. The purpose here is to see whether there are distinct features in the interaction of hydrogen with pearlite and its convoluted structure. Given that pearlite contains two phases, the fracture mechanism in its embrittled state may include both ductile and cleavage mechanisms, Figure 11.1.

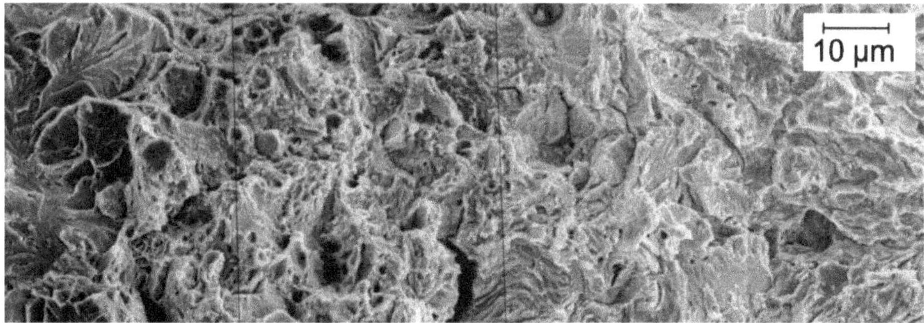

Figure 11.1 Fe-0.85C-0.6Mn-0.26Si wt% steel with $\sigma_y = 600$ MPa, $S_I \approx 0.3$ μm, axisymmetric tensile sample with a gentle, smooth notch, tested at a slow strain-rate while immersed into a $pH = 12.5$ (alkaline) solution with a constant electro-chemical potential. Fracture propagation was from the left to the right of the image. The sample illustrates a mixture of ductile and brittle failure. Reproduced from Toribio [7], with minor adaptation under the CCBY 4.0 DEED licence.

[1]There is another 'burning reaction' in which a pair of ionised hydrogen atoms, i.e., protons, fuse to produce a helium nucleus, with an accompanying loss of mass that manifests as energy, thus powering the Sun.

DOI: 10.1201/9781032631981-11

11.1 ABILITY TO ABSORB AND RETAIN HYDROGEN

The solubility of hydrogen in ferrite or cementite depends on the nature of the hydrogen in contact with the steel. The equilibrium concentration of hydrogen in ferrite exposed to hydrogen gas at one-atmosphere pressure and ambient temperature is just $x_{\mathrm{H}}^{\alpha \mathrm{H}_2} = 1.6 \times 10^{-8}$. The concentration when the steel encounters hydrogen ions, $x_{\mathrm{H}}^{\alpha \mathrm{H}^+}$, can be much greater when using electrochemical or immersion hydrogen-charging techniques [8, 9]. For example, when submerged into H_2S saline-solutions $x_{\mathrm{H}}^{\alpha \mathrm{H}^+} = 1.5 \times 10^{-4}$ ($\equiv 100\,\mathrm{mm}^3\,\mathrm{g}^{-1}$), Fig. 11.2. Some of this hydrogen may be 'trapped' at defects such as interfaces, dislocations or other discontinuities in the lattices of the component phases. The location and nature of the hydrogen (H or H_2) can be determined by monitoring the amount released from the steel as the temperature is ramped up from ambient. As will be seen later, it is now possible to image the atoms to verify their location and concentration.

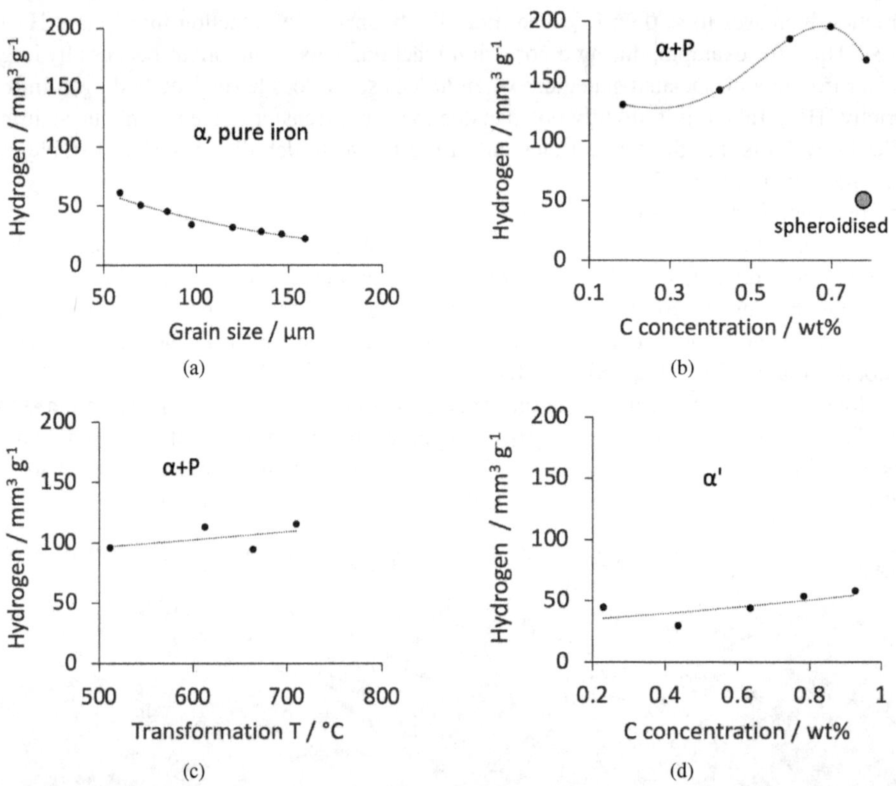

Figure 11.2 Hydrogen content of undeformed samples after immersion in a H_2S saline-solution for 8 h, followed by measurement of hydrogen released during heat treatment at 680 °C for 20 min. Note that $100\,\mathrm{mm}^3\,\mathrm{g}^{-1} \equiv 1.5 \times 10^{-4}$ mol fraction. Selected data from [10, 11]. (a) Pure iron in its ferritic condition. (b) Mixtures of ferrite and pearlite as a function of the average carbon concentration of the steel. Also included is a datum from spheroidised pearlite. (c) A hypoeutectoid steel (0.59C wt%) transformed partially into allotriomorphic-α and then cooled to generate pearlite at the temperature indicated. (d) Martensite as a function of the average carbon concentration of the steel.

Hydrogen atoms locate preferentially at octahedral interstices in cementite [12] (Figure 11.3), which is unlike ferrite where the tetrahedral interstices are favoured, although the enthalpies of solution of hydrogen in α and θ are similar. The barrier Q for hydrogen diffusion in cementite is an order of magnitude greater then that in ferrite $Q_{\mathrm{H}}^{\theta} = 0.8\,\mathrm{eV}$, $Q_{\mathrm{H}}^{\alpha} = 0.09\,\mathrm{eV}$, so the passage of hydrogen through cementite will be much slower. This is because the carbon atoms in cementite hinder the relaxation of the structure, thus lowering the volume available for hydrogen through a narrow channel of iron atoms. These two factors have been taken to indicate that the properties of cementite are unlikely to be significantly affected by diffusible hydrogen.

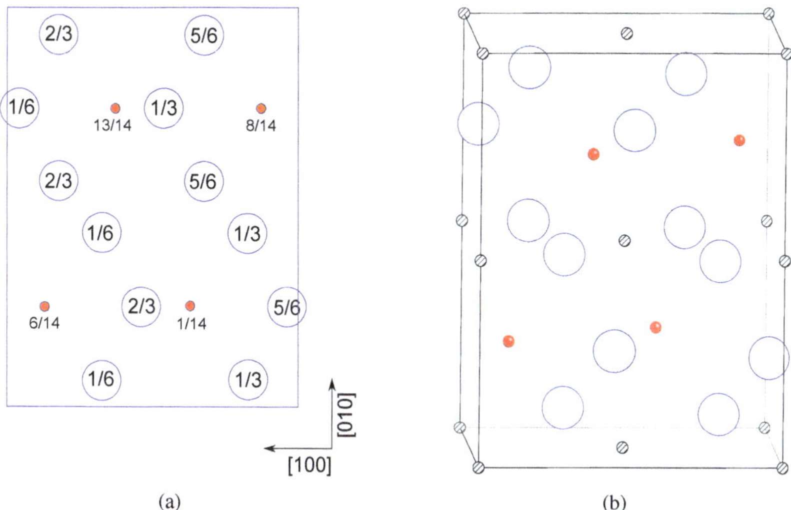

Figure 11.3 (a) Structure of cementite, marked with the fractional height normal to the diagram. The Fe and C atoms are blue and red, respectively. (b) The octahedral interstices are represented by the small shaded circles. These are the sites preferentially occupied by hydrogen in cementite [12].

In the discussion that follows, the measurements of hydrogen content depend both on its ability to penetrate the particular microstructure and on whether it encounters traps. When not pinned down at defects, the hydrogen can escape from the steel before experimental measurements are implemented, one reason why pearlitic, patented rods are allowed a rest period before being sold.

In pure iron, the ability to absorb and retain hydrogen decreases as the ferrite grain size becomes larger, i.e., as the interfacial area per unit volume ($S_V^{\alpha\theta}$) decreases, indicating that the primary traps are the grain boundaries, Figure 11.2a. Ferrite-pearlite steels though, appear to be better able to retain hydrogen, Figure 11.2b – this could be because the α/θ interfaces trap hydrogen or because of slow diffusion through cementite which with its 'lamellar' structure renders the migration pathway rather tortuous. The absorption of hydrogen into the steel also becomes more difficult for the same reason. That hydrogen is not trapped at α/θ interfaces, is revealed in atom-probe experiments. Figure 11.4 shows that the segregation of deuterium in the proximity of the interfaces develops only in deformed pearlite due to localised-strains introduced there as a consequence of mechanical incompatibilities between the two phases.

The hypothesis that the cementite lamellae are barriers to hydrogen diffusion is supported by the observation that spheroidisation greatly reduces the amount of hydrogen retained in the steel, Figure 11.2b.

In hypoeutectoid steel containing about $V_V^{\alpha} = 0.22$ with the residue being pearlite, the hydrogen retained seems insensitive to the temperature at which pearlite forms (Figure 11.2c), presumably because the ferrite provides a diffusion short-circuit.

These observations on steels are consistent with studies of spheroidal cast-iron using secondary ion mass spectroscopy on deuterium-charged specimens. The evolution of hydrogen during heating began first from ferrite and then at temperatures in excess of 200 °C from pearlite. The additional feature is that graphite traps held on to the hydrogen until temperatures in excess of 300 °C were reached [14].

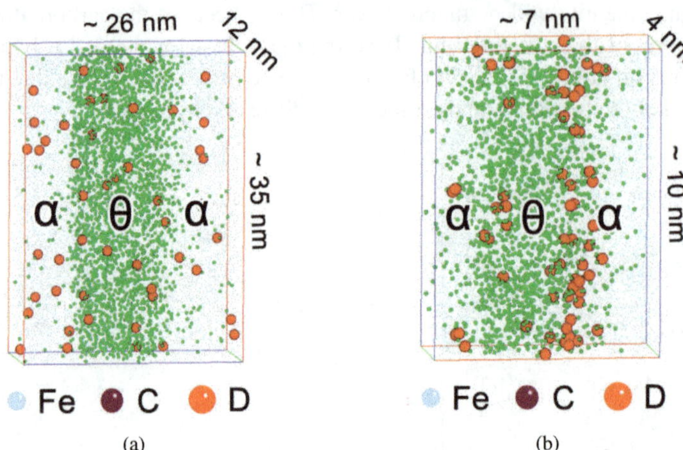

Figure 11.4 Fe-0.53C-1.48Si-0.7Mn-0.76Cr wt% steel transformed during continuous cooling. The atom-probe experiments were conducted on small samples that were charged with deuterium by immersing in a NaOH+D_2O solution. (a) Undeformed pearlite showing the distribution of deuterium atoms across a cementite lamella. (b) Deformed pearlite (total of 88% reduction of area), showing deuterium atoms located in the proximity of the θ/α interfaces. Images courtesy of Taisuke Sasaki, reproduced with permission of Elsevier from Li et al. [13].

With martensite, (Figure 11.2d), the net retention of hydrogen is similar to that of pure, ferritic iron. This is surprising because martensite contains many more defects. However, hydrogen uptake depends both on the sinks present in the structure and on how these reduce the effective diffusion coefficient of hydrogen into the steel [15]. This may explain why martensite has a hydrogen concentration comparable to that of ferrite.

11.2 HYDROGEN IN BANDED MICROSTRUCTURES

Banded α+P steel is illustrated in Figure 6.2, highlighting the characteristic anisotropy that facilitates specific mechanisms of hydrogen cracking [11, 16]. Industrial work in this area has focused on pipelines for the fossil fuel industries, pressurised hydrogen containers etc., which can fail by parting along the bands [17]. Chemical concentration gradients inherited from the solidification process become elongated along the principal hot-deformation direction. Those regions with a lower than average concentration of substitutional solutes first transform into α on cooling, thus partitioning carbon into the solute-rich regions, which then evolve into the pearlite bands.

Banding can be mitigated during large-scale manufacture using accelerated cooling from the austenitic state so that transformations are suppressed to lower temperatures where $|\Delta G^{\gamma\rightarrow\alpha}|$ is greater. As a result, transformation is stimulated at *all* locations, not just in the substitutional-solute depleted regions. This means that the partitioning of carbon as the ferrite grows is not localised into bands, resulting in a more uniform structure without large deviations in hardness (Figure 11.5) [18–20]. Hydrogen-induced cracking ceases to be a problem because the hardness in all regions becomes less than about 250 HV [18].

Homogeneous structures in low-carbon steels are less prone to both hydrogen-induced cracking and sulphide stress-corrosion cracking, where the latter occurs when stressed steel is exposed to wet H_2S, as encountered in the oil industry. This causes the generation of nascent hydrogen by the reaction [21]

$$\mathrm{Fe} + \mathrm{H_2S} \xrightarrow{\mathrm{H_2O}} \mathrm{FeS} + 2\mathrm{H}.$$

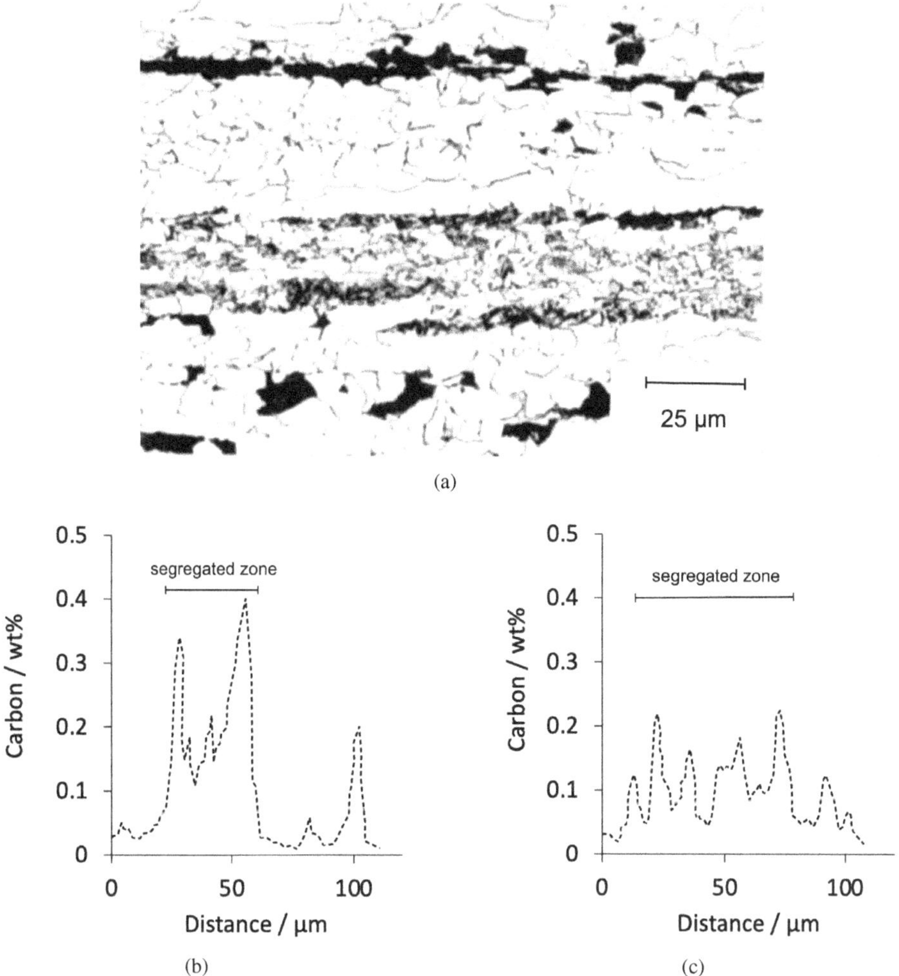

Figure 11.5 (a) The effect on microstructure, of chemical segregation along the mid-thickness of heavy gauge plate. Distribution of carbon concentration in the segregated zone for (b) conventional control-rolled and (c) rapidly cooled steel plates of composition Fe-0.082C-0.28Si-1.33Mn-0.007P-0.009S-0.04Nb-0.09V wt%. Micrograph and data courtesy of H. Tamehiro.

The hydrogen then embrittles the steel in the normal manner [22]. The H_2S at the same time poisons the evolution of H_2 gas, so the focus of the reactions becomes charging the steel with atomic hydrogen. In low-carbon pipeline steels, a bainitic microstructure is more resistant to these problems than one containing allotriomorphic ferrite [23]. This again is a reflection of the greater $|\Delta G^{\gamma\rightarrow\alpha}|$ at the temperatures where bainite forms, which stimulates more uniform transformation in spite of the chemical concentration gradients within the austenite.

The passage of hydrogen through uniform mixtures of α+P becomes more sluggish as the fraction of pearlite is increased (Figure 11.6a). The *apparent* diffusivity[2] becomes anisotropic in a banded microstructure, with the greatest resistance to transport lying normal to the bands [16, 24]. Most attribute this to the α/θ interfaces somehow interfering with the migration of hydrogen, but direct observations [13] indicate otherwise; rather, $D_{\mathrm{H}}^{\theta} \ll D_{\mathrm{H}}^{\alpha}$. The calculated enthalpy of solution of

[2]The diffusion of hydrogen is impeded by traps [15] – hence the adjective *apparent*.

hydrogen, relative to molecular hydrogen, is similar for cementite and ferrite, which could indicate that in a $\theta + \alpha$ phase mixture, the solubility of hydrogen will be similar for both phases (hence the uniform distribution of deuterium in Figure 11.4a).

The degree of anisotropy in diffusion is related to the severity of banding and the quantity of pearlite. X70 pipeline steel has just 0.074C wt% so there isn't a discernible difference in the flux of hydrogen along different directions within the steel [25]. A quantitative treatment of this would be useful, perhaps by analogy with the work on duplex stainless steel which typically exhibit a banded $\alpha + \gamma$ microstructure where the tortuosity of the two-phase dispersion is accounted for [26].

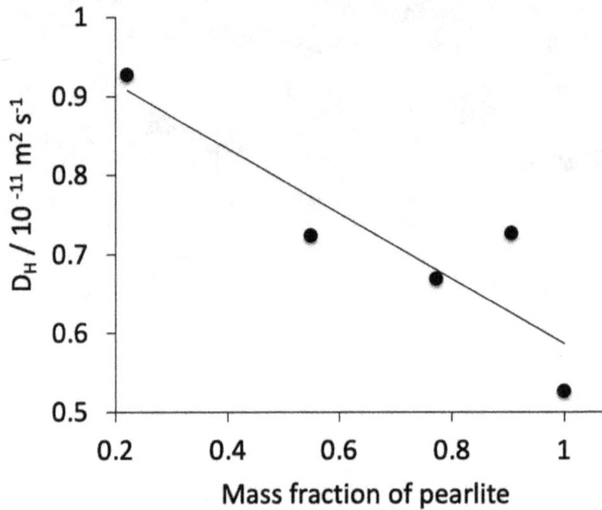

Figure 11.6 Apparent diffusivity of hydrogen at room temperature, through uniform mixtures of ferrite and pearlite. Data from Chan and Charles [11], with the pearlite fractions derived from the stated carbon concentrations of the steels studied. The original D_H values [11] have been reduced by two orders of magnitude following discussion with the author [27].

Experiments comparing a α+P and fully pearlitic microstructures have shown that the permeability of hydrogen is much reduced in the latter, to an extent that cannot be explained by trapping effects. If instead it is assumed that most of the hydrogen flux must be through the ferrite, then the path for the hydrogen becomes much more tortuous in the fully pearlitic microstructure. The length of the diffusion path is increased by a factor of about 30 compared with the α+P microstructure [28]. The main flux is through the ferrite because of $D_H^{\theta} \ll D_H^{\alpha}$, making cementite in this context, impenetrable to hydrogen.

11.3 PICKLING

Hot-processed steels usually are coated with oxide; *pickling* refers to a chemical method of removing this oxide prior to cold-processing. It was during attempts to remove the oxide by immersing steel wires into hydrochloric or sulphuric acid, that led Johnson in 1875 to discover hydrogen embrittlement, when he noticed an 'extraordinary decrease in toughness and breaking-strain of the iron so treated' [8]. It is the reaction between the steel and the acid that introduces hydrogen atoms into the steel.

As a consequence, hot-rolled pearlitic wire rods are baked at about 150 °C for 15 min after pickling, and a prolonged period is set aside to allow some of the hydrogen to diffuse out before drawing into smaller diameters. If, after this cold work, the steel is left in a state of internal residual-stress combined with dissolved hydrogen, the wire can undergo delayed spontaneous-fracture without the application of external stress [29]. Microscopic cracking that generates acoustic emissions is followed by their longitudinal splitting into arced halves.

11.4 HYDROGEN IN STEEL WIRES

The susceptibility of steels to hydrogen embrittlement can be assessed by electrolytically charging with hydrogen and then testing in tension at a slow strain-rate. The concentration of hydrogen in the steel is always minute, so to do damage, it must be able to accumulate at stress concentrations. The steel only fails in a brittle fashion if the testing procedure is slow enough to permit the hydrogen to concentrate at specific locations over the time-scale of the experiment.

Electrolytic charging forces hydrogen atoms into the steel; any not located at deep traps within the iron leave the steel via the free surface, a process that is rapid at high temperatures. Pearlitic wires and strands in ropes have small diameters that facilitate the escape of the mobile hydrogen. The efficacy of the traps in holding back the hydrogen atoms is assessed using thermal desorption analysis in which, a hydrogen-charged specimen is heated at a constant rate to ever greater temperatures while the gas that evolves is characterised quantitatively using chromatography. A plot of desorption rate versus temperature then contains peaks which can be interpreted to obtain the binding energy of hydrogen to specific traps. The location of the peak on the temperature scale depends on the heating rate which typically is in the range 100-200 $^\circ\mathrm{C\,h^{-1}}$; the number density of traps can also influence the peak position [e.g., 30]. In those circumstances, a low-temperature peak occurring at $\lesssim 100\,^\circ\mathrm{C}$ would be associated with diffusible hydrogen that eventually escapes from the steel, even at ambient temperature. For a given set of test conditions, higher temperature peaks correspond to more effective traps for hydrogen so they can be useful in mitigating embrittlement [22].

Pearlitic steel wires ($\varepsilon_\mathrm{p} = 1.91$) that are charged with hydrogen tend to show two desorption peaks during heating at 100 $^\circ\mathrm{C\,h^{-1}}$, the first at 100 °C representing diffusible hydrogen that embrittles and the second at 300 °C corresponding to that which is trapped, i.e., innocuous as long as it remains bound [31]. Ageing the drawn wire at 100 °C or 300 °C for ten minutes in each case, improved the resistance to hydrogen embrittlement [32]. The strength of the wire was increased by the annealing, attributed originally to strain ageing phenomena (Section 13.7), but later to fine precipitation observed in the deformation-induced carbon-rich regions of the ferrite. The effect, therefore, of the 100 °C anneal is simply to allow diffusible hydrogen that is weakly trapped to migrate out of the steel. The 300 °C anneal reduced the strength which makes the steel less susceptible to embrittlement [8].

Superior hydrogen resistance of deformed pearlite

Stronger steels are more prone to hydrogen embrittlement [8, 33]. The average concentration $\overline{x}_\mathrm{H}^\alpha$ is minute, of the order of 10^{-6} so it is difficult to understand how it can affect the properties in the dramatic way that it does. We have seen that an appropriate stress field attracts hydrogen, causing it to concentrate at critical locations. The stress that can be applied to a stronger steel is larger, so the driving force for hydrogen migration to a crack tip increases, explaining the essence of the strength effect.

However, notched tensile tests show that deformed pearlite is more resistant to embrittlement than a quenched and tempered martensitic version with identical strength in the absence of hydrogen, and identical chemical composition, (Fe-0.62C-2.02Si-0.23Mn-1.01Cr wt%) Figure 11.7a [34]. This is because hydrogen-charged elongated pearlite, during tensile deformation, develops cracks that are parallel to the principal loading-axis, which means that gross fracture is prevented until the cracks link laterally. It also is possible that longitudinal cracks permit the diffusible hydrogen to escape from the specimen; cracks are known to trap hydrogen [35].

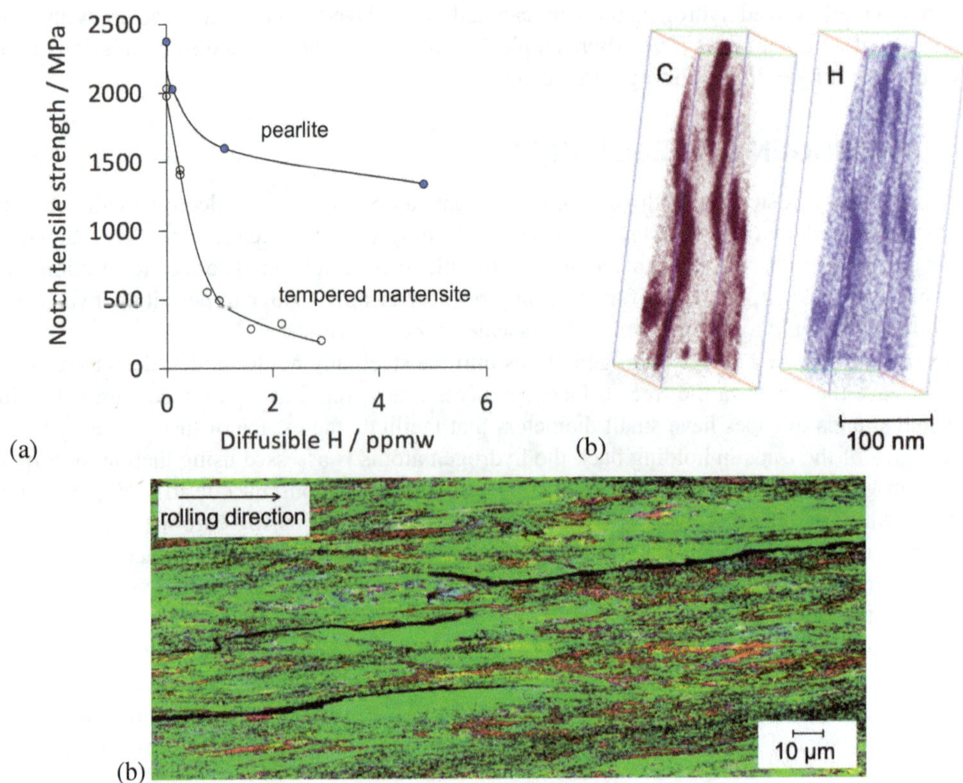

Figure 11.7 (a) Slow strain-rate, notched tensile tests on pearlite deformed using caliber rolling from a 40 mm bar to a 14 × 14 mm square-sectioned rod, and from the same steel in the quenched and tempered (500 °C) martensitic condition. Selected data from Ueji et al. [34]. (b) Atom-probe tomograph of drawn wire showing anisotropic distribution of hydrogen that correlates with that of carbon. Image courtesy of Baptiste Gault. (c) Cracks shown in black, in a hydrogen-charged sample identical to that in (a), following the rolling direction. Orientation image, adapted from Ueji et al. [34], with the permission of ISIJ International.

Drawn pearlite has a strongly anisotropic structure. Studies on microscopic, notched cantilevers of drawn pearlite that are loaded while being charged with hydrogen, have revealed that the cracking occurs parallel to the α/θ interfaces rather than normal to their orientations [36]. The critical stress intensity for hydrogen-induced crack propagation along the interfaces was estimated at just $1.7\,\mathrm{MPa\,m^{1/2}}$. This might not surprise given the tendency for hydrogen to segregate in the vicinity of the α/θ interfaces in deformed pearlite [13]. Samples tested in air, without *in situ* hydrogen charging, did not reveal anisotropy in crack propagation behaviour.

Limited observations suggest that hydrogen in deformed pearlite ($\varepsilon_{\mathrm{p}} = 3.7$) tends to partition to the carbon-rich regions associated with cementite that has decomposed due to plastic deformation and slight indications of hydrogen at α/θ interfaces [37]. Thus, the distribution of hydrogen becomes anisotropic, its location correlating with the corresponding anisotropic distribution of carbon, Figure 11.7b. This may contribute to the tendency for cracks to propagate parallel to the drawing direction. Figure 11.7c illustrates such cracks.

11.5 STATIC FRACTURE AND RESIDUAL STRESS

Pre-stressed pearlitic-steel rods used for reinforcing concrete suffer from environmentally-assisted failure if they become exposed, because of hydrogen infusion from corrosion reactions. The steel can then fracture at a constant stress after an often unpredictable amount of time. Experiments to assess the susceptibility of the rod involve dead-loading while the sample is immersed in a corrosion cell containing an ammonium thiocyanate solution; the time to failure is then monitored as a function of the applied stress.

Reinforcing rods are fabricated by cold-drawing down 12 mm diameter rods down to ≈ 5 mm, which leaves an axial, macroscopic tensile-stress in the surface shell of the wire (p. 286). A tensile residual stress is obviously a disadvantage because it must limit what can be applied before failure follows. Figure 11.8 shows data from two sets of experiments, the first for as-drawn rods that contained surface tensile residual-stresses of 604 MPa, and another set for stress-relieved rods where with just 176 MPa of tension [38].[3] If the residual stresses are added to the applied stresses, then the data for the two kinds of samples are seen to be broadly similar, emphasising the fact that a residual tensile-stress exacerbates hydrogen embrittlement.

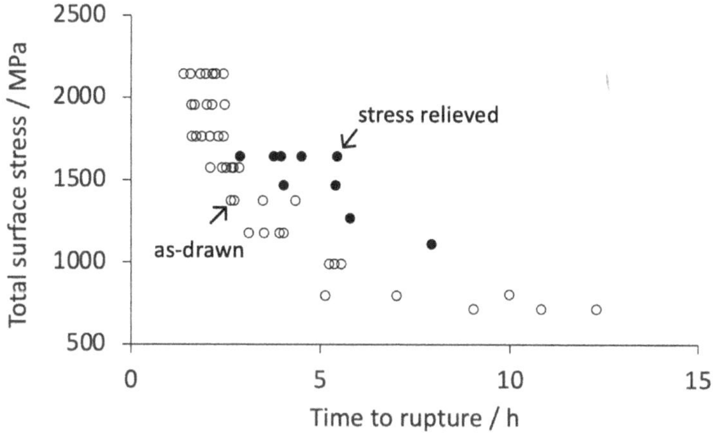

Figure 11.8 The time to rupture for steel samples immersed in ammonium thiocyanate. The raw data from Elices [38] have been modified by adding the measured tensile residual stresses, to generate the total stress.

REFERENCES

1. H. Cavendish: 'XIX. Three papers, containing experiments on factitious air', *Philosophical Transactions of the Royal Society of London*, 1766, **56**, 141–184.
2. G. E. R. Lloyd: Aristotle: the growth and structure of his thought: Cambridge, U. K.: Cambridge University Press, 1968.
3. H. W. Woolley, R. B. Scott, and F. Brickwedde: 'Compilation of thermal properties of hydrogen in its various isotopic and ortho-para modifications', *Journal of Research of the National Bureau of Standards*, 1948, **41**, 379–475.
4. C. Minot, and C. Demangeat: 'The iron–hydrogen system: Lattice location of hydrogen, heat of formation, and hydrogen–hydrogen binding energy', *The Journal of Chemical Physics*, 1987, **86**, 2161–2167.

[3]The tensile properties were otherwise similar with the 0.1% proof strength being 1548 MPa for the as-drawn sample, and 1571 MPa for the stress-relieved specimen [38].

5. E. M. Burbidge, G. R. Burbidge, W. A. Fowler, and F. Hoyle: 'Synthesis of the elements in stars', *Reviews of Modern Physics*, 1957, **29**, 547–650.
6. H. K. D. H. Bhadeshia, and R. W. K. Honeycombe: Steels – structure, properties and design: 5th edition ed., London, U.K.: Elsevier Butterworth-Heinemann, 2024.
7. J. Toribio: 'Hydrogen-assisted microdamage of eutectoid pearlitic steel in the presence of notches: The tearing topography surface', *Metals*, 2023, **13**, 1365.
8. W. H. Johnson: 'On some remarkable changes produced in iron and steel by the action of hydrogen and acids', *Proceedings of the Royal Society of London*, 1875, **23**, 168–179.
9. S. L. I. Chan, M. Martinez-Madrid, and J. A. Charles: 'Evaluation of three different test methods for charging hydrogen into iron', *Metals Technology*, 1983, **10**, 464–470.
10. S. L. I. Chan: 'Hydrogen trapping ability of steels with different microstructures', *Journal of the Chinese Institute of Engineers*, 1999, **22**, 43–53.
11. S. L. I. Chan, and J. A. Charles: 'Effect of carbon content on hydrogen occlusivity and embrittlement of ferrite–pearlite steels', *Materials Science and Technology*, 1996, **2**, 956–962.
12. E. J. McEniry, T. Hickel, and J. Neugebauer: '*Ab initio* simulation of hydrogen-induced decohesion in cementite-containing microstructures', *Acta Materialia*, 2018, **150**, 53–58.
13. Z. Li, T. Sasaki, R. Ueji, Y. Kimura, A. Shibata, T. Ohkubo, and K. Hono: 'Role of deformation on the hydrogen trapping in the pearlitic steel', *Scripta Materialia*, 2024, **241**, 115859.
14. K. Takai, Y. Chiba, K. Noguchi, and A. Nozue: 'Visualization of the hydrogen desorption process from ferrite, pearlite, and graphite by secondary ion mass spectrometry', *Metallurgical & Materials Transactions A*, 2002, **33**, 2659–2665.
15. L. S. Darken, and R. P. Smith: 'Behaviour of hydrogen in steel during and after immersion in acid', *Corrosion*, 1949, **5**, 1–16.
16. S. A. Golovanenko: 'Steels for the gas and oil industry', *Journal of Materials for Energy Systems*, 1980, **1**, 66–78.
17. J. S. Laws, V. Fraick, and J. McConnell: 'Hydrogen gas pressure vessel problems in the M-1 facilities': Tech. Rep. CR-1305, National Aeronautics and Space Administration, Cleveland, Ohio, USA, 1969.
18. H. Tamehiro, R. Habu, N. Yamada, H. Matsuda, and M. Nagumo: 'Properties of large diameter line pipe steel produced by accelerated cooling after controlled rolling', In: P. D. Southwick, ed. *Accelerated Cooling of Steel*. 1985:401–414.
19. H. Tamehiro, T. Takeda, S. Matsuda, K. Yamamoto, and N. Okumura: 'Effect of accelerated cooling after controlled rolling on hydrogen induced cracking resistance of line pipe steel', *Transactions of the Iron and Steel Institute of Japan*, 1985, **25**, 982–988.
20. M. K. Graf, H. G. Hillenbrand, and P. A. Peters: 'Accelerated cooling of plate for high-strength large-diameter pipe', In: P. D. Southwick, ed. *Accelerated Cooling of Steel, TMS AIME*. 1985:165–180.
21. R. D. Kane: 'Roles of H_2S in behaviour of engineering alloys', *International Metals Reviews*, 1985, **30**, 291–301.
22. H. K. D. H. Bhadeshia: 'Prevention of hydrogen embrittlement in steels', *ISIJ International*, 2016, **56**, 24–36.
23. M.-C. Zhao, Y.-Y. Shan, F. R. Xiao, K. Yang, and Y. H. Li: 'Investigation on the H_2S-resistant behaviors of acicular ferrite and ultrafine ferrite', *Materials Letters*, 2002, **57**, 141–145.
24. H.-L. Lee, and S. L.-I. Chan: 'Hydrogen embrittlement of AISI 4130 steel with an alternate ferrite/pearlite banded structure', *Materials Science & Engineering A*, 1991, **142**, 193–201.
25. A. J. Haq, K. Muzaka, D. P. D. an A. Calka, and E. V. Pereloma: 'Effect of microstructure and composition on hydrogen permeation in X70 pipeline steels', *International Journal of Hydrogen Energy*, 2013, **38**, 2544–2556.
26. A. Turnbull, and R. B. Hutchings: 'Analysis of hydrogen diffusion in a two-phase alloy', *Materials Science & Engineering A*, 1994, **177**, 161–171.

27. S. L. I. Chan: 'Personal communication', 2020: Correction to apparent diffusivity data.
28. C. Forot, E. Legrand, E. Roguet, J. Creus, and J. Kittel: 'Impact of cementite tortuosity on hydrogen diffusion in pearlitic steels', In: *Earth, Water, Fire, Air, Corrosion happens everywhere*. Graz, Austria: The European Corrosion Congress, 2015:1–8.
29. N. K. Mukhopadhyay, G. Sridhar, N. Parida, S. Tarafder, and V. R. Ranganathan: 'Hydrogen embrittlement failure of hot dip galvanised high tensile wires', *Engineering Failure Analysis*, 1999, **6**, 253–265.
30. W. Song, J. Von Appen, P. Choi, R. Dronskowski, D. Raabe, and W. Bleck: 'Atomic-scale investigation of ε and θ precipitates in bainite in 100Cr6 bearing steel by atom probe tomography and ab initio calculations', *Acta Materialia*, 2013, **61**(20), 7582–7590.
31. K. Takai, and R. Watanuki: 'Hydrogen in trapping states innocuous to environmental degradation of high-strength steels', *ISIJ International*, 2003, **43**, 520–526.
32. D. Hirakami, T. Manabe, K. Ushioda, K. Noguchi, K. Takai, Y. Hata, S. Hata, and H. Nakashima: 'Effect of aging treatment on hydrogen embrittlement of drawn pearlitic steel wire', *ISIJ International*, 2016, **56**, 893–898.
33. R. P. Frohmberg, W. J. Barnett, and A. R. Troiano: 'Delayed failure and hydrogen embrittlement in steel': Tech. Rep. 54-320, Wright-Patterson Air Force Base, Ohio, U. S. A., 1954.
34. R. Ueji, Y. Kimura, and T. Inoue: 'Preferable resistance against hydrogen embrittlement of pearlitic steel deformed by caliber rolling', *ISIJ International*, 2022, **62**, 368–376.
35. W. Solano-Alvarez, E. J. Song, D. K. Han, D. W. Suh, and H. Bhadeshia: 'Cracks in martensite plates as hydrogen traps in a bearing steel', *Metallurgical & Materials Transactions A*, 2015, **46**, 665–673.
36. K. Tomatsu, T. Amino, T. Chida, S. Uji, M. Okonogi, H. Kawata, T. Omura, N. Maruyama, and Y. Nishiyama: 'Anisotropy in hydrogen embrittlement resistance of drawn pearlitic steel investigated by in-situ microbending test during cathodic hydrogen charging', *ISIJ International*, 2018, **58**, 340–348.
37. A. J. Breen, L. T. Stephenson, B. Sun, Y. Li, O. Kasian, D. Raabe, M. Herbig, and B. Gault: 'Solute hydrogen and deuterium observed at the near atomic scale in high-strength steel', *Acta Materialia*, 2020, **188**, 108–120.
38. M. Elices: 'Influence of residual stresses in the performance of cold-drawn pearlitic wires', *Journal of Materials Science*, 2004, **39**, 3889–3899.

12 Pearlite in rail steels

The bold idea of metal wheels running on smooth metal-rails has evolved dramatically since wooden rails were first implemented at collieries some two centuries ago [1]. It does seem extraordinary that un-lubricated metal-on-metal locomotion can deliver comfort, speed and efficiency in a reliable and routine manner. Thanks to George Stephenson an extraordinarily courageous engineer, the 'Father of the railways' [2].

Rails are large objects, both in cross-section and length, weighing some 30-75 $kg\,m^{-1}$. They must be capable of mass manufacture at a reasonable cost and of withstanding large cyclic-contact stresses, braking friction, wear, vibrations and must be weldable. Rail production today is the archetypal near-net manufacturing process with zero waste or finish machining. Continuous cast rectangular blooms $\approx 410 \times 320$ mm can directly be led into a hot-rolling process, thus avoiding the need for reheating. They are deformed while in their hot austenitic state into the required profile using rolls that are shaped so that the cross-section evolves during multiple passes, Figure 12.1a.

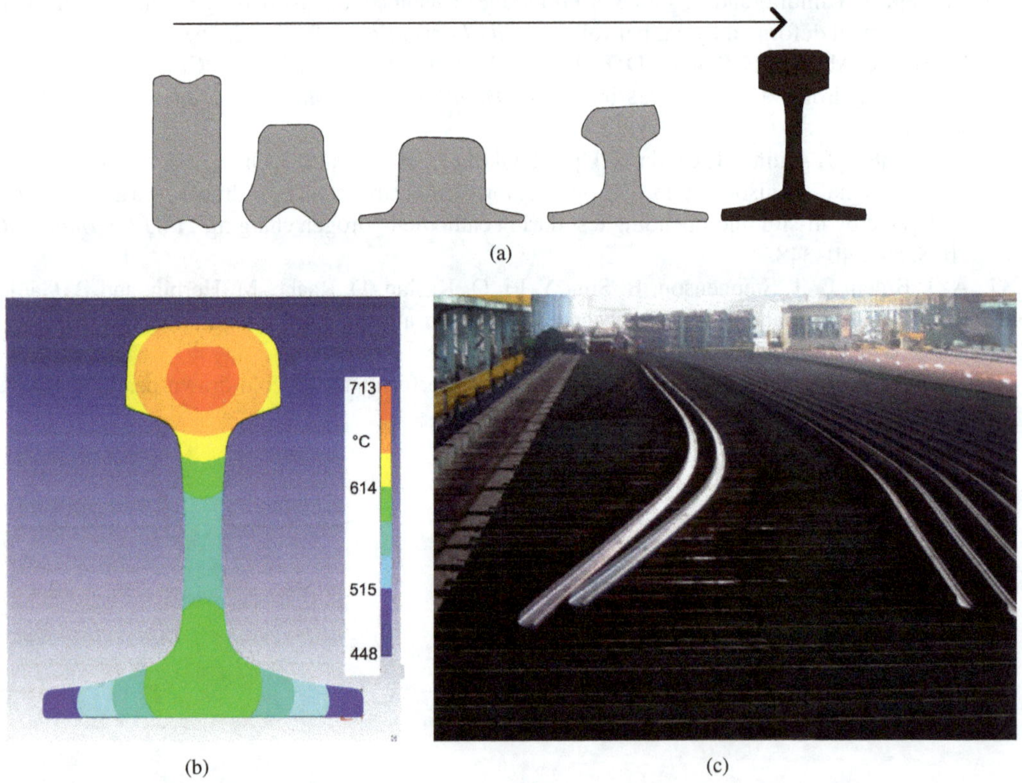

Figure 12.1 (a) Evolution of the rail-section during hot-rolling. Only a few of the 15 or so steps are illustrated. (b) Calculated temperature field of heavy rail (Fe-0.74C-0.76Si-0.82Mn wt%) following cooling for 800 s after rolling, when the transformation to pearlite is almost complete. (c) Because the cooling is not uniform, the rails bend as the temperature decreases. The rails on the right are relatively hot. (b,c) are adapted from Chen et al. [3] with permission of Springer Nature.

Because of the asymmetrical cross-section at the end of the rolling process, the rail does not cool uniformly, causing it to curve while laid sideways on a cooling bay Figure 12.1b,c. Limited

DOI: 10.1201/9781032631981-12

cold-deformation is therefore implemented by passing it through a series of driven rollers that are located in an alternating sequence above and below the rail. The top rolls of the straightening machine are profiled to the required crown shape and the loads imparted are sufficient to make marginal, final modifications to the shape of the crown.

12.1 RAIL COMPOSITION

The majority of steel used in the manufacture of rails relies on pearlite and the steels have a simple chemical composition, Figure 12.2. They usually contain $\lesssim 0.6$Si wt% because of the deoxidation requirements during steelmaking or for a modest contribution to solid-solution hardening ($w_{\mathrm{Si}} \times 106$ MPa [4]). Manganese is common in concentrations between 0.5-1.2 wt% because it allows pearlite to form over a range of cooling rates from grains of γ that are in a recrystallised form.

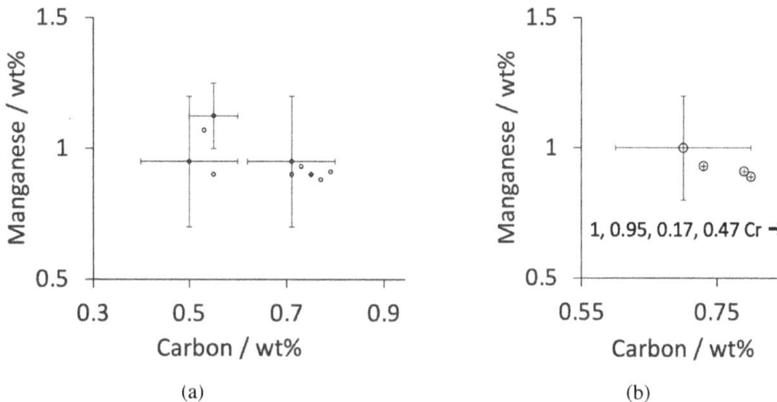

Figure 12.2 Aspects of the chemical compositions of commercially available rails. The bars represent specification ranges; points without the bars are actual reported compositions. (a) Rail containing deliberate additions of only C, Mn, Si. (b) Rails containing deliberate additions of C, Mn, Si, Cr . The point on the far left contains 1Cr wt% whereas that on the far right has 0.47Cr wt%. Data from a compilation in [5].

The carbon concentration tends to be close to the nominal eutectoid-concentration (p. 166) although in the case of rails destined for less onerous applications, it can be significantly smaller. The manganese is necessary because it imparts hardenability, thus allowing finer pearlite to form at greater undercoolings below T_{E}, and over the range of cooling rates expected from the non-uniform cooling of the bulk of the rail. Figure 12.3a shows that without manganese, the austenite would decompose rapidly at high temperatures, making it impossible to achieve the undercooling associated with a fine interlamellar spacing within the pearlite. The concentration of manganese cannot be so large that the growth rate of pearlite is retarded to an extent that leaves the residual austenite susceptible to transformation into high-carbon bainite.

A typical chromium addition does not enhance the hardenability much and hardly affects S_{I}, Equation 8.3. It is knownthat chromium partitions into the cementite during the pearlite transformation (Figure 8.22) [6]. Chromium-enriched cementite is harder, has a greater elastic modulus and is more resistant to abrasive wear and spalling, Figure 12.3b [7, 8]; empirical experiments on rail steels confirm this role or chromium [9, 10].

Rail steels must be capable of transforming completely into pearlite when air-cooled at reasonable rates following hot-rolling. A typical rail-steel will contain about 0.8 wt% of Mn and not much of any other hardenability-enhancing solutes. Depending also on the austenite grain size, this can lead to a fully pearlitic microstructure if the cooling rate is maintained $\lesssim 4\,^{\circ}\mathrm{C\,s^{-1}}$ [12], confirmed

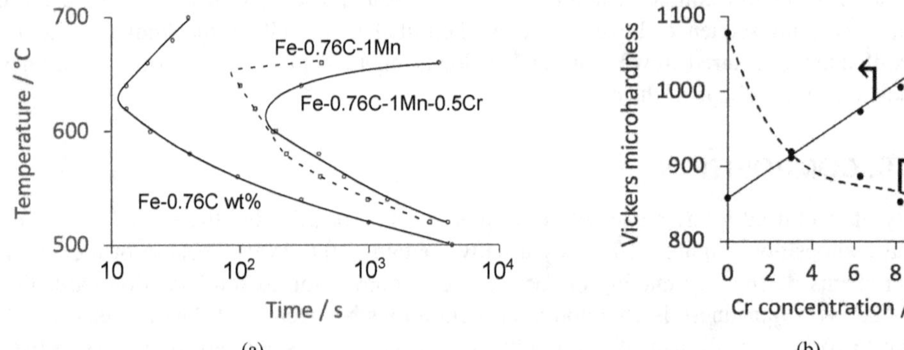

Figure 12.3 (a) Calculated time-temperature-transformation diagrams for the initiation of pearlite [11]. (b) Hardness and pin-on-disc wear loss of polycrystalline cementite as a function of its chromium concentration. Adapted using data from Zheng et al. [7].

by the continuous cooling transformation diagram in Figure 12.4a. At greater Mn concentrations (≈ 1 wt%), the bainite-start temperature is $B_S \approx 520\,°\mathrm{C}$ [13] so pearlite should consume all of the austenite if bainite is to be avoided; in Figure 12.4a, the cooling rate $1\,°\mathrm{C\,s^{-1}}$ meets this condition whereas $5\,°\mathrm{C\,s^{-1}}$ does not.

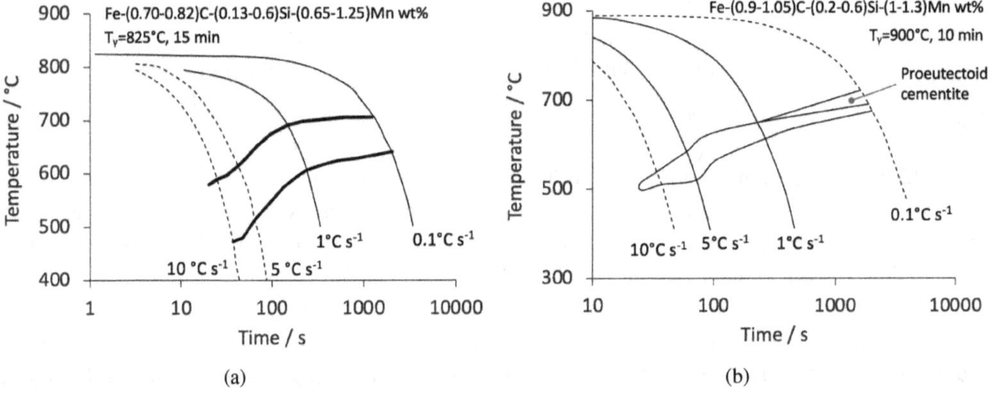

Figure 12.4 Continuous cooling transformation diagrams for some rail steels, the exact compositions are not available, nor the austenite grain sizes, in the original publications. In all cases, dashed cooling curves lead to ultimate microstructures that are not fully pearlitic, whereas the other such curves lead only to pearlite. (a) Adapted using selected data from Yuryev et al. [14]. (b) Hypereutectoid rail steel. Adapted using selected data from Chen et al. [15].

Hyper- or hypo-eutectoid steels must be transformed at a temperature $T < T_E$ within the Belaiew-Hultgren region (Figure 8.17) in order to generate a pearlitic microstructure that is free from proeutectoid phases. Fig. 12.4b shows for a hypereutectoid rail-steel, the range of cooling rates available to avoid the precipitation of detrimental proeutectoid-θ at the austenite grain surfaces.

12.2 ELEMENTARY MECHANICAL PROPERTIES OF RAILS

The strength of pearlitic rails is *determined* primarily by the interlamellar-spacing; for a rail of composition Fe-0.64C-1.4Mn-0.29Si wt%, $S_I \approx 0.15\,\mu\mathrm{m}$ [16] the σ_y given by Equation 4.8 comes to 450 MPa [17] with only ≈ 83 MPa due to solution-strengthening of α_P by Mn, Si [4].

If a steel has the eutectoid carbon concentration then proeutectoid phases are absent even in binary Fe-C alloys. Adding hardenability enhancing solutes such as manganese allows pearlite to be generated at lower temperatures, at the cooling rates achievable across the cross-section of a rail weighting some $60\,\mathrm{kg\,m^{-1}}$. An increase in the cooling rate by using air or mist blasts on the hot rail, can refine S_I, particularly at the rail head which bears the brunt of the contact with the wheel.

The stress-strain behaviour in tension of rail steels is similar to other air-cooled pearlitic steels, exhibiting continuous yielding (p. 52). The tensile properties vary across the section of the rail. The head of the rail can be much stronger if the interlamellar spacing there has been refined by accelerated cooling, Figure 12.5, where key properties are summarised. The yield strength and hardness are both linear functions of $S_I^{-1/2}$; the softer rails see service in less-demanding applications whereas the hardest ones are applied when the curvature of the track has a radius $\lesssim 3000$ m [18], because the forces on the wheel-rail interface then become larger, capable of inducing localised plasticity and shape change. These forces scale with the axle-load, necessitating harder rails to avoid excessive wear when transporting heavy goods.

It is difficult to be explicit about the tensile elongation required to successfully implement rail steels, and oddly, a greater elongation is specified for less-strong rails; lower values of $9 \rightarrow 10\,\%$ are accepted for harder rails, presumably because larger ductility simply cannot be achieved, Figure 12.5d. A part of measured elongation sometimes reflects the inclusion content of the steel because non-metallic particles such as oxides and sulphides nucleate voids that ultimately link to cause gross failure. A large number density of such particles leads to a lower elongation because the linking occurs more rapidly with closely spaced voids. Modern rails have reduced impurity contents so inclusions become less of an issue. The surface of a rail undergoes plastic shear during service, leading to an accumulation of strain. Cracks initiate when the ductility there becomes exhausted so there must be a tolerable extent of ductility, whether that should be measured in tension or in compression experiments [19]; the stress system during rolling contact is approximately a combination of in-phase uniaxial compression and torsion [20].

Cracks in rails usually are initiated at the surface due to fatigue; it is important that these cracks do not reach a critical size for rapid propagation as defined by the fracture toughness, Figure 12.5e. There is a routine maintenance procedure whereby the surface of the rail is ground down to remove any fatigue damage while maintaining the head profile.

There is a complication in the application of fracture mechanics to rail scenarios. The loading conditions and crack length are time-dependent so a dynamic fracture toughness may be more relevant; the dynamic toughness for a particular rail steel has been reported to be $\approx 10\,\mathrm{MPa\,m^{1/2}}$ smaller [21].

The structure and mechanical properties do vary with depth from the working surface, given that the cooling rate is not uniform, but they can be different along the length of the rail when placed on the cooling platform after hot rolling [29, 39–41]. This applies also to heat-treated rails [42]. It has even been argued that a heat treatment process needs to be developed that allows hardness gradients to develop on the rail-head, such that lateral wear can be reduced [43].

12.2.1 HEAT TREATMENT OF RAILS

The aim of heat treatment is to condition the properties of the rail head. The rolled rail may be reheated to $\approx 900\,^\circ\mathrm{C}$ for 50 min and then passed through a chamber containing a series of symmetrically placed air jets, Figure 12.6, which illustrates the hardness distribution following this forced cooling [44]. The microstructure obtained was fully pearlitic with a hardness of ≈ 380 HV, compared with the ordinary hardness of the same rail of 350 HV [45].

Induction heating followed by cooling using jets of air-water or polymer-water mixtures can be used to strengthen the rail surface [46, 47]. Voestalpine has developed a commercial process in which the whole length of the hot-rolled rail is dipped into the heat-treatment bath immediately

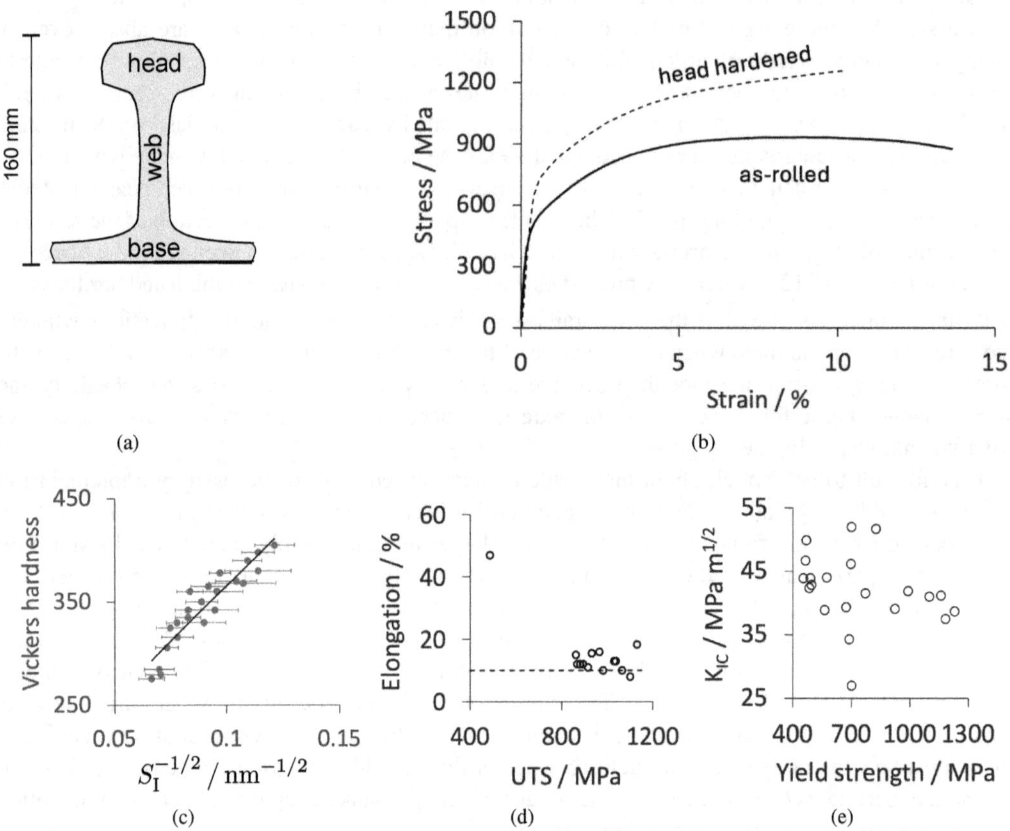

Figure 12.5 (a) Terminology used in defining positions within a rail cross-section. A typical rail may be about 160 mm in height. (b) Stress-strain curves from tensile tests conducted on the head of as-hot-rolled rail with the dashed curve representing a head-hardened rail. Compiled using data from [22, 23]. The elongation data have sometimes been corrected to remove the elastic part of the σ-ε curve, which often includes machine deflection. The dashed line refers to a 10% elongation often quoted in specifications. (c) Hardness as a function of $S_I^{-\frac{1}{2}}$, using rail steel data due to Perez-Unzueta and Beynon [24]. (d) Compilation of strength and elongation data on a variety of rail steels [22, 23, 25–32]. The data come from individual tensile tests conducted on the head of the rail, rather than the minimum specified values that often are quoted in the literature. (e) Fracture toughness of rail heads as a function of the yield strength. Compilation of data from [33–38].

after rolling. This results in a fine, relatively uniform pearlite, which enhances many of the properties required for high-performance rails [48]. All heat treatments that require separate heating after the rail has cooled, or induction heating as an off-line process, reduce productivity. Modern practice is to conduct accelerated cooling as the hot rail emerges from the rolling process [47], with any straightening applied after this in-line heat treatment.

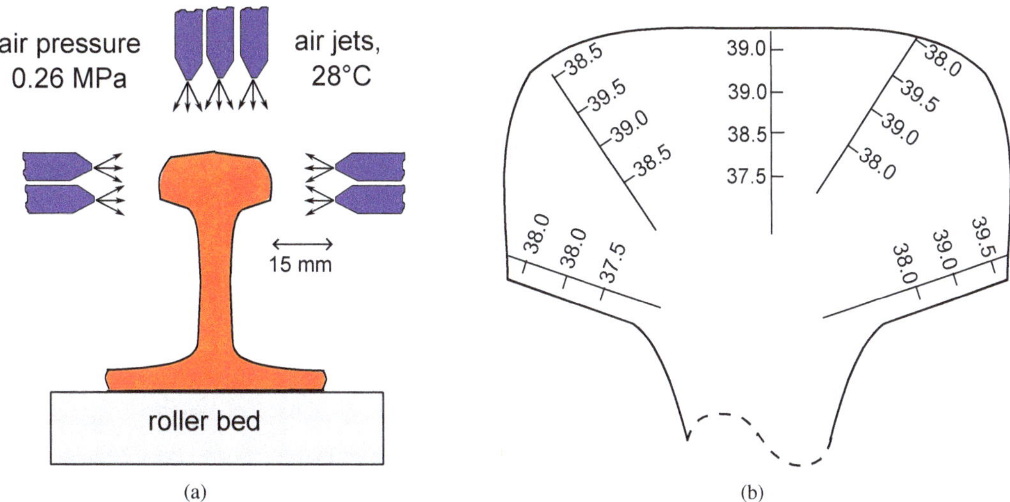

Figure 12.6 (a) How the rail-head might be heat treated after austenitisation, with relatively rapid cooling implemented using air jets as the hot rail passes through an enclosed cooling chamber. (b) Distribution of Rockwell C hardness in rail-head. The composition of the rail is Fe-0.75C-0.85Mn-0.6Si-0.084V wt%, with the commercial designation U75V. A Rockwell C hardness of 39 is equivalent to 376 HV. Selected data from Hao et al. [44].

12.3 WEAR OF RAILS

The un-lubricated, abrasive wear of rails involves the rolling/sliding motion of a wheel against the rail, leading to considerable shear deformation at the surface. This hardens the surface leading to the detachment of wear debris so the rail progressively loses mass and given time, can alter the shape of the contact region. Many experiments indicate that the wear rate correlates with the hardness (Figure 12.7a) but the problem is more complicated than the Archard Law [49] would indicate with an inverse dependence of the rate on hardness. For example, the structure below the contact surface is greatly sheared by the relative motion of the wheel and rail, the depth of the deformation decreasing with hardness, as shown in Figure 12.7.

The shear deformation causes a partial fragmentation and thinning of the cementite, and more importantly, to its alignment parallel to the contact surface.[1] Thin cementite-lamellae are more capable of accommodating deformation so they readily align with the contact surface, leading to an enhancement of wear-resistance [24]. This effect is much more prominent than in martensitic or bainitic rail-steels.

The one exception is carbide-free bainitic steel which not only better resists wear but also is the only structure which at the same time reduces wear on the wheel [5, 50, 51]. The absence of carbides avoids fragmentation debris of the type associated with the abrasive wear of ordinary rails.

Although most correlations rely on the properties of pearlite before it is subjected to wear, the hardness at the wear surface increases sharply with the number of cycles in a disc-on-disc wear test [52]. The ability of the steel to harden therefore may be relevant in the interpretation of abrasion data.

[1]This, as noted elsewhere (p. 328) is a common feature of the lamellar structure when subjected to large deformations.

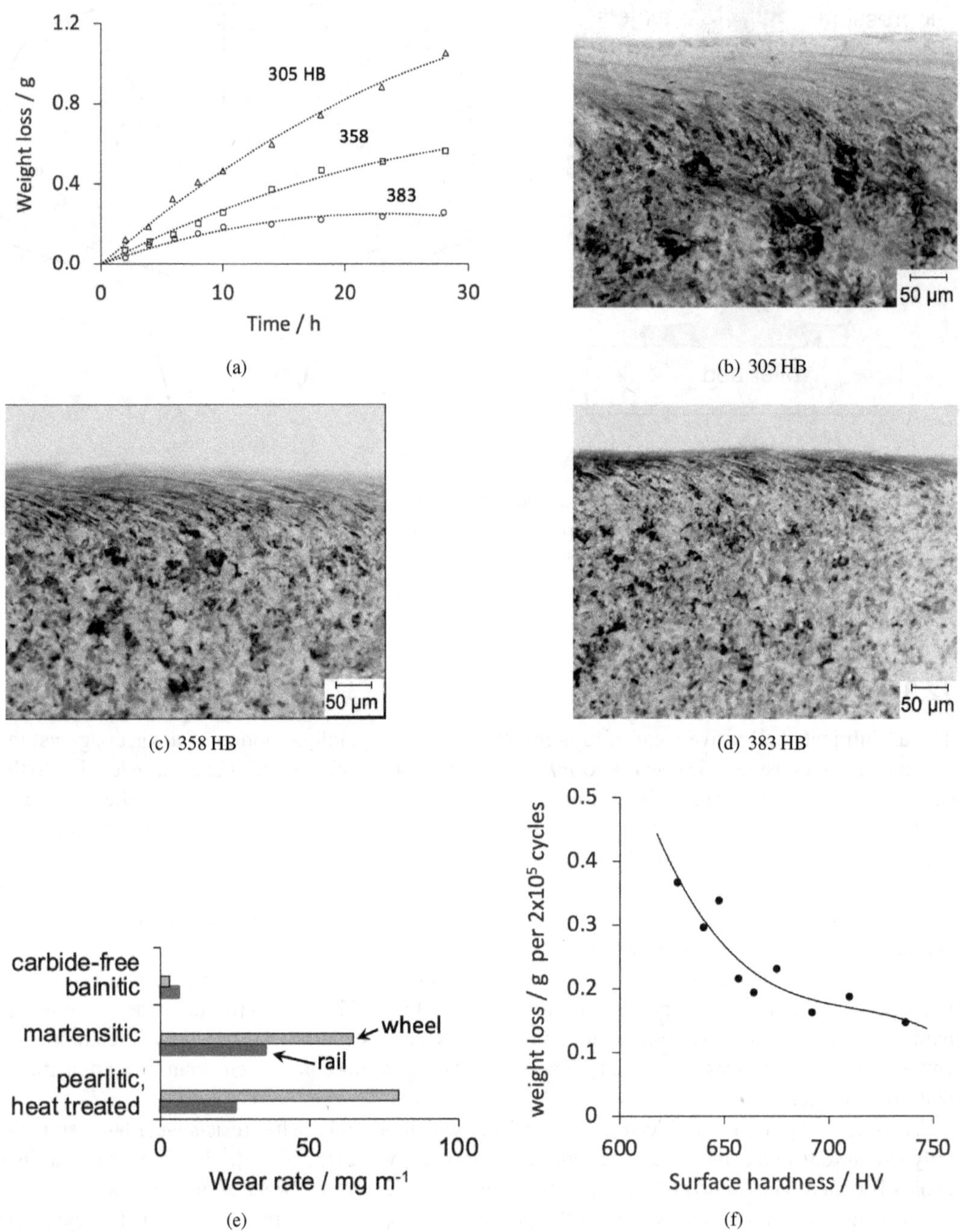

Figure 12.7 Wear data for rail steel that has been austenitised and transformed into pearlite with different interlamellar spacings (263, 161, 104 nm) in order to achieve a variety of hardnesses [53]. (a) Weight loss data generated by rotating a wheel steel at 300 rpm against the sample at 270 rpm with a contact load of 3.2 kN and a speed differential giving a 10% slip ('creepage'). (b,c,d) Cross sections of the rail steel following wear-testing, to show the sheared regions, the depths of which decrease as the initial hardness increases. Images courtesy of Professor Wu Run, with the contrasts inverted here for clarity [53]. (e) Comparison of wear rates [50]. (f) Disc-on-disc wear data using a 640 MPa contact stress and a slip of 20%. On the vertical axis, the origin is at 10^5 cycles, and 0.5 corresponds to 7×10^5 cycles. The pre-test hardness range was only 385 → 395 HV. Adapted using selected data from Ueda et al. [52].

12.4 ROLLING CONTACT FATIGUE OF RAILS

The stress field created when two frictionless bodies are pressed together into elliptical contact whilst avoiding plasticity was studied by Hertz [54, 55]. The most interesting prediction was that the shear stress τ peaks at a certain depth below the surface, Figure 12.8. So the repeated motion of the moving body, for example railway wheels on rails, will cause the shear stress to oscillate and therefore lead to fatigue damage accumulating below the rail surface. This eventually can lead to spalling if left unchecked.

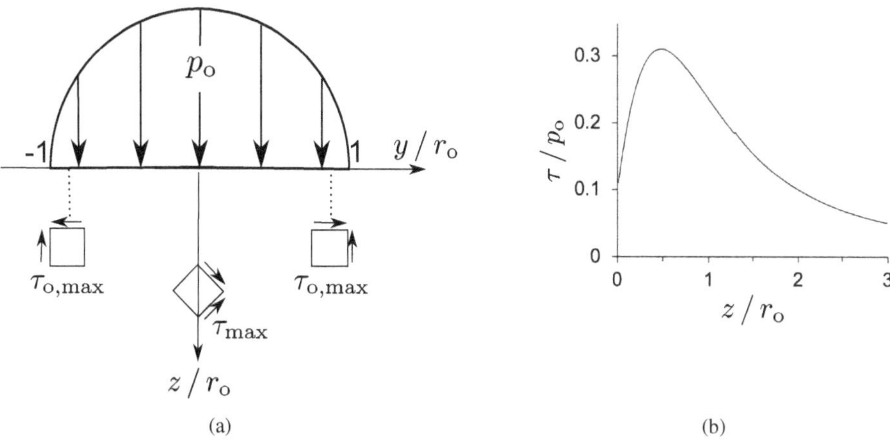

Figure 12.8 Hertzian contact. with the coordinate z is along the depth direction whereas y is parallel to the rolling direction. (a) Normalised stresses as a function of depth z/r_o along the z axis ($y = 0$). (b) The maximum shear stress a due to contact over an area of radius r_o due to a load p_o.

Combinations of rolling contact and sliding can also lead to surface fatigue-crack initiation [56], a mechanism pertinent to rail tracks where sliding features more prominently. Such cracks subsequently propagate through the sub-surface zone with the highest mode II shear stress.

Unlike bearings, localised plasticity caused by friction with the rolling stock is significant in the case of rails, making the Hertzian approach based on elasticity too simplistic. Fatigue cracks can grow along shear bands associated with the surface deformation layer, Figure 12.9.

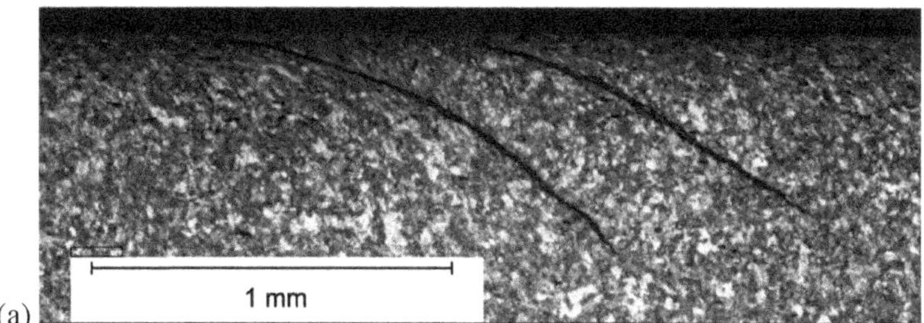

Figure 12.9 Plasticity at the contact surface, caused by rolling-sliding. Rolling contact fatigue cracks following the directional shear deformation. Image reproduced from Trummer et al [57] with the permission of Elsevier.

The deformation accumulates at the surface as the sum of small shear strain increments ($\sum_1^N \gamma_o$) of plastic strain over N cycles as the wheels traverse a specific region of track. For a pearlitic rail with a hardness $\approx 270\,\mathrm{HV}$ an empirical description of this ratcheting is as follows [58]:

$$\begin{aligned} \gamma_o &= 8.3275\times10^{-8}\exp\{0.024867\mu_f p_o\} \\ \text{and}\quad \frac{d\gamma}{dN} &= \gamma_o b N^{b-1} = 1.1023-0.00127644\mu_f p_o, \end{aligned}$$

where μ_f is the friction coefficient and the product $\mu_f p_o$ is the maximum shear stress, p_o is the contact pressure, and $b\approx 0.5$ is a fitting parameter. It follows that the rate of accumulation of shear strain decreases as N increases.

Experiments where pearlite is rubbed against pearlite using torsion indicate a friction coefficient under dry conditions of $\approx 0.6\rightarrow 0.7$ and $\approx 0.3\rightarrow 0.45$ when wet [59]. With dry ball-on-disc sliding $\mu_f\approx 0.36$ [60]. The actual value of μ_f depends on many factors including surface roughness, which will not be constant. The coefficient will therefore evolve during service, starting off at a small value and then reaching a steady-state magnitude indicated above.

There has been no investigation on how the structure of pearlite might be engineered to reduce the friction coefficient. However, an increase in strength, whether that is achieved by the usual method of refining S_I or with the help of interphase vanadium-carbide precipitation in α_P [61] should help in both respects, as long as the rail has a smooth surface and level aspect when first placed into service. Some friction is necessary to achieve traction during locomotive acceleration or deceleration.[2]

12.5 WHITE-ETCHING LAYERS

White-etching surface layers are common on railway tracks where rolling-contact and sliding damage occurs, on machined surfaces and in general where steel surfaces rub together [62]. The discussion on this has focussed on whether the interaction between the wheel and rail leads to temperatures sufficiently high to austenitise the steel, in which case the white layer might consist of severely deformed martensite; the alternative is that the initial microstructure is mechanically alloyed to produce carbon supersaturated ferrite. The white character comes from the fine scale of the structure which is homogeneous relative to the virgin material.

Early transmission electron microscopy and calculations indicated that the transient temperature does not exceed Ae_1, with the conclusion that the white layer represents severely deformed ferrite enriched in carbon because the cementite is forced into solution by mechanical alloying [63]. Recent finite element estimates that include friction contributions suggest otherwise, that it *is* possible for transients to exceed Ae_3. When combined with characterisation using the atom-probe, the results suggest that the white-layers consist of fine martensite, Figure 12.10 [64]. The cementite lamellae dissolve so the carbon ends up in the martensite and at defects, but the manganese and silicon atoms do not relocate from the original position of the the now-dissolved cementite. This is because heating into the austenite phase field is so rapid, and the time spent there is so small, that substitutional solutes are unable to diffuse over significant distances.

[2] A magnetic levitation train has no friction but brakes using the same magnets but in mutual repulsion mode.

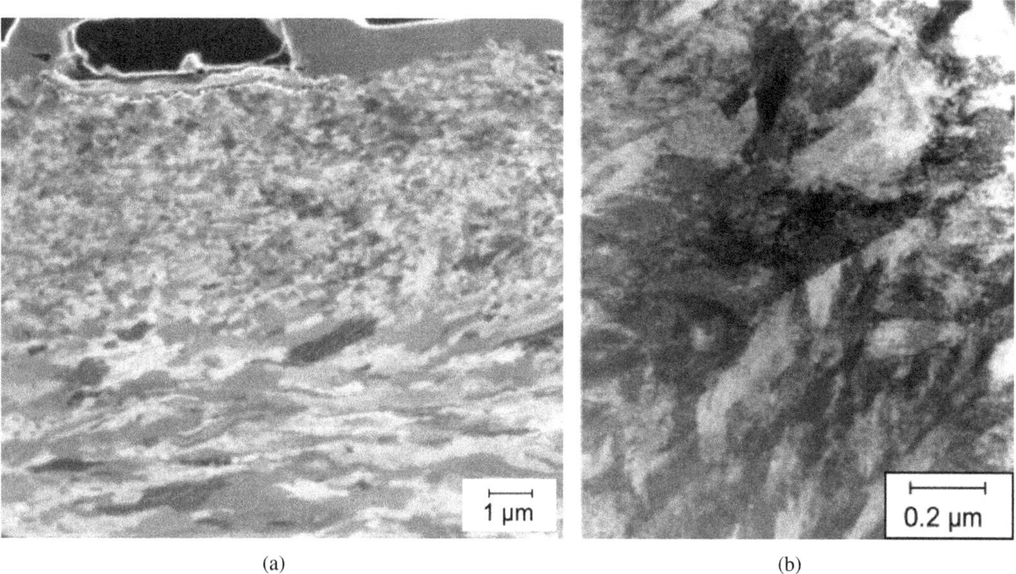

Figure 12.10 High-speed rail steel (Fe-0.7C-0.93Mn-0.23Si wt%) that has experienced service.(a) The region at the top with the fine structure is the white-etching layer observed in an optical microscope – the underlying structure is fine, less-deformed pearlite. (b) Martensite in the white-etching layer. Images reproduced from Takahashi et al. [64] with permission of Elsevier.

12.6 RAILWAY WHEELS

Each railway wheel weighs about 500 kg and can reach 900 kg for freight trains, Figure 12.11a. They are shaped to safely guide the train on the track while subjected to large tangential and normal forces during operation. The wheel is made by heating a cylindrical blank and compression forging it into the approximate shape before punching out the centre where the shaft eventually fits. While still hot, it is then roll-forged to create the precise shape of the flange before cooling the rim with water and then leaving it to slowly cool in air. This makes the rim harder than the central regions which therefore better resist crack propagation. The later cooling of the core also puts the rim into a state of compression. The wheel is then machined into its final dimensions and profile before service. The distribution of hardness and the fraction of proeutectoid-α is illustrated in Figure 12.11b. The plasticity and shear deformation associated with service changes the hardness profile. That for the flange-face is illustrated in Figure 12.11c but the data are representative also for parts of the running band [65].

Rolling-contact fatigue is the dominant damage mechanism [65]. The surface is during maintenance, machined to remove a few millimetres that contain signs of damage; this can be repeated until the minimum diameter permitted is reached. Proeutectoid-α is weaker than pearlite – hence, it accommodates plastic strain better than pearlite. Consequently, it's ductility rapidly becomes exhausted. It is argued that a greater V_V^α therefore increases susceptibility to rolling-contact fatigue, and ferrite elongated within the sheared surface can provide a pathway for the easy propagation of cracks [66].

If a wheel is prevented from rotating while the train is in motion, flat surfaces are created as it slides along the rail, causing complex damage, including the formation of martensite due to transient heating and cooling [67, 68]. The high-carbon martensite is inherently brittle. Modern braking systems on passenger trains have essentially eliminated the locking of wheel sets during breaking, thus preventing the temperature of the steel from exceeding Ac_1 [69] and therefore minimising the risk of obtaining martensite.

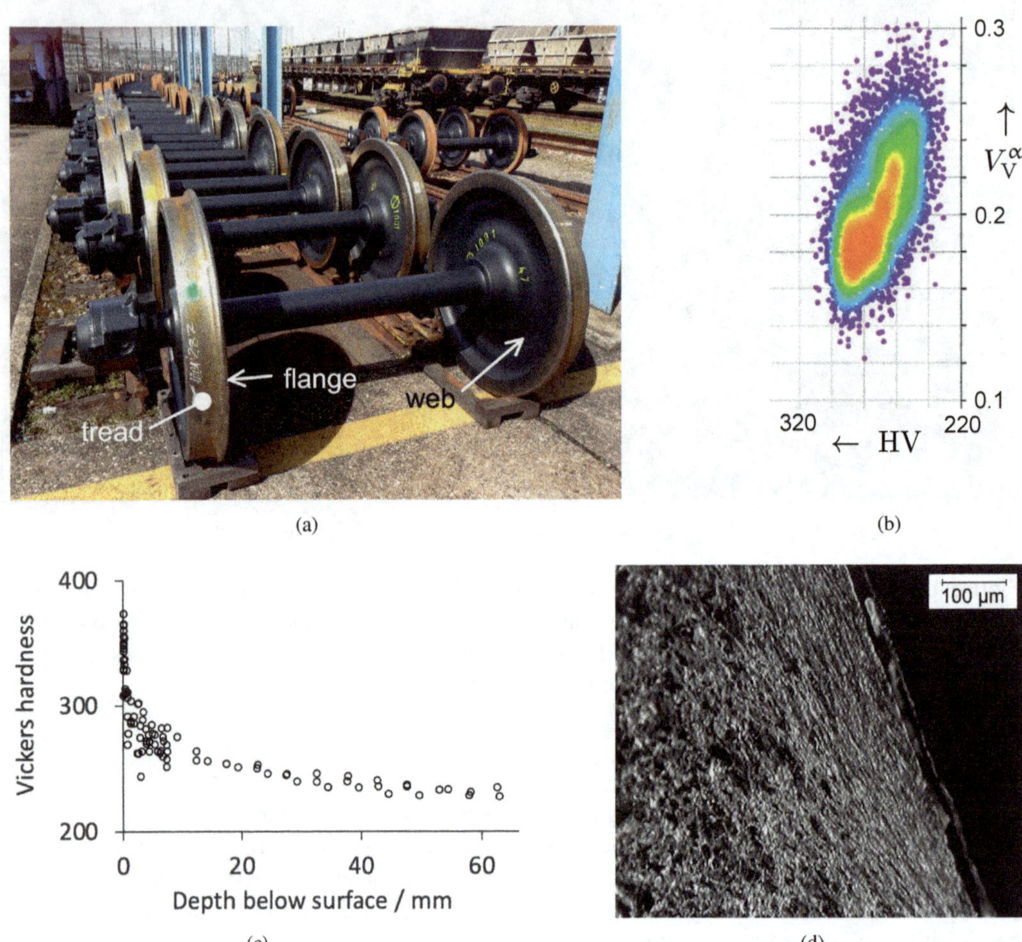

(a) (b) (c) (d)

Figure 12.11 (a) Railway wheel sets illustrating the scale and engineering design. Permission-free image from Wikimedia. (b) Vickers microhardness and proeutectoid-α distribution in the rim of a wheel (Fe-0.5C-0.76Mn-0.34Si-0.22Cr-0.15Ni wt%) that has not experienced service. The colours represent the density of points, red being the greatest. Adapted from Liu et al. [70] with permission of Elsevier. (c,d) Data from wheel that has served for 270 Mm since last refurbished. Mixture of hardness and microhardness data from the flange face as a function of depth below the surface, and micrograph of the shear deformation at that face. Adapted under CC BY license from [65].

The wheels typically are hypoeutectoid steels (Table 12.1) with a structure containing proeutectoid-α ($V_V^\alpha \approx 0.05 \rightarrow 0.1$) and pearlite ($S_I \approx 0.15\,\mu$m), although quantities of bainite have also been reported [71]. Given the size of the wheels, the details of the structure will vary with depth from the surface of the as-manufactured wheel, and service increases the hardness in areas that encounter contact with the rail and hence undergo plastic deformation [65].

Table 12.1
Some chemical compositions and some properties of steels used in the manufacture of railway wheels. The apparent fracture toughness K_Q refers to a measured value that does not quite satisfy the test-sample dimensional requirements, that ensure plane-strain conditions. The reduction of area and elongation columns in [72] were interchanged in the original publication.

C wt%	Mn wt%	Si wt%	Cr wt%	σ_y MPa	σ_{UTS} MPa	Elongation %	ΔK_{th} MPa $\sqrt{m}$	da/dN m cycle^{-1}	K_Q MPa $\sqrt{m}$	Ref.
0.49	0.75	0.34		610	911	14	7.4	$1.8 \times 10^{-9}\Delta K^{3.3}$	89 ± 6	[71]
0.65	0.63	0.26		659	1035	15	7.1	$5.6 \times 10^{-9}\Delta K^{3.0}$	58 ± 7	[71]
0.74	0.80	0.34		730	1140	15	7.2	$1.8 \times 10^{-9}\Delta K^{3.4}$	48 ± 4	[71]
0.63	0.84	0.88		687	1148	14	7.5	$8.6 \times 10^{-9}\Delta K^{2.9}$	49 ± 3	[71]
0.52	0.73	0.26	0.25	610	945	20				[72]
0.57	0.73	0.26	0.26	615	965	19				[72]
0.52	0.93	0.93	0.04	650	985	23				[72]
0.58	0.71	0.22	0.04	576	868	12				[73]

REFERENCES

1. N. Wood: A Practical Treatise on Rail-roads, and Interior Communication in General: 3rd ed., London, U.K.: Longman, Orme, Brown, Green, & Longmans, 1838.
2. Anonymous: 'George Stephenson (1781-1848)': In: D. Winterbone, and K. Moore, eds. *An Engineering Arhcive*. London, U.K.: Institution of Mechanical Engineers, 1997: 5.
3. L. Chen, M. Yang, Y. Xu, Z. Zhang, Z. Li, and F. Zhang: 'Effects of cooling on bending process of heavy rail steel after hot rolling', *Metallography, Microstructure, and Analysis*, 2016, **5**, 196–206.
4. W. C. Leslie: 'Iron and its dilute substitutional solid solutions', *Metallurgical Transactions A*, 1972, **3**, 5–23.
5. W. Solano-Alvarez, and H. K. D. H. Bhadeshia: 'Steels for rails', *Progress in Materials Science*, 2024, **146**, 101313 (pp. 1–64).
6. J. Chance, and N. Ridley: 'Chromium partitioning during isothermal transformation of a eutectoid steel', *Metallurgical Transactions A*, 1981, **12A**, 1205–1213.
7. B. Zheng, Z. Huang, J. Xing, and Y. Xiao: 'Two-body abrasion resistance of cementite containing different chromium concentrations', *Journal of Materials Research*, 2016, **31**, 655–662.
8. B. Zheng, Z. Huang, J. Xing, and X. Fan: 'Three-body abrasive wear behavior of cementite with different chromium concentrations', *Tribology Letters*, 2016, **61**, 1–13.
9. J. Dearden: 'The wear of steel rails: A review of the factors involved', *Proceedings of the Institution of Civil Engineers*, 1954, **3**, 456–481.
10. A. Dobuzhskaya, G. Galitsyn, G. Yunin, E. Polevoi, and A. Yunusov: 'Effect of chemical composition, microstructure and mechanical properties on the wear resistance of rail steel', *Steel in Translation*, 2020, **50**, 906–910.
11. H. K. D. H. Bhadeshia: 'A thermodynamic analysis of isothermal transformation diagrams', *Metal Science*, 1982, **16**, 159–165.
12. B. L. Bramfitt: 'Accelerated cooling of rail', *Iron & Steelmaker*, 1991, **18**, 33–41.
13. K. F. Rodrigues, G. M. M. Mourao, and G. L. Faria: 'Kinetics of isothermal phase transformations in premium and standard rail steels', *Steel Research International*, 2021, **92**, 2000306.

14. A. Yuryev, N. Kozyrev, R. Shevchenko, A. Mikhno, and O. Gutak: 'Phase transformations occurring during the formation of a welded joint from rail steel', In: *E3S Web of Conferences*, vol. 330. France: EDP Sciences, 2021:02007.
15. P. Chen, L. Chen, J. Xu, S. Huang, and Z. Xia: 'Formation mechanism of pearlite during thermal cycling in U75V steel rail repaired by laser directed energy deposition', *Journal of Laser Applications*, 2021, **33**, 101–106.
16. V. M. Ferreira, M. G. Mecozzi, R. H. Petrov, and J. Sietsma: 'Microstructure development of pearlitic railway steels subjected to fast heating', *Materials & Design*, 2022, **221**, 110989.
17. H. K. D. H. Bhadeshia, and A. R. Chintha: 'Critical assessment: the strength of undeformed pearlite', *Materials Science and Technology*, 2022, **38**, 1291–1299.
18. P. Pointner, A. Hierg, and J. Jaiswal: 'Definitive guidelines on the use of different rail grades': Tech. Rep. TIP5-CT-2006-031415, D4.1.5GL, Chalmers University, Sweden, 2006.
19. J. Jaiswal: 'Private communication to H. K. D. H. Bhadeshia', 2024: Elongation in rails.
20. K. Burkart, H. Bomas, R. Schroeder, and H.-W. Zoch: 'Rolling contact and compression-torsion fatigue of 52100 steel with special regard to carbide distribution', In: *Advances in rolling contact fatigue strength testing and related substitute technologies*. West Conshohocken, PA, USA: ASTM International, 2012:218–236.
21. D. H. Stone: 'An introduction to the fracture mechanics of railroad rails': In: *Railroad Track Mechanics and Technology*. Oxford, U.K.: Pergamon Press, 1978:353–368.
22. T. Bandula-Heva, and M. Dhanasekar: 'Failure of discontinuous railhead edges due to plastic strain accumulation', *Engineering Failure Analysis*, 2014, **44**, 110–124.
23. P. Boonsukachote, S. Kingklang, and V. Uthaisangsuk: 'Modelling of mechanical properties of pearlitic rail steel', *Key Engineering Materials*, 2019, **798**, 3–8.
24. A. J. Perez-Unzueta, and J. H. Beynon: 'Microstructure and wear resistance of pearlitic rail steels', *Wear*, 1993, **162-163**, 173–182.
25. P. Liu, Y. Quan, and G. Ding: 'Dynamic mechanical characteristics and constitutive modeling of rail steel over a wide range of temperatures and strain rates', *Advances in Materials Science and Engineering*, 2019, **2019**, 6862391.
26. K. Iwafuchi, Y. Satoh, Y. Toi, and S. Hirose: 'Fatigue property analysis of rail steel based on damage mechanics', *Quarterly Report of RTRI*, 2004, **45**, 203–209.
27. A. Kapoor, J. H. Beynon, D. Fletcher, and M. Loo-Morrey: 'Computer simulation of strain accumulation and hardening for pearlitic rail steel undergoing repeated contact', *The Journal of Strain Analysis for Engineering Design*, 2004, **39**, 383–396.
28. B. Panda, R. Balasubramaniam, and A. Moon: 'Microstructure and mechanical properties of novel rail steels', *Materials Science and Technology*, 2009, **25**, 1375–1382.
29. Y. Lan, G. Zhao, Y. Xu, C. Ye, and S. Zhang: 'Effects of quenching temperature and cooling rate on the microstructure and mechanical properties of U75V rail steel', *Metallography, Microstructure, and Analysis*, 2019, **8**, 249–255.
30. J. Scutti, R. Pelloux, and R. Fuquen-Moleno: 'Fatigue behavior of a rail steel', *Fatigue & Fracture of Engineering Materials & Structures*, 1984, **7**, 121–135.
31. F. Yu, P.-Y. B. Jar, and M. Hendry: 'Effect of temperature on deformation and fracture behaviour of high strength rail steel', *Engineering Fracture Mechanics*, 2015, **146**, 41–55.
32. H. Aglan, and Y. Gan: 'Fatigue crack growth analysis of a premium rail steel', *Journal of Materials Science*, 2001, **36**, 389–397.
33. Y. Ochi, and A. McEvily: 'An evaluation of the fatigue crack growth and fracture toughness characteristics of rail steels', *Engineering Fracture Mechanics*, 1988, **29**, 159–172.
34. L. P. Moreira, T. G. Viana, L. B. Godefroid, G. L. de Faria, and L. C. Cândido: 'Effect of microstructure on fracture toughness and fatigue crack growth resistance of pearlitic steels for railroad application', In: *Proceedings of 23rd ABCM International Congress of Mechanical Engineering*. Rio de Janeiro, Brazil: ABCM, 2015:1–8.

35. Z. Sajuri, N. A. Alang, N. A. A. Razak, and M. Aziman: 'Fracture toughness and fatigue crack growth behavior of rail track material', *Key Engineering Materials*, 2011, **462**, 1109–1114.
36. J. W. Seo, S. J. Kwon, H. K. Jun, and D. H. Lee: 'Evaluation of fatigue and fracture characteristics of high-speed rail material', *Journal of the Korean Society for Precision Engineering*, 2017, **34**, 861–866.
37. H. A. Aglan, and M. Fateh: 'Fatigue crack growth and fracture behavior of bainitic rail steels.', *Journal of Mechanics of Materials and Structures*, 2007, **2**, 335–346.
38. S. Dhua, A. Ray, S. Sen, M. Prasad, K. Mishra, and S. Jha: 'Influence of nonmetallic inclusion characteristics on the mechanical properties of rail steel', *Journal of materials engineering and performance*, 2000, **9**, 700–709.
39. V. A. Shilov, D. L. Shvarts, and E. O. Skosar: 'Aspects of the rolling of long rails on a universal rail-beam mill', *Metallurgist*, 2016, **60**, 260–266.
40. D. Nesterov, N. Razin'kova, L. Chernyakova, and B. Kisil: 'Structure and mechanical properties of rails type r65 bulk-hardened in oil', *Metal Science and Heat Treatment*, 1991, **33**, 331–336.
41. A. Khourshid, Y. Gan, and H. Aglan: 'Microstructure origin of strength and toughness of a premium rail steel', *Journal of materials engineering and performance*, 2001, **10**, 331–336.
42. M. Jabłońska, F. Lewandowski, B. Chmiela, and Z. Gronostajski: 'Advanced heat treatment of pearlitic rail steel', *Materials*, 2023, **16**, 6430.
43. V. Sinel'nikov, and G. Filippov: 'Technological aspects of improving the quality and service properties of railroad rails', *Metallurgist*, 2001, **45**, 403–407.
44. K. Hao, W. Di, and X.-m. Zhao: 'Surface temperature change of U75V 60 kg/m heavy rail during heat treatment', *Journal of Iron and Steel Research, International*, 2013, **20**, 33–67.
45. H. Wu, S. Huang, Z. Li, P. Chen, H. Tan, and Z. Xia: 'Optimization of microstructure and properties in U75V steel rail cladding layers manufactured by laser melting deposition and laser shock peening', *Optics & Laser Technology*, 2023, **163**, 109436.
46. D. Nesterov, V. Sapozhkov, N. Levchenko, and V. Dubrov: 'Heat treatment of rail steel using induction heating', *Metal Science and Heat Treatment*, 1990, **32**, 589–594.
47. K. Saeki, and K. Iwano: 'Progress and prospects of rail for railroads': Tech. Rep. 105, Nippon Steel & Sumitomo Metal Technical Report, Japan, 2013.
48. R. Heyder, and G. Girsch: 'Testing of HSH® rails in high-speed tracks to minimise rail damage', *Wear*, 2005, **258**, 1014–1021.
49. J. F. Archard: 'Friction between metal surfaces', *Wear*, 1986, **113**, 3–16.
50. J. K. Yates: 'Innovation in rail steel', *Science in Parliament*, 1996, **53**, 2–3.
51. H. K. D. H. Bhadeshia: 'Steels for rails': In: *Encyclopedia of Materials Science*. Pergamon Press, Oxford, Elsevier Science, 2007:1–7.
52. M. Ueda, K. Uchino, and A. Kobayashi: 'Effects of carbon content on wear property in pearlitic steels', *Wear*, 2002, **253**, 107–113.
53. J. Fei, G. Zhou, J. Zhou, X. Zhou, Z. Li, D. Zuo, and R. Wu: 'Research on the effect of pearlite lamellar spacing on rolling contact wear behavior of U75V rail steel', *Metals*, 2023, **13**, 237.
54. H. Hertz: 'Über die berübrung fester elastiscber Körper (on the contact of elastic solids)', *J. reine und angewandte Mathematik*, 1882, **92**, 156–171.
55. K. L. Johnson: Contact Mechanics: Cambridge, U. K.: Cambridge University Press, 1985.
56. W. E. Littmann: 'The mechanism of contact fatigue (sp-237)', In: *Interdisciplinary approach to the lubrication of concentrated contacts*. New York, USA: NASA, 1970:309–377.
57. G. Trummer, C. Marte, S. Scheriau, P. Dietmaier, C. Sommitsch, and K. Six: 'Modeling wear and rolling contact fatigue: Parametric study and experimental results', *Wear*, 2016, **366**, 71–77.
58. X. Su, and P. Clayton: 'Ratchetting strain experiments with a pearlitic steel under rolling/sliding contact', *Wear*, 1997, **205**, 137–143.
59. C. Jessop, and J. Ahlström: 'Friction between pearlitic steel surfaces', *Wear*, 2019, **432**, 102910.

60. S. H. Lee, J. Y. Kang, H. N. Han, K. H. Oh, H. C. Lee, D. W. Suh, and S. J. Kim: 'Variant selection in mechanically-induced martensitic transformation of metastable austenitic steel', *ISIJ International*, 2005, **45**, 1217–1219.
61. S. A. Parsons, and D. V. Edmonds: 'Microstructure and mechanical properties of medium carbon ferrite-pearlite steel microalloyed with vanadium', *Materials Science and Technology*, 1987, **3**, 894–904.
62. T. S. Eyre, and A. Baxter: 'The formation of white layers at rubbing surfaces', *Tribology*, 1972, **5**, 256–261.
63. S. B. Newcomb, and W. M. Stobbs: 'A transmission electron microscope study of the white etching layer on a rail head', *Materials Science & Engineering A*, 1984, **66**, 195–204.
64. J. Takahashi, K. Kawakami, and M. Ueda: 'Atom probe tomography analysis of the white etching layer in a rail track surface', *Acta Materialia*, 2010, **58**, 3602–3612.
65. P. Molyneux-Berry, C. Davis, and A. Bevan: 'The influence of wheel/rail contact conditions on the microstructure and hardness of railway wheels', *The Scientific World Journal*, 2014, **2014**, 1–17.
66. J. E. Garnham, and C. L. Davis: 'The role of deformed rail microstructure on rolling contact fatigue initiation', *Wear*, 2008, **265**, 1363 – 1372.
67. J. Ahlström, and B. Karlsson: 'Microstructural evaluation and interpretation of the mechanically and thermally affected zone under railway wheel flats', *Wear*, 1999, **232**, 1–14.
68. D. Nikas, Y. Zhang, and J. Ahlström: 'Effect of annealing on microstructure in railway wheel steel', In: *IOP Conference Series: Materials Science and Engineering*, vol. 1249. IOP Publishing, 2022:012059.
69. K. Cvetkovski, J. Ahlström, and B. Karlsson: 'Monotonic and cyclic deformation of a high silicon pearlitic wheel steel', *Wear*, 2011, **271**, 382–387.
70. Z. Liu, L. Yang, G. Zhang, L. Zhao, Q. Shao, D. Huang, C. Zhu, Y. Wang, X. Shen, Z. Yang, et al.: 'Correlation of microstructure and hardness distribution of high-speed train wheels under original and service statuses', *Engineering Failure Analysis*, 2024, , 107994.
71. M. Faccoli, A. Ghidini, and A. Mazzù: 'Changes in the microstructure and mechanical properties of railway wheel steels as a result of the thermal load caused by shoe braking', *Metallurgical and Materials Transactions A*, 2019, **50**, 1701–1714.
72. D. Zeng, L. Lu, Y. Gong, N. Zhang, and Y. Gong: 'Optimization of strength and toughness of railway wheel steel by alloy design', *Materials and Design*, 2016, **92**, 998–1006.
73. C. Suetrong, and V. Uthaisangsuk: 'Investigations of fatigue crack propagation in ER8 railway wheel steel with varying microstructures', *Materials Science and Engineering A*, 2022, **840**, 142980.

13 Pearlitic wire

Strong wires based on cold-drawn eutectoid steels are used routinely to manufacture ropes for use in composites such as automotive tyres, musical instruments, etc., some examples of which are illustrated in Figure 13.1. Most of the applications are safety-critical, for example the winding ropes on heavy-lifting cranes, suspension bridges, ski-lifts and elevators, guide ropes and transportation systems for deep mine-shafts. Safety barriers on highways often use wire ropes for energy-absorption during glancing impacts. Commercial body-armour incorporating 3 mm diameter steel wire provides ballistic and stab protection. The annual production of steel wires is probably in excess of 30 million tonnes with international trade amounting to about £2 billion.

Figure 13.1 A few applications of cold-worked pearlitic steel wire. (a) Tyre cord, with strength in the range 2750-4000 MPa. (b) Bicycle spokes. These can be galvanised or polyethylene terephthalate plastic coated. (c) Thermoplastic reinforced with 2800 MPa continuous steel wire fabric, produced using compression or injection moulding. Used in energy-absorbing applications in the automotive industries. (d) The Akashi-Kaikyo Bridge in Japan, the longest single-span suspension bridge supported by steel cables some 2 GPa in strength. Images a-c courtesy of Bekaert via Pascal Antoine; image-d courtesy of Professor Nobutaka Yurioka.

There are clever engineering aspects to the assembly of wires and ropes, Figure 13.2. The individual wires in strands, and strands in ropes, are preformed into helices before assembly to facilitate

DOI: 10.1201/9781032631981-13

an even distribution of the load in the final assembly and to reduce the chance of kinking [1].

The *lay* of the rope is defined as the distance over which a strand makes a complete turn around the rope and there are different ways in which the strands can be combined to form the rope. In the popular regular lay, the wires are twisted in one direction and the strands in the opposite sense. In the final assembly, the wires appear to run parallel to the rope, Figure 13.2. Ropes with regular lay are relatively flexible, easier to splice and resist crushing.

With Lang's lay, the wires and strands have an identical sense of twist, i.e., the wires are laid in the same direction as the lay of the strands; they present a greater wear-surface when in use, for example, in hoists. The Lang lay is more susceptible to kinking; such cables are appropriate for lifting applications whereas the regular lay performs better in scenarios where bending features. There may be a central core made of synthetic material or steel, to prevent abrasion between the wires of adjacent strands. For a given diameter of rope, an increase in the number of constituent wires increases the overall fatigue resistance of the rope [2]. Fatigue takes many forms in stranded rope: due to tension, torsion, bending and fretting. It can be exacerbated by concurrent corrosion.

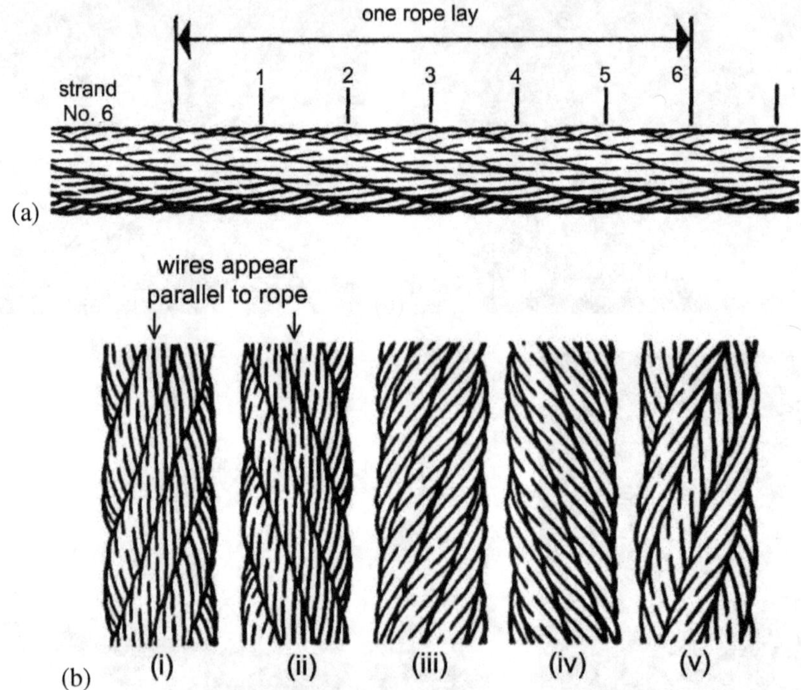

Figure 13.2 (a) A six-strand rope and the definition of a *lay*. (b) Illustration of a variety of lays. (i) Right regular lay; (ii) left regular lay; (iii) right Lang's lay; (iv) left Lang's lay; (v) cross lay. The right and left versions of the rope are designed to compliment the drums on which the ropes may be wound, given that the drums also have a handedness as they rotate about their axes. Images reproduced from Pelmear [1] by permission of Oxford University Press.

Figure 13.3 is an elegant example of the use of steel ropes and a giant steel-ball to form a tuned damper that stops the tallest building in Taiwan from swaying by elastic deformation under the influence of wind or tremors due to tectonic plate movements. The building has 101 stories and is 509.2 m tall so the accumulated elastic strain can lead to large deflections at the very top, which would be unacceptable in terms of practicality and can cause nausea. The steel ropes used to suspend the heavy ball from the 91st to the 87th floor are made of cold-drawn pearlite.

Figure 13.3 Regular right-lay steel rope as a part of a system of ropes that suspend a gigantic steel ball weighing 660 tonnes, the world's largest tuned mass damper that keeps the skyscraper Taipei 101, from swaying in the wind or during earthquakes. (a) Steel ball and rope. (b) Enlarged image of rope to show that the wire strands are aligned with the major axis of the rope, i.e., the vertical direction in the diagram. Images courtesy of Jessie Lin of Taipei 101, via the good offices of Professor J. R. Yang

13.1 HOT-ROLLED WIRE ROD

Wire production begins with hot-rolled rod, typically 16 mm in diameter, followed by dry-drawing down to a diameter of ≈ 5 mm. This wire is then transformed into fine pearlite by one of three processes that restore the undeformed state. In *patenting*, the red-hot steel in its austenitic condition enters a bath of lead at temperatures between 450-500 °C. The production process is continuous with wire speeds of the order of $100\,\mathrm{m\,s^{-1}}$, so the lead bath must have sufficient thermal mass to ensure that the temperature is maintained at the desired value. The bath may be several metres in length – Figure 13.4 illustrates the scale of the process, which ensures efficient heat transfer, resulting in pearlite that has a fine interlamellar spacing ($S_\mathrm{I} \approx 50$nm). However, lead is no longer considered to be safe so fused-salt baths are substituted, such as mixtures of sodium and potassium compounds. The most modern processes involve a continuous production line from the hot-rolling process directly to the salt bath treatment.

In the Stelmor process, the hot-rolled rod at a temperature of about 1050 °C emerges from the last stand on the mill at speeds of 12-42 $\mathrm{m\,s^{-1}}$. As it travels a distance of about 30 m, water sprays bring its temperature down to ≈ 800 °C, which helps avoid the coarsening of the austenite grain structure and minimises surface oxidation [3]. It is then laid in a continuous spiral on a conveyor belt, Figure 13.5, where it is subjected to forced air cooling that suppresses the precipitation of proeutectoid cementite whilst at the same time transforming the austenite into fine pearlite. The cooling rate is usually $> 15\,^\circ\mathrm{C\,s^{-1}}$ but this may vary depending on the position on the conveyor belt, which may not matter if the desired microstructure is obtained and if the rod is to be processed further, e.g., wire

Figure 13.4 (a) An industrial patenting production-line, illustrating the scale of the process. (b) The dry-drawn wires entering the production-line furnace for austenitisation. (c) The patented wires emerging from the process. Images courtesy of Bekaert via Pascal Antoine.

drawing. The Stelmor conveyor usually is limited to a maximum of 22 mm diameter rod because larger cross-sections become stiff after cooling, making it difficult to form compact coils.

The use of lead is avoided in the Stelmor process, but strength is sacrificed by about 100-200 MPa when compared with patented steel. This is because S_I increases, from 69 to 82 nm [4]. The effect of this difference is exaggerated during cold-drawing because finer cementite dissolves more readily during the deformation [5]. To overcome this, heat transfer can be improved by replacing forced-air cooling by a mist of fine droplets of water that impinge on the steel [6]. This not only increases the strength but also the resistance to delamination by longitudinal fracture when the wire is twisted. The mechanism of delamination is described in Section 13.6, but Figure 13.6 gives an indication of the role of the process used in generating pearlite in the wire rod, on the tendency for delamination. The enhanced strength with both the lead-patented and the mist-cooled samples, is because the processes induce fine pearlite. Galvanised lead-patented wire has been shown to have better torsional ductility than similarly treated Stelmor wire [4].

The ductility of patented steel suffers if harder phases other than pearlite, such as bainite, are present in the microstructure, or if the pearlite is coarse, in which case shear cracks in the thick

Figure 13.5 Stelmor conveyor line. The coils are positioned on the belt to allow better control of the cooling conditions. Image courtesy of Steeltec AG.

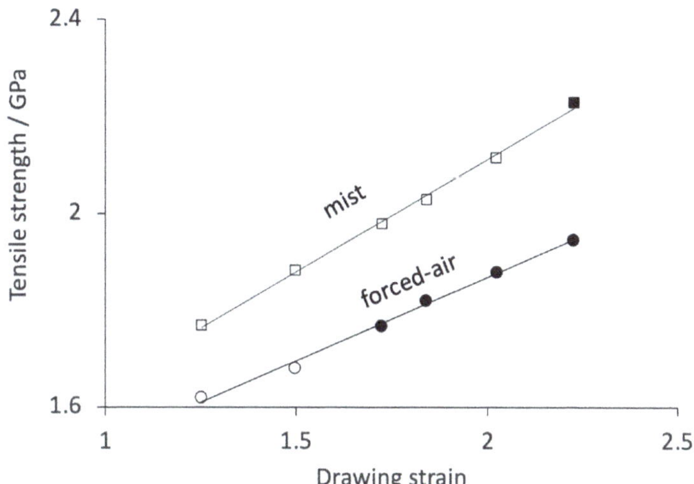

Figure 13.6 Mechanical data on wire rods that were transformed into pearlite using the Stelmor process, followed by drawing to the strains indicated. The cooling on the conveyer was implemented using either water mist or forced air. The filled points represent samples exhibiting delamination during torsion testing. Selected data from [6].

cementite lamellae compromise ductility even though the strength is reduced [7]. A fine and uniform pearlitic microstructure ensures drawability, i.e., whether the material has the ductility to cope with the imposed drawing strains. It is assumed sometimes that increased ductility in patented wire can enhance the torsional ductility in drawn wire, but the evidence is lacking.

13.2 WIRE DRAWING

The hot-rolled rod has to be pickled in order to remove scale, prior to passage through the sequence of cold-drawing dies. The reduction of the diameter in each die is about 20-25%, but the details depend on the dimensions and properties required, Table 13.1.

Table 13.1
Examples of steps in the cold-drawing of pearlitic steel. Selected data from [8].

	Wire diameter / mm								Final properties	
	step 0	step 1	step 2	step 3	step 4	step 5	step 6	step 7	σ_y / GPa	RA / %
1	12.11	10.80	9.81	8.94	8.22	7.56	6.98		1.57	44
2	11.03	9.90	8.95	8.21	7.49	6.80	6.26	5.40	1.49	23

Fine wire ($\phi = 0.2$ mm) can achieve strength levels 3-4 GPa whereas galvanised wire (5 mm diameter) for modern suspension-bridges typically has a strength of 1.4-2.0 GPa depending on the initial heat treatment to achieve the pearlitic state and the subsequent extent of the cold-drawing operation. In research, small samples have been created with strengths in excess of 6 GPa. Although there are subtle variations in the chemical compositions of wires (Table 13.2), the mechanical properties of commercial wires are determined largely by their processing.

The interlamellar spacing will naturally decrease during wire drawing in a manner compatible with the external dimensions as the wire proceeds through the drawing die. Many pearlite colonies will not have their lamellae aligned to the drawing direction, making the cementite undergo shape-contortions, buckling and fracture that induces some form of alignment along the principal deformation direction, Figure 13.7. But given the three-dimensional structure of a pearlite colony, even those lamellae which on two-dimensional sections appear to be aligned along the deformation axis, will have their connectivity reduced by damage accumulation. The deformation therefore is by no means homogeneous within the sample. Experiments suggest that the spacing decreases in proportion to the patented-wire diameter z_D [21], so following Embury [12],

$$\frac{1}{\overline{S}_I} = \frac{1}{\overline{S}_{I_o}} \frac{z_{D_o}}{z_D} \tag{13.1a}$$

where z_{D_o} and S_{I_o} represent initial values at zero strain, with the important assumption that the pearlite lamellae are aligned to the drawing direction so S_I is likely to be overestimated. The plastic strain along the drawing direction is given by

$$\varepsilon_p = \ln\left\{\frac{A_o}{A}\right\} \equiv 2\ln\left\{\frac{z_{D_o}}{z_D}\right\} \tag{13.1b}$$

where the initial cross-sectional area of the rod is A_o and its instantaneous area is A. By comparing these two equations, it follows that

$$\frac{1}{\overline{S}_I} = \frac{1}{\overline{S}_{I_o}} \exp\left\{\frac{\varepsilon_p}{2}\right\}. \tag{13.1c}$$

This can now be substituted into an equation of the form of the Hall-Petch relationship (Equation 4.17) to obtain the flow stress as a function of the drawing strain:

$$\sigma = \sigma^f + \frac{k_\varepsilon}{\sqrt{\overline{S}_I}} = \sigma^f + \frac{k_\varepsilon}{\sqrt{\overline{S}_{I_o}}} \exp\left\{\frac{\varepsilon_p}{4}\right\}. \tag{13.1d}$$

Table 13.2

Chemical compositions, wt%, of steels used in the production of wire. Blank entries represent data missing in the original publications.

Alloy	Reference	C	Si	Mn	P	S	Other
1	[9]	1.06	0.20	0.3	0.003	0.003	0.21Cr
2	[10]	1.01	0.22	0.33	0.008	0.004	0.23Cr
3	[9]	0.96	0.19	0.31	0.002	0.001	0.20Cr
4	[11]	0.98	0.21	0.41			0.20Cr
5	[11]	0.97	0.20	0.40			0.79Ni
6	[11]	0.96	0.19	0.39			
7	[10]	0.94	1.24	0.32	0.005	0.004	0.23Cr
8	[12]	0.93	0.20	0.37	0.02	0.03	
9	[13]	0.93	0.19	0.32			0.22Cr
10	[10]	0.93	0.22	0.33	0.006	0.004	0.23Cr
11	[14]	0.92	1.3	0.5			0.3 Cr
12	[15]	0.92	0.22	0.48	0.014	0.008	
13	[15]	0.91	0.20	0.31	0.006	o.007	0.20Cr
14	[16]	0.90	1.27	0.49	0.008	0.002	0.30Cr
15	[17]	0.82	0.22	0.49	0.004	0.004	
16	[18]	0.82	0.23	0.79			0.03Al
17	[18]	0.82	0.98	0.78			0.04Al
18	[19]	0.82	0.19	0.75	0.017	0.015	
19	[20]	0.82	0.23	0.70			
20	[20]	0.82	0.17	0.70			
21	[5]	0.81	0.22	0.41	0.004	0.004	
22	[20]	0.70	0.26	0.45			

It is assumed that the controlling barriers to plastic flow are the α/θ interfaces and resistance from dislocations generated during deformation. Both of these parameters will be a function of the instantaneous interlamellar spacing S_{I}. An empirical multiplier was used to account for the inclination of the slip plane relative to the interface but is avoided here. After all, a similar parameter is unnecessary in the conventional analysis of $\sigma = f\{S_{\mathrm{I}}\}$; furthermore, it would require five independent slip-systems to operate in the ferrite in order to maintain structural integrity.

Figure 13.8a shows experimental data plotted in a manner consistent with Equation 13.1d; the slope of the line is 1779 MPa.[1] The constant k_ε should equal k_{HP} at zero strain; Equation 13.1d then reduces to the Hall-Petch relationship. However, k_ε is based on the change in interlamellar spacing during plastic deformation so it includes, for example, the changes in strength due to dislocation generation. Therefore, in general, $k_\varepsilon \neq k_{\mathrm{HP}}$. This is clear from the fact that the slope of the plot $(= k_\varepsilon/\sqrt{S_{\mathrm{I}_0}})$ in Figure 13.8a is 1779 MPa, but should be just 660 MPa if only S_{I} controls strength.

The Hall-Petch Equation 4.17 gives reasonable estimates of the undeformed wire-rod using S_{I_0} so the problem lies in applying this equation to plastically deformed pearlite. The dislocation density

[1]Data from swaged samples [22] are omitted from Figure 13.8 because the deformations involved during swaging are complex and cyclic. Measurements of S_{I} reported in [12, 20, 21] are omitted because they do not account for stereological effects.

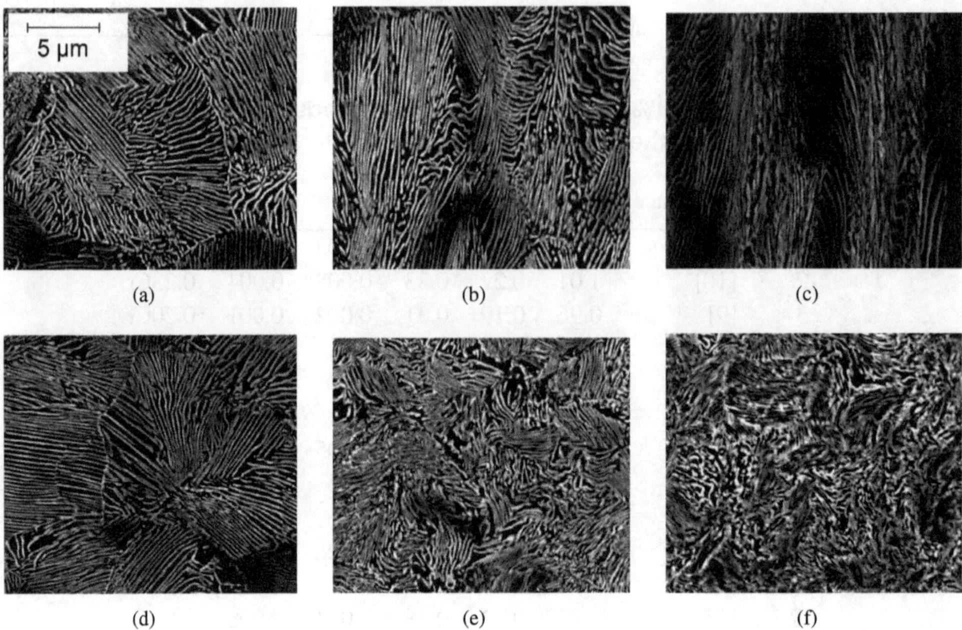

Figure 13.7 Microstructure of Fe-0.79C-0.21Si-0.68Mn-0.22Cr wt% pearlitic steel wire as a function of the drawing strain. Longitudinal sections (drawing direction vertical): (a) Hot-rolled 11 mm diameter rod following processing in the Stelmor line, $\varepsilon_p = 0$; (b) $\varepsilon_p = 0.774$; (c) $\varepsilon_p = 1.566$. Cross-sections: (d),(e),(f) with the same ε_p as (a),(b),(c), respectively. Selected micrographs adapted from Toribio et al. [8], reproduced under the CC-BY license http://creativecommons.org/licenses/by/4.0/

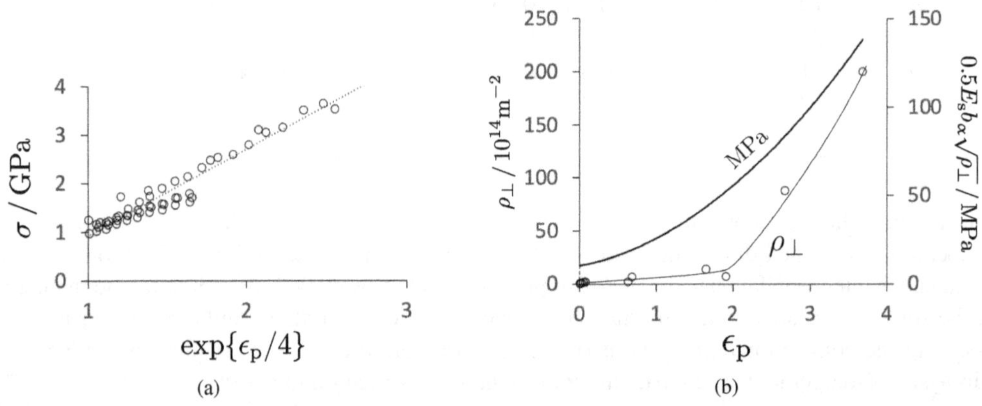

Figure 13.8 (a) The flow stress as a function of the drawing strain for pearlite generated by isothermal transformation. Selected data from [12, 20, 21]. The data cover a range of interlamellar spacings from apparent $S_{\mathrm{I_o}}$ =65-81 nm. (b) The dislocation density and its contribution to strengthening as a function of the cold-drawing strain. Data from [19, 23–25]. Comprehensive $\rho_\perp$ data in Table 4.1.

of the ferrite and perhaps that of the cementite undoubtedly increases with plastic strain; this is not accounted for in Embury and Fisher model [12]. However, the magnitude of dislocation strengthening is rather small, as illustrated in Figure 13.8b; other explanations could include the dissolution of cementite at large strains and the alignment of the microstructure along the drawing direction.

A more sophisticated model has been developed [26], but a simpler version using methods common in crystallography, is presented here. A deformation matrix **S** that represents axisymmetric tension (i.e., die-drawing) can be multiplied by a direction vector **u** to examine how it changes into a new vector **v** as follows [27, 28]:

$$(\mathrm{X}\ \mathbf{S}\ \mathrm{X})[\mathrm{X};\mathbf{u}] = [\mathrm{X};\mathbf{v}]$$

$$\text{in expanded form} \begin{pmatrix} 1/\sqrt{S_{33}} & 0 & 0 \\ 0 & 1/\sqrt{S_{33}} & 0 \\ 0 & 0 & S_{33} \end{pmatrix} \begin{bmatrix} u_1 \\ u_2 \\ u_3 \end{bmatrix} = \begin{bmatrix} v_1 \\ v_2 \\ v_3 \end{bmatrix} \qquad (13.2)$$

with the orthonormal basis X defined by the unit vectors **x**, **y** and **z**, where the latter is parallel to the drawing direction, so $S_{33} \geq 1$, where S_{ii} are principal distortions, i.e., the ratios of the final to initial length of a unit vector parallel to the coordinate concerned. In terms of the true drawing-strain, $S_{33} = \exp\{\varepsilon\}$. To consider the change in lamellar spacing and orientation, it is necessary to consider plane normals defined in the reciprocal basis X*. The initial unit normal is designated **h** which after deformation becomes **k**:

$$(\mathbf{h};\mathrm{X}^*)(\mathrm{X}^*\ \mathbf{S}^{-1}\ \mathrm{X}^*) = (\mathbf{k};\mathrm{X}^*) \qquad (13.3)$$

$$\text{in expanded form } (h_1 \quad h_2 \quad h_3) \begin{pmatrix} \sqrt{S_{33}} & 0 & 0 \\ 0 & \sqrt{S_{33}} & 0 \\ 0 & 0 & 1/S_{33} \end{pmatrix} = (k_1 \quad k_2 \quad k_3). \qquad (13.4)$$

Since the interlamellar spacing will be related to the reciprocal of the magnitude of **k**, the fractional change in S_I is given by the ratio $|\mathbf{h}|/|\mathbf{k}|$. The calculations in Table 13.3 show that lamellae parallel to the drawing axis do not change in orientation ($\mathbf{h}\angle\mathbf{k} = 0$) but S_I is reduced exactly by the fraction that the wire diameter is reduced.

Table 13.3

Some calculations using Equation 13.4, showing the change in the reciprocal vector normal to the lamella plane, the fractional change in interlamellar spacing ($|\mathbf{h}|/|\mathbf{k}|$) and the new orientation of the lamellae as defined by k. Data to a maximum of three significant figures for brevity. The angle ∠ is in degrees.

S_{33}	**h**	**k**	$\lvert\mathbf{h}\rvert/\lvert\mathbf{k}\rvert$	$\mathbf{h}\angle\mathbf{k}$
2	$[1\,0\,0]$	$[\sqrt{2}\,0\,0]$	$\frac{1}{\sqrt{2}}$	0
2	$[\frac{1}{\sqrt{2}}\,\frac{1}{\sqrt{2}}\,0]$	$[1\,1\,0]$	$\frac{1}{\sqrt{2}}$	0
2	$[\frac{1}{\sqrt{2}}\,0\,\frac{1}{\sqrt{2}}]$	$[1\,0\,\frac{1}{2\sqrt{2}}]$	0.942	25.5
2	$[\frac{1}{\sqrt{3}}\,\frac{1}{\sqrt{3}}\,\frac{1}{\sqrt{3}}]$	$[0.817\,0.817\,0.289]$	0.840	21
2	$[0\,0\,1]$	$[0\,0\,\frac{1}{2}]$	2	0
10000	$[\frac{1}{\sqrt{2}}\,0\,\frac{1}{\sqrt{2}}]$	$[70.71\,0\,\approx 0]$	0.014	45.0

Lamellae that lie normal to the drawing axis will be associated with an increase in S_I in proportion to the longitudinal drawing strain. Other orientations lead to different fractional changes in the interlamellar spacing but the lamellae all tend to rotate so that their normals become more parallel to the wire cross sections. An extreme example is when $S_{33} = 10,000$ where **k** becomes parallel to

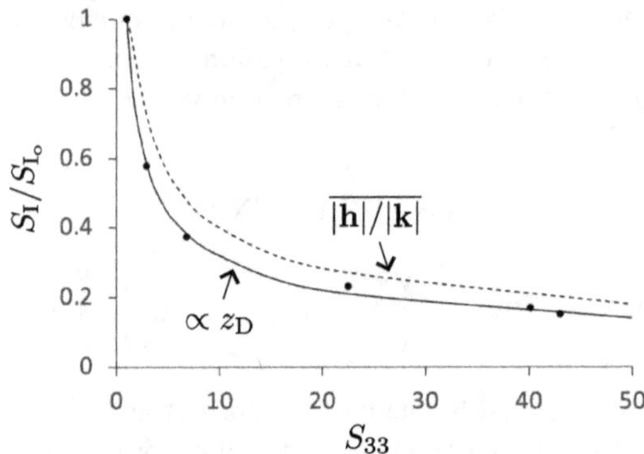

Figure 13.9 Normalised interlamellar spacing (S_I/S_{I_o}) versus $S_{33} = \exp\{\varepsilon\}$ which is the principal distortion along the drawing axis. The spacing taken to be proportional to the instantaneous wire-diameter is represented by a continuous line. When cementite lamellae are at first orientated at random relative to the drawing axis, the mean interlamellar-spacing of all these orientations is given by the dashed line [29] assuming all orientations have equal presence within the pearlite. The points are from [21] but the method of measurement and estimates of uncertainty are not available.

a radial direction – all orientations in the plane defined by **x** and **y** (Table 13.3). As Langford stated [26], 'at infinite elongation, all lamellae will be completely reoriented...' to be parallel to the tensile axis.

Equation 13.4 can be applied to estimate what happens to S_I for all possible lamellar orientations relative to the drawing axis, implemented in a computer program [29]. Figure 13.9 compares the case where S_I is proportional to the wire diameter according to Equation 13.1a, and where this assumption is not made. Setting the spacing to be proportional to the diameter underestimates the actual spacing but the experimental data are consistent with $S_I \propto z_D$; the explanation may lie in the rapid alignment of lamellae to be parallel to the drawing axis as a function of strain. Figure 13.10 shows how an initial uniform distribution of orientations of cementite lamellae ends up with all lying parallel to the drawing direction even though the deformation is modelled as homogeneous.

Equations 13.1a and 13.2 represent homogeneous deformation. The real process may be more complex – after all, fairly planar lamellae turn into the cross-sectional microstructure illustrated in Figure 13.11a. But perhaps the complexities can be neglected given that gross mechanical properties depend on parameters such as S_I. Even though the drawn structure appears contorted, it remains the case that straight segments of the lamellae point in all radial directions as indicated in Figure 13.11b. The deformed lamellae formed in wire drawing can be visualised as curved, thin slabs, with centres of curvature towards the mid-point of the wire and long axes parallel to the drawing direction. The mechanism of the curled microstructure is discussed further in Section 13.2.3.

There are some useful empirical equations for estimating the proof and ultimate strengths of pearlitic wire containing 0.78-0.80 wt% carbon, as a function of the drawing strain ($\varepsilon_p \leq 1.3$) [8]:

$$\sigma_y\,/\,\text{GPa} \approx 0.432\varepsilon_p + 0.758 \pm 0.1 \tag{13.5}$$

$$\sigma_{UTS}\,/\,\text{GPa} \approx 0.370\varepsilon_p + 1.244 \pm 0.04. \tag{13.6}$$

All of the individually recorded stress-strain curves used to derive these equations showed continuous yielding. The fractional reduction of area correlates rather badly with the drawing strain, Figure 13.12. The ductility exhibits a minimum as a function of ε_p because the fracture mode changes from smooth at zero ε_p to one containing an increasing proportion of radial fissures at large ε_p. The fissures represent an energy-absorbing mechanism that may enhance the ultimate fracture-strain.

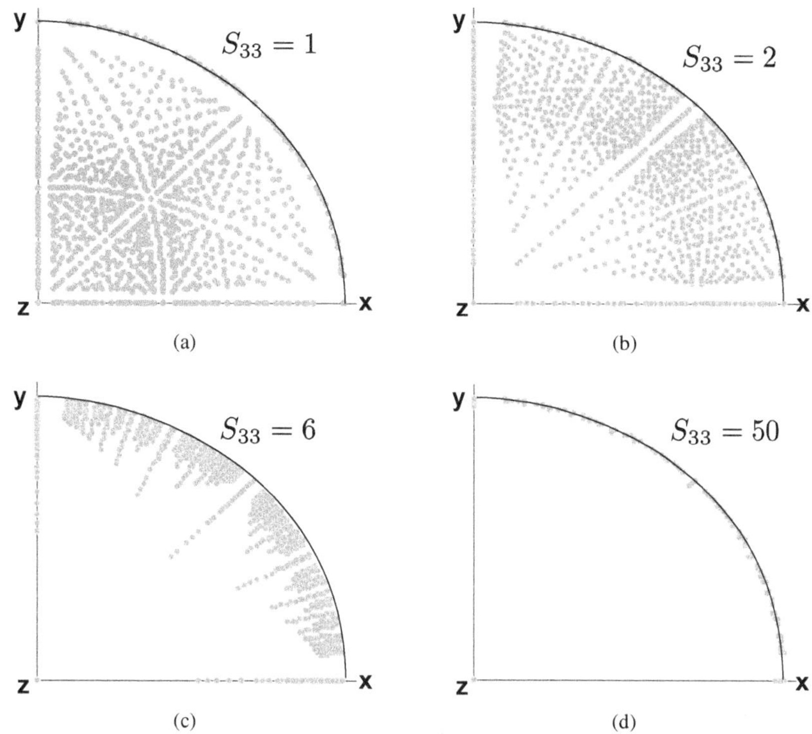

Figure 13.10 Quadrants of equal-angular stereographic projections showing the distribution of poles defining the normals to cementite lamellae within a colony, as a function of the drawing distortion S_{33} during wire drawing. **x** and **y** are mutually perpendicular radial directions and **z** is along the drawing axis. (a) Undeformed, showing the initial distribution of normals to the cementite lamellae. (b-d) Reorientation of the poles with increasing S_{33}, until eventually, all the lamellae are essentially parallel to **z**, though with their normals in all radial directions. The poles therefore all lie on the perimeter of the stereographic projection.

13.2.1 STRENGTH AND RELIABILITY

If, say 100 tensile tests are carried out on 100 cm length of drawn pearlitic wire, there will be a distribution of breaking loads; if the length is doubled then the distribution will be different and with a smaller mean breaking load. This is because the chance of finding a defect that initiates failure increases as the size does. In typical cables, the reduction in strength due to this weakest-link effect is of the order of 10%. A further reduction of about 8% occurs when there is a large number of parallel fibres in the cable, because the individual wires will not be mechanically identical [30].

13.2.2 ANISOTROPY OF STRENGTH AND TOUGHNESS

The anisotropy described here covers two aspects: the direction of the applied uniaxial-stress, whether it is tensile or compressive, and strength when compared along orthogonal directions of the wire.

The drawing operation renders both the structure and crystallography, directional, even though the material is polycrystalline. Experiments by Fujita et al. [31][2] show that the 0.1% proof-strength measured along the drawing direction **z** is different in compression and tension, Figure 13.13a,

[2]The chemical composition of the pearlitic wire studied by Fujita et al. was not stated in the original paper [31] but is assumed from their later paper [32].

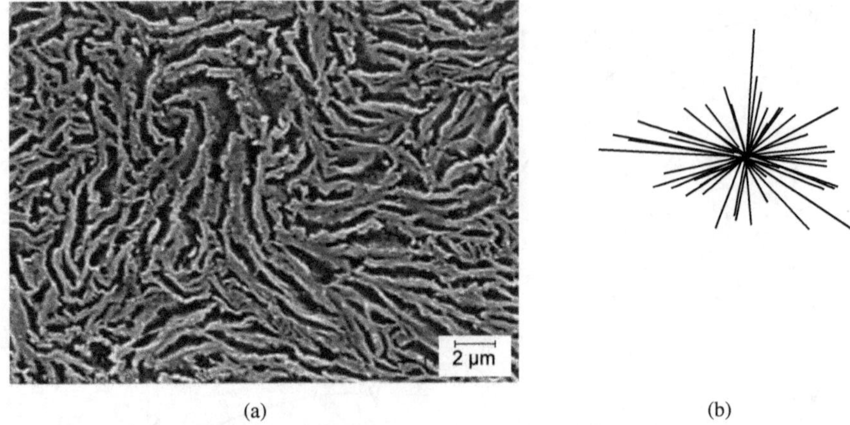

Figure 13.11 (a) Cross-section of a pearlitic wire showing the contortions of the cementite lamellae, generated when the steel is subjected to a wire-drawing reduction of 20% in one pass. Micrograph reproduced from Zelin [21] with permission from Elsevier. (b) Approximately straight segments from (a) showing that the lamellae point essentially in all radial directions.

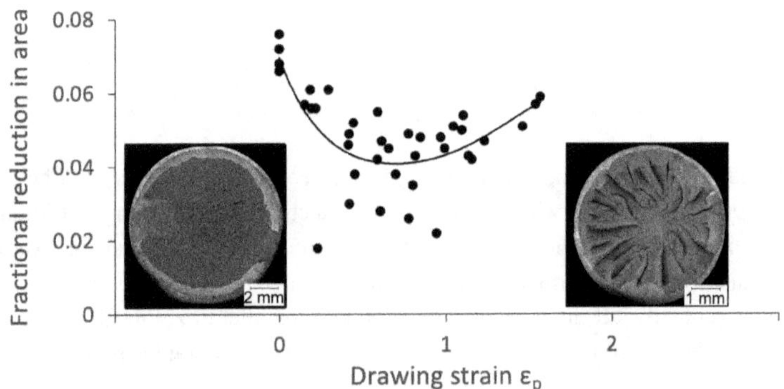

Figure 13.12 Fractional reduction in area following tensile fracture of Fe-0.79C-0.21Si-0.68Mn-0.22Cr wt% pearlitic steel wire as a function of the drawing strain. The curve is a polynomial fit which correlates with the data with a coefficient of 0.71. The images on the left and right correspond the fracture surfaces observed following ε_p of 0 and 1.566, respectively. Fractographs and data due to Toribio et al. [8], adapted and reproduced under the CC-BY license http://creativecommons.org/licenses/by/4.0/

because of compressive residual micro-strains parallel to **z**, within α_p [33–37]. These oppose tensile loading but assist compression, but there are difficulties in invoking micro-strains to explain the observed anisotropy.

In residual-stress terminology, type II stresses [38] are associated with micro-strains; they affect local deformation, such as the operation of specific slip systems within an individual grain. The strength anisotropy discovered by Fujita et al. is *macroscopic*. Type I tensile residual stresses extend over large distances in the surface layer of a drawn wire, with balancing compression in the core (p. 286). And even with the micro-strains, any compression within the ferrite will be balanced by tension within the cementite; in a wire drawn to $\varepsilon_p = 1.96$, the stress within the cementite is tensile at 2000 MPa whereas that in the ferrite is compressive at −140 MPa [33]. Any role of crystallographic texture has been ruled out by an experiment in which the wire was annealed in order to eliminate the residual strains but retain the texture.

The difference in tension and compression might be attributed simply to the Bauschinger effect [39], where the back stresses associated with deformation in one direction eases yield in the reverse direction. However, the experiments of Fujita et al. are on separate samples for the longitudinal and transverse tensile tests, so the Bauschinger back-stress explanation does not apply.

The second effect, Figure 13.13b, shows that σ_{UTS} is greater along the drawing direction than in the transverse orientation. This could be related to texture but the reasons are not obvious.

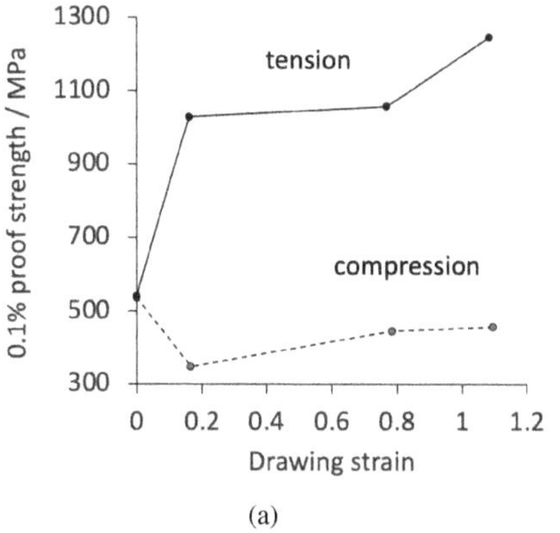

(a)

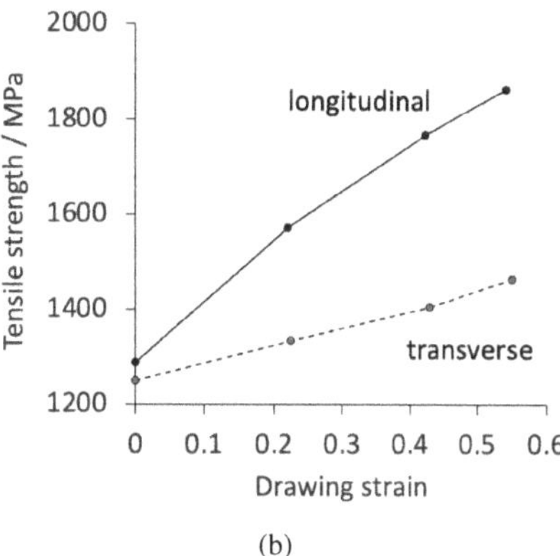

(b)

Figure 13.13 Pearlitic wire of probable composition Fe-0.81C-0.19Si-0.51Mn wt%. (a) The 0.1% proof strength measured using rod samples that are loaded along the drawing direction, in both tension and compression. (b) Tensile strength measured along the drawing and transverse directions using very small tensile specimens. Adapted using selected data from [31].

Whereas toughness is not a defining parameter in rope designs based on bundles of thinner wires, cracks advance parallel to the drawing direction **z** is easier than across the wire, with K_{IC} being more than twice as large when fracture propagation is normal to **z** [40]. Between $-200 \rightarrow 25\,^{\circ}$C, the toughness of a 7 mm diameter cold-drawn wire is $20 < K_{IC}^{\text{longitudinal}} < 40\,\text{MPa}\,\text{m}^{1/2}$ but $40 < K_{IC}^{\text{transverse}} < 95\,\text{MPa}\,\text{m}^{1/2}$ [40]. This may indicate a weakness in the α/θ interface in the drawn condition, defects parallel to the drawing direction due to the fragmentation of cementite, oriented defects due to the incompatibility in the plastic deformation rates of the two phases, etc., but a precise mechanistic explanation is missing.

13.2.3 MECHANISM OF CURLING

The first analytical explanation of the mechanism by which the curled microstructure illustrated in Figure 13.11 evolves, is due to Hosford [41], based on the deformation of a single crystal of ferrite with its $z \parallel [011]_\alpha$ direction aligned to the drawing direction, $x \parallel [100]_\alpha$ and $y \parallel [0\bar{1}1]_\alpha$. In ferrite, slip occurs by *pencil glide* whereby a shear direction $\langle 111 \rangle_\alpha$ is shared by a number of slip planes including $\{110\}_\alpha$, $\{112\}_\alpha$, and $\{123\}_\alpha$. It is assumed that slip is on the system $\langle 11\bar{1} \rangle_\alpha \{112\}_\alpha$ so when the drawing direction is $z \parallel [011]_\alpha$, specific slip systems must operate. The slip directions $[111]_\alpha$ and $[\bar{1}11]_\alpha$ are favourably oriented to accommodate extension along z but the deformation is then anisotropic in the x-y plane. To ensure axisymmetric flow, the $[\bar{1}11]_\alpha$ and $[1\bar{1}1]_\alpha$ would need to activate:

$$\mathbf{P}_1: \quad [111]_\alpha(\bar{2}11)_\alpha$$
$$\mathbf{P}_2: \quad [\bar{1}11]_\alpha(\bar{2}11)_\alpha$$
$$\mathbf{P}_3: \quad [1\bar{1}1]_\alpha(\bar{2}\bar{1}1)_\alpha$$
$$\mathbf{P}_4: \quad [\bar{1}\bar{1}1]_\alpha(2\bar{1}1)_\alpha.$$

$z \parallel [011]$ $y \parallel [01\bar{1}]$ $x \parallel [100]$

Shear on $\mathbf{P}_1$ and $\mathbf{P}_2$ can produce elongation along the wire axis z but the line at the intersection of their slip planes would remain invariant, i.e., there would be no strain recorded along the y-axis, with a contraction along $x \parallel [100]_\alpha$. The slip planes of $\mathbf{P}_3$ and $\mathbf{P}_4$ are normal to the wire axis so slip on those systems does not elongate the wire. Increments in the normal and shear plastic-strains are related by the Schmid factor $\mathrm{M} = \cos\phi \cos\lambda$, where the normal to the slip plane lies at an angle ϕ to the axis of interest and λ is the angle between the axis and the slip direction [42–45]:

$$\mathrm{d}\varepsilon_{\mathrm{p},j} = \sum_i \mathrm{M}_{j,\mathbf{P}_i} \times \mathrm{d}\gamma_{\mathrm{p},\mathbf{P}_i} \tag{13.7}$$

where j represents the x, y or z axis in turn. For the wire drawing scenario described above where only $\mathbf{P}_1$ and $\mathbf{P}_2$ contribute to strain along z, Hosford expresses Equation 13.7 as:

$$\mathrm{d}\varepsilon_{\mathrm{p},z} = \frac{\sqrt{2}}{3}(\mathrm{d}\gamma_{\mathrm{p},\mathbf{P}_1} + \mathrm{d}\gamma_{\mathrm{p},\mathbf{P}_2}) \tag{13.8a}$$

where $\mathrm{M}_{z,\mathbf{P}_1} = \mathrm{M}_{z,\mathbf{P}_2} = \sqrt{2}/3$. After correcting a numerical error in [41], the strain increment along x is

$$\mathrm{d}\varepsilon_{\mathrm{p},x} = -\frac{\sqrt{2}}{3}(\mathrm{d}\gamma_{\mathrm{p},\mathbf{P}_1} + \mathrm{d}\gamma_{\mathrm{p},\mathbf{P}_2}) + \frac{\sqrt{2}}{6}(\mathrm{d}\gamma_{\mathrm{p},\mathbf{P}_3} + \mathrm{d}\gamma_{\mathrm{p},\mathbf{P}_4}) \tag{13.8b}$$

and that along y is given by:

$$\mathrm{d}\varepsilon_{\mathrm{p},y} = -\frac{\sqrt{2}}{3}(\mathrm{d}\gamma_{\mathrm{p},\mathbf{P}_3} + \mathrm{d}\gamma_{\mathrm{p},\mathbf{P}_4}). \tag{13.8c}$$

Hosford then considered two cases, the first in which only $\mathbf{P}_1$ and $\mathbf{P}_2$ are active with zero deformation along the y-axis which lies on both the slip planes, so $\mathrm{d}\varepsilon_{\mathrm{p},x} = -\mathrm{d}\varepsilon_{\mathrm{p},z}$. It follows that the shear required to generate a unit strain along z is $(\mathrm{d}\gamma_{\mathrm{p},\mathbf{P}_1} + \mathrm{d}\gamma_{\mathrm{p},\mathbf{P}_2})/\mathrm{d}\varepsilon_{\mathrm{p},z} = 3/\sqrt{2}$. This deformation is not axisymmetric, so a second case was considered where all four slip systems are activated with the axial symmetry condition that $\mathrm{d}\varepsilon_{\mathrm{p},x} = \mathrm{d}\varepsilon_{\mathrm{p},y} = -(1/2)\mathrm{d}\varepsilon_{\mathrm{p},z}$. When this is substituted into the above equations, it can be demonstrated that $\sum_{i=1}^{4} \mathrm{d}\gamma_{\mathrm{p},\mathbf{P}_i} = (3/\sqrt{2}) \times (3/2)$. In other words, the amount of slip needed for axisymmetric deformation is one and a half times greater than with the invariant-line strain $\mathbf{P}_1\mathbf{P}_2$, leading to the conclusion that there would be a tendency for grains in fibre-textured samples to deform by invariant-line strains, causing them in turn to develop elliptical shapes normal to the fibre axis. To maintain strain compatibility (space-filling), 'neighbouring grains would need to bend around one another', explaining the curling effect.

There is another way of dealing with homogeneous deformations. Each of the shears $\mathbf{P}_i$ can be represented as a 3×3 deformation matrix [p. 156, 45]:

$$(\mathrm{X}\ \mathbf{P}\ \mathrm{X}) = \mathbf{I} + s[\mathrm{X};\mathbf{d}](\mathbf{p};\mathrm{X}^*) \tag{13.9}$$

where $\mathbf{I}$ is the identity matrix, s is the magnitude of the shear, $\mathbf{d}$ is the unit shear direction and $\mathbf{p}$ is the unit normal to the shear plane. The symbol X denotes the frame of reference with X^* being the corresponding reciprocal basis.[3] The orthonormal basis X is defined by the x, y and z axes of the

---

[3] $[\mathrm{X};\mathbf{d}]$ is a 3×1 column matrix and $(\mathbf{p};\mathrm{X}^*)$ a 1×3 row matrix. Their product therefore results in a 3×3 matrix.

wire. It follows that for $s = 0.1$

$$(\mathrm{X}\,\mathbf{P}_1\,\mathrm{X}) = \begin{pmatrix} 0.95286 & 0.02357 & 0.02357 \\ -0.04714 & 1.02357 & 0.02357 \\ -0.04714 & 0.02357 & 1.02357 \end{pmatrix} \quad (\mathrm{X}\,\mathbf{P}_2\,\mathrm{X}) = \begin{pmatrix} 0.95286 & -0.02357 & -0.02357 \\ 0.04714 & 1.02357 & 0.02357 \\ 0.04714 & 0.02357 & 1.02357 \end{pmatrix}$$

and using $s = -0.05$ to ensure axisymmetric deformation,

$$(\mathrm{X}\,\mathbf{P}_3\,\mathrm{X}) = \begin{pmatrix} 1.02357 & 0.01179 & -0.01179 \\ -0.02357 & 0.98821 & 0.01179 \\ 0.02357 & 0.01179 & 0.98821 \end{pmatrix} \quad (\mathrm{X}\,\mathbf{P}_4\,\mathrm{X}) = \begin{pmatrix} 1.02357 & -0.01179 & 0.01179 \\ 0.02357 & 0.98821 & 0.01179 \\ -0.02357 & 0.01179 & 0.98821 \end{pmatrix}$$

giving the total deformation

$$\mathbf{P}_1 \times \mathbf{P}_2 \times \mathbf{P}_3 \times \mathbf{P}_4 = \begin{pmatrix} 0.95408 & 0.00172 & 0.00273 \\ 0.00355 & 1.02635 & 0.07238 \\ 0.00577 & 0.07237 & 1.02635 \end{pmatrix}$$

The effect of any of these deformations on a vector $\mathbf{u}$ can be calculated as follows:

$$(\mathrm{X}\,\mathbf{P}\,\mathrm{X})[\mathrm{X};\mathbf{u}] = [\mathrm{X};\mathbf{v}] \tag{13.10}$$

where $\mathbf{v}$ is the vector resulting from the operation of $\mathbf{P}$ on an initial vector $\mathbf{u}$. Table 13.4 shows what happens to unit vectors parallel to the principal axes of the wire due to combinations of the deformations considered in the Hosford analysis. Perhaps the most important point to note is the the pairs $\mathbf{P}_1\mathbf{P}_2$ and $\mathbf{P}_3\mathbf{P}_4$ cause opposing deformations in the x-y plane. For example, in the former case, the x-axis is contracted while it is expanded in with the latter deformation.

The methods are obviously idealistic. The wire texture does not equal a single-crystal, the role of cementite is neglected, work hardening is not considered, it is unlikely that each of the four deformations act in unison, nor will they have equal shears at any instant in the drawing operation. Each of the deformations will encounter different resistance because the microstructure becomes anisotropic during drawing. Therefore, any *momentarily unequal* opposing strains in the x-y plane may lead to plastic bending that is constrained by the die, and hence the curls. However, affine transformations such as those considered by Hosford and the matrix method cannot reproduce the curls; some heterogeneity is essential, underlined also by the fact that the plastic strain distribution across the cross section of a wire during drawing is not expected to be uniform [46].

Figure 13.14 shows how grain curling can be predicted using finite element modelling, beginning with a structure in which a $\langle 011\rangle_\alpha$ fibre texture already exists. The huge advantage is that heterogeneous deformation on a microscopic scale is accounted for. The simulation illustrated is in two dimensions for the sake of clarity, with biaxial compression implemented to alter the diameter of the wire; the original work includes also a three-dimensional simulation that reveals the development of the texture and the curling of grains but for obvious reasons, is more difficult to visualise [47]. The lines marked β are parallel to $\langle 0\bar{1}1\rangle$ which remains invariant to $\mathbf{P}_1\mathbf{P}_2$ in the Hosford model. Thus $\beta = \langle 0\bar{1}1\rangle$ in grain 'I' is nearly uncompressed but the $\langle 100\rangle$ direction normal to it is compressed preferentially. In order to accommodate this deformation in grains II and III, grain I has to curl. Therefore, it is the anisotropic deformation of grains in the cross-sectional plane, and accommodation between grains, that forces the curling.

Similar results have been obtained by Sevillano et al. [48], with their emphasis on internal-stress anisotropy leading to large stresses on the $\{100\}$ planes parallel to the wire axis, which were assumed to contribute to the delamination mechanism through cleavage. However, it is now known that delamination failure is ductile although a post-mortem examination of the fracture would reveal a smooth, sheared surface – torsion continues after delamination, so the crack faces are twisted into contact thus rubbing together in shear mode [49, 50]. A normal failure without delamination exhibits anisotropic ductile-dimples sheared along the torsion direction, Figure 13.15.

Table 13.4
Extensions, contractions and rotations of unit vectors parallel initially to the principal axes of the wire, as a result of the deformations described by Hosford [41]. The combination of four invariant-planes strains can therefore lead to axisymmetric deformation.

Initial vector	Deformation	Final vector	Magnitude after deformation
$[0\ \frac{1}{\sqrt{2}}\ \frac{1}{\sqrt{2}}]_z$	$\mathbf{P}_1\mathbf{P}_2$	[0.00314 0.77691 0.77691]	1.0987
$[0\ \overline{\frac{1}{\sqrt{2}}}\ \frac{1}{\sqrt{2}}]_y$	$\mathbf{P}_1\mathbf{P}_2$	$[0\ \overline{\frac{1}{\sqrt{2}}}\ \frac{1}{\sqrt{2}}]$	1.0000
$[1\,0\,0]_x$	$\mathbf{P}_1\mathbf{P}_2$	[0.91016 0.00444 0.00444]	0.9101
$[0\ \frac{1}{\sqrt{2}}\ \frac{1}{\sqrt{2}}]_z$	$\mathbf{P}_3\mathbf{P}_4$	$[0\ \frac{1}{\sqrt{2}}\ \frac{1}{\sqrt{2}}]$	1.0000
$[0\ \overline{\frac{1}{\sqrt{2}}}\ \frac{1}{\sqrt{2}}]_y$	$\mathbf{P}_3\mathbf{P}_4$	[0.00078 −0.67455 0.67455]	0.9540
$[1\,0\,0]_x$	$\mathbf{P}_3\mathbf{P}_4$	[1.04825 −0.00111 0.00111]	1.0483
$[0\ \frac{1}{\sqrt{2}}\ \frac{1}{\sqrt{2}}]_z$	$\mathbf{P}_1\mathbf{P}_2\mathbf{P}_3\mathbf{P}_4$	[0.00314 0.77691 0.77691]	1.0987
$[0\ \overline{\frac{1}{\sqrt{2}}}\ \frac{1}{\sqrt{2}}]_y$	$\mathbf{P}_1\mathbf{P}_2\mathbf{P}_3\mathbf{P}_4$	[0.00071 −0.67455 0.67456]	0.9540
$[1\,0\,0]_x$	$\mathbf{P}_1\mathbf{P}_2\mathbf{P}_3\mathbf{P}_4$	[0.95408 0.00354 0.00577]	0.9540

13.3 CRYSTALLOGRAPHIC TEXTURE

During the deformation of a polycrystalline material, crystals that initially are randomly oriented tend to rotate as their individual slip systems attempt to comply with applied stresses. The original random distribution of orientations becomes less so. The deformed polycrystalline-material is then said to be *crystallographically textured*, in a state somewhere between an isotropic steel and a single crystal.

A convenient method for communicating texture in rolled sheet is by stating the set $\{h\,k\,l\}\langle u\,v\,w\rangle$, i.e., the planes which lie roughly parallel to the rolling plane, and direction in the rolling plane that tends to be parallel to the rolling direction. The overall texture can be represented as the sum of components:

$$\text{texture} = \sum_i \lambda_i \{h\,k\,l\}_i \langle u\,v\,w\rangle_i$$

where λ represents the weighting given to a particular type of texture. With textured wires, the crystals tend to share a common direction along the drawing axis with other equivalent directions equally distributed such that there is rotational symmetry about the drawing axis. Cold-drawn pearlitic wires have the $\langle 110\rangle_\alpha$ fibre texture because these directions tend to rotate progressively towards the wire axis during deformation. For example, at the early stages of drawing, the texture is reported to consist of the components $\langle 110\rangle_\alpha\{001\}_\alpha$ and $\langle 110\rangle_\alpha\{\bar{1}11\}_\alpha$; an increase in drawing strain changes the surface texture to $\langle 110\rangle_\alpha\{1\bar{1}0\}_\alpha$ with the core at $\langle 110\rangle_\alpha\{\bar{1}11\}_\alpha$ [52]. All of these share the $\langle 110\rangle_\alpha$ fibre.

Similar information is scarce for the orientations of the cementite, with one exception [53] that shows little change in cementite texture during shear deformation (p.45). The particular orthorhombic lattice of cementite means that many lattice planes have close Bragg angles. As a result, even if

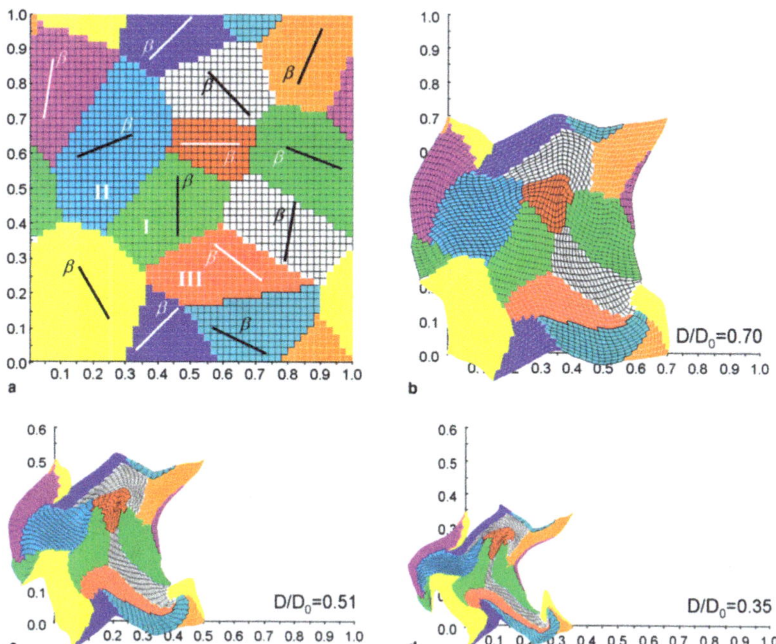

Figure 13.14 Finite element simulation of the biaxial compression of a body-centred cubic material in the form of a wire with a $\langle 110\rangle$ fibre texture prior to the compression. The ratios of the wire diameter to initial diameter are indicated. The line *beta* in each grain is along the $\langle\bar{1}10\rangle$ direction. Image reproduced from Očenášek et al. [47] with permission from Elsevier.

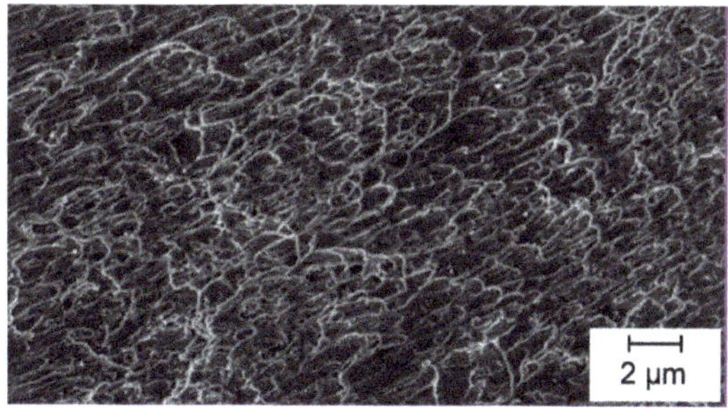

Figure 13.15 Fracture surface of a pearlitic wire that has been subjected to torsion to the point of failure. Shows ductile dimples that are elongated by torsion. Image courtesy of Aparna Singh, more details in [51].

the cementite has a preferred orientation in drawn pearlitic wire, it gives the appearance of isotropy because of overlapping diffracted intensity from a variety of planes [54]. Coarse cementite (tens of micrometres) in ancient swords made by forging [55] exhibits a strong texture in which $(010)_\theta$ planes align parallel to broad plane of the blade; the strains involved in wire drawing are much greater than in forging so the cementite there must exhibit some sort of texture that has yet to be revealed.

The α-texture may vary across the section of the wire; a 1.2 mm diameter drawn-wire exhibited the $\langle 110\rangle_\alpha$ fibre texture, but when chemically thinned to 1 mm, the texture lost circular symmetry and hence could be described as $\langle 110\rangle_\alpha\{1\bar{1}0\}_\alpha$ (circular or cylindrical texture), meaning that most ferrite grains have a $\{110\}_\alpha$ plane tangent to the surface of the wire [54, 56]. The fibre texture appears at the surface and core of the wire, whereas the circular texture is between these two, and strongest in the annular ring located about 0.8 of the wire radius (Figure 13.16a), corresponding to the position of the shear maximum during drawing [57]. The crystallographic fibre-axis and the drawing axis are not perfectly aligned, which is expected given the random orientation of ferrite prior to deformation.

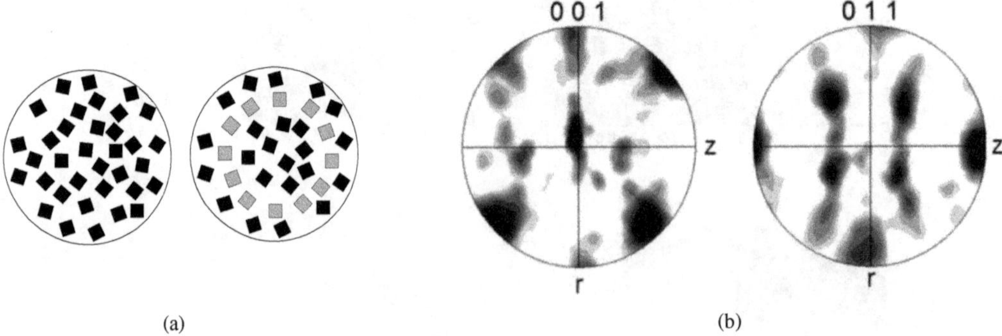

Figure 13.16 (a) Schematic texture on a cross-section of a pearlitic steel wire; each square represents an α unit-cell. The fibre texture on the left has the normal to each of the squares aligned along the drawing direction, but the edges of the squares are randomly oriented. On the right is a region below the surface, where the edges of the grey squares are radially aligned (circular texture). The core and surface regions retain the fibre texture. (b) $\{001\}_\alpha$ and corresponding $\{011\}_\alpha$ pole figure from Fe-0.9C-0.22Si-0.33Mn-0.23Cr wt% pearlitic steel subjected to a drawing strain of 3.5. The $\langle 011\rangle_\alpha$ directions are strongly aligned to the wire axis ('z') indicating a fibre texture, but it is not possible to rule out the circular texture. Reproduced from Jamoneau et al. [58] with permission from Elsevier.

An example of texture developed in pearlite after drawing the wire to a strain of about 3.5 is presented in Figure 13.16b, which illustrates the 001_α and 011_α pole figures. If it is assumed that ferrite slips on $\langle\bar{1}10_\alpha\rangle\{112\}_\alpha$ slip system, then two of the slip planes would contain the $[\bar{1}10]_\alpha$ wire axis. This may be important in explaining the formation of shear bands during torsion, as described later (p. 293).

Although the texture has been described in terms of the fibre and circular components, other components are possible, and the variation across the cross section of the wire depends on process design [57]. With die-angles in the range 9-12° in a 21 pass process ($\varepsilon_p = 3.5$), by far the strongest texture is $\langle 110\rangle_\alpha$ fibre parallel to **z**, with the strength decreasing with depth from the surface, Figure 13.17. A second weak component, $\langle 112\rangle_\alpha$ parallel to **z**, is also present. When the die angle is increased to 16-18° in the final three stages, the texture obtained in no longer the fibre or circular texture; instead, $\langle 112\rangle_\alpha$ increases with depth, consistent with the greater shearing associated with the larger die-angles.

13.4 MACROSCOPIC RESIDUAL STRESSES DUE TO WIRE DRAWING

Residual stresses in a body are those not needed to maintain an equilibrium between the body and its environment [38]. The strains during wire drawing are not uniform because the wire surface is in contact with that of the die, whereas the core is not. This leads to Type I residual stresses that vary continuously across the cross-section of the wire. Type II residual stresses are due to differences in deformation behaviour of adjacent grains, either due to crystallographic orientation or crystal structure. They are therefore characterised over a microstructural scale. Type III are stresses within a material on an atomic scale; we shall not be concerned with such stresses in the context of pearlitic

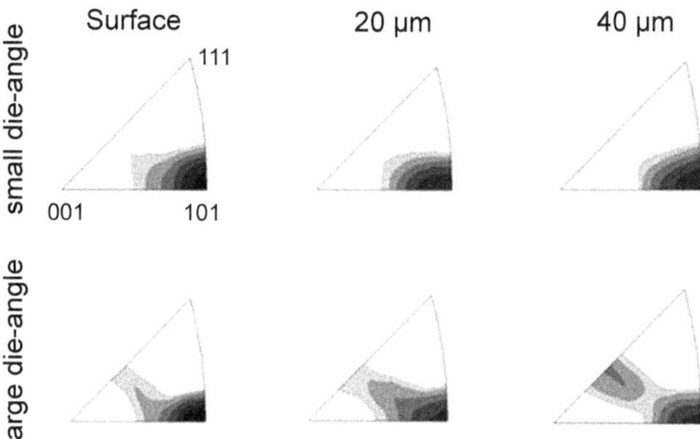

Figure 13.17 Inverse pole figures showing the drawing direction with respect to crystallographic indices, for pearlitic wires drawn to a strain of 3.5, using two different die angles in the final three stages of the process. Adapted from Jamoneau et al. [57], where much more detail is available, with the permission of Wiley Materials. ©WILEY-VCH Verlag GmbH & Co. KGaA, Weinheim.

wires. Figure 13.18a shows schematically how the Types I and II stresses are related in pearlite, with the macrostress calculated by summing the microstresses within the two phases after weighting them by their respective volume fractions.

Type II stresses often are characterised using neutron diffraction, which when it penetrates the steel does not pick up the balancing macrostress, whereas X-ray diffraction that has a shallow penetration helps determine the combined effects of Type I and Type II stresses, i.e., the total stress at a location. Using both techniques helps to decipher the contributions [59], as illustrated in Figure 13.18b where it is assumed that the microstress in cementite can be derived by balancing against that in ferrite, taking into account the phase fractions.

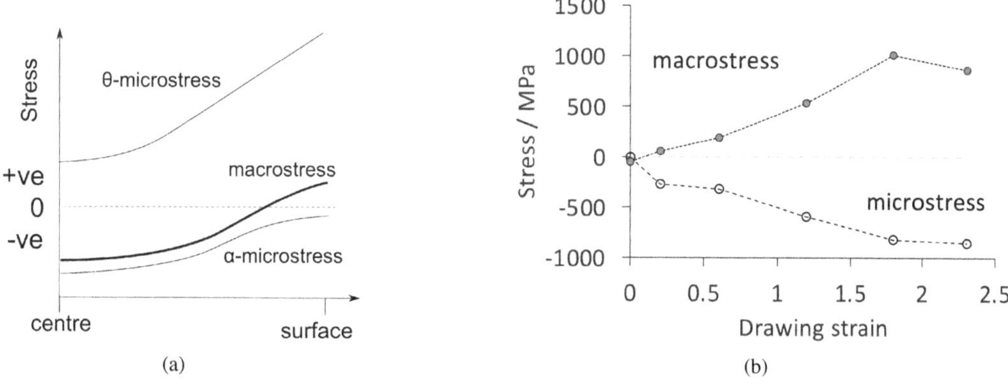

Figure 13.18 (a) Difference between Type I (macro) and Type II (micro) stresses in pearlite, across the section of a cold-drawn pearlitic wire. The macrostress is given by a summation of the microstresses of the two phases, scaled by their respective volume fractions. Actual data are available in [60]. (b) Measured residual stresses within the ferrite of a drawn, eutectoid steel wire. The α-microstresses are determined using X-ray diffraction and the α-macrostress calculated using both neutron and X-ray diffraction. Selected data from Kriska et al. [59].

There are observations and simulations of the macroscopic residual-stress along the drawing direction of pearlitic wire, showing that it is tensile in the surface regions and compressive at the core, Figure 13.19a. This is because of friction with the die that constrains the deformation at the surface of the wire. In Figure 13.19b, (i) is an imaginary operation in which the core is elongated more than the outer cylinder which is in contact with the die. In the second imaginary operation (ii), the surface cylinder and core are forced to the same length, by compression the latter and pulling the former, leading to the observed state of axial residual-stress.

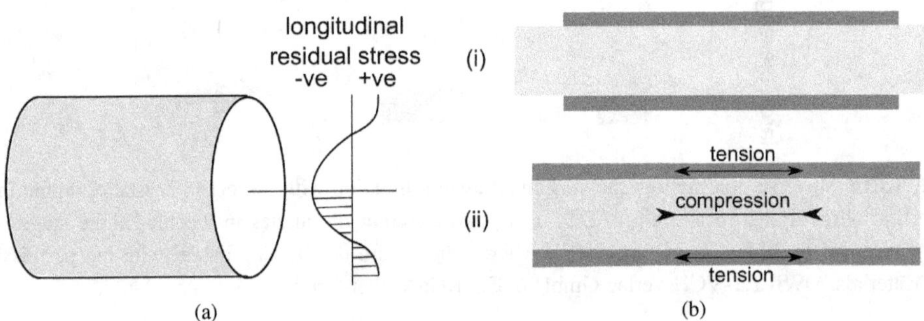

Figure 13.19 (a) Distribution of residual stress in the longitudinal direction of a drawn wire, with tension at the surface and compression within the core. (b) Schematic explanation of the development of residual stress.

The macroscopic stress distribution can be determined using X-ray or neutron diffraction from gauge volumes of the sample. Within those volumes, these techniques measure the residual strains from the lattice spacings within each phase and then calculate an overall stress by scaling in proportion to the phase fractions present. The average strains measured within each gauge volume represent a combination of the macroscopic stress and type II microstrains but the latter cancel when scaled by the volume fractions of the phases involved. The total macrostress then is given by [61]:

$$\sigma_{ij}^{\text{macro}} = (1 - V_V^\theta)\overline{\sigma_{ij}^\alpha} + V_V^\theta \overline{\sigma_{ij}^\theta} \tag{13.11}$$

where $\overline{\sigma_{ij}}$ represent the average stress-tensors derived from the corresponding strain tensors and elastic constants.

In an interesting set of experiments [62, 63] 20 mm diameter bars of completely ferritic and a completely pearlitic, annealed steels, were reduced from 20 mm diameter to 18 mm diameter ($\varepsilon_p = 0.21$) by cold drawing and the residual stress distributions determined. The purpose of the ferritic steel was simply to establish whether simulations can reproduce the data observed in a single phase material; notwithstanding that, the measured axial residual-stress distributions for both steels are similar, Figure 13.20. They reveal a tensile stress towards the surface of the rods and a compressive stress towards the core. The similarity between the two steels may indicate that it is the deformation of the ferrite, i.e., the major phase within pearlite, that is key in determining the residual stress distribution.

Many other results confirm the existence of a tensile residual-stress at the surface drawn wires [64]. The magnitude of that stress scales with drawing strain, reaching 600 MPa at $\varepsilon_p = 1.8$ [65] and 1.1 GPa at $\varepsilon_p = 2.4$ [59].

Ateinza et al. [63] used 20 mm initial diameter and a small drawing strain of 0.21 to characterise the stress distribution. It would be interesting to see whether smaller diameters and larger strains could lead to a homogenisation of stresses, possibly a cancellation of the +ve and −ve residual stresses.

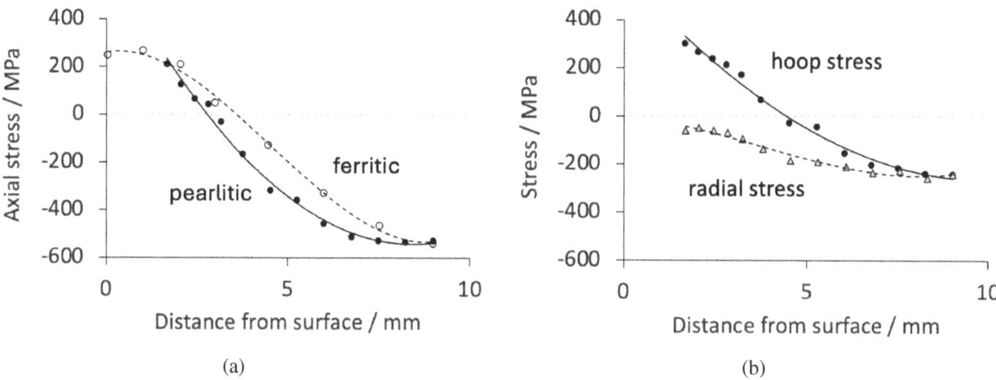

Figure 13.20 Fully ferritic, or fully pearlitic rods reduced from $20 \rightarrow 18$ mm diameter by drawing deformation. (a) Observed distribution of macroscopic residual-stress along the longitudinal direction of a drawn wire, with tension at the surface and compression within the core. (b) Hoop and radial components of residual stress plots for the pearlitic steel. Adapted using data from Atienza et al. [62, 63].

13.5 DUCTILITY OF WIRES

It might be assumed that the cold-drawn wires used to make ropes would benefit from greater ductility; the elongation to failure can be as small as $\approx 2\%$ (Figure 13.21), comparable to the elastic strain of 1.9% when a wire is loaded to 4 GPa. In a long rope, the elastic elongation to failure is the dominant component of elongation [30].

Plasticity associated with elongation must of course be limited in strong materials, but elongation can decrease when the wire diameter is reduced. In copper, ductility decreases as wires are made thinner because of the localisation of slip and the greater influence of slip steps on the surface on stress concentration [66]. There is some evidence to show that the ductility of pearlitic wires decreases with diameter, when the strength is quite similar [67]:

wire diameter / mm	σ_y / MPa	σ_{UTS} / MPa	% elongation
2.5	1857	1972	9.1
1.8	1858	2021	6.2
1.6	1870	2044	3.1

Low tensile-ductility pearlitic wires have been in service for many decades, so their minimal elongation must be tolerable. Ropes contain many wires which means that the failure of a few does not necessarily compromise structural integrity. The probability of failure is reduced by limiting the loads on the ropes through safety factors as large as two. Indeed, the ductility of individual wires is often not listed in tables describing rope properties [e.g., 68]. On the other hand, damage due to inter-wire stresses in spiral strands can be mitigated by increasing ductility.

Annealing drawn wire can boost ductility but at the expense of strength. To avoid this, patented rod is drawn down from 12.2 to 5.10 mm in nine stages, but with intermediate treatments of 280 °C for 10 min prior to each of the final drawing stages 8 and 9 followed by a final anneal at 415 °C for 15 min. This substantially increases the elongation from about 1% to just over 5%, Figure 13.22. The elongation-strength data are included in Figure 13.21 where they are seen to follow the general trend, in which case the complication of adding annealing treatments may not be justified.[4]

[4]The elongation data reported in [72] have been corrected to remove elastic deformation due to the sample and machine.

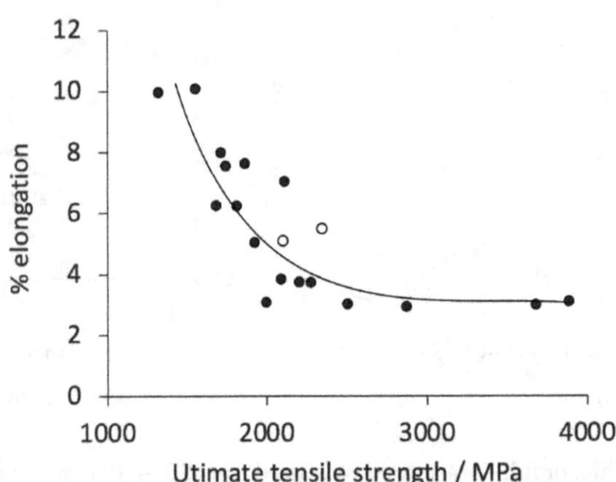

Figure 13.21 The elongation to failure of drawn pearlitic wires. Data assembled from [14, 69]. The open points are from [70], for later discussion. Further data available from Gondo et al., as a function of the drawing reduction per pass [71].

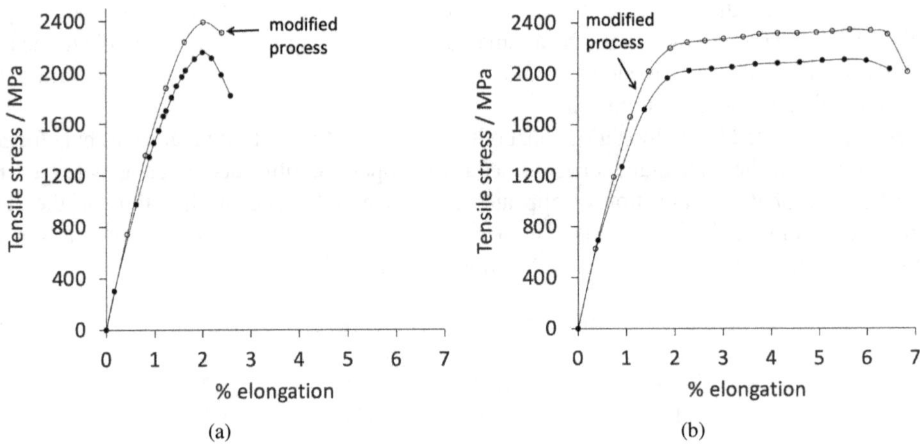

Figure 13.22 The filled points represent conventionally drawn pearlite, whereas the open points are from a modified process in which a low-temperature anneal (280 °C, 10 min) is applied before each of the final two drawing stages. (a) Properties after the drawing operations were completed. (b) Following an anneal at 415 °C for 15 min after the drawing operations were completed. Selected data from Zhou et al. [72].

13.6 TORSIONAL DUCTILITY AND DELAMINATION

Individual strands are twisted together into larger diameter ropes for bridge cables, stranded tyre-cords or coiled springs. Uniformly twisting a drawn wire such that the accumulated shear strain is ≈ 1 makes negligible difference to its tensile strength or ductility, although the proof strength decreases [73]. A larger degree of twisting can cause *delamination*, i.e., longitudinal splitting of the wire driven by torsional stresses. The splitting is accompanied by a sudden reduction in the torque recorded. Failure in this manner is different from torsional ductility which depends on the number of turns to fracture, irrespective of the mode of fracture. It is important to note that delamination can occur during uniaxial tensile or compressive loading.

The plastic strains during a torsion test, typically carried out at a rate $\dot{\gamma} = 0.6\,\mathrm{s}^{-1}$, can be much greater than during tensile testing because the plastic instability that causes necking does not occur. Even though the shape of the specimen is preserved until fracture, the deformation across the cross-section of the sample is not uniform.[5] For a solid, isotropic specimen, the maximum shear strain experienced at the surface is [74, 75]:

$$\gamma_{\max} = \frac{r_o \theta}{L}, \tag{13.12a}$$

where L is the length and r_o the radius of the sample subjected to torsion, and θ is the twist angle in radians. The corresponding maximum shear stress at large plastic strains is [75]

$$\tau_{\max} = \frac{1}{2\pi r_o^3}\left(\frac{\theta}{L}\frac{\mathrm{d}M_\mathrm{T}}{\mathrm{d}(\theta/L)} + 3M_\mathrm{T}\right) \tag{13.12b}$$

where $M_\mathrm{T} = 2\pi \int_0^{r_o} \tau r^2\,\mathrm{d}r$ is the torsional moment. A typical torsion-test curve is illustrated in Figure 13.23; there is considerable strain recorded following the onset of plasticity, but the ultimate shear-strength in torsion (1270 MPa) is naturally smaller than that in tension (2240 MPa). According to the von Mises yield criterion, (Equation 4.3), $\sigma_\mathrm{y} = \sqrt{3}\tau_\mathrm{y}$, so a difference is to be expected. However, the comparison here is with the ultimate tensile strength, not the yield strength. A part of the explanation must then depend on the work hardening behaviour, which during torsion testing is most intense in the vicinity of the surface, whereas the entire volume of the gauge length in principle undergoes identical work hardening during tensile testing. The difference emphasises the heterogeneous nature of the plastic deformation during torsion testing. The ratio $\sigma_\mathrm{UTS}/\tau_\mathrm{UTS}$, where the denominator is the shear stress when the ultimate tensile strength is reached, is found experimentally to be in the range 1.76-2.64 [14, 76].

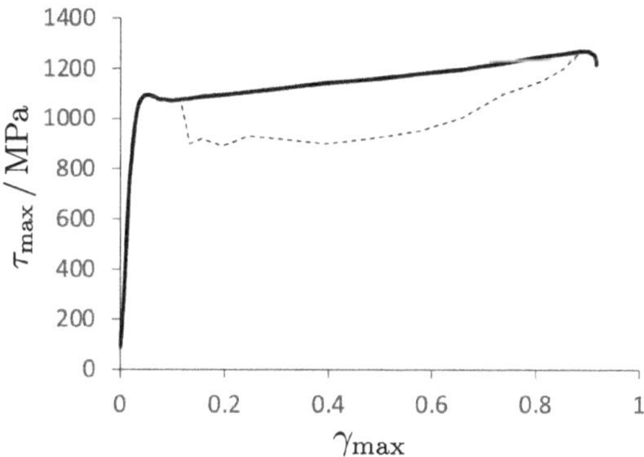

Figure 13.23 Shear stress and strain felt at the surface of a pearlitic wire subjected to a drawing strain of 1.95. The steel composition is Alloy 11, Table 13.2. The slight peak at the beginning of torsion is because of the change in the strain path from drawing to twisting [77]. Selected data from Jung et al. [14]. The dashed line indicates schematically, the behaviour expected if delamination occurs, with a sudden drop in $\tau_{\max}$, after which the longitudinal split is held together by the mechanical forces of torsion.

Whereas Equation 13.12b assumes that the shape and size of the torsion specimen is constant, it is possible to for length of the sample to change due to the development of anisotropy in the material, for example from changes in crystallographic texture. In such cases [78],

$$\tau_{\max} = \frac{1}{2\pi r_o^3}\left(\frac{\theta}{L}\frac{\mathrm{d}M_\mathrm{T}}{\mathrm{d}(\theta/L)} + 3M_\mathrm{T}\right)\left[1 - \frac{1}{2}\left(3\varepsilon_z + \frac{\theta}{L}\frac{\mathrm{d}\varepsilon_z}{\mathrm{d}\{\theta/L\}}\right)\right]^{-1} \tag{13.13}$$

[5]It sometimes is assumed that the centre of the specimen is elastically loaded, but at the large plastic strains associated with torsion testing spread the deformation from the surface inwards.

where ε_z is the axial strain, and given volume conservation, $\varepsilon_x = \varepsilon_y = -0.5\varepsilon_z$. The shear strain becomes

$$\gamma_{max} = r_o(1+\varepsilon_x)\frac{\theta}{L}. \tag{13.14}$$

The axial strain in general will be a small fraction of the shear strain.

13.6.1 MECHANISM OF DELAMINATION

Figure 13.23 shows what to expect during a torsion test when the wire splits longitudinally, causing a sudden drop in the stress indicating irreversible damage. The drop is followed by a somewhat steady stress level as the crack propagates along the wire. The final increase in stress with continued torsion is because the split parts are held together by the continued twisting, but the wire has by then lost its mechanical integrity. A number of mechanisms have been proposed for the delamination.

Fragmentation of cementite

Delamination is a ductile failure mode, beginning when small holes that decorate shear bands parallel to the wire axis and a radial direction link [e.g., 58, 77]. Anything that assists void nucleation, such as an increased number density of cementite particles resulting from fragmentation of cementite during the wire drawing operation, should promote delamination.

Coarse pearlite is more susceptible to delamination than the finer variety when the comparison is made on the same steel [17]. Cold-deformation can fragment cementite lamellae resulting in an increase in the locations where voids can initiate. Lamellae that are not well-aligned with the drawing axis become bent through large angles, making them more susceptible to fragmentation. Delamination cracks have been found along arrays of voids at the cementite fragments [17]. It would, in this context, be instructive to study delamination in aligned pearlite generated by growth in a temperature gradient, which should make it less susceptible to fragmentation if the lamellae are set all along the drawing direction [79, 80].

It is easier to initiate voids at the larger particles associated with coarse pearlite, because they pose greater obstacles to the flow ferrite (p. 95). The lamellar structure is not disrupted at relatively low strains (e.g., $\varepsilon_p = 1.13$) where delamination is not observed [16]. Wire subjected to drawing strain of 2.38 and annealed at 500 °C for 1 h has a structure consisting of linear arrays of spheroidal cementite but does not exhibit delamination. The annealing reduces the strength from 1900 MPa to 1550 MPa, which may facilitate the flow of ferrite around the cementite particles.

There also is a decrease in the work-hardening capacity of the material when lamellae change into linear arrays of spheroidal particles during annealing of drawn wire [14]. The reduced work hardening capacity is because the mean-free-path for dislocation motion in ferrite increases Figure 13.24. It may reduce the susceptibility to delamination, although the wire normally would not be used in an annealed condition due to the loss of strength.

Heat generation during drawing

The design of the drawing sequence can help reduce delamination. The wire can become hot during the final stages of high-speed wire drawing causing an increase strength (Section 13.7) with an accompanying loss in ductility [82]. Implementing a greater part of the drawing strain in the earlier passes helps reduce the temperature of the wire in the ultimate stages of drawing, thus eliminating delamination [82].

This technique is supported by the fact that annealing the wire at 150 °C, 1 h introduces delamination that was absent in samples tested prior to the heat treatment [50]. The heat treatment causes the excess carbon in the ferrite, originating from the mechanical dissolution of cementite during drawing, to precipitate as fine carbides (of unknown structure). This precipitation is localised at the sites of the original cementite, may assist void formation and hence delamination.

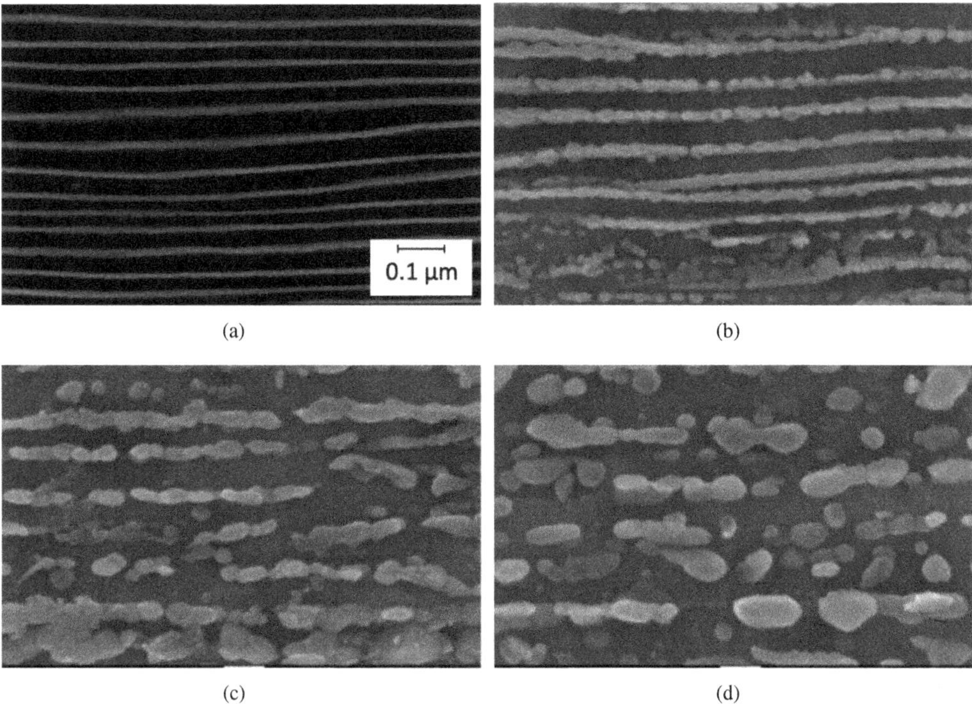

Figure 13.24 The effect of annealing on the structure of pearlitic wire drawn to a strain of 1.95 (Alloy 11, Table 13.2). The aligned cementite lamellae disintegrate into linear arrays of somewhat spheroidal particles, leaving gaps for the penetration of dislocations within the ferrite. (a) In the as-drawn condition. (b) Following annealing at 500 °C for 30 s, (c) 15 min, (d) 1 h. Images reproduced from Jung et al. [14] under the terms of the CC BY 4.0 license [81].

Residual stress and delamination

A small drawing reduction (≈ 0.02) in the final pass of the production process for thin wires reduces tensile residual stresses in the axial and tangential directions, and can put it into compression [83]. While the original work related to the fatigue of tyre cord, it is known that such a final drawing pass eliminates the early onset of delamination in pearlitic wires that had a total drawing strain of ≈ 3.5, Figure 13.25. This emphasises the role of the surface in initiating longitudinal cracks and why there is considerable scatter in the locations where delamination initiates because of variations in surface condition. In controlled experiments, the location of delamination can be controlled by introducing a small hole (0.2 mm diameter on a sample 160 mm long) through the wire sample [50].

Texture effects and shear bands

Cold-drawing makes the wire mechanically anisotropic, in part due to the evolution of crystallographic texture. This leads to grain curling across the cross-section of the wire, but the reduced constraint at the wire surface can cause it to dip locally – these perturbations then elongate during drawing to produce longitudinal grooves [58].

The stress during torsion concentrates at the grooves, thus localising shear at strain rates that are much greater than in the surrounding material. This causes adiabatic heating and the recrystallisation of the structure into very fine grains. Delamination cracks initiate there and subsequently propagate by a repeat of the sequence of recrystallisation and fracture within. An examination of the delamination cracks shows them to be decorated beneath the crack surface by the fine grains,

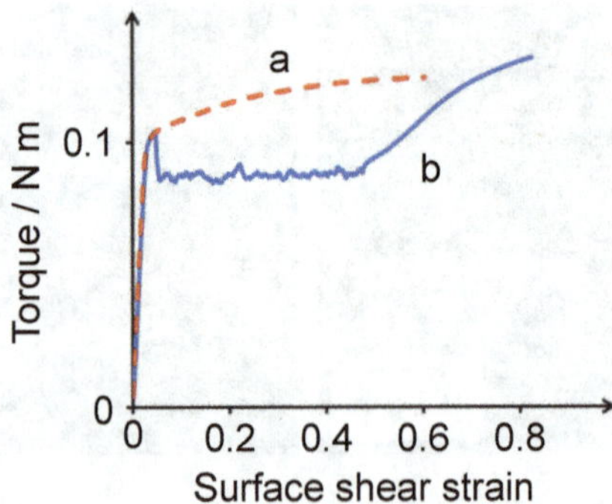

Figure 13.25 Torsion tests on 0.3 mm diameter, drawn pearlitic wire ($\varepsilon_p = 3.5$). (a) Wire that had received an additional drawing pass with a true strain of just 0.02. (b) Wire without the additional pass, exhibiting early failure by delamination. Figure reproduced from Jamoneau et al. [58] with permission from Elsevier.

Figure 13.26.

It has been considered that the fibre texture of the drawn wire makes it easier for it to shear along the axis than on the cross-section. This is said to be the core reason why delamination occurs along the longitudinal axis, not by a brittle mechanism, but by the rapid localisation of shear due to the texture. This interpretation is somewhat weakened by the fact that the texture changes quite significantly as torsion progresses [84].

Figure 13.26 Orientation map showing a thin layer of fine grains below the delamination crack that formed during torsion testing. The fine grains form within shear bands prior to fracture, through a process of adiabatic heating and recrystallisation driven by localised, high-strain rate deformation. Boundaries are defined by misorientations $> 15^\circ$. Image reproduced from Jamoneau et al [58] with permission from Elsevier.

Wires that have a significant circular crystallographic-texture component where $\{110\}_\alpha$ tangent to the wire surface (p. 284), are more susceptible to failure by delamination when compared against those with only a $\langle 110 \rangle_\alpha$ fibre texture [77]. The circular component is usually less than 10% of the total fibre component, making it difficult to isolate the relationship between texture types and the tendency to delaminate. Furthermore, to change the texture requires manipulation of the drawing conditions, which may lead to alterations in other factors such as the distribution of plastic strain across the section.

Wires have been produced using non-circular drawing, in which dies that have elliptical holes, the major axes of which are rotated in sequence, until the final stage where a circular die is used to produce a round-sectioned wire [85]. The circular component of texture decreased from 8% to just 1% for conventionally drawn wire and non-circularly drawn wire, respectively, with a concomitant increase in delamination resistance. The magnitude and distribution of plastic strains across the cross-section of the wires changes, as illustrated in Figure 13.27, so strain homogenisation may be a contributing factor to better delamination performance. The lack of circular symmetry about the wire axis in Figure 13.27b is striking.

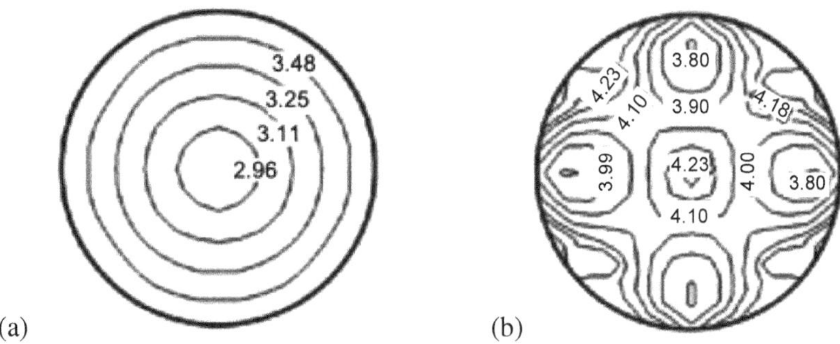

Figure 13.27 Calculated effective plastic strains, which is the strain obtained by integrating the incremental strains experienced by each material point. (a) Wire drawn through a sequence of circular dies. (b) Wire drawn through a combination of dies with circular and elliptical holes. Images reproduced from Baek et al. [85] with permission from Elsevier.

Initiation of strain localisation

The continuity of the cementite is disrupted during wire drawing. Because the fragmentation process depends on the lamellar orientation relative to the applied strain, a hetrogeneous microstructure of cementite results.

Cementite dissolves progressively during wire drawing, in part because the carbon prefers to be at dislocations rather than in Fe_3C. There is a suggestion that this partitioning of carbon promotes delamination. In a comparison of dry- and wet-drawn pearlitic wires, delamination in torsion occurred after a much smaller drawing strain in the former case. Atom-probe measurements show that there was a greater rate of carbon partitioning into the ferrite (as θ dissolves) in the wires drawn without lubrication, because the temperature rise during the deformation accelerated the process [86]. Cause and effect for this mechanism are not established, but wires with comparable textures and residual stress gradients show easier delamination when the ferrite is harder and the concentration gradients of carbon are greater [87]. It is not clear why only a small proportion of samples that are tested identically exhibit delamination.

In further experiments in the same study [87] undeformed pearlitic wire-rods were studied using top-hat shear deformation to reveal that the maximum shear strains occurred when the pearlite lamellae were at a large angle to the wire-rod axis (Figure 13.28). The initial microstructure prior to wire drawing may as a consequence affect delamination. There has not been any serious effort to control this alignment in a commercial scenario, for example by generating pearlite in a temperature gradient (so-called forced-velocity pearlite) [79, 80].

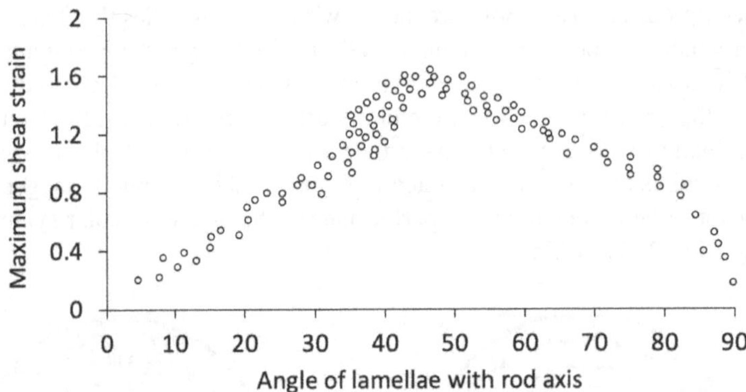

Figure 13.28 The maximum shear strain recorded in top-hat shear tests in individual colonies of pearlite in wire rod as a function of the apparent angle (degrees) made by lamellae with the rod axis. Adapted using data from Durgaprasad et al. [87].

13.7 CHANGES DUE TO ANNEALING

Much of the work done in the deformation of the steel during wire drawing is dissipated as heat. An additional contribution comes from the friction between the steel and the wire surface. The heat increases the temperature of the steel, the details of which depend on the design and parameters of the drawing process [82], but typical die-exit temperatures are 150-200 °C. Indeed, because of the heat generated, no significant difference is found between the elementary mechanical properties of wires where the starting wire-temperature was −196 °C or ambient [88]. The increase in temperature heat treats the deformed structure and is sufficient, for example, to change the hardness of cementite from 960 to 890 HV due to a 120 °C increment above ambient [89, 90]. It may influence the diffusible-hydrogen content of the wire by enhancing the rate at which the steel degases, which might explain the well-known resistance of as-drawn wire to embrittlement [19]. However, the low-temperature ageing is not sufficiently potent to influence dislocation recovery, Figure 13.29a.

Annealing of pearlitic wire (Alloy 18, Table 13.2 $\varepsilon_p = 1.91$) at 100 °C for 10 min does not alter the dislocation density, crystallography or general microstructure when compared with the as-drawn condition, but the tensile strength nevertheless increases from 1940 $\rightarrow$ 2030 MPa [19]. Annealing at 300 °C for 10 min revealed some dislocation recovery and polygonisation but the tensile strength nevertheless increased to 2040 MPa.

These increments in strength have been attributed to the segregation of carbon to dislocations, i.e., the classical strain-ageing phenomenon where dislocations become anchored by Cottrell atmospheres. The ageing temperature need not be low if the duration of the heat-treatment is short. strain ageing does not necessarily lead to the discontinuous yielding during tensile testing, mobile dislocations persist or are generated by coherency strains at the α/θ interfaces. Furthermore, the ultimate tensile strength increases in concert with the proof strength. There is evidence from electrical resistivity measurements that some cementite may dissolve during low-temperature ageing, because there is a reduction in free energy when carbon moves from cementite to dislocation cores [92, 93]. This might explain the increment in strength on the transition from elastic to plastic deformation, but not of the *ultimate tensile strength*, which could only be explained by a sustained increase in the number density of (unknown) barriers to dislocation motion. And the increase in both σ_y and σ_{UTS} also occurs after annealing for 5 min at 75-325 °C, for pearlite subjected to just 0.016 of plastic 'pre-strain' prior to ageing [94], though not in the absence of the pre-strain [92].

Figures 13.30a,b illustrate the problem. When a partly strained tensile specimen is unloaded, aged and reloaded, the yield point recurs but once the dislocations are liberated, the stress-strain

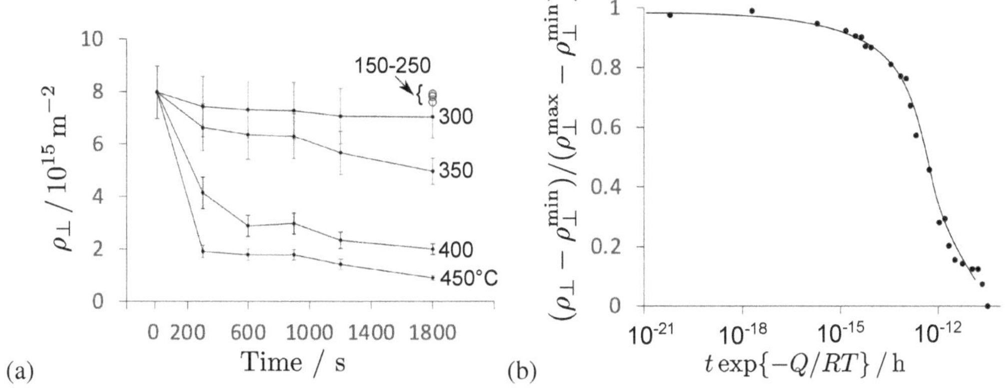

Figure 13.29 (a) Changes in dislocation density due to annealing temperature and time, for Fe-0.82C-0.2Si-0.5Mn wt% steel patented at 560 °C and cold-drawn to $\varepsilon_p = 3$ with the temperature rise during drawing limited to < 100 °C. Data from Chen et al. [91]. (a) Rationalisation of the data plotted in (a) in terms of the kinetic strength of the heat treatment, with $Q = 190\,\mathrm{kJ\,mol^{-1}}$, with the time in seconds and T is the absolute temperature.

"

curve follows the original path. This is similar to observations on single-crystals of iron-containing small concentrations of interstitials [95]. In polycrystalline low-carbon steel that is strained and then aged at 60 °C, there is no change in the flow curve beyond the yield phenomena, as long as the ageing time is limited to about 200 min, but prolonged ageing causes precipitation or clustering that enhances the work hardening rate, thus explaining the increase in σ_{UTS} [97]. It is conceivable that precipitation occurs during the ageing of pearlitic wire, in the carbon-enriched ferrite where cementite existed prior to wire drawing, even though the temperatures involved are low. The tempering of high-carbon martensite is established to cause carbon-clustering and precipitation < 100 °C [98]. Such effects can in principle explain the increase in σ_{UTS} in strain-aged pearlitic wire.

The role of carbides precipitated during the low-temperature annealing of wires in enhancing the flow stress, suggested originally by Wilson and Russell [97], has now been verified, Figures 13.30c,d [96]. The tomograph also shows how the substantial carbon-dispersal fields form in the ferrite, with the shape reminiscent of the original cementite lamellae.

During the high-temperature annealing, the reduction in hardness is more rapid when the initial plastic strain is large, as might be expected given the larger associated defect density and hence the driving force for change, Figure 13.31. If annealing reduces the overall dislocation density, then the volume fraction of cementite must increase in proportion; this most likely occurs by the growth of existing cementite particles rather than the nucleation of new ones.

Vacancy clusters

Severe deformation naturally creates defects within the pearlite. But aside from dislocations, the study of other defects such as vacancies has been limited. Positrons have a greater life at vacancy defects because they encounter fewer electrons there than in the bulk. Figure 13.32 shows how the ratio of $\approx$ 10-vacancy clusters in pearlitic wire drawn to a strain of 3 changes during annealing for 30 min [99]. Smaller vacancy clusters or even individual vacancies were not measured, either because their association with carbon made positron annihilation ineffective, or because they rapidly agglomerate into clusters. Using precipitate dispersion strengthening theory with dislocations cutting the vacancy clusters, the strength increment due to vacancy clusters at $\varepsilon_p = 3$ has been estimated to be in the range 137-204 MPa [100].

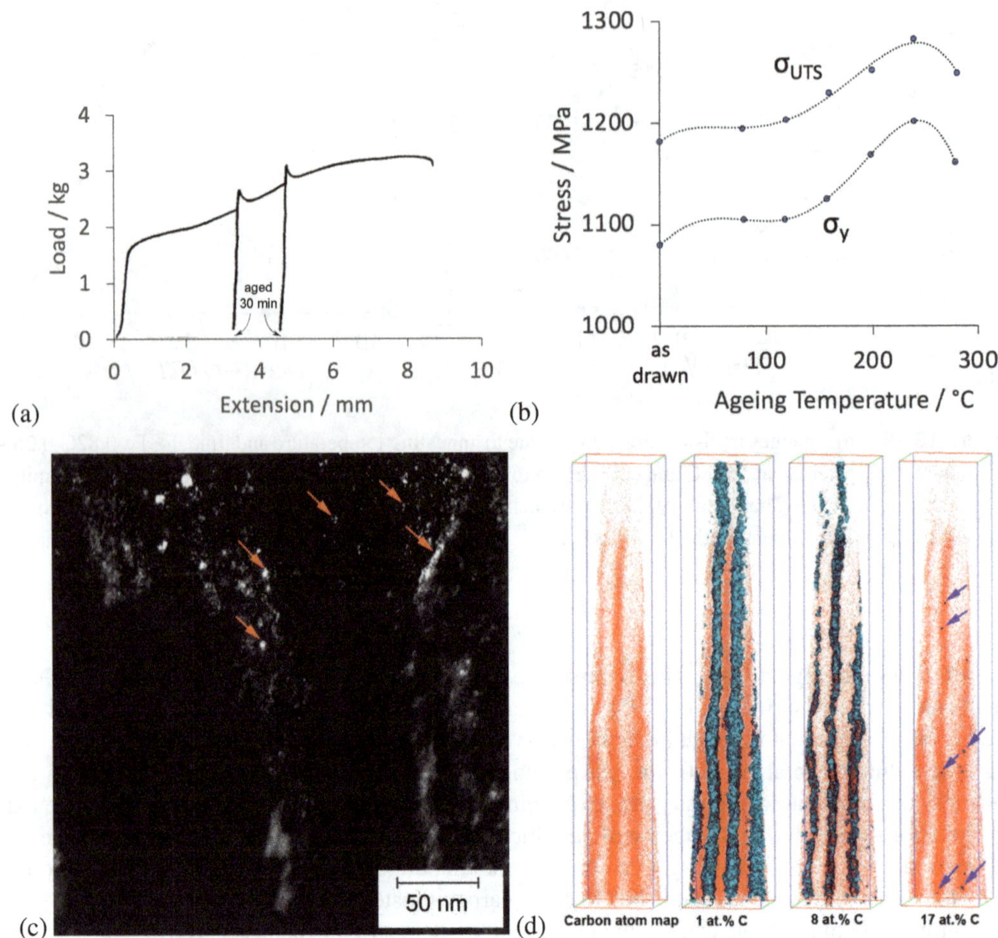

Figure 13.30 (a) Strain-ageing behaviour of single crystal iron to tensile deformation, unloading, ageing and then reloading. Adapted from Holden [95] with the permission of AIP Publishing. (b) Data showing how both σ_y and σ_{UTS} of cold-drawn pearlitic wire changes with ageing at a variety of temperatures for 5 min in each case. Selected data from Yamada [92]. (c) Fine precipitation of carbides in cold-drawn pearlitic wire ($\varepsilon_p = 4.5$) annealed at 100 °C for 12 h. (d) The same wire as (c), showing the distribution of carbon characterised using atom-probe tomography. The arrows indicate positions where the carbon concentration peaks, identified as carbides. Images (c,d) reproduced from Wei et al. [96] with the permission from Elsevier.

Rapid heat-treatment, $T > Ac1$

Although not directly related to wire steel, rapid annealing experiments on a 75% cold-rolled eutectoid steel (Fe-0.85C-0.77Mn-0.17Si-0.2Mo wt%) at 730-820 °C for just 5-60 s are revealing [101]. The strength the pearlite was increased by ≈ 750 MPa by cold-rolling relative to the undeformed state with a reduction in the interlamellar spacing due to the rolling strain (Figure 13.33a). This strain-induced refinement in the interlamellar spacing does not explain the increase in strength, as is evident from the Hall-Petch line plotted in Figure 13.33a, based on Equation 4.17. The majority of the strength increase must therefore be attributed to the increase in dislocation density, particularly within the ferrite.

The short-duration heat treatments on the cold-rolled state have the effect of annealing the ferrite. Some ferrite also transforms into austenite as the intercritical annealing temperature is increased, but

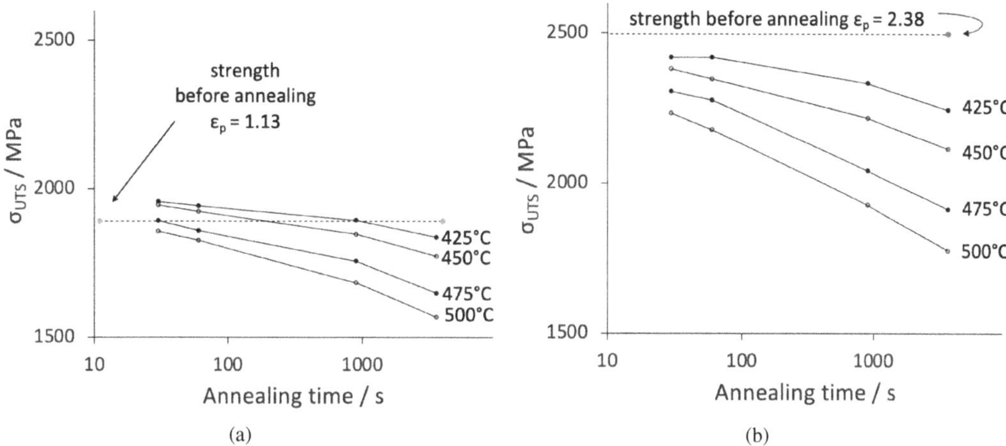

Figure 13.31 The strength of cold-drawn eutectoid steel, Alloy 14, Table 13.2 as a function of annealing time at the temperature indicated. (a) Drawn to a strain of 1.13 followed by annealing. (b) Drawn to a strain of 2.38 followed by annealing. Adapted using data from Joung et al. [16].

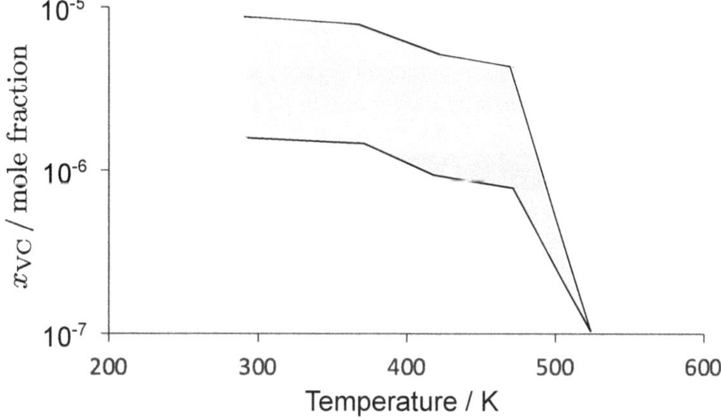

Figure 13.32 Positron annihilation experiments for the wire as in Figure 13.29, showing the change in the mole fraction of vacancy clusters during annealing at the temperature indicated for 30 min. The term x_{VC} refers to the ratio of the number of moles of vacancy clusters to that of iron; each cluster has about ten vacancies. The uncertainty (shaded) is associated with the interpretation of the positron data. Adapted using data from Chen et al. [99].

the cementite that is oriented along the rolling direction, apparently does not. This reverted austenite transforms back into ferrite, with a much reduced defect density, during cooling [102]. The oriented cementite structure is therefore retained in spite of the excursion into the austenite phase field. The continuous yielding typical of virgin pearlitic microstructures changes to σ-ε curves exhibiting Lüders strains and prominent upper and lower yield points, Figure 13.33b. This is because the cold-deformed ferrite now contains a recovered dislocation structure that will have lost its coherency strains with cementite; it is the coherency strains in pearlite when it first forms that generates the continuous yielding behaviour (p. 52). The Lüders strain decreases as the heat-treatment temperature is increased, because some of the cementite will dissolve in γ, resulting in the ever-increasing quantities of virgin pearlite during cooling to ambient temperatures.

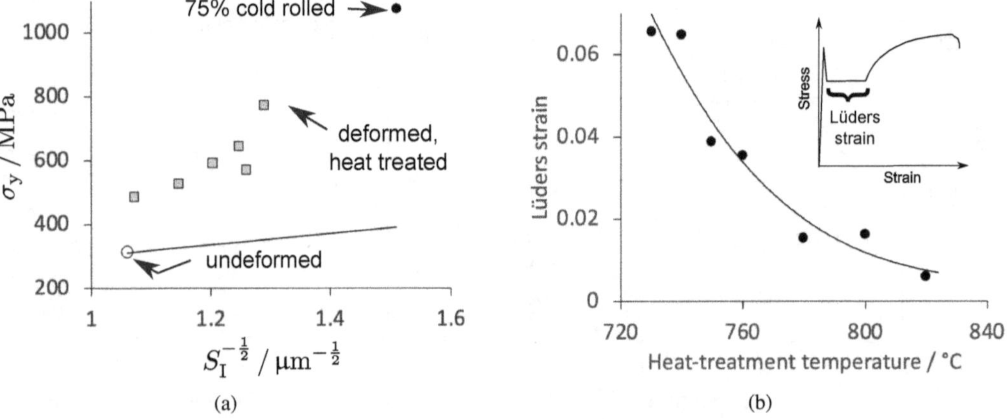

Figure 13.33 (a) Eutectoid steel, transformed into pearlite at 705 °C, air cooled and then cold-rolled to a reduction of 75%. Samples then intercritically annealed as described in the text. The line represents the Hall-Petch relationship for undeformed pearlite (Equation 4.17). (b) Lüders strains observed on deformed and rapidly-heated samples. Adapted using data from Querales and Byrne [101].

Electric treatment

High-frequency pulses of electrical current can induce dramatic changes in the structure and properties of materials [103]; this is illustrated for a drawn pearlitic wire in Figure 13.34 [104]. It is

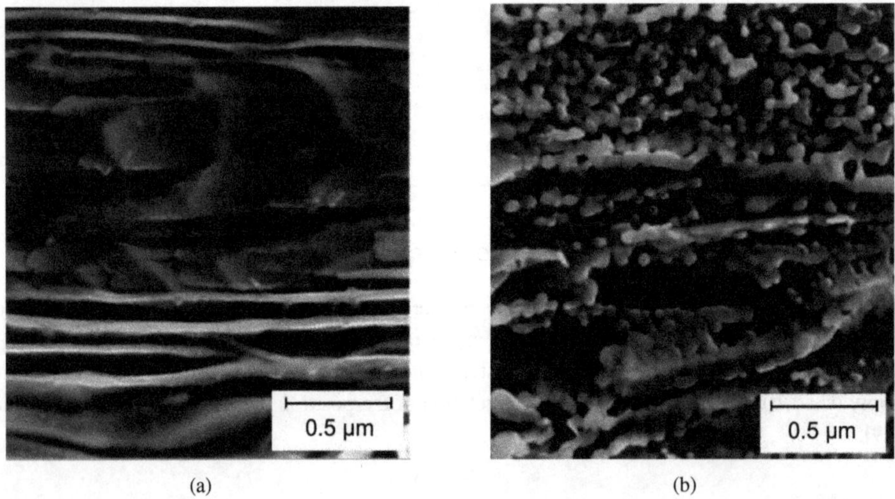

Figure 13.34 Cold-drawn pearlitic steel wire with an 80% reduction and approximate composition Fe-0.8C-0.2Si-0.5Mn wt%, subjected to a peak current density of $9 \times 10^9\,\mathrm{A\,m^{-2}}$ with a single pulse of duration of 150 μs. (a) In the cold drawn state. (b) The same following the application of electropulsing. Reproduced from Qin et al. [104] under the CCBY license.

believed that the observations are not related to electromigration *per se*, which increases atomic mobility, because electropulsing has been found in stainless steels to *suppress* precipitation reactions [104]. The reasons for the electrically-induced structural changes are not understood, but the following phenomena have been discussed *qualitatively*:

- that the pulsing changes the thermodynamic free energy of the system. This could be tested (not yet been done) using phase diagram calculations to predict the effect of electropulsing, in a manner akin to calculations for the effects of pressure and magnetic fields on phase stability.
- Joule heating due to the applied current density is considered unimportant [105]. The pearlitic wire experiments of Qin et al. involve a current density of about $< 14\,\mathrm{A\,mm^{-2}}$, a pulse duration of 80 μs, pulse frequency 100 Hz over a period of 10 min, resulting in an average temperature increment of just $\approx 25\,°\mathrm{C}$ in experiments [106]. Much greater current densities and pulse durations ($18.6\,\mathrm{kA\,mm^{-2}}$, 800 μs) can lead to calculated temperature increments of $\Delta T \approx 750\,°\mathrm{C}$ [107]. For a pulse duration Δt,

$$\Delta T = \frac{\int_0^{\Delta t} j^2 \rho_e \, \mathrm{d}t}{C_P \rho} \tag{13.15}$$

  where j is the current density, ρ_e is the electrical resistivity, C_P the heat capacity at constant pressure and ρ the density.
- Stress effects caused by thermal changes, but the timescales of the pulsing and the fact that the cementite lamellae spheroidise means that diffusion must play a role – stress alone cannot change the shape of pearlite.
- Heterogeneities due to the skin effect whereby the electrical current density varies as a function of depth from the surface of the sample.

The mechanical properties of electropulsed pearlitic wire are not particularly different from those without such treatment [106].

13.8 DEFORMATION-INDUCED DISSOLUTION OF CEMENTITE

Cementite in steel can become unstable with respect to ferrite that is rich in dislocations. This is because a carbon atom fits better within the strain field of a dislocation [108], leading to a reduction in energy by ≈0.5 eV when it is located within a distance b off the dislocation core. The corresponding energy reductions when the carbon is in the lattices of ε or θ carbides are about 0.27 and < 0.5 eV respectively [109, 110]. Plastic deformation can therefore stimulate the carbides to dissolve with the carbon migrating to the dislocation-rich ferrite. Figure 13.35a illustrates the case for cementite as a function of the $\rho_\perp$ and the number of carbon atoms in a single plane of iron atoms threaded by the dislocation line. Each line represents the concentration of carbon that can be absorbed by the dislocations.

Figure 13.35b shows how the volume fraction of cementite in a eutectoid steel (Alloy 21, Table 13.2) evolves during cold-drawing. The thinner cementite in the 570 °C sample dissolves more rapidly. For both samples, the amount of carbon introduced into the ferrite at the expense of the cementite, at a maximum drawing strain of 3.2, can be estimated at $w_C^\alpha = 0.27$. From Figure 13.8, the dislocation density within the ferrite at a strain of 3.2 is $\approx 1.36 \times 10^{16}\,\mathrm{m^{-2}}$. Experimental observations using the atom-probe indicate that there are $N_C \approx 21$ carbon atoms per iron-atom plane [111]. Using these data, the point plotted in Figure 13.35a shows that the dislocations can in principle accommodate more than the 0.27C wt%, consistent with the observed fraction of cementite that dissolves.

Other data [112] on hypereutectoid pearlite drawn to strains of up to 5.1, showing the dissolution of cementite using at atom probe, can similarly be explained in terms of the dislocation density. Precise agreement is not expected [113] because the capacity of dislocations to attract carbon decreases when they are closely spaced. In a particularly interesting study where the X-ray diffraction patterns of high-C martensite and of drawn pearlite were compared, the former exhibited the tetragonality expected from Bain-strain induced ordering of carbon atoms, whereas the latter did not because the carbon atoms there are primarily located at dislocation cores [114].

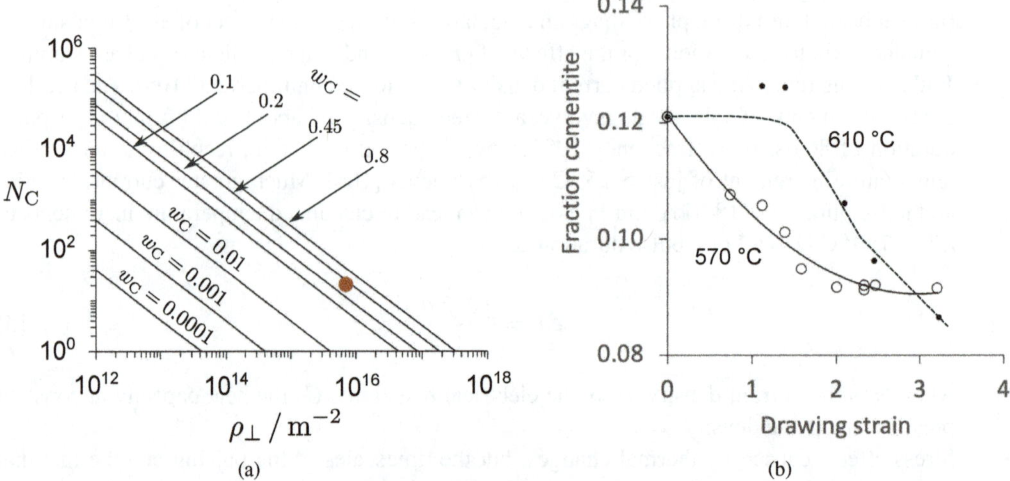

Figure 13.35 (a) The number N_C of carbon atoms segregated at a dislocation, per plane of iron atoms normal to a dislocation line, versus the dislocation density, for a variety of carbon concentrations. Each line represents the combination of N_C and $\rho_\perp$ that would make the carbon indicated more stable at dislocations than in cementite. The dislocations are assumed to consist of an equal mixture of screw and edge varieties. After Kalish and Cohen [110]. The point plotted is described in the text. (b) Dissolution of cementite during the cold drawing of eutectoid steel, as determined using Mössbauer spectroscopy. The temperatures represent that at which the pearlite was generated prior to deformation. The original data from Nam et al. [5] have been adjusted to allow the two curves to begin with a common origin at the expected equilibrium fraction of cementite at zero strain, since they all originate from the same steel.

The atomic mechanism of deformation-induced cementite dissolution has an analogy in mechanical alloying [115], i.e., the mechanical mixing of particles of θ and α until eventually, the particles mix on an atomic scale. During the process, the interface may lose definition and the composition of the cementite may change with size.

Figure 13.36 shows that the concentration of carbon varies 'smoothly' on tracing a path through ferrite towards a lamella of cementite and back into ferrite on the other side [13]. The atom probe technique in this case does not reveal the crystal structure, so the crystallography of the regions that have less than 25 at% carbon is not clear. Microscopy and X-ray diffraction of wire drawn to $\varepsilon_p = 2$ have demonstrated that the crystalline order within the cementite is compromised, with some regions apparently becoming amorphous [116]. The cementite in the vicinity of the α/θ interface can exhibit amorphous regions in such wires [72], perhaps not surprising given its state of dissolution during deformation. These observations are limited by the volume of material examined at high resolution; the cementite cannot all be in an amorphous state. Extracted cementite from wire drawn to a strain as large as 3.47 still exhibits periodicity; there are at least five coincidences between diffraction peaks from the severely drawn wire and that which was not deformed, Figure 13.36c. There may exist more such coincidences, rendered unclear because of peak broadening due to the deformed state.

There is another reason why cementite can dissolve when the steel is severely deformed. The plastic deformation thins the cementite by elongation, so increasing the θ/α interfacial area per unit volume; the cementite may also be refined by the deformation cutting the particles as they pass through [118]. Both of these factors reduce its stability until it is more energetically favourable for the cementite to dissolve.

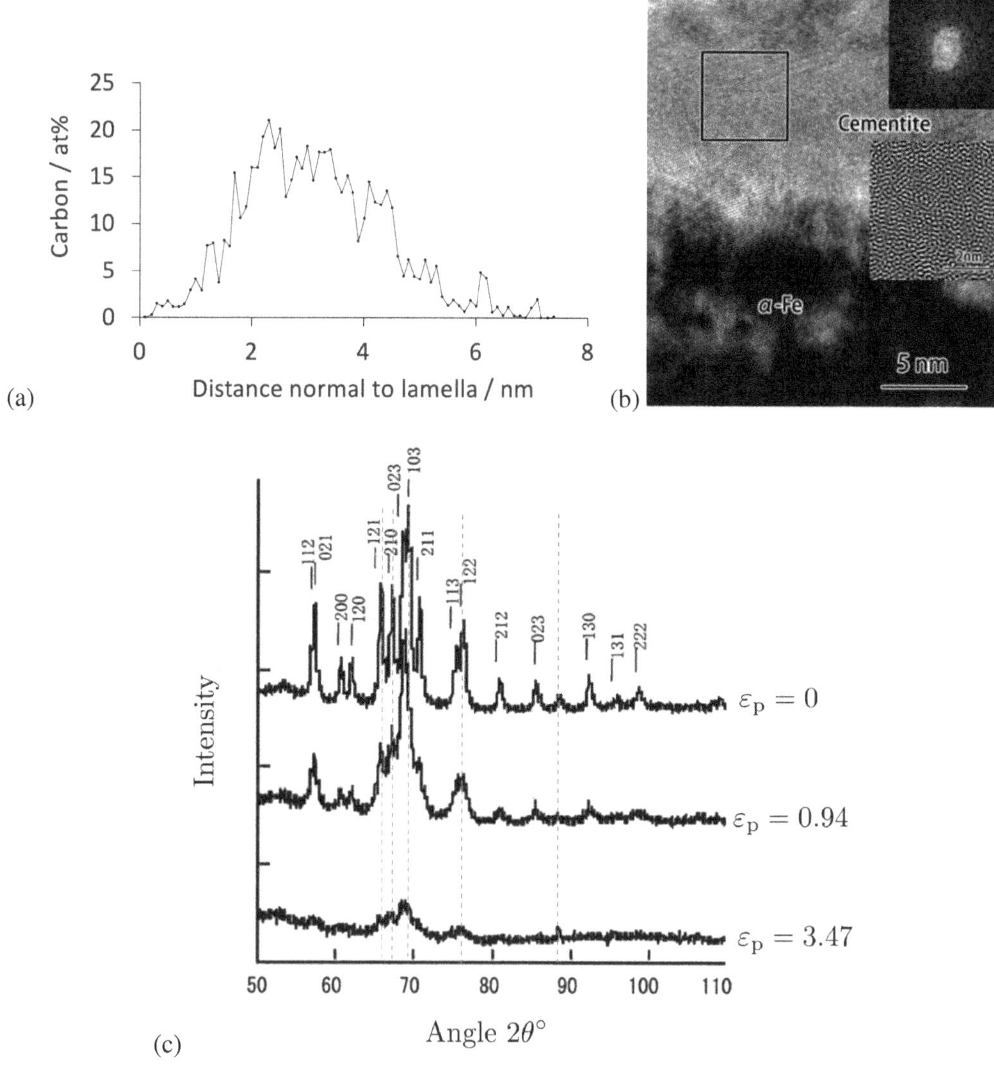

Figure 13.36 (a) An atom probe concentration-profile for carbon across a cementite lamella in Alloy 9, Table 13.2. The steel was cold-drawn to $\varepsilon_p = 3.5$. Data adapted from Danoix et al. [13]. (b) Micrograph showing the amorphous state of cementite in wire drawn to $\varepsilon_p = 0.2$. The insets show the fast Fourier transform and its inverse, confirming the amorphous state in the cementite region examined. Image reproduced from Zhou et al. [116], with the permission from Elsevier. (c) X-ray diffraction from extracted cementite. Image reproduced with minor adaptations, from Borchers et al. [117], with the permission of Elsevier.

13.9 HIGH-CARBON, HYPEREUTECTOID WIRES

The amount of carbon that can be added to steel destined for wire production is not limited by the eutectoid concentration 0.76 wt%, because with sufficient undercooling, fully pearlitic microstructures can be achieved even in hypereutectoid steels (Chapter 7). However, the tendency to form proeutectoid cementite networks at the austenite grain surfaces increases, usually with an accompanying loss of ductility. The carbon concentration determines the volume fraction of cementite and therefore, the strength of the wire.

Any factor that leads to a reduction in the interlamellar spacing and therefore the mean free path $z_\perp$ for dislocation motion, increases the work hardening rate. For drawn wire, this translates into a stronger wire.

Most of the compositions listed in Table 13.2 strictly are hypereutectoid, but the purpose here is to focus on carbon concentrations that are close to 1 wt%, where problems due to the formation of proeutectoid cementite and the lack of the ductility that is needed during the drawing process can become problematic. The way to avoid proeutectoid cementite in such steels is to cool sufficiently rapidly before the steel reaches the patenting temperature. Figure 13.37 maps experimental data on the carbon concentration versus the cooling rate at the centre of a typical hot-rolled rod, with domains where only pearlite forms in hypereutectoid steel. The cooling rate achieved in a Stelmor line can be 10-15 °C s^{-1} so it is possible for a 1.1 wt% steel to transform completely into pearlite at the patenting temperature (575 °C) [9].

Based on this principle, Ochiai et al. designed a hypereutectoid steel of composition Fe-0.96C-0.2Si-0.3Mn-0.2Cr wt%, with the Mn and Cr concentrations kept small in order to avoid chemical segregation effects that can lead to martensite formation. The pearlite was generated by transformation at 575 °C. After drawing down to 0.04 mm diameter ($\varepsilon_p = 9.8$) from the initial 5.5 mm, an extraordinary $\sigma_{UTS} = 5.7$ GPa was recorded with excellent resistance to delamination during torsion testing, primarily because the microstructure did not contain bainite or martensite; this has the effect of increasing the work-hardening rate of the steel, which is known to enhance the resistance to torsional failure. The ultimate strength on drawing down to 1 mm ($\varepsilon_p = 3.4$) was just over 3 GPa, which is comparable to σ_{UTS} data reported for almost identical composition (Alloy 4, Table 13.2), transformation temperature and ε_p, by Song et al. [119].

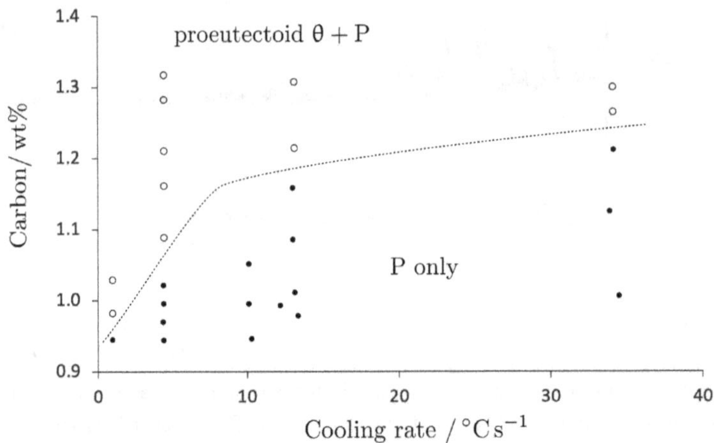

Figure 13.37 The carbon concentration versus the cooling rate of a 5.5 mm rod at its centre, indicating whether or not proeutectoid cementite forms before the pearlite reaction. The steel also contains 0.2Si and 0.5Mn wt%. Adapted using data from Ochiai et al. [9].

13.10 GALVANISING OF STEEL WIRE

A number of intermetallic compounds form by the rapid reaction of steel dipped into a Zn-rich liquid at about 450 °C during the galvanising process. The coating therefore is not homogeneous, consisting instead of multiple phases that become richer in zinc with distance normal to the original steel surface. The initial thickness of the coating in the case of steel wires is of the order of 30 μm, which naturally will decrease during the drawing operation, not simply because of the spreading of the zinc over a greater surface, but also because some of the zinc may be stripped by the mechanical forces experienced as the wire progresses through the dies.

The modern galvanising bath usually contains aluminium, nickel, bismuth and tin as deliberate additions in small concentrations. The additions influence the morphology, kinetics and structure of the coating. Larger concentrations of aluminium ($\approx 10\,\text{wt}\%$) in the predominantly zinc bath have been shown to dramatically improve the resistance of pearlitic wire used for bridge construction, in both humid and wet environments; the mechanism involves the formation of more compact and adherent corrosion products when compared with just zinc galvanisation [120].

Steel wires are usually cleaned by pickling before galvanising in molten zinc at 450 °C. They may then be drawn in the galvanised condition, with the zinc accommodating the drawing strain and may even act as a solid lubricant. Nevertheless, the coating may suffer roughening and wear during the forming operations, and in some cases peel (delaminate). The presence of oxides on the surface of the wire can lead to a non-uniform coating-thickness with variable adhesion to the steel. In particular, silicon within the steel can have an adverse effect on galvanised coatings by promoting more brittle Fe-Zn compounds [121]. Steels containing sufficient concentrations of silicon tend to form a low melting temperature oxide called fayalite (Fe_2SiO_4), which then penetrates both the steel and any other oxide to form a mechanical key. As a consequence, routine descaling operations fail to remove all traces of the oxide, making subsequent galvanising imperfect.

Galvanising is in effect a short heat-treatment, which does cause the as-drawn wire to lose some strength. It is, nevertheless, a necessary process when manufacturing cables for marine bridges. An increase in the silicon concentration of the steel helps to mitigate any loss of strength, Figure 13.38, because of the well-known effect that it has on retarding spheroidisation (p. 176).

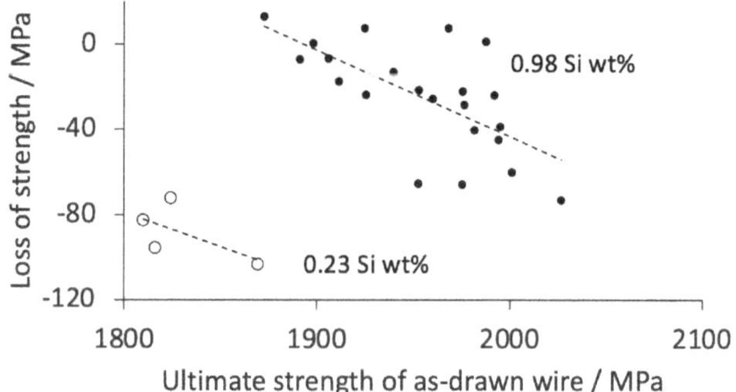

Figure 13.38 Loss of strength due to hot-dip galvanising. Selected data from Takahashi et al. [18].

REFERENCES

1. P. L. Pelmear: 'Manufacturing processes', *Journal of the Society of Occupational Medicine*, 1976, **26**, 100–101.
2. K. Kumar, D. Goyal, and S. S. Banwait: 'Effect of key parameters on fretting behaviour of wire rope: A review', *Archives of Computational Methods in Engineering*, 2020, **27**, 549–561.
3. R. D. Morales, A. Lopéz, and I. M. Olivares: 'Mathematical simulation of the Stelmor process', *Ironmaking and Steelmaking*, 1991, **18**, 129–138.
4. C. Ciganik, J. Speer, K. Findley, and W. van Raemdonck: 'Effects of post-drawing surface heat treatments on the strength and torsional ductility of high strength steel wire', *Materials Science Forum*, 2018, **941**, 1790–1795.

5. W. J. Nam, C. M. Bae, S. J. Oh, and S. J. Kwon: 'Effect of interlamellar spacing on cementite dissolution during wire drawing of pearlitic steel wires', *Scripta Materialia*, 2000, **42**, 457–463.
6. Anonymous: 'High tensile steel wire rods applying mist cooling process': Tech. Rep. JFE Technical Report 15, JFE, Japan, 2010.
7. K.-S. An, S. W. Jeong, H. J. Bea, and W. J. Nam: 'Formation of abnormal structures and their effects on the ductility of eutectoid steel', *Metals and Materials International*, 2016, **22**, 995–1002.
8. J. Toribio, F. J. Ayaso, B. González, J.-C. Matos, D. Vergara, and M. Lorenzo: 'Tensile fracture behavior of progressively-drawn pearlitic steels', *Metals*, 2016, **6**, 114.
9. I. Ochiai, S. Nishida, H. Ohba, and A. Kawana: 'Application of hypereutectoid steel for development of high strength steel wire', *Tetsu-to-Hagané*, 1993, **79**, 1101–1107.
10. C. Ciganik: 'Influence of processing and composition on the strength and torsional ductility of high strength steel wire': Master's thesis, Colorado School of Mines, Colorado, USA, 2017.
11. H. R. Song, E. G. Kang, and W. J. Nam: 'Effect of alloying elements on work hardening behavior in cold drawn hyper-eutectoid steel wires', *Materials Science and Engineering: A*, 2007, **449**, 1147–1150.
12. J. D. Embury, and R. M. Fisher: 'The structure and properties of drawn pearlite', *Acta Metallurgica*, 1966, **14**, 147–159.
13. F. Danoix, D. Julien, X. Sauvage, and J. Copreaux: 'Direct evidence of cementite dissolution in drawn pearlitic steels observed by tomographic atom probe', *Materials Science & Engineering A*, 1998, **250**, 8–13.
14. J. Y. Jung, K. S. An, P. Y. Park, and W. J. Nam: 'Effects of wire drawing and annealing conditions on torsional ductility of cold drawn and annealed hyper-eutectoid steel wires', *Metals*, 2020, **10**, 1043.
15. H. Tashiro, and T. Tarui: 'State of the art for high tensile strength steel cord': Tech. Rep. 88, Nippon Steel Corporation Technical Report, Japan, 2003.
16. S. W. Joung, U. G. Kang, S. P. Hong, Y. W. Kim, and W. J. Nam: 'Aging behavior and delamination in cold drawn and post-deformation annealed hyper-eutectoid steel wires', *Materials Science & Engineering A*, 2013, **586**, 171–177.
17. C. M. Bae, W. L. Nam, and C. S. Lee: 'Effect of interlamellar spacing on the delamination of pearlitic steel wires', *Scripta Materialia*, 1996, **35**, 641–646.
18. T. Takahashi, T. Tarui, and S. Konno: 'Development of hightensile strength wire of 180 kgf/mm^2 for bridge cable', *Steel Construction Engineering*, 1994, **1**, 119–126.
19. D. Hirakami, T. Manabe, K. Ushioda, K. Noguchi, K. Takai, Y. Hata, S. Hata, and H. Nakashima: 'Effect of aging treatment on hydrogen embrittlement of drawn pearlitic steel wire', *ISIJ International*, 2016, **56**, 893–898.
20. V. T. L. Buono, B. M. Gonzalez, T. M. Lima, and M. S. Andrade: 'Measurement of fine pearlite interlamellar spacing by atomic force microscopy', *Journal of Materials Science*, 1997, **32**, 1005–1008.
21. M. Zelin: 'Microstructure evolution in pearlitic steels during wire drawing', *Acta Materialia*, 2002, **50**, 4431–4447.
22. V. M. Kardonskii, G. V. Kurdjumov, and M. D. Perkas: 'Effect of the size and shape of cementite particles on the structure and properties of deformed steel', *Metal Science and Heat Treatment*, 1966, **6**, 62–68.
23. M. Dollar, I. M. Bernstein, and A. W. Thompson: 'Influence of deformation substructure on flow and fracture of fully pearlitic steel', *Acta Metallurgica*, 1988, **36**, 311–320.
24. Y. Guo, Z. Li, C. Yao, K. Zhang, F. Lu, K. Feng, J. Huang, M. Wang, and Y. Wu: 'Microstructure evolution of Fe-based nanostructured bainite coating by laser cladding', *Materials and Design*, 2014, **63**, 100–108.

25. X. Zhang, A. Godfrey, X. Huang, N. Hansen, and Q. Liu: 'Microstructure and strengthening mechanisms in cold-drawn pearlitic steel wire', *Acta Materialia*, 2011, **59**, 3422–3430.
26. G. Langford, and M. Cohen: 'Calculation of cell size strengthening of wire-drawn iron', *Metallurgical Transactions*, 1970, **1**, 1478–1480.
27. Q. Zhu, C. M. Sellars, and H. K. D. H. Bhadeshia: 'Quantitative metallography of deformed grains', *Materials Science and Technology*, 2007, **23**, 757–766.
28. H. K. D. H. Bhadeshia: Geometry of Crystals: 2nd edition. Institute of Materials, 2001.
29. H. K. D. H. Bhadeshia: 'Axisymmetric tensile deformation of randomly oriented pearlite': http://www.msm.cam.ac.uk/map/steel/programs/spacing.html, 2024.
30. M. H. Faber, S. Engelund, and R. Rackwitz: 'Aspects of parallel wire cable reliability', *Structural Safety*, 2003, **25**, 201–225.
31. T. Fujita, Y. Yamada, and T. Yamada: 'Anisotropy of high carbon steel wire', *Tetsu-to-Hagané*, 1973, **59**, S255.
32. Y. Yamada, T. Yokoyama, T. Yamada, T. Fujita, and S. Kinoshita: 'Effects of nitrogen and drawing schedule on the tensile properties of high carbon steel wire', *Tetsu-to-Hagané*, 1978, **64**, 420–429.
33. K. van Acker, J. Root, P. van Houtte, and E. Aernoudt: 'Neutron diffraction measurement of the residual stress in the cementite and ferrite phases of cold-drawn steel wires', *Acta Materialia*, 1996, , 4039–4049.
34. T. Suzuki, Y. Tomoga, A. Moriai, N. Minakawa, and Y. Morii: 'Anisotropic residual stresses in a pearlitic steel after tensile deformation', In: *Advanced Technology in Experimental Mecahnics*. Tokyo, Japan: Japan Society of Mechanical Engineers, Materials and Mechanics Division, 2003:OS04W0082.
35. T. Suzuki, Y. Tomota, M. Isaka, A. Noriai, N. Minakawa, and Y. Morii: 'Strength anisotropy and residual stress in drawn pearlite steel wire', *ISIJ International*, 2004, **44**, 1426–1430.
36. M. Kumagai, S. Sato, S. Suzuki, M. Imafuku, H. Tashiro, and S. Ohya: 'Residual stress analysis of cold-drawn pearlite steel wire using white synchrotron radiation', *ISIJ International*, 2015, **55**, 1489–1495.
37. T. Akada, R. Ueji, M. Mitsuhara, S. Yamasaki, and M. Tanaka: 'Plastic anisotropy in yield stress of drawn pearlitic steels', *Materials Science and Engineering: A*, 2024, **898**, 146380.
38. P. J. Withers, and H. K. D. H. Bhadeshia: 'Residual stress part 1 – measurement techniques', *Materials Science and Technology*, 2001, **17**, 355–365.
39. J. Bauschinger: 'Ueber die veranderung der elasticitatsgrenge und der festigkeit des eisens und stahls durch strecken und quetschn, durch erwarmen und abkuhlen und durch oftmal wiederholte beanspruchung (about the change in elastic limit and the strength of iron and steel through stretching and squeezing ...)', *Mitteilungen aus dem Mechanisch-Technischen Laboratorium der K. Technischen Hochschule in Munchen*, 1886, **13**, 1–115.
40. M. Elices: 'Influence of residual stresses in the performance of cold-drawn pearlitic wires', *Journal of Materials Science*, 2004, **39**, 3889–3899.
41. W. F. Hosford, Jr: 'Microstructural changes during deformation of [011] fiber-textured metals', *Transactions of the Metallurgical Society of the AIME*, 1964, **230**, 12–15.
42. E. Schmid, and W. Boas: Plasticity of Crystals (translated from the 1935 edition of Kristalplastizitaet): London, U.K.: F. A. Hughes and Co., 1950.
43. J. W. Christian: 'Deformation by moving interfaces', *Metallurgical Transactions A*, 1982, **13**, 509–538.
44. A. Kelly, and K. M. Knowles: Crystallography and Crystal Defects: 2nd ed., New York, USA: John Wiley & Sons, Inc., 2012.
45. H. K. D. H. Bhadeshia: Geometry of Crystals, Polycrystals and Phase Transformations: Florida, USA: CRC Press, Taylor and Francis, 2017.

46. R. K. Chin, and P. S. Steif: 'A computational study of strain inhomogeneity in wire drawing', *International Journal of Machine Tools & Manufacture*, 1995, **35**, 1087–1098.
47. J. Očenášek, M. Rodriguez Ripoll, S. M. Weygand, and H. Riedel: 'Multi-grain finite element model for studying the wire drawing process', *Computational Materials Science*, 2007, **39**, 23–28.
48. J. Gil Sevillano, D. González, and J. M. Mariínez-Esnaola: 'Heterogeneous deformation and internal stresses developed in BCC wires by axisymmetric elongation', *Materials Science Forum*, 2007, **550**, 75–84.
49. M. Tanaka, T. Manabe, T. Morikawa, and K. Higashida: 'Nature of delamination cracks in pearlitic steels', *Scripta Materialia*, 2016, **112**, 32–36.
50. M. Tanaka, T. Manabe, T. Morikawa, and K. Higashida: 'Mechanism behind the onset of delamination in wire-drawn pearlitic steels', *Tetsu-to-Hagané*, 2019, **105**, 155–162.
51. V. N. Khiratkar, K. Mishra, and A. Singh: 'Deformation and microstructural evolution of nanostructured pearlite under tension versus torsion', *Steel Research International*, 2021, **92**, 2000526.
52. S. Gondo, R. Tanemura, R. Mitsui, S. Kajino, M. Asakawa, K. Takemoto, K. Tashima, and S. Suzuki: 'Relationship between mesoscale structure and ductility of drawn high carbon steel wire', *Materials Science & Engineering A*, 2021, **800**, 140283.
53. C. S. A. da Silva, B. R. C. Saraiva, L. Novotný, P. W. C. Sarvezuk, M. Masoumi, C. C. Silva, L. F. G. Herculano, J. L. Cardoso, H. F. G. de Abreu, and M. Béreš: 'Texture and lattice strain evolution in a pearlitic steel during shear deformation: an in situ synchrotron X-ray diffraction study.', *International Journal of Plasticity*, 2024, **?**, https://doi.org/10.1016/j.ijplas.2024.104083.
54. T. Montesin, and J. Heizmann: 'Evolution of crystallographic texture in thin wires', *Journal of Applied Crystallography*, 1992, **25**, 665–673.
55. M. R. Barnett, A. Sullivan, and R. Balasubramaniam: 'Electron backscattering diffraction analysis of an ancient wootz steel blade from central india', *Materials Characterization*, 2009, **60**, 252–260.
56. P. van Houtte: 'Crystallographic texture in drawn wires: measurement, modeling and effect on mechanical properties', *Wire Journal International*, 2012, **45**, 58–68.
57. A. Jamoneau, J.-H. Schmitt, and D. Solas: 'Measurement of texture gradient in heavily cold-drawn pearlitic wires', *Advanced Engineering Materials*, 2018, **20**, 1700279.
58. A. Jamoneau, D. Solas, J. Bourgon, P. Morisot, and J.-H. Schmitt: 'Strain localization and delamination mechanism of cold-drawn pearlitic steel wires during torsion', *Materials Science & Engineering A*, 2021, **814**, 141222.
59. M. Kriška, J. Tacq, K. van Acker, M. Seefeldt, and S. van Petegem: 'Neutron and X-ray diffraction study of residual and internal stress evolution in pearlitic steel during cold drawing', *Journal of Physics: Conference Series*, 2012, **340**, 012101.
60. J. M. Atienza, J. Ruiz-Hervias, and M. Elices: 'Improved cold-drawn eutectoid steel wires based on residual stress measurement and simulation: Part 2. optimization of mechanical properties', *Wire Journal International*, 2009, **42**, 52–63.
61. M. L. Martinez-Perez, F. J. Mompean, J. Ruiz-Hervias, C. R. Borlado, J. M. Atienza, M. Garcia-Hernandez, M. Elices, J. Gil-Sevillano, R. L. Peng, and T. Buslaps: 'Residual stress profiling in the ferrite and cementite phases of cold-drawn steel rods by synchrotron X-ray and neutron diffraction', *Acta Materialia*, 2004, **52**, 5303–5313.
62. J. Atienza, and M. Elices: 'Influence of residual stresses in the stress relaxation of cold drawn wires', *Materials and Structures*, 2004, **37**, 301–304.
63. J. M. Atienza, J. Ruiz-Hervias, M. L. Martinez-Perez, F. J. Mompean, M. Garcia-Hernandez, and M. Elices: 'Residual stresses in cold drawn pearlitic rods', *Scripta Materialia*, 2005, **52**, 1223–1228.

64. S. He, A. van Ball, S. Y. Li, P. van Houtte, F. Mei, and A. Sarban: 'Residual stress determination in cold drawn steel wire by FEM simulation and X-ray diffraction', *Materials Science & Engineering A*, 2003, **346**, 101–107.
65. F. Yang, J. Q. Jiang, Y. Wang, C. Ma, F. Fang, K. L. Zhao, and W. Li: 'Residual stress in pearlitic steel rods during progressively cold drawing measured by X-ray diffraction', *Materials Letters*, 2008, **62**, 2219–2221.
66. G. Khatibi, R. Stickler, and B. Weiss: 'Strength and ductility of thin Cu wires', *International Journal of Materials Research*, 2022, **94**, 749–753.
67. A. Durgaprasad, S. Giri, S. Lenka, S. Kundu, S. Mishra, S. Chandra, R. D. Doherty, and I. Samajdar: 'Microstructures and mechanical properties of as-drawn and laboratory annealed pearlitic steel wires', *Metallurgical & Materials Transactions A*, 2017, **48**, 4583–4597.
68. J. M. Walton: 'Developments in steel cables', *Journal of Constructional Steel Research*, 1996, **39**, 3–19.
69. C. Borchers, and R. Kirchheim: 'Cold-drawn pearlitic steel wires', *Progress in Materials Science*, 2016, **82**, 405–444.
70. L. Zhou, F. Fang, M. Kumagi, E. Pickering, and X. Zhang: 'A modified pearlite microstructure to overcome the strength-plasticity trade-off of heavily drawn pearlitic wire', *Scripta Materialia*, 2022, **206**, 114236.
71. S. Gondo, S. Suzuki, M. Asakawa, K. Takemoto, K. Tashima, and S. Kajino: 'Improvement of ductility with maintaining strength of drawn high carbon steel wire', *Key Engineering Materials*, 2016, **716**, 32–38.
72. L. Zhou, F. Fang, M. Kumagai, E. Pickering, and T. Yu: 'Structure restoration and coarsening of nanocrystalline cementite in cold drawn pearlitic wire induced by low temperature annealing', *Scripta Materialia*, 2022, **215**, 114696.
73. N. Guo, B. Luan, and Q. Liu: 'Influence of pre-torsion deformation on microstructures and properties of cold drawing pearlitic steel wires', *Materials and Design*, 2013, **50**, 285–292.
74. A. Nadai: Theory of Flow and Fracture of Solids: 3rd ed., New York, USA: McGraw-Hill, 1950.
75. G. E. Dieter: Mechanical Metallurgy: 3rd ed., McGraw-Hill, 1986.
76. J. Gil Sevillano: 'A twist on heavily drawn wires', *Wire Journal International*, 2011, **44**, 58–70.
77. I. Lefever, U. D'Haene, W. van Raemdonck, E. Aernoudt, P. van Houtte, and J. Gil Sevillano: 'Modeling of the delamination of high strength steel wire', *Wire Journal International*, 1998, **31**, 90–95.
78. H. Jin, W.-Y. Lu, J. W. Foulk, J. Ostien, S. L. Kramer, and A. Jones: 'Solid cylinder torsion experiment for large shear deformation': Tech. Rep. SAND2020-2415C, Sandia National Laboratories, U.S.A., 2020.
79. G. F. Bolling, and R. H. Richman: 'Forced velocity pearlite', *Metallurgical Transactions*, 1970, **1**, 2095–2104.
80. D. D. Pearson, and J. D. Verhoeven: 'Forced velocity pearlite in high purity Fe-C alloys: Part 1. experimental', *Metallurgical Transactions A*, 1984, **15**, 1037–1045.
81. 'Creative commons attribution (CC BY) license': last accessed 30th November 2021: URL https://creativecommons.org/licenses/by/4.0/.
82. S. K. Lee, D. C. Ko, and B. M. Kim: 'Pass schedule of wire drawing process to prevent delamination for high strength steel cord wire', *Materials and Design*, 2009, **30**, 2919–2927.
83. P. F. Willemse, and B. P. Naughton: 'Effect of small drawing reductions on residual surface stresses in thin cold-dravvn steel vvire, as measured by X-ray diffraction', *Materials Science and Technology*, 1985, **1**, 41–45.
84. N. Guo, B. Luan, and Q. Liu: 'Influence of torsion deformation on textures of cold drawing pearlitic steel wires', *Acta Metallurgica Sinica (English Letters)*, 2015, **28**, 707–714.

85. H. M. Baek, S. K. Hwang, H. S. Joo, Y.-T. Im, I.-H. Son, and C. M. Bae: 'The effect of a non-circular drawing sequence on delamination characteristics of pearlitic steel wire', *Materials and Design*, 2014, **62**, 137–148.
86. N. Maruyama, T. Tarui, and H. Tashiro: 'Atom probe study on the ductility of drawn pearlitic steels', *Scripta Materialia*, 2002, **46**, 599–603.
87. A. Durgaprasad, S. Giri, S. Lenka, S. K. Sarkar, A. Biswas, S. Kundu, S. Mishra, S. Chandra, R. D. Doherty, and I. Samajdar: 'Delamination of pearlitic steel wires: The defining role of prior-drawing microstructure', *Metallurgical & Materials Transactions A*, 2018, **49**, 2037–2047.
88. D. Wei, X. Min, X. Hu, Z. Xie, and F. Fang: 'Microstructure and mechanical properties of cold drawn pearlitic steel wires: Effects of drawing-induced heating', *Materials Science & Engineering A*, 2020, **784**, 139341.
89. M. Umemoto, Z. G. Liu, H. Takaoka, M. Sawakami, K. Tsuchiya, and K. Masuyama: 'Production of bulk cementite and its characterization', *Metallurgical & Materials Transactions A*, 2001, **32**, 2127–2131.
90. H. K. D. H. Bhadeshia: 'Cementite', *International Materials Reviews*, 2020, **65**, 1–27.
91. H. Chen, Z. Yang, C. Zhang, K. Zhu, and S. van der Zwaag: 'On the transition between grain boundary ferrite and bainitic ferrite in Fe–C–Mo and Fe–C–Mn alloys: The bay formation explained', *Acta Materialia*, 2016, **104**, 62–71.
92. Y. Yamada: 'Static strain aging of eutectoid carbon steel wires', *Transactions of the Iron and Steel Institute of Japan*, 1976, **16**, 417–426.
93. V. T. L. Buono, M. S. Andrade, and B. M. Gonzalez: 'Kinetics of strain aging in drawn pearlitic steels', *Metallurgical & Materials Transactions A*, 1998, **29**, 1415–1423.
94. I. P. Kemp, G. Pollard, and A. N. Bramley: 'Static strain aging in high carbon steel wire', *Materials Science and Technology*, 1990, **6**, 331–337.
95. A. N. Holden, and F. W. Kunz: 'Note on the strain aging of iron single crystals', *Journal of Applied Physics*, 1952, **23**, 799.
96. D. Wei, L. Li, X. Min, F. Fang, Z. Xie, and J. Jiang: 'Microstructure and mechanical properties of heavily cold drawn pearlitic t steel wires: Effects of low temperature annealing', *Materials Characterization*, 2019, **153**, 108–114.
97. D. V. Wilson, and B. Russell: 'The contribution of precipitation to strain ageing in low carbon steels', *Acta Metllurgica*, 1960, **8**, 468–479.
98. G. R. Speich: 'Tempering of low-carbon martensite', *Transactions of the Metallurgical Society of AIME*, 1969, **245**, 2553–2564.
99. Y. Z. Chen, G. Csiszár, J. Cizek, X. H. She, C. Borchers, Y. J. Li, F. Liu, and R. Kirchheim: 'Defect recovery in severely deformed ferrite lamellae during annealing and its impact on the softening of cold-drawn pearlitic steel wires', *Metallurgical & Materials Transactions A*, 2016, **47**, 726–738.
100. Y. Z. Chen, G. Csiszár, J. Cizek, S. Westerkamp, C. Borchers, T. Ungár, S. Goto, F. Liu, and R. Kirchheim: 'Defects in carbon-rich ferrite of cold-drawn pearlitic steel wires', *Metallurgical & Materials Transactions A*, 2013, **44**, 3882–3889.
101. A. Querales, and J. G. Byrne: 'The effect of rapid annealing on the mechanical behaviour of cold rolled eutectoid steel', In: P. Haasen, V. Gerold, and G. Kostorz, eds. *Proceedings of the 5th International Conference on the Strength of Metals and Alloys, ICMSA 5.* Oxford, U. K.: Pergamon Press, 1979:921–926.
102. A. Querales, and J. G. Byrne: 'Improvement of fatigue life of 1080 steel by thermomechanical processing', *Fatigue of Engineering Materials and Structures*, 1979, **1**, 371–382.
103. O. A. Troitskii: 'Electromechanical effect in metals', *JETP Letters (English version)*, 1969, **10**, 11–14.

104. R. S. Qin, A. Rahnama, W. J. Lu, X. F. Zhang, and B. Eliott-Bowman: 'Electropulsed steels', *Materials Science and Technology*, 2014, **30**, 1040–1044.
105. X. R. Chu, S. X. Lin, Z. M. Yue, J. Gao, and C. S. Zhang: 'Research of initial dynamic recrystallisation for AZ31 alloy with pulse current', *Materials Science and Technology*, 2015, **31**, 1601–1606.
106. O. Omoigiade, A. Haldar, and R. Qin: 'Macroscopic characterization of mechanical properties in electric current treated dry drawn high strength wires', *MRS Advances*, 2017, **2**, 963–974.
107. X. L. Wang, Y. B. Wang, Y. M. Wang, B. Q. Wang, and J. D. Guo: 'Oriented nanotwins induced by electric current pulses in cu–zn alloy', *Applied Physics Letters*, 2007, **91**, 163112.
108. A. H. Cottrell, and B. A. Bilby: 'Dislocation theory of yielding and strain ageing of iron', *Proceedings of the Physics Society A*, 1949, **62**, 49–62.
109. L. S. Darken, and R. W. Gurry: Physical Chemistry of Metals: New York, USA: McGraw-Hill, 1953.
110. D. Kalish, M. Cohen, and S. A. Kulin: 'Strain tempering of bainite in 9Ni-4Co-0.45C steel', *Journal of Materials*, 1970, **5**, 169–183.
111. J. Wilde, A. Cerezo, and G. D. W. Smith: 'Three-dimensional atomic-scale mapping of a Cottrell atmosphere around a dislocation in iron', *Acta Materialia*, 2000, **43**, 39–48.
112. K. Hono, M. Ohnuma, M. Murayama, S. Nishida, A. Yoshie, and T. Takahashi: 'Cementite decomposition in heavily drawn pearlite steel wire', *Scripta Materialia*, 2001, , 977–983.
113. D. Kalish, and M. Cohen: 'Structural changes and strengthening in the strain tempering of martensite', *Materials Science & Engineering*, 1970, **6**, 156–166.
114. Y. Shiota, Y. Tomota, A. Moriai, and T. Kamiyama: 'Structure and mechanical behavior of heavily drawn pearlite and martensite in a high carbon steel', *Metals and Materials International*, 2005, **11**, 371–376.
115. A. Y. Badmos, and H. K. D. H. Bhadeshia: 'Evolution of solutions', *Metallurgical & Materials Transactions A*, 2000, **28**, 2189–2193.
116. L. Zhou, F. Fang, X. Zhou, Y. Tu, Z. Xie, and J. Jiang: 'Cementite nano-crystallization in cold drawn pearlitic wires instigated by low temperature annealing', *Scripta Materialia*, 2016, **120**, 5–8.
117. C. Borchers, T. Al-Kassab, S. Goto, and R. Kirchheim: 'Partially amorphous nanocomposite obtained from heavily deformed pearlitic steel', *Materials Science & Engineering A*, 2009, **502**, 131–138.
118. J. Languillaume, G. Kapelski, and B. Baudelet: 'Cementite dissolution in heavily cold drawn pearlitic steel wires', *Acta Materialia*, 1997, **45**, 1201–1212.
119. H. Song, R. Shi, Y. Wang, and J. J. Hoyt: 'Simulation study of heterogeneous nucleation at grain boundaries during the austenite-ferrite phase transformation: comparing the classical model with the multi-phase field nudged elastic band method', *Metallurgical and Materials Transactions A*, 2017, **48**, 273–2738.
120. T. Jarwali, and S. Nakamura: 'Anti-corrosion performance of bridge strands consisting of steel wires galvanised with zinc–aluminium alloy', *Structure and Infrastructure Engineering 12 (2016) 682-694*, 2016, **12**, 682–694.
121. J. K. Singh, and D. D. N. Singh: 'Studies on defects in galvanised coating on steel wire', *Indian Journal of Chemical Technology*, 2012, **19**, 361–365.

14 Miscellaneous

14.1 AUSTENITISATION

The heat treatment required to generate austenite is not usually an important issue in production scenarios because the size of the objects manufactured tends to be large so they have to be held at a high temperature for a considerable period in order to homogenise the temperature. In modern facilities, the steel that emerges red-hot from the casting process can be fed directly into the processing lines, thus saving the energy and handling costs associated with reheating furnaces. Austenite formation may be an issue in the heat-affected zones of welds where coarse γ-grain sizes can lead to detrimental microstructures on cooling, which is why structural steels are designed with carbon equivalents that are small enough to avoid such problems. For the same reason, fully pearlitic steels rarely are welded. Nevertheless, unravelling of the mechanisms of austenite formation during heating can be interesting.

14.1.1 NUCLEATION OF AUSTENITE

When mixtures of proeutectoid-α and pearlite are heated above Ae_1, austenite tends to nucleate preferentially at the α-P or pearlite-colony boundaries [1, 2]. Nucleation at colony boundaries prevails in hypereutectoid steels [3, 4]. The lamellar boundaries within pearlite are also potential nucleation sites. These locations are illustrated schematically in Figure 14.1a.

There is little known about the crystallography of austenite formation from pearlite or its spheroidised form, other than that it sometimes is possible for a Kurdumov-Sachs type orientation relationship to develop when γ nucleates at α/α grain boundaries [5].

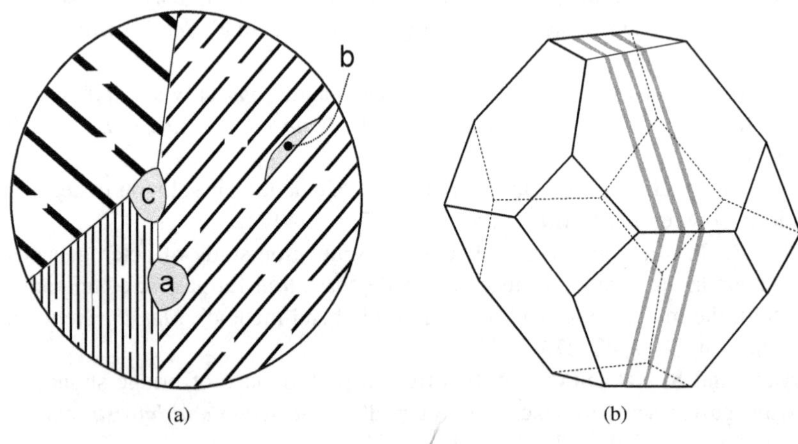

Figure 14.1 (a) Variety of nucleation sites within a fully pearlitic steel. ‘a, b, c’ represent austenite particles. (b) Representation of a pearlite colony as a tetrakaidecahedron. Diagrams adapted from Roósz et al. [3].

Consider the nucleation of γ at the favoured locations-c (Figure 14.1).[1] If a colony of pearlite is treated as a truncated octahedron (tetrakaidecahedron) of side length z_o (Figure 14.1), and if these

[1]A factor neglected is that colonies within a nodule of pearlite are not strongly misoriented, Chapter 2. After all, a whole nodule containing several colonies is essentially a bicrystal. The focus perhaps, should be on nodule boundaries.

DOI: 10.1201/9781032631981-14

objects are arranged to fill all space, then the total length of edges per unit volume (L_V) where the colonies meet is

$$L_V = \frac{36z_o}{4}\frac{1}{11.31z_o^3} \propto \frac{1}{z_o^2} \tag{14.1}$$

because each tetrakaidecahedron of volume $11.31z_o^3$ [6] has 36 edges but each edge is shared by four such objects. But both α and θ play a role in the γ nucleation process so the number of intersections of cementite lamellae with the tetrakaidecahedron edges becomes important, with this quantity proportional to S_I^{-1}. It follows that the nucleation rate per unit volume would vary as $\dot{N}_V \propto z_o^{-2} S_I^{-1}$:

$$\begin{aligned} \dot{N}_V &\propto \frac{1}{z_o^2 S_I} \exp\left\{\frac{-G^*}{RT}\right\} \\ &\approx 10^{-4} \exp\left\{-\frac{25.73}{T - T_E}\right\} \text{mm}^{-3}\,\text{s}^{-1}, \end{aligned} \tag{14.2}$$

an equation (Figure 14.2) that should be regarded as empirical because it does not follow the form expected from nucleation theory, making it impossible to assess for example, an activation energy.[2]

The measurements illustrated in Figure 14.2 are unique but don't lend themselves to rigorous interpretation based on nucleation theory, in part because they cover a narrow range of temperatures. There are numerous attempts at modelling the formation of austenite from pearlite or from mixtures of allotriomorphic ferrite and pearlite. However, none adequately justify the assumptions on which they are based, do not nurture a closure between experiment and theory other than perhaps a fitting to volume fractions, nor do they lead to generic outcomes or applicability. What Roberts and Mehl pointed out > 80 years ago in their paper [1] on austenite formation [1], that 'no rigorous quantitative mathematical treatment is as yet possible', remains true today.

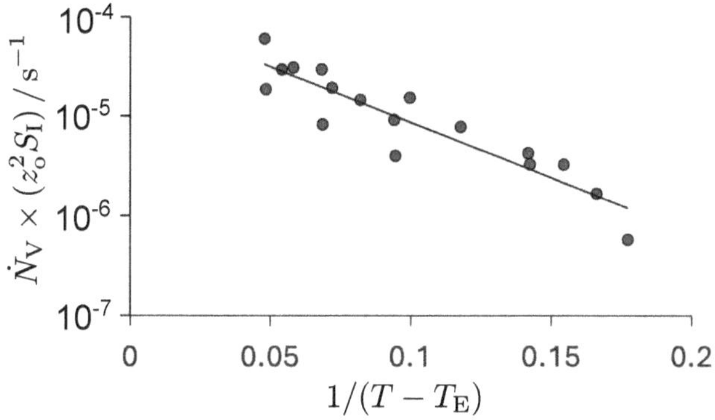

Figure 14.2 Plot of the nucleation rate of austenite when a eutectoid steel (Fe-0.78C-0.35Si-0.55Mn wt%) is heated to a variety of temperatures $T > T_E$. Selected data from Roósz et al. [3].

When banded proeutectoid-α+P microstructures (Chapter 6) are heated, the austenite first forms along the pearlite bands, thus replicating the original anisotropy [1]. The solidification-induced chemical segregation that is responsible for such microstructures is not eliminated by typical austenitisation procedures, so the ferrite and pearlite bands reappear when the fully austenitic steel is cooled appropriately to ambient temperatures [7, 8].

Figure 14.3 shows some exceptional micrographs related to the nucleation and growth of austenite. It is possible to see a tiny amount of austenite that has formed at the enlarged termination of a

[2]This presentation differs from [3] where the square of the term $z_o^2 S_I$ was used without justification.

cementite lamella at its junction with proeutectoid-α. There is a similar region of γ that has formed inside the pearlite colony but Figure 14.3c is a vivid illustration that the genesis of austenite must predominantly be at the proeutectoid-α/P boundary with subsequent almost allotriomorphic growth towards the centre of the pearlite colony.

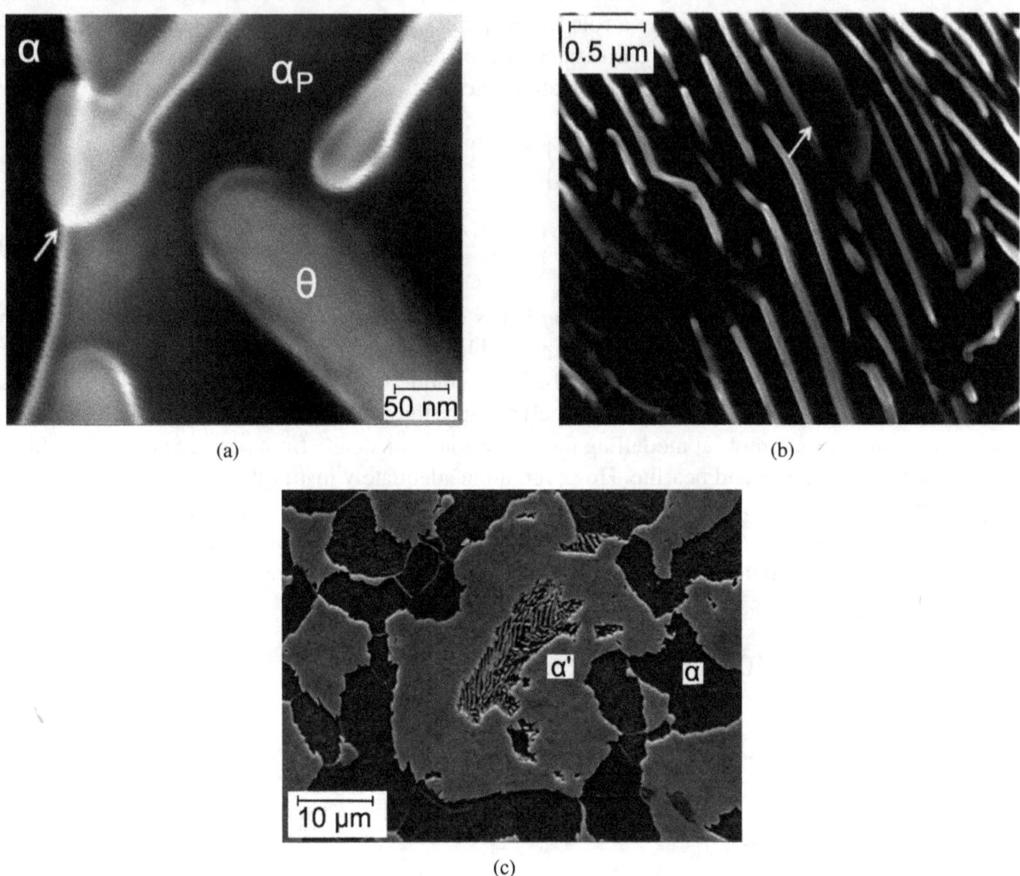

Figure 14.3 Hypoeutectoid steels with an initial α+P structure heated at 1500 °C s^{-1} to 750 °C and then quenched to ambient temperature. (a) Fe-0.17C-10.22Si-1.08 wt%. The arrow highlights the nucleation of austenite at the proeutectoid-α-P boundary. (b) The arrow shows austenite nucleated within pearlite for the same steel as in (a). α_P signifies ferrite that is a constituent of pearlite. (c) Fe-0.45C-0.26Si-0.63Mn wt%. Illustrates the inward growth of austenite (now martensite) into a region of pearlite. Micrographs reproduced from Cerda et al. [9] with permission of Elsevier.

In spheroidised steels, cementite particles located at α boundary triple points are more potent nucleation sites than intragranular particles, because the former leads to the elimination of some of the α/α boundaries [10–12]. The pearlite can become annealed during continuous heating – when cold-rolled hypoeutectoid steels with a rolling reduction of 80% are heated at 300 K s^{-1}, the pearlite within spheroidises first followed by the formation of austenite in the regions that are rich in cementite particles [13, 14].

14.1.2 GROWTH OF AUSTENITE

Austenite penetrates more into pearlite than into any proeutectoid-α that may be present in the original microstructure, because the pearlite is a rich source of carbon.

The initial rapid growth of γ into pearlite often is treated assuming carbon diffusion-controlled interface advance, though the problem is not well-defined because there are two phases that have to be consumed as the austenite grows. Consider the carbon concentration distribution illustrated in Figure 14.4, representing the thickening of austenite along $\pm z$ normal to the pearlite lamellae with the austenite nucleated at the θ/α interface and growing into both phases. As the austenite grows isothermally as a function of time t into the cementite, the rate at which it absorbs carbon must equal that carried away from the θ/γ interface by diffusion:

$$\frac{dz^\theta}{dt}(c^{\theta\gamma} - c^{\gamma\theta}) \approx -D_C^\gamma \frac{c^{\gamma\alpha} - c^{\gamma\theta}}{z^\alpha - z^\theta} \qquad \text{for } 0 \leq z^\theta \leq S_I \times \frac{0.76 - 0.02}{6.67 - 0.02} \tag{14.3a}$$

where the numerical term on the right hand side is the fraction of cementite in pearlite of eutectoid composition, D_C^γ is the diffusion coefficient of carbon in austenite, z^θ and z^α represent the positions of the moving interfaces, between which the gradient on the right is defined and assumed to be constant. The distance $z^\alpha - z^\theta$ will initially be less than S_I.

The diffusion coefficient of carbon in ferrite [15] is much greater than in austenite [16];

$$D_C^\gamma\{750^\circ\mathrm{C}, 0.76\mathrm{C\,wt\%}\} = 0.16 \times 10^{-11}\,\mathrm{m^2\,s^{-1}}$$
$$D_C^\alpha\{750^\circ\mathrm{C}\} = 0.16 \times 10^{-9}\,\mathrm{m^2\,s^{-1}}$$

so assuming that the gradient of concentration in the ferrite is uniform,

$$\frac{dz^\alpha}{dt}(c^{\gamma\alpha} - c^{\alpha\gamma}) = D_C^\alpha \frac{c^\alpha - c^{\alpha\gamma}}{z^* - z^\alpha} \tag{14.3b}$$

where $z^* \approx S_I \times (6.67 - 0.76)/(6.67 - 0.02)$ is the thickness of the ferrite lamella, calculated by applying the lever rule at T_E, and assuming that there is no austenite growing from the opposite side of the lamella. Equation 14.3b can integrated to show that

$$z^* z^\alpha - \frac{(z^\alpha)^2}{2} = D_C^\alpha \frac{c^\alpha - c^{\alpha\gamma}}{c^{\gamma\alpha} - c^{\alpha\gamma}} t \qquad \text{for } z^\alpha \leq z^*. \tag{14.3c}$$

Equations 14.3a and 14.3b are coupled. When solved using a stepwise numerical procedure [17], they show that the transformation of the cementite and of ferrite is essentially simultaneous in the sense that both phases are replaced by austenite at roughly the same instant.

It is worth emphasising that the theory presented assumes diffusion-controlled growth, but given the very small starting carbon concentration in the α_P prior to heating, it could be the case that the growth into ferrite is interface controlled. Diffusion-controlled growth also means that local equilibrium is satisfied at all the interphase-interfaces [p.217, 18].

Substitutional solutes such as manganese or chromium can affect the growth rate of austenite by orders of magnitude when compared against Fe-C alloys [19]. This is related to the partitioning of these solutes between the austenite and ferrite once the cementite has essentially dissolved [12]. The cementite in particular, if subjected to annealing, can absorb a lot of manganese prior to reaching a temperature where austenite formation begins.

14.1.3 OVERALL KINETICS OF TRANSFORMATION TO AUSTENITE

The change in the fraction of austenite during the heating of a hypoeutectoid steel initially is rapid, becoming more gradual as the temperature continues to increase (Figure 14.6a). This transition

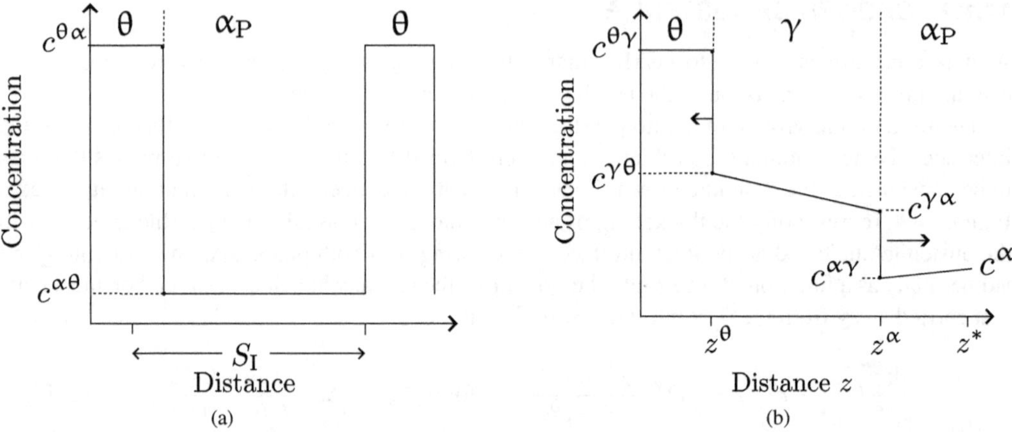

Figure 14.4 (a) The distribution of concentration prior to heating. S_I is the interlamellar spacing. (b) Schematic representation of the growth of austenite in a direction $\pm z$ which is normal to the pearlite lamellae. z^{θ} and z^{α} represent the time-dependent locations of the θ/γ and α/γ interfaces respectively. The concentration c^{α} is that of the ferrite at ambient temperature, now superheated to $T > Ae_1$.

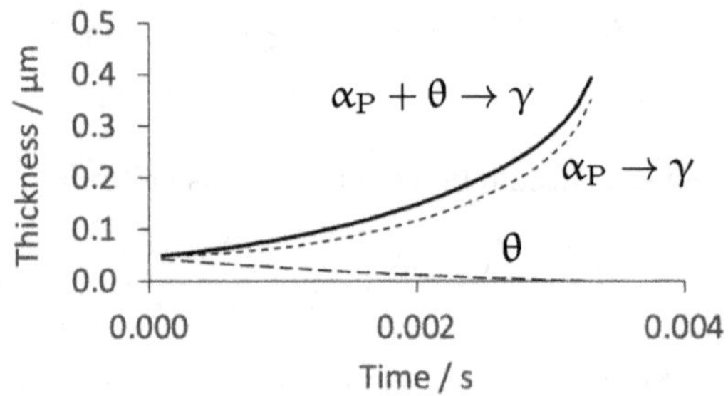

Figure 14.5 Solution to Equations 14.3a and 14.3b for a eutectoid Fe-0.76C wt% binary steel, representing isothermal transformation at 750 °C with $S_I = 0.4\,\mu$m. The growth of the austenite is normal to the notional lamella plane. The θ curve represents the thickness of the cementite lamella as a function of time, $\alpha_P \rightarrow \gamma$ the change in the thickness of austenite at the expense of ferrite, and the net thickness of the austenite at an instant is represented by $\alpha_P + \theta \rightarrow \gamma$. The calculations do not include nucleation so they begin with an assumed finite thickness of austenite.

occurs when $V_V^{\gamma} \approx V_V^{P}\{\text{ambient}\}$, where the pearlite fraction is that in the original microstructure prior to heating. The para-equilibrium fraction of austenite and that measured during continuous heating essentially follow the same trend with increasing temperature, with the sharp change in slope corresponding to the point where pearlitic part of the microstructure in the hypoeutectoid steel is consumed. The change in the gradient of the function $V_V^{\gamma}\{T\}$ is therefore, largely a thermodynamic effect. The plot in Figure 14.6b shows a similar change in the slope of the driving force versus temperature, with a much smaller gradient once all the cementite has dissolved. Kinetic phenomena simply shift the function $V_V^{\gamma}\{T\}$ to a small superheat of ≈ 15 °C in the case illustrated. The driving force is defined as $\Delta G^{\alpha+\theta\rightarrow\gamma} = G^{\gamma} - G^{\alpha+\theta}$ or $\Delta G^{\alpha\rightarrow\gamma} = G^{\gamma} - G^{\alpha}$ depending on the phase(s) available for transformation.

The isothermal formation of austenite at $T > Ae_1$ can be represented by a time-temperature-transformation diagram. But because both diffusion coefficients and the driving force for austenite formation increase with temperature, so the rate of transformation increases monotonically with

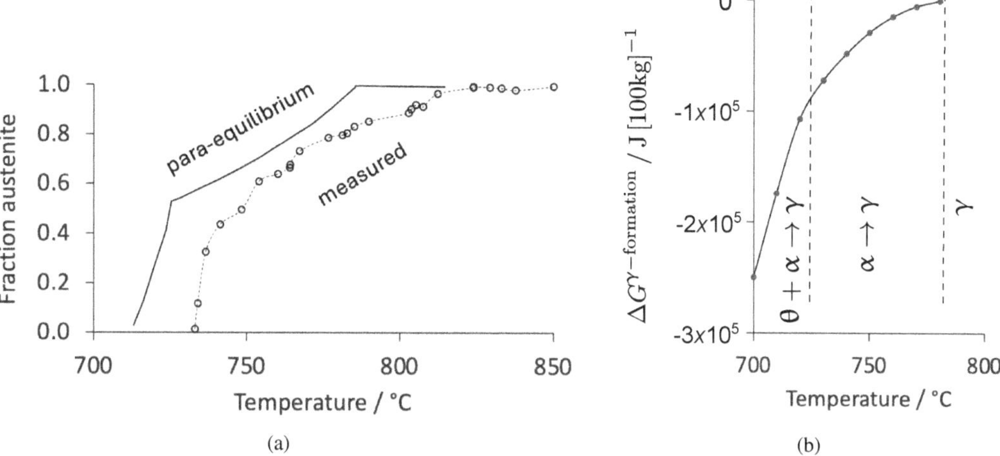

Figure 14.6 (a) A comparison of the calculated paraequilibrium fraction of austenite against that measured using X-ray diffraction when the hypoeutectoid steel is heated at $10\,^{\circ}\mathrm{C\,min}^{-1}$. The steel has the composition Fe-0.36C-0.66Mn-0.31Si-0.23Cu-0.18Cr wt%, with $V_V^P = 0.56$ at ambient temperature. Adapted using data from Savran et al. [20]. (b) Calculated equilibrium driving force for austenite formation for the same steel as in (a). These calculations used MTDATA [21], with the permitted phases being γ, α and θ. The phases indicated are those that the calculations permit at equilibrium.

superheat, i.e., there is no classical 'C' curve. Figure 14.7a illustrates this for a eutectoid steel which can transform completely into austenite at all temperatures above Ae_1. This is not the case for hypoeutectoid steels where the equilibrium fraction of austenite increases from zero at Ae_1 to 1 at Ae_3. So although the curve representing the beginning of transformation shows an increasing rate with superheat, that for achieving the equilibrium fraction of austenite shows a 'C'-curve tendency. This is because the driving force is small close to Ae_1 whereas the fraction to be achieved is large at temperatures approaching Ae_3. Figure 14.7b illustrates this, admittedly with the starting microstructure at ambient temperature being martensitic, but this may not affect the essence of the late C-curve phenomenon.

14.2 RAPID HEATING OF PEARLITE

There are circumstances where pearlite may be heated into the γ phase field so rapidly that solutes including carbon cannot diffuse readily.

Samples of Fe-0.18C-0.28Si-0.9Mn wt% hypoeutectoid steel exposed to an energetic electron-beam (1.4 MeV accelerator) experience excursions in surface temperatures up to $\approx 1200\,^{\circ}\mathrm{C}$ over a period of just a second and then cool back to $< 500\,^{\circ}\mathrm{C}$ within another couple of seconds [23, 24]. Figure 14.8 shows that some regions at the surface transform back into fine ferrite, but the original pearlitic regions become martensitic. During the heating cycle, the pearlite transforms first into carbon-rich austenite (γ_{rich}), followed by the ferrite changing into carbon-poor austenite (γ_{poor}) as the superheat increases. The speed of the process does not permit chemical homogenisation, even of carbon when the region is fully austenitic. As a result, during cooling, the transformation sequence would be $\gamma_{\mathrm{poor}} \rightarrow \alpha$ followed by $\gamma_{\mathrm{rich}} \rightarrow \alpha'$. The martensite in Figure 14.8b has a microhardness of 750 HV whereas the ferrite is as soft as 170 HV. A martensite hardness of 720 HV corresponds to a carbon concentration of 0.62 wt% [25], almost identical to the composition of the original pearlite.

The boundaries between the α'-regions and ferrite are diffuse, so there may well exist gradients of carbon concentration and therefore of microstructure there. There are indirect observations that

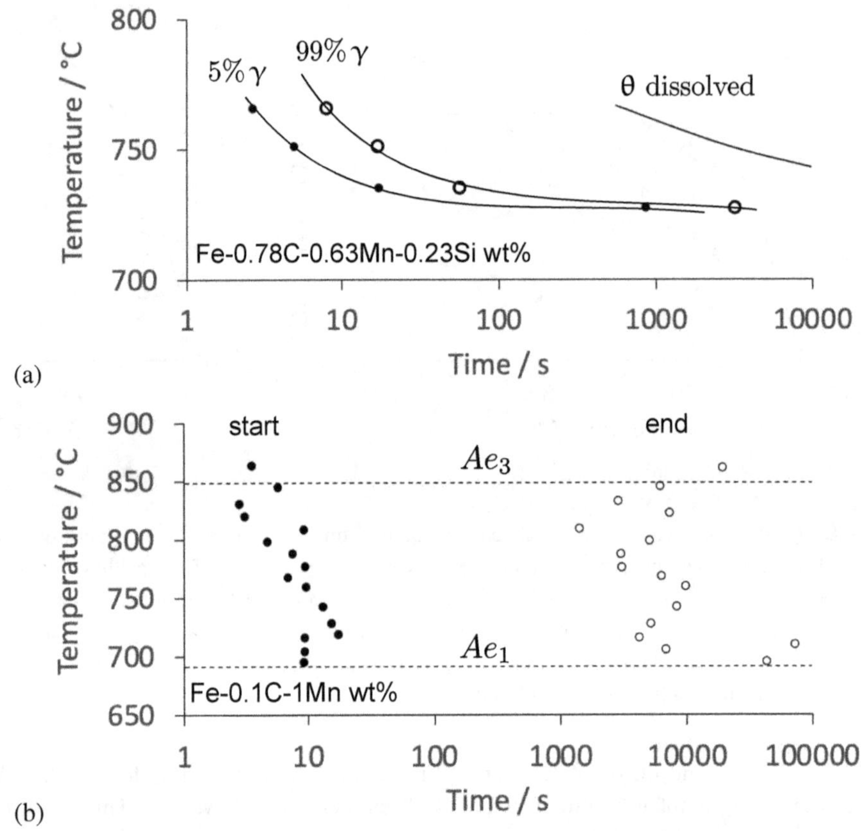

Figure 14.7 (a) Time-temperature-transformation diagram for isothermal transformation to austenite from an initially pearlitic microstructure in an essentially eutectoid steel. The curve on the right is when all of the cementite is dissolved but the carbon in the austenite is not necessarily homogeneously distributed. Selected data from Roberts and Mehl [1]. (b) TTT diagram for austenite formation in a hypoeutectoid steel beginning with a martensitic microstructure. Note that there is an order of magnitude difference in the time scales of these plots. Selected data from Lenel [22].

support the existence of such gradients and simulations of the rapid heating of pearlite predict them [26].

Rapid heating ($1000\,K\,s^{-1}$) using induction, followed by quenching from the peak temperature, of Fe-0.99C-0.39Mn-0.25Si-1.38Cr wt% steel has demonstrated the heterogeneous distribution of carbon in the austenite generated [27]. Regions close to incompletely dissolved cementite transformed into twinned martensite whereas distant locations contained dislocated martensite. While not rigorously true, larger carbon concentrations favour internally twinned martensite because of the reduced martensite-start temperature [28]. This is because a twinning lattice-invariant deformation during martensitic transformation leads to a more glissile interface at low temperatures, than the corresponding slip mode [29].

Rapid heating experiments on hypoeutectoid steels using electrical current pulses reveal a substantial loss of strength at the point where the pearlitic regions transform into austenite [30, 31]. This loss is greater when the pearlite fraction is large. This may be relevant to machining phenomena, where the strain rates and transient temperatures can be very large [30]; the deformed austenite would have different transformation characteristics [32].

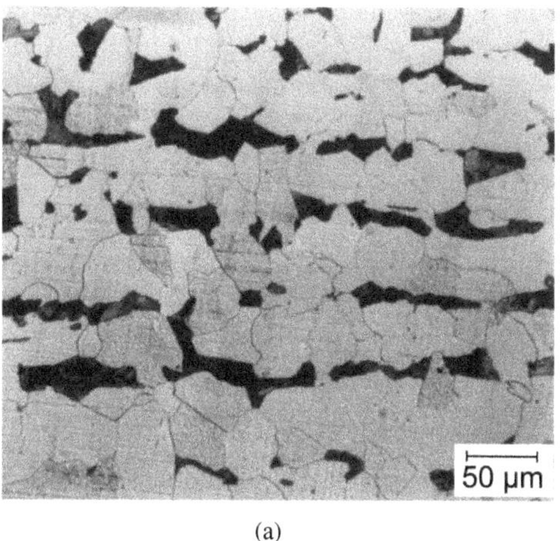

(a)

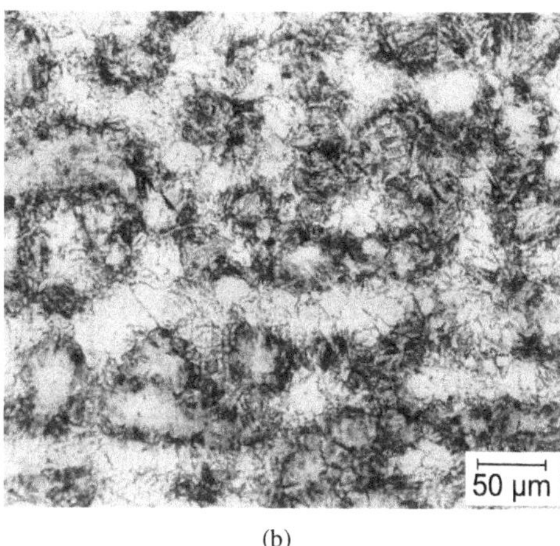

(b)

Figure 14.8 (a) The original α+P microstructure. (b) The structure at the surface irradiated with electrons. Images from [23] with the permission of Springer Nature.

14.3 AUSTENITE AND GHOST PEARLITE

Retained austenite can be beneficial to mechanical properties but it may require expensive alloying-elements to ensure its stability at ambient temperature and to reap the benefits of transformation-induced plasticity [33]. To avoid this, a technique has been developed in which cementite that is not at equilibrium in tempered martensite, is enriched in manganese by partitioning from the surrounding ferrite. Reheating this mixture to a temperature where austenite can form, followed by quenching leaves the Mn-rich prior-cementite regions as stable islands of austenite, while the rest of the Mn-depleted austenite transforms [34, 35].

The phases within pearlite, when it first forms, do not have their equilibrium compositions (p. 200); this is confirmed in microanalysis experiments [36–39]. When annealed, the structure slowly tends towards equilibrium with substitutional solutes partitioning between the α_P and θ. Figure 14.9 shows a microanalytical trace across a thin plate of cementite in Fe-1Cr-1.04Mn-0.84C wt% steel transformed into pearlite at 630 °C. The concentration of chromium at the centre is less than in the vicinity of the α_P/θ interface on either side, because of partitioning from $\alpha_P \rightarrow \theta$ following transformation, as the system meanders towards equilibrium.

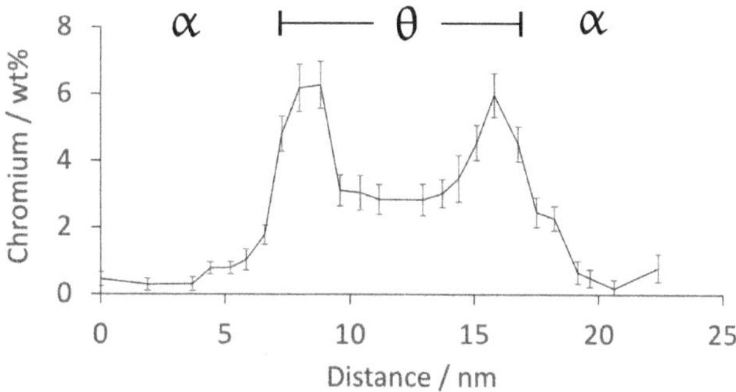

Figure 14.9 Composition profile across a cementite lamella with pearlite, adapted using data from Garratt-Reed et al. [40].

Manganese is of particular interest because it is present in most eutectoid steels and much prefers to be in cementite. It also is a cheap and yet a strong austenite-stabilising solute. Table 14.1 shows that the equilibrium concentration of manganese in cementite is greatest at low temperatures.

Table 14.1
The calculated [21] equilibrium weight percentages of manganese in cementite and ferrite within a eutectoid steel of average chemical composition Fe-2Mn-0.8C wt%.

Temperature / K	$w_{\mathrm{Mn}}^{\theta\alpha}$	$w_{\mathrm{Mn}}^{\alpha\theta}$	Temperature / °C
800	13.67	0.42	527
840	13.07	0.50	567
880	12.19	0.63	607
920	11.22	0.76	647
960	10.18	0.90	687

In an experiment (Figure 14.10) where pearlite was generated in Fe-0.39C-3.69Mn wt% at 570 °C and held there for 6 h, the Mn/(Mn+Fe) atom ratio reached $u_{\mathrm{Mn}}^{\theta} \approx 20$ compared with the depleted value in ferrite at $u_{\mathrm{Mn}}^{\alpha} \approx 2$ [41–43]. When this was heated to 750°C $> Ae_3$, and then quenched, the lamellae of Mn-rich austenite substituted for the positions where cementite previously existed. The intervening Mn-depleted regions became low-carbon martensite; it is interesting that when tempered the martensite did not precipitate carbides, as explained below.

The sample reached a fully austenitic condition above Ae_3. The activity of the carbon in that austenite homogenises by uphill diffusion [44]. Its concentration in the Mn-rich regions increases at the expense of the Mn-poor regions (Figure 6.4). Equation 6.1b applied to the data from [41] indicates that the Mn-poor regions of the γ should contain only 0.08C wt% as a_{C}^{γ} becomes uniform by uphill diffusion. The reason, therefore, why the tempering of low-Mn α' does not precipitate cementite is because it only contains 0.08C wt%, much of which is likely to be trapped at dislocations in the martensite [45], where it is more stable than cementite [46].

Figure 14.10c shows that u_{Mn}^{θ} is reduced during the short excursion into the austenite phase field, with a corresponding increase in u_{Mn}^{α}, indicating sufficient atomic mobility to render some homogenisation.

It is extraordinary that the structure that develops on cooling resembles pearlite but does not contain cementite; hence the term *ghost pearlite*, Figure 14.10. The claim is that an elongation of $\approx 11\%$ at an ultimate strength of ≈ 2 GPa is achieved, the ductility in part due to the presence of the retained austenite in the microstructure. This is a good combination when it is considered that the ghost pearlite is generated without the cold deformation that gives exceptional strength to pearlitic steel wires (Chapter 13). The lamellar austenite can undergo deformation-induced martensitic transformation though at a slower rate than γ generated by first spheroidising the cementite [47].

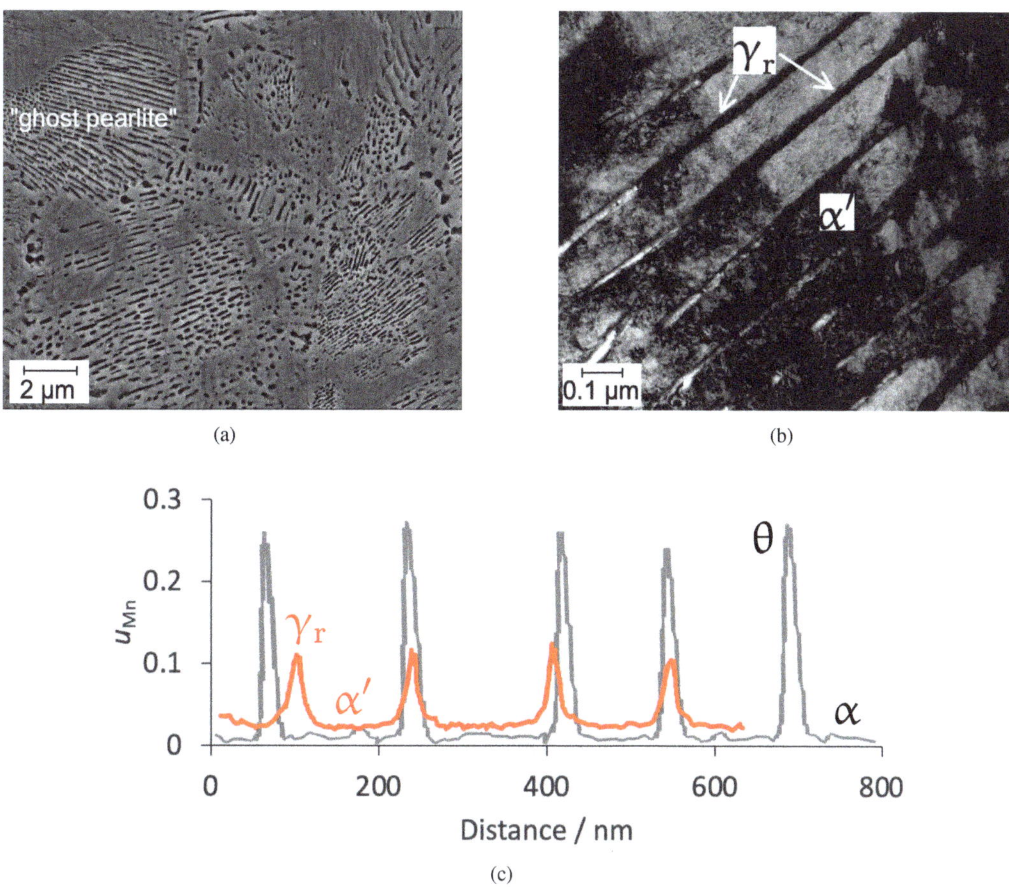

Figure 14.10 Fe-0.39C-3.69Mn wt% austenitised at 800 °C, 10 min, held at 570 °C, 6 h, cooled to room temperature. Then heated to 750 °C, 30 s, quenched. (a) A microstructure that appears to be pearlite but is not. Therefore labelled as 'ghost pearlite'. (b) It consists in fact of alternating lamellae of retained austenite (γ_r) and low-carbon martensite. Images courtesy of Zhiping Xiong and Xingwang Cheng based on [41]. (c) The Mn/(Mn+Fe) atom ratio in the pearlite after annealing the 570 °C treatment (in grey) and in the ghost pearlite (in red) consisting of retained austenite and martensite following the 750 °C treatment. Data digitised and re-plotted using data from [41], for the purposes of superposition, though the horizontal origins of the two plots are not intended to correlate.

14.4 FREE-MACHINING STEELS

Most free-machining steels are not entirely pearlitic, usually with carbon concentrations < 0.5 wt%. The mechanical properties required from such steels often form a secondary selection-criterion to the ease of machining [48]. They would not normally be used in circumstances where fatigue or toughness feature prominently in engineering design.

Machinability is best defined by cost, that is determined by the quality required, the machining time, the wear and tear of associated equipment, whether the component is to be mass produced or is a limited edition, and the cost of machining relative to that of the component itself.[3] The focus

[3]Machine learning models take cutting speed, feed rate, cutting depth, tool wear parameters, material type and type of cutting tool as inputs. There are, of course, many more variables, such as machine rigidity, but machine learning relies on the availability of sufficient data on each variable, and missing variables reflect in the noise of the calculation model.

here is on the components of the steel microstructure that mitigate the cost of machining, assuming that all the specified properties are maintained.

Metal cutting involves localised deformation and fracture. The surface integrity of the component being machined, tool wear and tool forces depend on the ease with which layers of the material can be removed and on how the removed material detaches both from the substrate and from its contact with the cutting-tool.[4] Steels containing inclusions (e.g., MnS, Pb) facilitate fracture so the debris that is created becomes fragmented rather than continuous, Figure 14.11. There may be a limited role of the inclusion providing lubrication between the steel and the tool if it is sufficiently plasticised during the machining process, in which case it can help reduce the forces on the tool. Lead particles are established in this role but MnS in free-machining steel serves primarily to provide crack or void initiation sites.

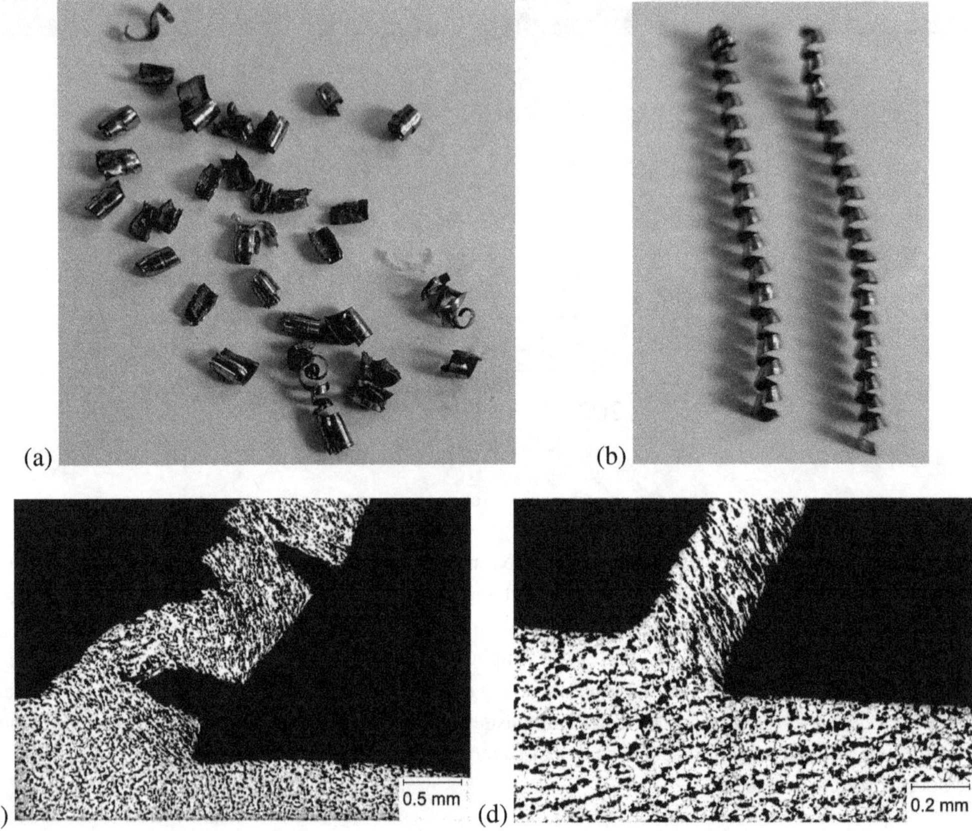

Figure 14.11 Machining debris. (a) Fragmented swarf from free-machining steel. (b) Continuous swarf from ordinary steel. Photographs courtesy of Paul Nicholls and Paul Stokes. (c,d) Morphology of machining debris from re-sulphurised steel, showing both continuous and fragmented chips. Micrographs from Sidjanin and Kovac [50], copyright © Institute of Materials, Minerals and Mining, reprinted by permission of Taylor & Francis Ltd, www.tandfonline.com on behalf of the Institute of Materials, Minerals and Mining.

Some of the debris detached from the steel being machined, can in principle become pressure welded to the cutting tool, leading to the so-called *built-up edge* on the tool. This changes the local

---

[4]To quote a classic [p. 20, 49], 'the ultimate cause for a tool giving out when cutting metal is the dullness or wear of the tool produced by the rubbing or pressure of the chip upon the lip surface of the tool'.

geometry of the tool and therefore, can compromise the surface finish and tool life. Pearlite does not adhere to the tool as much as ferrite so its presence in the microstructure reduces the tendency to generate built-up edges [51].

The inclusions may typically be sulphides, bismuth or in rare circumstances, graphite. Lead, though particularly effective, is not recommended because of its toxicity. Sulphur seems to be the most common free-machining addition to the steel. During conventional steelmaking, the amount of sulphur in the molten steel decreases with the progressive reduction of its carbon concentration. After tapping, *re-sulphurising* is implemented with the help of a slag on the liquid that does not absorb sulphur, with elemental sulphur added deliberately. In modern practice, the steel normally may be produced using scrap that is melted in an electric arc furnace, placed into a ladle furnace for deoxidation (using FeMn, FeSi) and removal of sulphur using a mixture of CaO and CaFe while being aerated with argon bubbling from the bottom of the ladle. Sulphur is then added using wire containing a core of elemental sulphur, followed by CaSi injection [52]. The final structure will consist of a mixture of ferrite and pearlite, together with MnS inclusions. Manganese sulphide has the advantage over the harder inclusions that sometimes are added to enhance machinability, that it's softness and limited-plasticity reduces wear on the tools [48].

The deoxidation practice before desulphurisation determines the type and complexity of the MnS precipitates. There are contradictory reports but recent research suggests that in ferrite-pearlite steels, an oxygen concentration less than ≈ 0.01 wt% leads to single-phase MnS precipitates which are elongated in shape following hot deformation. Alloys richer in oxygen yield more rounded MnS precipitates that are in contact with harder oxides such as $MnO\text{-}Al_2O_3$, $MnO\text{-}SiO_2$ and $2MnO\text{-}SiO_2$. The mixed precipitates are associated with a deterioration in tool wear [53].

Typical chemical compositions of free-machining α+P steels are listed in Table 14.2; the carbon concentration depends on the mechanical properties required and there is sufficient manganese to combine with sulphur to form MnS, while at the same time ensuring that the correct microstructure is obtained following processing. The silicon concentration is kept to a minimum because it is known to be detrimental to machinability. The quantity of sulphur needed depends on the combination of mechanical property and machinability requirements. A concentration greater than 0.2S wt% does not further reduce tool wear but does enhance chip formation and surface finish in medium-carbon steels [54].

The morphology and distribution of the MnS particles are determined by the steelmaking process in combination with the processing that follows.

Table 14.2

Chemical compositions of some free-machining α+P steels. *Mean composition from published range.

C	Si	Mn	Cr	Ni	Cu	S	P	Other	Reference
0.16	0.05	1.18	0.17	0.18	0.38	0.27	0.01		[50]
0.07		1.35				0.40		0.08Bi, 0.15Sn	[55]*
0.40	0.14	1.17	0.09	0.05	0.38	0.15	0.02	0.02Sn	[54]*

14.5 BULK NANOSTRUCTURED PEARLITIC STEELS

The term *nanostructure* is reserved for structures containing an exceptional interfacial area per unit volume, i.e., when the spacing between interfaces is below about 30 nm. [56, 57]. The term *bulk* refers here to objects that can be large in all three dimensions. It is safe to say that bulk, nanostructured pearlitic steels do not exist in the undeformed condition.

Nanostructured pearlite can be generated by phase transformation without calling on plastic deformation. The quite remarkable, finest pearlite spacing ever achieved in commercial, eutectoid steel is via the patenting process (Chapter 13). To recap, patenting is a continuous process whereby the steel, in its austenitic state, proceeds through a molten lead-alloy bath so transformation temperatures in the range 510-570 °C are achieved routinely, with corresponding values of S_{I} in the range 28-44 nm [58]. The steel naturally is in the form of wire-rods.

In what follows, the alloy designation refers to Tables 14.3 and 14.4. To achieve fine pearlite in plates that are not of an appropriate shape or size for the patenting process, requires the transformation temperature to be suppressed by other means. The formation of bainite would need to be avoided. The driving force for transformation can be enhanced by alloying with cobalt or aluminium. Alloy 1 adopts the latter approach with two interlamellar spacings generated by isothermal transformation at 600 and 700 °C, because the estimated bainite-start temperature was 550 °C [59]. The properties achieved are not remarkable in the context of many commercially available steels but it interesting that the fracture toughness (unlike the total elongation) of the stronger sample with the smaller interlamellar spacing is somewhat greater than that with the coarser S_{I}. In both cases, the fracture surfaces following tensile testing indicated rough, quasicleavage type fracture and pearlite nodule sizes were shown to be identical for both samples; the elongation preceding fracture was uniform. Given that the process of fracture in pearlite begins with the cracking of cementite (p. 95), thin-cementite may be better suited to resist the propagation of brittle fracture across pearlite colonies or nodules.

Greater strength can in principle be achieved in cheaper steel without compromising the elongation, by increasing the carbon concentration, eliminating manganese and adding aluminium. This change in substitutional solute content results in an increase in the driving force for transformation, permitting the refinement of interlamellar spacing during continuous cooling transformation (Alloy 2, [60]). It is likely that the toughness is reduced by the greater fraction of cementite.

An alternative is to boost the driving force using a magnetic field during transformation, because this promotes ferromagnetic phases. Ferrite has a Curie temperature $T_{\mathrm{C}}^{\alpha} = 769\,^{\circ}\mathrm{C}$, which exceeds the temperature at which pearlite is expected. Cementite too is ferromagnetic but only below about $T_{\mathrm{C}}^{\theta} = 186\,^{\circ}\mathrm{C}$ [18, 61] so it does not respond to the magnetic field at typical pearlite formation temperatures. Alloy 3 was designed originally for the production of nanostructured bainite, but when subjected to a 30 T magnetic field during continuous cooling from austenite at $60\,^{\circ}\mathrm{C\,min^{-1}}$, pearlite with $S_{\mathrm{I}} \approx 50$ nm is generated instead [62]. In fact, a fully pearlitic nanostructure can be generated in Alloy 3 simply by cooling at the slower rate of $< 6\,^{\circ}\mathrm{C\,min^{-1}}$ from austenite; the interlamellar spacing achieved is even finer at 44 nm and yet, the data in Table 14.4 indicate that the hardness is less than the pearlite obtained at the greater cooling rate (30T field). This may be related to the short transformation time increasing the dislocation density, whereas slow cooling would anneal the pearlite after formation (p. 71). When cooled at an even slower rate of $1.2\,^{\circ}\mathrm{C\,min^{-1}}$, the structure exhibits a degree of spheroidisation (Figure 14.12); the fine S_{I} makes the pearlite more susceptible to spheroidisation (Chapter 9) resulting in a hardness of only 355 HV.

In Alloy 4, nanostructured pearlite was generated by cooling at $6\,^{\circ}\mathrm{C\,min^{-1}}$ [63]; this particular experiment demonstrates that fine pearlite can be achieved in a bulk, undeformed steel that exclusively contains common solutes, during continuous cooling transformation at reasonable cooling rates. The pearlite was monitored to form over the temperature range 662-639 °C. The mechanical properties, however, have not been characterised adequately.

The data in general indicate that there is no commercial advantage to be had from the steels listed in Table 14.3 because the properties measured can easily be obtained in much cheaper martensitic or bainitic steels.

Table 14.3
Compositions (wt%) of 'nanostructured', fully pearlitic steels.

Alloy	C	Si	Mn	Cr	Ni	Mo	Co	Al	V	Reference
1	0.63	1.44	0.44	1.08	0.04	0.24	3.84	1.06		[59]
2	0.90	0.86	0.01	1.51				0.86		[60]
3	0.78	1.60	2.02	1.01		0.24	3.87	1.37		[62]
4	0.79	1.59	1.94	1.33		0.3			0.11	[63]
5	0.7	0.99	1.81	0.36		0.3	3.11	2.13		[64]
5	0.7	0.99	1.81	0.36		0.3	3.11	2.13		[64]

Table 14.4
Some properties of the steels described in Table 14.3. The cooling rate ($\dot{T}$) is not stated for those alloys transformed isothermally. Where necessary, hardness data have been converted into Vickers values for comparison purposes. Many of the data listed come from non-standard, small specimen tests; the elongation data are therefore probably optimistic.

Alloy	Hardness / HV	σ_y / MPa	σ_{UTS} / MPa	ε_t	K_{IC} / MPa m$^{\frac{1}{2}}$	S_I / nm	$\dot{T}$ / °C min^{-1}	Reference
1		1075	1289	8	42 ± 0.9	63		[59] [a]
1		790	1153	14	35 ± 0.2	110		[59] [a]
2	410	857	1487	10		128	40	[60]
2	430	880	1575	11		96	60	[60]
2	460	1329	1670	11		86	100	[60] [b]
2	470	1342	1685	12		82	120	[60] [b]
3	484					≈ 50	60	[62][c]
3	423					44	6	[63]
3	375					66	3	[63]
4	401					81	6	[63]
5		524	868	29		168		[64]
5		615	1032	24		86		[64]

[a] The strength data from [58] were transposed in their Table 2 and the reported elongations included the elastic response from the testing system, both corrected here.

[b] σ_{UTS} here corresponds to the fracture strength, which was reached before a maximum in the σ-ε plot. The σ_y values for alloy 2 were not stated in the original publication but have been deduced from further data provided by Professor Zhinan Yang.

[c] Transformed under influence of 30T magnetic field.

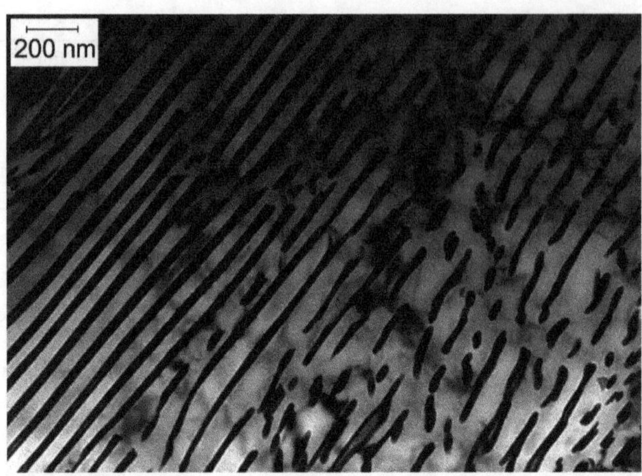

Figure 14.12 Alloy 3 cooled at 1.2 °C min^{-1} from its austenitic state. There is evidence for the early stages of spheroidisation that occurs during cooling following transformation. Transmission electron micrograph.

14.6 SEVERE DEFORMATION

14.6.1 BALL MILLING AND BALL DROP-TEST

During ball milling, a pearlitic-steel powder is placed into a rotating cylinder containing relatively large steel balls, usually made from an ordinary bearing steel such as 52100, which contains 1C-1.5Cr wt%. The balls repeatedly impact the particles which are intensely deformed as a consequence, with strain rates approaching $10^4\,s^{-1}$. The process first causes a nanocrystalline layer to form at the powder surfaces (hardness $\approx$ 11 GPa), delineated sharply from the underlying work-hardened region (hardness $\approx$ 5 GPa) [65], Figure 14.13a. With continued milling, the deformation spreads through each particle with possible dissolution of the cementite. It is not clear why the boundary between the nanocrystalline and work-hardened region is so sharp, but considerable porosity is visible in the hardened region near the interface, indicating a large strain incompatibility between the two regions due to the depth to which the impacting ball can influence the underlying structure. Transmission microscopy has shown that the nanocrystalline region actually has equiaxed grains in random orientations, indicating some sort of dynamic recrystallisation of the deformed structure.

A drop ball test is a more controlled simulation of the ball-milling operation, illustrated schematically in Figure 14.13b. The ball is impacted repeatedly on to the sample and in the case of pearlite, with strain rates comparable to ball milling; this leads to very similar outcomes as in ball milling [66].

14.6.2 EQUAL-CHANNEL ANGULAR DEFORMATION

Equal-channel angular processing (ECAP) [67] forces a billet through a die that has two intersecting channels such that it experiences severe deformation while preserving its shape, Figure 14.14a. The deformation involves shear but the plane and direction of the shear change as the material progresses through the bend in the die. Th extrusion can be repeated N times in order to accumulate the plastic strain within the material; the von Mises equivalent strain is given by [67]:

$$\varepsilon_{\mathrm{vM}} = \frac{2}{\sqrt{3}} N \cot\phi/2 \tag{14.4}$$

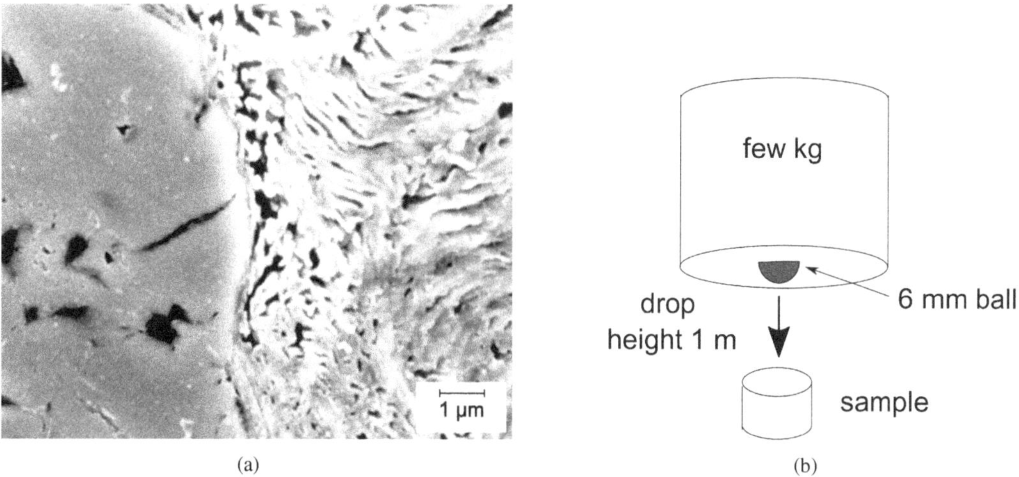

Figure 14.13 (a) Pearlite particle of overall composition Fe-0.89C wt%, ball milled for 360 ks. The region on the left is nanostructured and that on the right is simply work-hardened. Image courtesy of Professor Minoru Umemoto. (b) Schematic illustration of the drop ball test, adapted from [66].

Figure 14.14b shows the microstructure following an accumulated ECAP strain of 2. Some of the cementite has buckled and folded, perhaps due to the orientation of the lamellae being at a large angle to the principal deformation direction. ECAP enhances the ultimate strength, in this case from 950 MPa to $\approx$ 1700 MPa but the strength becomes orientation dependent and less reproducible [68].

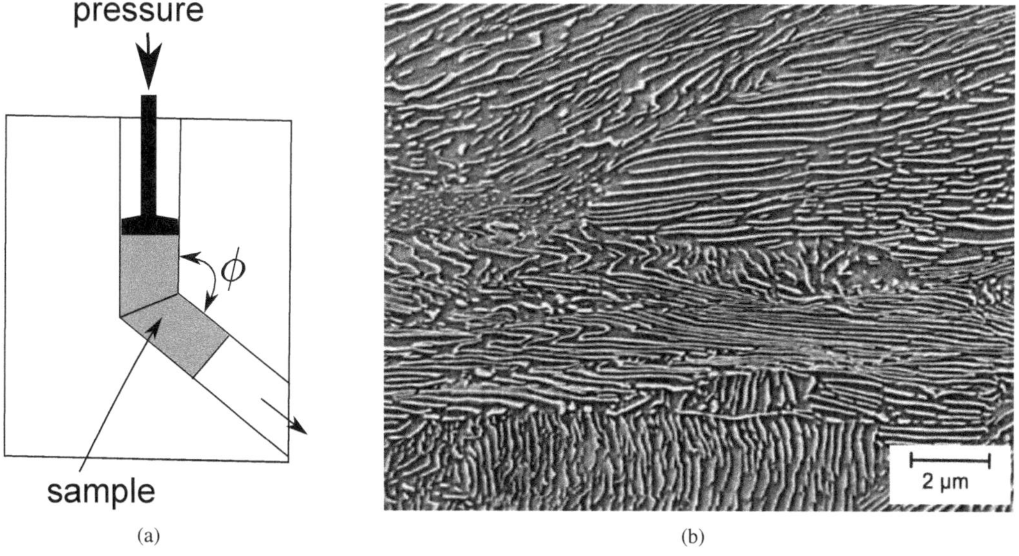

Figure 14.14 (a) Schematic illustration of the equal-channel angular processing die. (b) Microstructure of pearlitic steel following ECAP with three passes, with an accumulated $\varepsilon_{vM} = 2$. Micrograph adapted from Wetscher et al. [68], reproduced with the permission of Elsevier. Fe-0.76C-1Mn-0.37Si wt%

A variant of ECAP implements deformation with the material at elevated temperatures but within the $\alpha + \theta$ phase field. The purpose is to minimise macroscopic defects inherent in high-intensity cold-deformation, and stimulate dynamic recovery and recrystallisation. Spheroidised eutectoid-steel extruded at 650 °C, after two passes increases its yield strength from 290 to 590 MPa with slightly reduced elongation from 18 to 14%, the changes attributed to the refinement of the microstructure [69]. [5] Whilst interesting, similar or better properties are accessible in many commercial steels that are cheap to process. Similarly, ECAP treated α+P hypoeutectoid steels – for example, a Fe-0.15C-0.25Si-1.1Mn wt% α+P subjected to a total $\varepsilon_{vM} = 4$ and annealed subsequently into a spheroidised state did not exhibit important changes in the tensile properties relative to the unprocessed alloy [70].

14.6.3 HIGH-PRESSURE TORSION

Very large and shape-preserving plastic strains can be achieved in torsion, Figure 14.15a. The constrained sample is sheared by the action of torsion and compression, often referred to as high-pressure torsion because the vertical pressure on the disc sample is nominally in the several gigapascal range. The shear strain per rotation is the displacement $2\pi r$ divided by the thickness l, where r is the radial distance; it follows that the total shear strain for N rotations is given by [71]:

$$\gamma = 2\pi \frac{rN}{l} \qquad \text{with} \qquad \varepsilon_{vM} = \frac{\gamma}{\sqrt{3}}. \tag{14.5}$$

The shear strain is therefore zero at the centre but there may be additional deformation that is not accounted for, due to sliding between the sample and the torsion support or rotating member [72]. Experiments like these are curiosity-driven because the sample sizes are too small for applications, but the structural changes accompanying huge, cumulative shear-strains may be indicative of processes associated with rolling contact on rails. Nevertheless, high-pressure torsion cannot reproduce *damage* because of the constraint on the sample, so the information revealed is essentially about the structural and elementary-property changes due to severe shear.

Suppose that high pressure torsion is considered as a simple shear deformation; the pearlite spacing may then be little affected, but in practice, S_I has been found to decrease from 200 nm to 20 nm during torsion to a shear strain of 30 [73]. As pointed out by Utyashev et al. [74], Equations 14.4 and 14.5 are too simple to represent the processes that occur during severe deformation. The principal axes of the strain tensors rotate with respect to the sample frame, so the actual deformation implemented during high pressure torsion is more complex and of a greater magnitude than the simple shearing implied in Equations 14.4 and 14.5.

The strength measured parallel to the shear direction increases with the torsional strain but the ductility decreases [75]. Hardness data from a number of high-pressure torsion experiments on pearlitic steels are illustrated in Figure 14.15b; it is evident that the hardness before deformation determines the rest of the trajectory as a function of the von Mises strain. Also included are data from drawn pearlitic-wires to show that the hardness changes due to plastic strain are similar to those observed during high-pressure torsion once the hardness prior to deformation is taken into account; the comparisons are limited to the relatively low strains typical of wire drawing. Unlike wire drawing, it is not possible to estimate changes in S_I in the case of torsion experiments. For example, it can readily be demonstrated that the hardness does not follow the Embury and Fisher Equation 13.1d.

Fig. 14.15c,d shows the structure of the severely sheared samples. The two images are from the same sample (i.e., constant $\varepsilon_{vM} = 16$), indicating that in spite of the large shear implemented, the structure generated is not homogeneous. The lamellar alignment parallel to the shear strain is nevertheless largely retained, with the consequence that the toughness becomes anisotropic. In a sample deformed $\varepsilon_p = 15$, the fracture toughness decreased by an order of magnitude from 42 MPa m$^{1/2}$ to

[5] These elongations have been corrected from [69] to take account of the apparent extension prior to the onset of plasticity.

4.2 MPa m$^{1/2}$ as the fracture path changed from being normal to the lamellae, to propagation parallel to the lamellar structure [76]. The hardness of the sample remains remarkably uniform so the toughness variations are directly attributable to structural anisotropy.

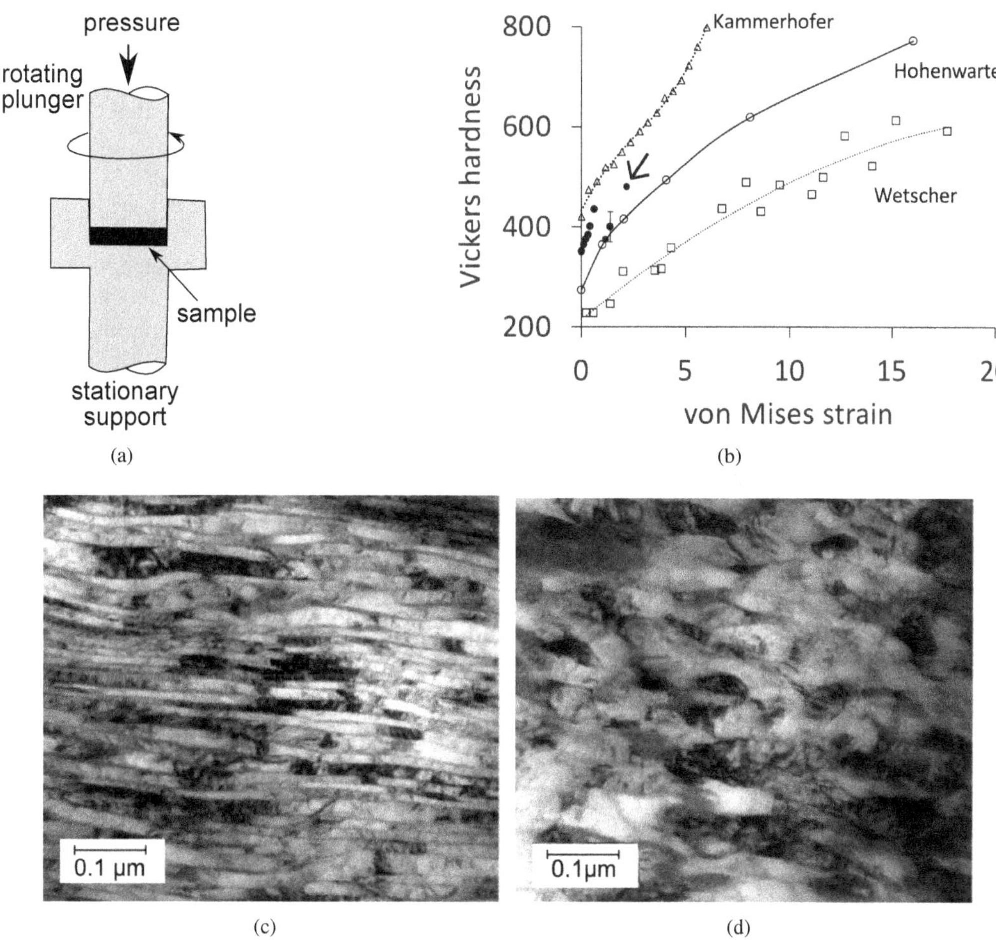

Figure 14.15 (a) Equipment used during constrained compression-torsion plastic-deformation of a disk 26 mm diameter and 8 mm thickness. (b) Compilation of hardness data (○,□) from pearlitic rail-steel (Fe-0.76C-0.35Si-1Mn wt%) subjected to high-pressure torsion [68, 77, 78]; error bars omitted for clarity but can be found in the original references. The filled points (•) highlighted with the arrow, are from drawn pearlitic wire [79–81]. (c,d) Fe-0.76C-0.35Si-1Mn wt% pearlitic steel subjected to $\varepsilon_{vM} = 16$ using torsion while the sample was pressed with a stress of 5.6 GPa. The two images show that the microstructure obtained is not entirely homogeneous. Micrographs adapted by permission from Springer Nature Customer Service Centre GmbH: published by Springer Nature, from Hohenwarter et al. [77] ©The Minerals & Materials Society of ASM International 2010.

High-pressure torsion leads to the alignment of the structure somewhat parallel to the shear plane so the fatigue crack growth rate becomes anisotropic [73, 82]. Cracks propagate much more rapidly parallel to the alignment when the comparison is made in the Paris law region, in samples subjected to $\gamma = 2$.

A comparison with drawn pearlitic wire is appropriate, even though the loading configuration is different from high-pressure torsion. It is established that crack advance is much easier parallel to the drawing direction than across the wire [83]. The cause is attributed to the weakening of α/θ interfaces in the drawn condition as defects are generated parallel to the drawing direction during the fragmentation of cementite (p. 281). However, during torsion to very large cumulative strains, it is likely that the cementite of the pearlite has entered solution.

A single sample of pearlite subjected to high-pressure torsion and fracture has been characterised using an atom probe [84], to reveal that fracture does indeed follow the α/θ interface[6] but it must be appreciated that studies like these examine an incredibly small volume of material. The outcome may or may not be representative of macroscopic fracture, in particular of the evolution of oriented damage.

14.7 CAVITATION EROSION

When a steel is in contact with liquid that undergoes a sudden local, transient change in pressure, a bubble forms and collapses when the initial stimulus is removed. This induces a shock wave that can be sufficiently intense to cause damage at the steel surface, either as a single event or due to repeated waves that lead to spalling of the type associated with fatigue processes. The pulse that locally reduces the liquid to below its vapour pressure can originate, for example, during turbulent flow, a momentary strike by a laser, or by ultrasonic vibration.

Samples of the same chemical composition (Fe-0.7C-0.24Si-1.12Mn wt%) but different microstructures, subjected to cavitation-erosion in a 3.5%-NaCl solution, show that allotriomorphic-α in a predominantly pearlitic sample increases the tendency for cavitation [85], with the ferrite observed directly to be more susceptible to attack. Uniform bainitic or tempered martensitic microstructures fared better, though untempered martensite did not, presumably because it is brittle.

14.8 ARTIFICIAL PEARLITE

Cementite can be manufactured in isolation using mechanical alloying [86] where different powders are placed in a ball mill so that repeated impact causes the mixing of the component powders, ultimately on a atomic scale [87]. Using this method cementite has been made in powdered form by mixing Fe, graphite and Mn in the correct proportion for $[Fe_{0.95}Mn_{0.05}]_{75}C_{25}$, followed by spark plasma sintering while under a stress of 50 MPa. The plasticity of cementite increases dramatically with deformation temperature, Figure 14.16a, due to superplasticity associated with the $\approx 0.5\,\mu m$ initial grain-size.

To create alternating layers of α and θ, cementite discs 6 mm diameter and 0.45 mm thickness, and low carbon Fe-0.15C wt% disks of identical diameter but double the thickness were stacked together and reduced 66% thickness by compression at 873 K and 100 MPa, resulting in the elegant structure shown in Figure 14.16b. This is worthy of further study in many respects, first because the each cementite layer, unlike ordinary pearlite, is polycrystalline with a micrometre grain size. What are the consequences of this of both the α/θ macroscopic interface plane, how does the deformation behaviour change since the object is not the normal bicrystal that is a pearlite colony, and how does the structure change if annealed for a prolonged period of time consistent with the spheroidisation of normal pearlite?

Obviously, the scale S_I of the artificial pearlite can be controlled by the compressive deformation imposed on the discs, but the spacing is of the order of 400 μm, much greater than in any pearlite generated by phase transformation. How then would the strength depend on S_I, would the Hall-Petch equation apply or is there something special about testing what is effectively a single colony of pearlite which should exhibit highly anisotropic strength represented by composite theory.

[6]The cementite at the interface may have been dissolved in part by the severe deformation, so the θ refers strictly to a high-carbon region in the vicinity of the original interface.

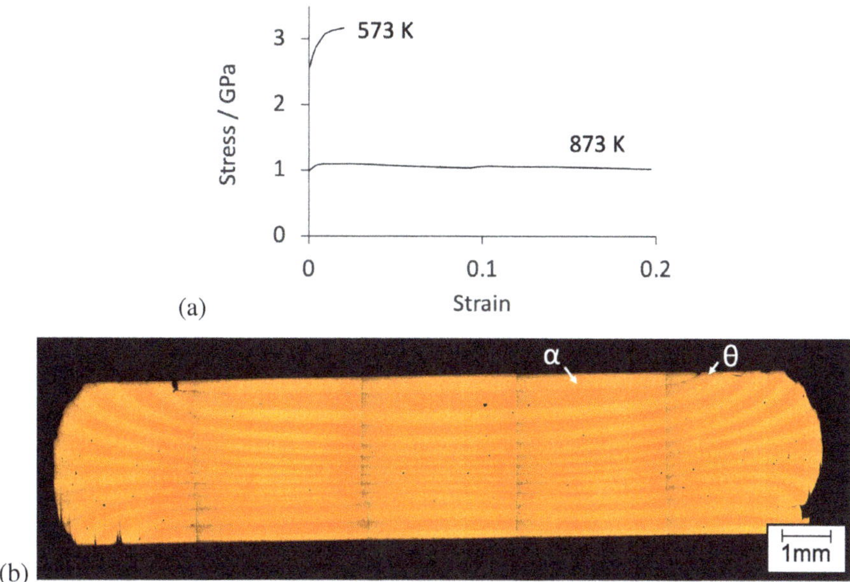

Figure 14.16 (a) The compressive strength of cementite as a function of temperature. Selected data from [88].(b) Artificial pearlite made of layers of ferrite (strictly Fe-0.15C wt%, so must contain some θ) and $[Fe_{0.95}Mn_{0.05}]_{75}C_{25}$. Image courtesy of Professor Minoru Umemoto.

14.9 ULTRASONIC ATTENUATION BY PEARLITE

The interest in ultrasonic attenuation mechanisms lies primarily on the interpretation of non-destructive testing. Sound can propagate in solids as longitudinal or transverse waves depending on whether the displacements are parallel or normal, respectively, to the direction of the wave. As it travels through the material, it may be scattered, for example, by grain boundaries, or some of its energy may be dissipated irreversibly as heat. Since a sound wave causes displacements, it is influenced by changes in the elastic modulus caused by grain orientations or the presence of more than one phase in the microstructure. Such 'heterogeneities' disturb the continuity of stress and displacement, leading to scattered waves that carry away some of the power of the incident wave, which is referred to as *attenuation* [89, 90].

The magnitude of the attenuation depends on the scale of the structure responsible for scattering to the wavelength (λ) and hence the frequency (ν) of the sound ($\nu = \text{velocity}/\lambda$). It follows that at low frequencies in the range 1-5 MHz ($\lambda = 300 \rightarrow 59$ m), pearlite with an interlamellar spacing of 10^{-6} m only slightly attenuates sound waves. The effect becomes significant at greater frequencies, Figure 14.17.

Pearlite is in this respect more heterogeneous than martensite and therefore more capable of scattering sound waves [91, 92]. Each martensite plate is elastically homogeneous and because each austenite grain can contain up to 24 different crystallographic variants of martensite, any anisotropy in modulus is minimal on the austenite grain size scale. The sound velocities are a little larger in pearlite than other microstructures (Table 14.5), attributed to density and effective modulus differences [92, 93].

Another technique that seems to be more effective in characterising structure is the monitoring of higher order harmonics of ultrasonic signals [95]. A sinusoidal longitudinal wave of amplitude A_1 moving along z in an isotropic material causes displacements u described by

$$u\{z,t\} = A_1 \sin\{kz - \omega t\}$$

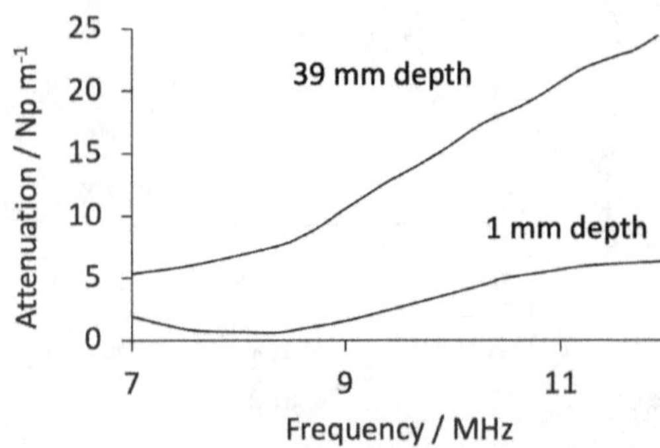

Figure 14.17 Measured attenuation of longitudinal sound waves as a function of depth into a pearlitic railway-wheel that had been heat treated to achieve a finer S_I at the surface. Selected data from [94]. The 'Np' refers to a Neper, which is equivalent to 8.68 decibels. The attenuation coefficient α plotted on the vertical axis is defined by $A = A_o \exp\{-\alpha z\}$ where A and A_o are the final and initial amplitudes respectively, and z is the depth.

Table 14.5

Ultrasonic sound velocities in Fe-0.39C-0.89Mn-0.31Si-1.13Cr-0.05Mowt% austenitised at 843 °C prior to transformation into a variety of microstructures. Selected data from [92].

	Density / kg m^{-3}	Velocity / m s^{-1}	
		Longitudinal	Transverse
Pearlite+proeutectoid-α	7813	5945	3252
Bainite	7809	5916	3236
Martensite	7776	5880	3195

if it is assumed that the displacements are strictly proportional to the product of the modulus and strain $E\varepsilon$. In this equation, ω is the angular frequency, t is the time, k is the wave number (number of wavelengths per unit length).

If the stress-strain relationship at a 'defect' in nonlinear,

$$\sigma = E\varepsilon(1 + b_{30}\varepsilon + \ldots)$$

then the travelling wave equation acquires additional harmonics[7], for example,

$$u\{z,t\} = \underbrace{A_1 \sin\{kz - \omega t\}}_{\text{first harmonic}} + \underbrace{\frac{A_1^2 k^2 \beta_e}{8} \cos\{2(kz - \omega t)\}}_{\text{second harmonic}}$$

where β_e is known as the nonlinear elastic coefficient that is a consequence of the nonlinear elastic behaviour at a 'defect'. It can be evaluated by measuring the amplitudes of the first and second

[7] A *harmonic* is a sinusoidal wave with a frequency that is an integral multiple of the fundamental frequency of the first harmonic

harmonic components for correlation against structure. It turns out that β_e is a much more sensitive parameter than the degree of attenuation or wave velocity in distinguishing the extent of spheroidisation of pearlite, Figure 14.18. In contrast, the attenuation and velocities changed by $< 5\%$.

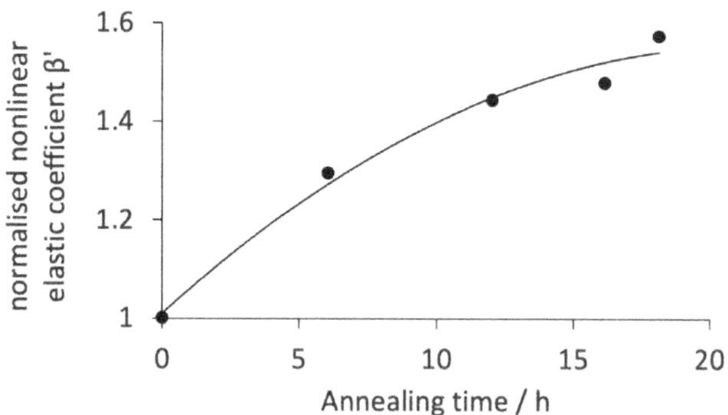

Figure 14.18 Fe-0.75C-0.96Mn-0.22Si wt% steel transformed into pearlite by cooling to 680 °C and held there for the time indicated, versus the nonlinear elastic coefficient β_e normalised to a value β' with an intrinsic value for the material. Adapted using data from Chen et al. [96]. A zero time corresponds to the initial lamellar structure whereas 18 h led to 'severe' spheroidisation.

REFERENCES

1. G. Roberts, and R. Mehl: 'The mechanism and the rate of formation of austenite from ferrite cementite aggregates', *Transactions of the American Society for Metals*, 1943, **31**, 613–650.
2. Z.-D. Li, G. Miyamoto, Z.-G. Yang, and T. Furuhara: 'Nucleation of austenite from pearlitic structure in an Fe-0.6 C-1Cr alloy', *Scripta Materialia*, 2009, **60**, 485–488.
3. A. Roósz, Z. Gacsi, and E. Fuchs: 'Isothermal formation of austenite in eutectoid plain carbon steel', *Acta Metallurgica*, 1983, **31**, 509–517.
4. D. Shtansky, K. Nakai, and Y. Ohmori: 'Pearlite to austenite transformation in an Fe-2.6 Cr-1C alloy', *Acta materialia*, 1999, **47**, 2619–2632.
5. U. Lenel, and R. Honeycombe: 'Morphology and crystallography of austenite formed during intercritical annealing', *Metal Science*, 1984, **18**, 503–510.
6. E. E. Underwood: Quantitative Stereology, chap. 4: Addison-Wesley Publication Company, 1970:90–93.
7. J. D. Verhoeven: 'A review of microsegregation induced banding phenomena in steels', *Journal of Materials Engineering and Performance*, 2000, **9**(3), 286–296.
8. J. Krawczyk, B. Pawłowski, and P. Bała: 'Banded microstructure in forged 18CrNiMo7-6 steel', *Metallurgy and Foundry engineering*, 2009, **35**, 45–53.
9. F. C. Cerda, I. Sabirov, C. Goulas, J. Sietsma, A. Monsalve, and R. Petrov: 'Austenite formation in 0.2% C and 0.45% C steels under conventional and ultrafast heating', *Materials & Design*, 2017, **116**, 448–460.
10. M. D. E. Robinson: 'The characteristics of the formation of austenite in eutectoid steel': Ph.D. thesis, University of British Columbia, Canada, 1947.
11. G. S. Ansell, P. J. Brofman, T. J. Nichol, and G. Judd: 'Effect of austenite strength on the transformation to martensite in Fe-Ni and Fe-Ni-C alloys', In: G. B. Olson, and M. Cohen, eds. *International Conference on Martensitic Transformations ICOMAT '79*. 1979:350–355.

12. Q. Lai, M. Gouné, A. Perlade, T. Pardoen, P. Jacques, O. Bouaziz, and Y. Bréchet: 'Mechanism of austenite formation from spheroidized microstructure in an intermediate fe-0.1 c-3.5 mn steel', *Metallurgical and Materials Transactions A*, 2016, **47**, 3375–3386.
13. H. Azizi-Alizamini, M. Militzer, and W. J. Poole: 'Austenite formation in plain low-carbon steels', *Metallurgical and Materials Transactions A*, 2011, **42**, 1544–1557.
14. D. Yang, E. Brown, D. Matlock, and G. Krauss: 'Ferrite recrystallization and austenite formation in cold-rolled intercritically annealed steel', *Metallurgical Transactions A*, 1985, **16**, 1385–1392.
15. R. B. McLellan, M. L. Rudee, and T. Ishibachi: 'The thermodynamics of dilute interstitial solid solutions with dual-site occupancy and its application to the diffusion of carbon in α-iron', *Transactions of the AIME*, 1965, **233**, 1939–1943.
16. H. K. D. H. Bhadeshia: 'Diffusion of carbon in austenite', *Metal Science*, 1981, **15**, 477–479.
17. H. K. D. H. Bhadeshia: 'Austenite growth in pearlite MAP_STEEL_AUSTENITEGROWTH': Computer program https://www.phase-trans.msm.cam.ac.uk/map/mapmain.html, Queen Mary University of London, University of Cambridge, 2024.
18. H. K. D. H. Bhadeshia: Theory of Transformations in Steels: London, U.K.: CRC Press, Taylor and Francis Group, 2021.
19. L. Karmazin, and J. Krejcí: 'The dependence of the austenitization kinetics on the type of initial spheroidized structure in low alloy steel', *Materials Science and Engineering: A*, 1994, **185**, L5–L7.
20. V. Savran, S. Offerman, and J. Sietsma: 'Austenite nucleation and growth observed on the level of individual grains by three-dimensional X-ray diffraction microscopy', *Metallurgical and Materials Transactions A*, 2010, **41**, 583–591.
21. NPL: 'MTDATA': Software, National Physical Laboratory, Teddington, U.K., 2006.
22. U. Lenel: 'TTT curves for the formation of austenite', *Scripta metallurgica*, 1983, **17**, 471–474.
23. D. Suh, S. Lee, and Y. Koo: 'Microstructural modification of plain carbon steels irradiated by high-energy electron beam', *Metallurgical and Materials Transactions A*, 1997, **28**, 637–647.
24. S. J. Choi, S.-J. Kwon, S.-H. Choo, and S. Lee: 'Phase analysis of surface layers irradiated with high-energy electron beam using mössbauer spectroscopy', *Materials Science and Engineering: A*, 1999, **265**(1-2), 208–216.
25. R. Blondeau, P. Maynier, J. Dollet, and B. Vieillard-Baron: 'Estimation of hardness, strength and elastic limit of C- and low-alloy steels from their composition and heat treatment', *Mémoires Scientifiques de la Revue de Métallurgie*, 1975, **72**, 759–769.
26. V. M. Ferreira, M. Mecozzi, R. Petrov, and J. Sietsma: 'Details of pearlite to austenite transformation in steel: Experiments and phase-field modeling', *Computational Materials Science*, 2023, **228**, 112368.
27. Y. Koizumi, T. Otsuka, Y. Minamino, T. Takayama, M. Ueyama, T. Daio, and S. Hata: 'Microstructures developed by super-rapid induction heating-and-quenching (SRIHQ) of Fe-1.4% Cr-1% C pearlitic steel', *Materials Science and Engineering: A*, 2013, **577**, 29–35.
28. P. M. Kelly, and J. Nutting: 'The martensite transformation in carbon steels', *Proceedings of the Royal Society of London*, 1960, **A259**, 45–58.
29. G. B. Olson, and M. Cohen: 'Theory of martensitic nucleation: a current assessment', In: M. S. H. I. Aaronson, D. E. Laughlin:, ed. *Proceedings of International Conference on Solid→Solid Phase Transformations*. Warrendale, Pennsylvania, USA: TMS-AIME, 1981:1209–1213.
30. S. Mates, M. Stoudt, and S. Gangireddy: 'Measuring the influence of pearlite dissolution on the transient dynamic strength of rapidly heated plain carbon steels', *JOM*, 2016, **68**, 1832–1838.
31. S. P. Mates, E. Vax, R. Rhorer, and M. R. Stoudt: 'Dynamic flow stress behavior of hypoeutectoid ferrite-pearlite steels under rapid heating', *Journal of Dynamic Behavior of Materials*, 2020, **6**, 246–265.

32. V. Khlestov, E. Konopleva, and H. McQueen: 'Effects of deformation and heating temperature on the austenite transformation to pearlite in high alloy tool steels', *Materials Science and Technology*, 2002, **18**, 54–60.
33. V. F. Zackay, E. R. Parker, D. Farh, and R. Busch: 'The enhancement of ductility in high-strength steels', *Quarterly Transactions of the ASM*, 1967, **60**, 252–259.
34. B. V. Rao: 'Ferrite-austenite dual phase steel': US Patent 4,544,422, 1985.
35. M. Grujicic, M. Buonanno, S. M. Allen, G. B. Olson, and M. Cohen: 'Heterogeneous precipitation of austenite for stabilization', In: G. B. Olson, M. Azrin, and E. S. Wright, eds. *Proc. 34th Sagamore Army Materials Research Conf. on Innovations of Ultrahigh-Strength Steel Technology*. 1987:527–593.
36. N. A. Razik, G. W. Lorimer, and N. Ridley: 'An investigation of manganese partitioning during the austenite-pearlite transformation using analytical electron microscopy', *Acta Metallurgica*, 1974, **22**, 1249–1258.
37. M. Miller, P. Beaven, R. Lewis, and G. Smith: 'Atom probe microanalytical studies of some commercially important steels', *Surface Science*, 1978, **70**, 470–484.
38. J. Chance, and N. Ridley: 'Chromium partitioning during isothermal transformation of a eutectoid steel', *Metallurgical Transactions A*, 1981, **12A**, 1205–1213.
39. H. K. D. H. Bhadeshia, A. Strang, and D. J. Gooch: 'Ferritic power plant steels: Remanent life assessment and the approach to equilibrium', *International Materials Reviews*, 1998, **43**, 45–69.
40. J. Garratt-Reed, T. D. Mottishaw, and G. D. W. Smith: 'First observations of solute redistribution at the pearlite growth front in alloy steels', In: *Institute of Physics Conference Series No. 68 (EMAG)*, vol. Chapter 9. London, U.K.: Institute of Physics, 1983:311–314.
41. Z. Xiong, D. Yang, H. Zhang, and X. Cheng: 'Carbide-free tempered martensite in fcc steels', *Journal of Materials Science & Technology*, 2024, **179**, 22–25.
42. D. Yang, C. Zhang, X. Cheng, and Z. Xiong: 'Lamellar pearlite as an initial microstructure for austenite reversion treatment', *Journal of Materials Engineering and Performance*, 2021, **30**, 1330–1339.
43. D.-z. Yang, Z.-p. Xiong, C. Zhang, G.-z. Feng, Z.-f. Cheng, and X.-w. Cheng: 'Evolution of microstructures and mechanical properties with tempering temperature of a pearlitic quenched and tempered steel', *Journal of Iron and Steel Research International*, 2021, **29**, 1393–1403.
44. L. S. Darken: 'Diffusion of carbon in austenite with a discontinuity of composition', *Transactions of the AIME*, 1949, **180**, 430–438.
45. A. H. Cottrell, and B. A. Bilby: 'Dislocation theory of yielding and strain ageing of iron', *Proceedings of the Physics Society A*, 1949, **62**, 49–62.
46. D. Kalish, and M. Cohen: 'Structural changes and strengthening in the strain tempering of martensite', *Materials Science & Engineering*, 1970, **6**, 156–166.
47. T. Abe, K. Fujikura, R. Nagashima, N. Nakada, and S. Yabu: 'Morphology dependence on mechanical stability of second-phase austenite in martensitic steels', *ISIJ International*, 2024, **64**, 421–429.
48. N. E. Luiz, and A. R. Machado: 'Development trends and review of free-machining steels', *Proceedings of the Institution of Mechanical Engineers, Part B: Journal of Engineering Manufacture*, 2008, **222**, 347–360.
49. F. W. Taylor: 'On the art of cutting metals', In: *Transactions of the American Society of Mechanical Engineers*, vol. 23. New York, USA: ASME, 1906:1–248.
50. L. Sidjanin, and P. Kovac: 'Fracture mechanisms in chip formation process', *Materials Science and Technology*, 1997, **13**, 4399–4444.
51. T. Akasawa, I. Fukuda, K. Nakamura, and T. Tanaka: 'Effect of microstructure and hardness on the machinability of medium-carbon chrome-molybdenum steel', *Journal of Materials Processing Technology*, 2004, **153-154**, 48–53.

52. C. Mapelli, W. Nicodemi, M. Vedani, and A. Zoppi: 'Control of inclusions in a resulphurised steel', *Steel Research*, 2000, **71**, 161–168.
53. H. Liu, and W. Chen: 'Effect of total oxygen content on the machinability of low carbon resulfurized free cutting steel', *Steel Research International*, 2012, **83**, 1172–1179.
54. D. J. Naylor, D. T. Llewellyn, and M. Keane: 'Control of machinability in medium-carbon steels', *Metals Technology*, 1976, **3**, 254–271.
55. Y. Kim, H. Kim, S. Y. Shin, K. Rhee, S. B. Ahn, D. L. Lee, N. J. Kim, and S. Lee: 'Formation mechanisms of cracks formed during hot rolling of free-machining steel billets', *Metallurgical & Materials Transactions A*, 2012, **43**, 882–892.
56. H. K. D. H. Bhadeshia: 'Nanostructured bainite', *Proceedings of the Royal Society of London A*, 2010, **466**, 3–18.
57. H. K. D. H. Bhadeshia: 'The first bulk nanostructured metal', *Science and Technology of Advanced Materials*, 2013, **14**, 014202.
58. J. A. Mohandesi, and M. Saadatmand: 'The optimization of interlamellar spacing in a nanopearlitic lead-patented hypoeutectoid steel wire', *Journal of Materials Engineering and Performance*, 2011, **20**, 1467–1473.
59. K. Mishra, and A. Singh: 'Effect of interlamellar spacing on fracture toughness of nanostructured pearlite', *Materials Science & Engineering A*, 2017, **706**, 22–26.
60. S. Liu, F. Zhang, Z. Yang, M. Wang, and C. Zheng: 'Effects of Al and Mn on the formation and properties of nanostructured pearlite in high-carbon steels', *Materials and Design*, 2016, **93**, 73–80.
61. A. Dick, F. Körmann, T. Hickel, and J. Neugebauer: '*Ab initio* based determination of thermodynamic properties of cementite including vibronic, magnetic, and electronic excitations', *Physical Review B*, 2011, **84**, 125101.
62. R. A. Jaramillo, S. S. Babu, G. M. Ludtka, R. A. Kisner, J. B. Wilgen, G. Mackiewicz-Ludtka, D. M. Nicholson, S. M. Kelly, M. Murugananth, and H. K. D. H. Bhadeshia: 'Effect of 30 Tesla magnetic field on transformations in a novel bainitic steel', *Scripta Materialia*, 2004, **52**, 461–466.
63. K. M. Wu, and H. K. D. H. Bhadeshia: 'Extremely fine pearlite by continuous cooling transformation', *Scripta Materialia*, 2012, **67**, 53–56.
64. V. N. Khiratkar, K. Mishra, and A. Singh: 'Deformation and microstructural evolution of nanostructured pearlite under tension versus torsion', *Steel Research International*, 2021, **92**, 2000526.
65. Y. Xu, M. Umemoto, and K. Tsuchiya: 'Comparison of the characteristics of nanocrystalline ferrite in Fe–0.89C steels with pearlite and spheroidite structure produced by ball milling', *Materials Transactions*, 2002, **43**, 2205–2212.
66. M. Umemoto, B. Huang, K. Tsuchiya, and N. Suzuki: 'Formation of nanocrystalline structure in steels by ball drop test', *Scripta Materialia*, 2002, **46**, 383–388.
67. V. M. Segal: 'Equal channel angular extrusion: from macromechanics to structure formation', *Materials Science & Engineering A*, 1999, **271**, 322–333.
68. F. Wetscher, R. Stock, and R. Pippan: 'Changes in the mechanical properties of a pearlitic steel due to large shear deformation', *Materials Science & Engineering A*, 2007, **445-446**, 237–243.
69. Y. Li, Y. Xiong, T. Zhou, H. Singh, J. Kömi, and M. Huttula: 'Impacts of warm equal-channel angular pressing on microstructure and mechanical properties of granular pearlitic steel', *Steel Research International*, 2021, **92**, 2000496.
70. D. H. Shin, S. Y. Han, K. T. Park, Y. S. Kim, and Y. N. Paik: 'Spheroidization of low carbon steel processed by equal channel angular pressing', *Materials Transactions*, 2003, **44**, 1630–1635.

71. R. Z. Valiev, Y. V. Ivanisenko, E. F. Rauch, and B. Baudelet: 'Structure and deformaton behaviour of Armco iron subjected to severe plastic deformation', *Acta Materialia*, 1996, **44**, 4705–4712.
72. R. Z. Valiev, R. K. Islamgaliev, and I. V. Alexandrov: 'Bulk nanostructured materials from severe plastic deformation', *Progress in Materials Science*, 2000, **45**, 103–189.
73. T. Leitner, G. Trummer, R. Pippan, and A. Hohenwarter: 'Influence of severe plastic deformation and specimen orientation on the fatigue crack propagation behavior of a pearlitic steel', *Materials Science & Engineering A*, 2018, **710**, 260–270.
74. F. Z. Utyashev, Y. E. Beygelzimer, and R. Z. Valiev: 'Large and severe plastic deformation of metals: Similarities and differences in flow mechanics and structure formation', *Advanced Engineering Materials*, 2021, **23**, 2100110.
75. N. Larijani, C. Kammerhofer, and M. Eke: 'Simulation of high pressure torsion tests of pearlitic steel', *Journal of Materials Processing Technology*, 2015, **223**, 337–343.
76. C. Kammerhofer, A. Hohenwarter, S. Scheriau, H. P. Brantner, and R. Pippan: 'Influence of morphology and structural size on the fracture behavior of a nanostructured pearlitic steel', *Materials Science & Engineering A*, 2013, **585**, 190–196.
77. A. Hohenwarter, A. Taylor, R. Stock, and R. Pippan: 'Effect of large shear deformations on the fracture behavior of a fully pearlitic steel', *Metallurgical & Materials Transactions A*, 2011, **42**, 1609–1618.
78. C. N. Kammerhofer: 'Influence of material properties on the in-service behavior of rails': Ph.D. thesis, University of Leoben, Austria, 2014.
79. A. Kisrane-Bouzidi, M. Zidani, M. C. Nebbar, T. Abid, A. L. Helbert, F. Brisset, and T. Baudin: 'Mechanical properties and texture evolution of high-carbon steel wires during wire drawing: Strand manufacturing', *International Journal of Engineering Research in Africa*, 2020, **49**, 130–138.
80. S. Sato, T. Shobu, K. Satoh, H. Ogawa, K. Wagatsuma, M. Kumagai, M. Imafuku, H. Tashiro, and S. Suzuki: 'Distribution and anisotropy of dislocations in cold-drawn pearlitic steel wires analyzed using micro-beam X-ray diffraction', *ISIJ International*, 2015, **55**, 1432–1438.
81. P. Kumar, N. P. Gurao, A. Haldar, and S. Suwas: 'Progressive changes in the microstructure and texture in pearlitic steel during wire drawing', *ISIJ International*, 2011, **51**, 679–684.
82. A. Leitner, A. Hohenwarter, and R. Pippan: 'Anisotropy in fracture and fatigue resistance of pearlitic steels and its effect on the crack path', *International Journal of Fatigue*, 2019, **124**, 528–536.
83. M. Elices: 'Influence of residual stresses in the performance of cold-drawn pearlitic wires', *Journal of Materials Science*, 2004, **39**, 3889–3899.
84. T. Leitner, S. Sackl, B. Völker, H. Riedl, H. Clemens, R. Pippan, and A. Hohenwarter: 'Crack path identification in a nanostructured pearlitic steel using atom probe tomography', *Scripta Materialia*, 2018, **142**, 66–69.
85. A. Rajput, J. Ramkumar, and K. Mondal: 'Cavitation behavior of various microstructures made from a C–Mn eutectoid steel', *Wear*, 2021, **486-487**, 204056.
86. J. S. Benjamin: 'Oxide dispersion strengthened ODS superalloys directional recrystallisation', *Metallurgical Transactions*, 1970, **1**, 2943–2951.
87. M. Umemoto, Z. G. Liu, H. Takaoka, M. Sawakami, K. Tsuchiya, and K. Masuyama: 'Production of bulk cementite and its characterization', *Metallurgical & Materials Transactions A*, 2001, **32**, 2127–2131.
88. M. Umemoto, Y. Todaka, and K. Tsuchiya: 'Mechanical properties of cementite and fabrication of artificial pearlite', *Materials Science Forum*, 2003, **426-432**, 859–864.
89. E. Papadakis: 'Physical acoustics and microstructure of iron alloys', *International Metals Reviews*, 1984, **29**, 1–24.

90. E. P. Papadakis: 'Ultrasonic attenuation caused by scattering in polycrystalline media', *Physical acoustics*, 2012, **4 (Part B)**, 269–328.
91. E. P. Papadakis, P. F. Sullivan, and D. J. Waltman: 'Ultrasonic attenuation and velocity in SAE 4150 steel': Tech. Rep. TR 143/37, Watertown Arsenal Laboratory (December 1961), Massachusetts, USA, 1961.
92. E. P. Papadakis: 'Ultrasonic attenuation and velocity in three transformation products in steel', *Journal of Applied Physics*, 1964, **35**, 1474–1482.
93. C. Gur Hakan, and B. O. Tuncer: 'Characterization of microstructural phases of steels by sound velocity measurement', *Materials Characterization*, 2005, **55**, 160–166.
94. H. Du, and J. A. Turner: 'Ultrasonic attenuation in pearlitic steel', *Ultrasonics*, 2014, **54**, 882–887.
95. Y. Xiang, M. Deng, and F.-Z. Xuan: 'Thermal degradation evaluation of HP40Nb alloy steel after long term service using a nonlinear ultrasonic technique', *Journal of Nondestructive Evaluation*, 2014, **33**, 279–287.
96. J. Chen, S.-J. Jin, and D. Qiao: 'Nondestructive evaluation of microstructure degradation of pearlitic steel', In: *2020 IEEE Far East NDT New Technology & Application Forum*. New York, USA: IEEE, 2020:27–30.

15 Other eutectoids

Pearlite contributes to the **49×10⁹ tonnes** of the steels in use today[1] – in brief, anyone making a train journey, using an elevator, or living in a building will experience, albeit unconsciously, this incredible coupling of phases. There are similar lamellar structures generated in other 'alloy' systems which are of little or no practical importance but are intriguing. Some of these, though not all, involve iron; the phase symbols used commonly, therefore, are not unique to a particular system.

15.1 DELTA EUTECTOID

There is another high-temperature eutectoid reaction common in steels that contain strong carbide-forming elements [1, 2], for example,

$$\underbrace{\delta}_{\text{Cubic } Im\bar{3}m} \rightarrow \underbrace{\gamma}_{\text{Cubic } Fm\bar{3}m} + \underbrace{M_{23}C_6}_{\text{Cubic } Fm\bar{3}m} \tag{15.1}$$

where the lattice type and space group are illustrated below each phase symbol. The reaction is referred to as δ-eutectoid in reverence to the parent phase δ-ferrite that evolves from the liquid steel. The eutectoid normally is associated with a small amount of residual δ that is otherwise surrounded by austenite. The structure produced is approximately lamellar with the cooperative growth of austenite and carbide at a common transformation front with the ferrite. The carbide is not limited to $M_{23}C_6$, as evident in Figure 15.1 for a steel with many strong carbide-forming solutes. The austenite within the colony of δ-eutectoid may decompose into martensite on cooling to ambient temperature.

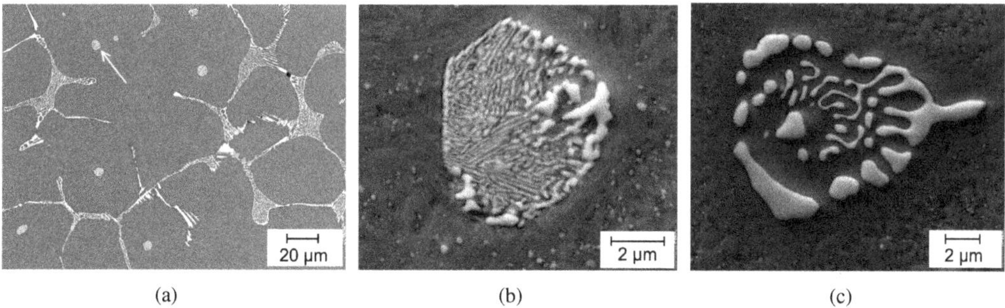

Figure 15.1 Structure of a tool steel. The exact composition was not specified, but from specifications, it is approximately Fe-0.9C-4.2Cr-0.25Mn-5Mo-0.3Si-6.1W-2V wt%. (a) General morphology – the arrow indicates the position of a δ-eutectoid colony. (b) δ-eutectoid where the carbide is M_2C. (c) δ-eutectoid where the carbide is M_6C. Images courtesy of Professor Alexander Chaus and Sahul [3] with permission of Elsevier.

The images in Figure 15.1 are only vaguely lamellar and there is no knowledge of their three-dimensional morphology. Much larger quantities of δ-eutectoid have been obtained by carburising

[1]This amounts to over 5×10^{38} atoms of iron, bearing in mind that there are only 10^{79} baryons (particles with mass) in the Universe, including dark matter but not considering the subdivision into quarks. The number of baryons is in English said to be between a vigintillion and a centillion. The number of carbon atoms in the iron that we use, is 2×10^{37}, somewhat under a duodecillion.

DOI: 10.1201/9781032631981-15

an initially Fe-14.9Cr wt% alloy at 900 °C, in which clear lamellar structures were observed between γ (which transformed into martensite lamellae on cooling) and $M_{23}C_6$ [4]. The carburisation increased the carbon concentration from zero to about 0.5 wt%. Subsequent similar experiments on a series of ternary alloys revealed clear evidence for the cooperative growth of the two product phases at a common transformation front into the δ-ferrite parent [5] although the detailed microstructure of the eutectoid depends on the substitutional solute [6]. There also exist kinetic analyses of the growth of such eutectoids accounting for boundary and volume diffusion mechanisms [6, 7]. Table 15.1 and Figure 15.2 illustrate some alloys in which the δ-eutectoid transformation has been observed. Many tool-steels also exhibit the structure [e.g., 8].

Table 15.1

Some alloys in which the δ-eutectoid reaction has been observed. The carbide listed is that within the eutectoid in association with austenite (which on cooling may transform into martensite). An extensive study of many alloys by Saito et al. [5, 6] is not included in this table for the sake of brevity, but the data are presented in Figure 15.2.

Composition / wt%	Temperature / °C	Carbide	Reference
Fe-0.2C-8.15Mo	900-1150	$(Fe_3Mo_3)C$	[1]
Fe-0.09C-0.40Mn-0.33Si-0.34Ni-17.20Cr	955	$M_{23}C_6$	[2]
Fe-0.17C-0.56Mn-0.46Si-0.35Ni-20.96Cr	982	$M_{23}C_6$	[2]
Fe-0.24C-0.46Mn-0.42Si-0.26Ni-24.85Cr	1093	$M_{23}C_6$	[2]
≈Fe-0.9C-4.2Cr-0.25Mn-5Mo-0.3Si-6.1W-2V	as cast	M_2C, M_6C	[3]
Fe-14.9Cr, carburised	900	$M_{23}C_6$	[4]
Fe-7Mo, carburised	1050	M_6C	[7]
Fe			

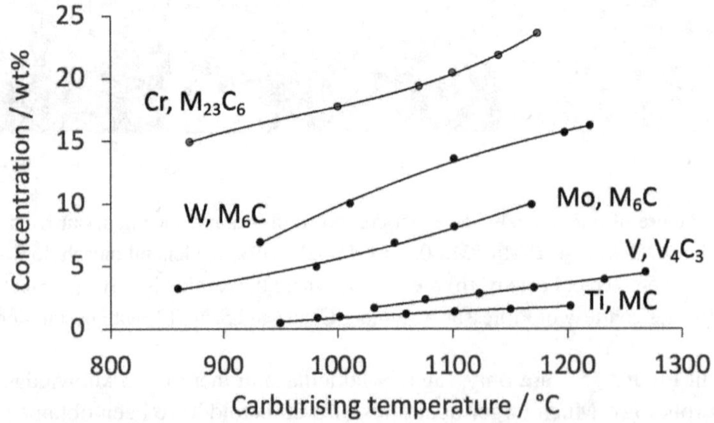

Figure 15.2 Binary alloys of iron exposed to carburising atmospheres in order to determine the eutectoid temperatures by examining the microstructures following exposure. The carbide associated with γ in the eutectoid is indicated. Each data point indicates a different alloy, though the eutectoid carbon concentration is not reported but is a function of the carburising temperature. Adapted using data from Saito et al. [6].

15.2 IRON-NITROGEN

The Fe-N system has a eutectoid (592 °C, 2.4N wt%) which forms a lamellar structure (Figure 15.3) on transforming from austenite

$$\underbrace{\gamma}_{\text{Cubic},Fm\bar{3}m} \rightarrow \underbrace{\alpha}_{\text{Cubic }Im\bar{3}m} + \underbrace{Fe_4N}_{\text{Cubic }Pm\bar{3}m} . \tag{15.2}$$

The nitride has a lattice parameter close to 0.379 nm and is the majority phase in the eutectoid, which has a nitrogen content of 5.88 wt% [9]. It is difficult to introduce large concentrations of nitrogen in iron so studies usually are based on thin foils that are gas-nitrided.

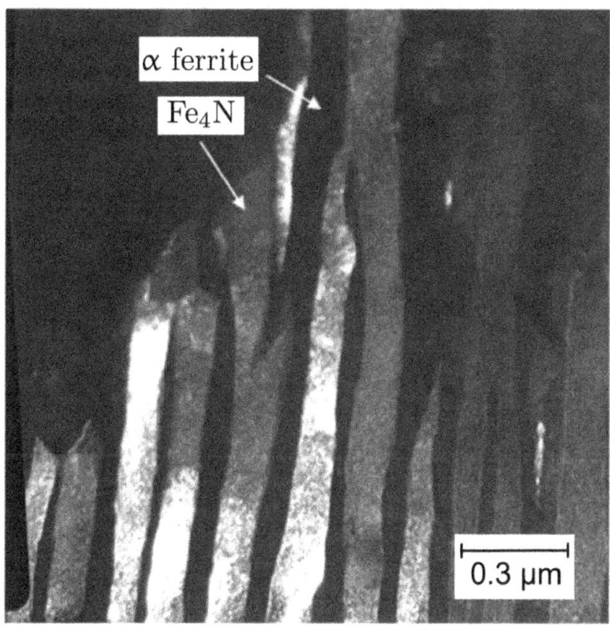

Figure 15.3 Dark-field transmission electron micrograph showing the Fe_4N as the dominant phase in its eutectoid mixture with ferrite. Image reproduced from Xiong et al. [9] with permission of Springer Nature

The Fe-N eutectoid mixture is referred to as *Braunite* after Hjalmar Braune who wrote on the influence of nitrogen on iron and steel during 1905 [10]. However, Braunite in fact was used before then, as the name of a silicate mineral, after Cammerath Braun of Gotha, Germany who pursued the study of mineralogy [11]. The mineral name remains in use today, so probably best to avoid it in the context of the Fe-N system.

15.3 ZIRCONIUM-IRON

The Zr-Fe system exhibits a eutectoid at a composition and temperature 4Fe at.% and ≈ 730 °C, respectively:

$$\underbrace{\beta}_{\text{Cubic }Im\bar{3}m} \rightarrow \underbrace{\alpha}_{\text{Hexagonal }P6_3/mmc} + \underbrace{Zr_3Fe}_{\text{Orthorhombic }Cmcm} , \tag{15.3}$$

with a clear lamellar microstructure, Figure 15.4. It is surprising how rapidly the reaction occurs, during quenching from the parent phase, in spite of the obvious need for substitutional atom diffusion [12]. The resulting phases are thought to have chemical compositions that deviate from equilibrium, but there is no adequate analysis of the kinetics of transformation. The phenomenon has been attributed in another system to favourable crystallography but that does not address the diffusion flux required to account for the large difference in the chemical compositions of the intermetallic phase and the other component of the eutectoid [13]. The idea that ordering occurs in the parent phase ahead of the transformation front also lacks evidence, which with modern techniques should be straightforward to establish.

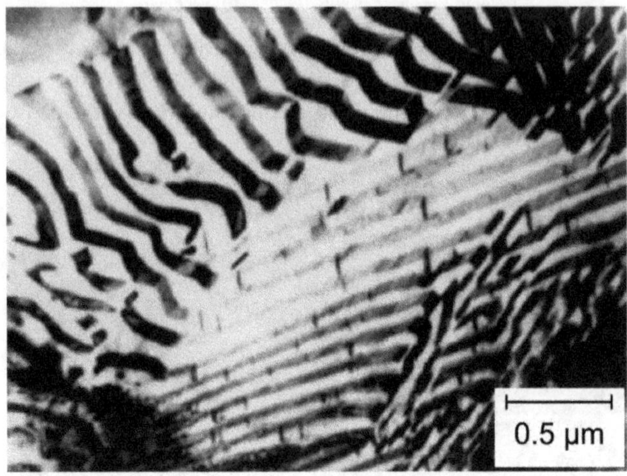

Figure 15.4 Eutectoid structure consisting of a mixture of α and Zr_3Fe in a near-eutectoid alloy sample water-quenched from 1000 °C. Reproduced from Kumar et al. [12] with permission of Elsevier.

15.4 TITANIUM-COPPER

The eutectoid temperature and composition for titanium-copper alloys is 790 °C and 7.1Cu wt%, respectively:

$$\underbrace{\beta}_{\text{Cubic } Im\bar{3}m} \rightarrow \underbrace{\alpha}_{\text{Hexagonal } P6_3/mmc} + \underbrace{\mathrm{Ti_2Cu}}_{\text{Tetragonal } I4/mmm} . \tag{15.4}$$

The lamellar microstructure is shown in Figure 15.5 with the intermetallic compound as the minor phase. The Young's modulus is a function of the volume fractions of the product phases, which opens up the possibility of adapting the material for medical applications. A low modulus mitigates stress shielding where implants reduce the load on adjacent bone, leading to an undesirable reduction in its density. A large modulus, on the other hand, can be useful in minimising elastic strains in devices such as dental bridges [14].

A fully-eutectoid microstructure can be generated even in hypereutectoid alloys such as Ti-12Cu wt% [13], simply by undercooling sufficiently below T_E so that the alloy falls in the Belaiew-Hultgren region of the binary phase-diagram (Figure 8.17). A larger copper concentration increases the fraction of the harder Ti_2Cu, thus boosting strength and tribological properties [15].

The eutectoid reaction in Equation 15.4 is not limited to binary alloys; it features in the classic Ti-6Al-4V alloy when copper added to induce a columnar to equiaxed solidification microstructure during additive manufacture. The copper works in this way by modifying constitutional supercooling during solidification. The eutectoid can form when two pieces of Ti-6Al-4V are joined by diffusion-brazing using a copper interlayer [16].

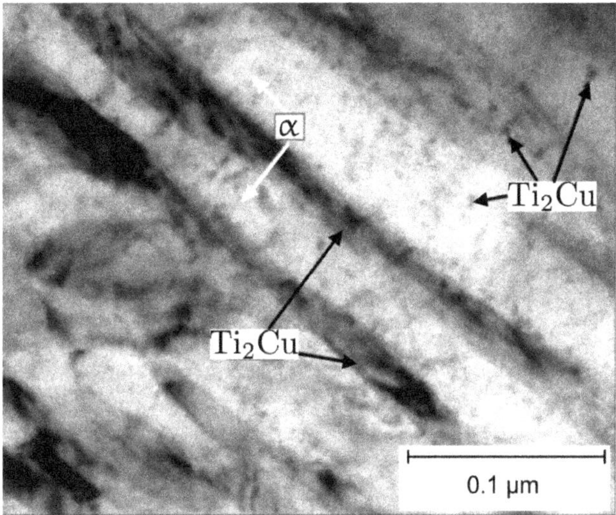

Figure 15.5 Lamellar eutectoid in Ti-7.1Cu wt%, generated during continuous cooling transformation. There are some independently precipitated small particles of Ti_2Cu highlighted at the top right, which presumably formed during cooling. Reproduced from Souza et al. [14] with permission of Elsevier.

15.5 Ag_2HgI_4 – Cu_2HgI_4

Ag_2HgI_4 is a good ionic conductor and to a lesser extent, so is Cu_2HgI_4 which in addition has electronic conductivity [17, 18]. The compounds change colour in their high (α) and low (β) temperature forms. The colour change is associated with order-disorder transformations which occur at 52 and 70 °C in the silver and copper based iodates, respectively. At ambient temperature, pure Ag_2HgI_4 and Cu_2HgI_4 have different tetragonal space-groups $I\bar{4}$ and $I\bar{4}2m$ respectively, whereas at high temperatures they are isomorphous with the cubic space-group $F\bar{4}3m$ [19]. Figure 15.6 shows the known part of the eutectoid phase diagram. Unfortunately, there do not seem to be any microstructural data revealing cooperative growth of the eutectoid from the parent phase.

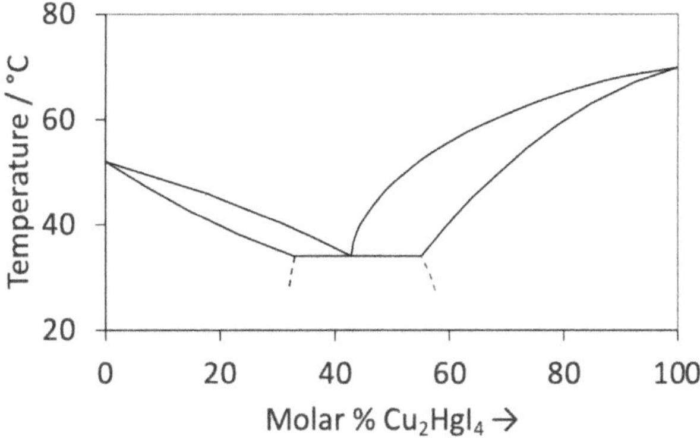

Figure 15.6 Phase diagram for Ag_2HgI_4-Cu_2HgI_4. The eutectoid composition is at 42.7 mol% Cu_2HgI_4 and 34 °C. Adapted using data from Friesel et al. [20].

15.6 IRON-ALUMINIUM

Iron-aluminium alloys involving intermetallic phases at one time attracted attention because of their low density; although strong at ≈ 1000 MPa yield strength, they were not found to be particularly ductile, especially when compared against commercially available iron-based alloys. A large aluminium concentration does instil oxidation resistance that can be exploited, for example, in combustion chambers. The system exhibits a eutectoid reaction at $T_E = 1103\,^\circ$C at the composition $Fe_{0.38}Al_{0.62}$ [21]:

$$\underbrace{Fe_5Al_8}_{\text{Cubic } I\bar{4}3m} \rightarrow \underbrace{FeAl}_{\text{Cubic } Pm\bar{3}m} + \underbrace{FeAl_2}_{\text{Triclinic } P\bar{1}} . \tag{15.5}$$

A directional eutectoid can be generated by allowing the transformation to occur in a temperature gradient, resulting in a composite of the two product intermetallic phases, Figure 15.7.

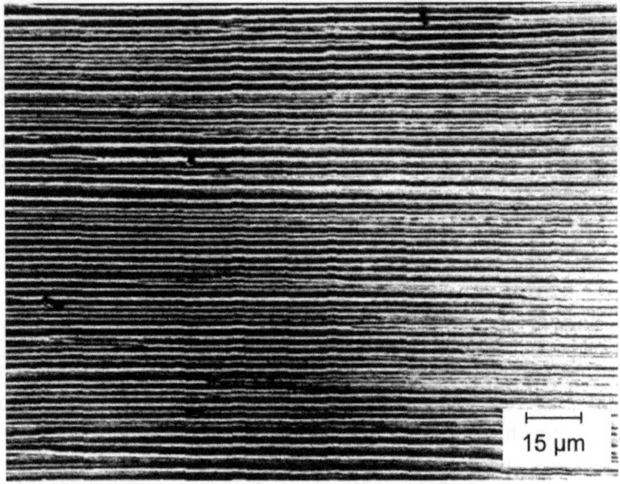

Figure 15.7 Longitudinal section of an iron-aluminium alloy of eutectoid composition, transformed into a directional structure by pulling it through an appropriate temperature gradient. The longitudinal direction is horizontal. Reproduced from Bastin et al. [22] with permission of Elsevier.

15.7 ALUMINIUM-ZINC

This is a particularly interesting system because there is a miscibility gap above the eutectoid temperature for the aluminium-zinc solid solution (γ) which leads to phase separation into two solid phases of identical structure but quite different compositions (Figure 15.8). For example, just before the eutectoid temperature of 277 °C, the γ phase separates into $\gamma_1\{16.5\,\text{at.\% Zn}\}$ and $\gamma_2\{59\,\text{at.\% Zn}\}$, where the latter has the exact eutectoid composition. Further cooling results in the eutectoid reaction:

$$\underbrace{\gamma_2}_{\text{Cubic } Fm\bar{3}m} \rightarrow \underbrace{\gamma_1}_{\text{Cubic } Fm\bar{3}m} + \underbrace{\beta}_{\text{Hexagonal } P6_3/mmc} \tag{15.6}$$

and γ_1 then becomes in equilibrium with β and evolves its composition accordingly, so is sometimes labelled α even though the crystal structure is unchanged.

Aluminium-zinc two-phase alloys can exhibit superplasticity, but not in the lamellar form.

15.8 FeO-IRON

The hot-processing of steel at temperatures in the austenite phase field creates surface oxides consisting of Fe_2O_3 which is the topmost layer, followed in decreasing oxygen to iron ratio by Fe_3O_4, FeO and the steel substrate. As the temperature of the steel is reduced to less than $\approx 570\,^{\circ}C$, the innermost layer FeO (strictly, non-stoichiometric $Fe_{1-x}O$) undergoes a eutectoid decomposition into mixture of Fe_3O_4 and Fe, Figure 15.9. The transformation influences the subsequent processing of the steel, but it occurs in two stages [24–26]:

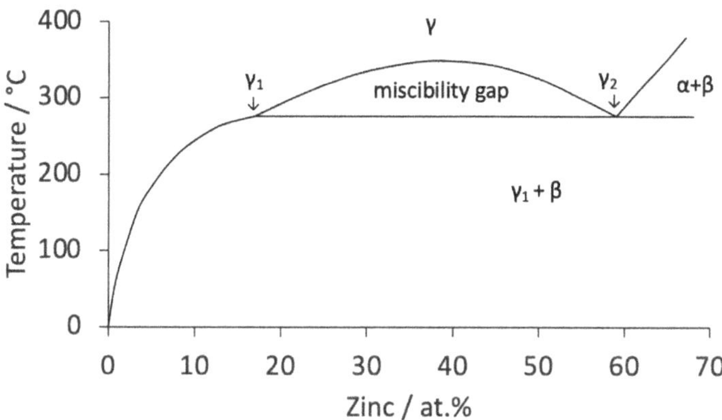

Figure 15.8 Part of the equilibrium Al-Zn phase diagram where all phases are solid. Adapted using a compilation and assessment by Murray [23].

$$\begin{aligned} (1-4x)\mathrm{Fe}_{1-y}\mathrm{O} &\rightarrow (y-x)\mathrm{Fe_3O_4} + (1-4y)\mathrm{Fe}_{1-x}\mathrm{O} \qquad \text{with } y > x \\ 4\mathrm{Fe}_{1-x}\mathrm{O} &\rightarrow \mathrm{Fe_3O_4} + (1-4x)\mathrm{Fe} \qquad \text{eutectoid} \end{aligned} \tag{15.7}$$

A time-temperature-transformation diagrams depicting the kinetics of the eutectoid transformation exhibit the classic 'C'-curves consistent with nucleation and growth phenomena, and the lamellar spacing follows the treatment described on p. 189 [27].

15.9 Ar-CO AND Ar-N

The argon and carbon monoxide system exhibits a eutectoid reaction:

$$\underbrace{\epsilon}_{\text{Trigonal } R3c} \rightarrow \underbrace{\gamma}_{\text{Ar-rich, Cubic } Fm\bar{3}m} + \underbrace{\chi}_{\text{CO-rich, Cubic } P2_13} \tag{15.8}$$

in which the eutectoid composition is $\approx 66\%$ CO and the eutectoid temperature 53 K [28]. Figure 15.10a was deduced using thermal analysis and X-ray diffraction. The ϵ was originally described as hexagonal close-packed but is better described as trigonal because of the way in which the molecules are arranged within the lattice. Orientational disorder of molecules would lead to a space group $P6_3/mmm$. In contrast, the argon-nitrogen phase diagram (Figure 15.10b) does not exhibit a eutectoid reaction. Is it possible to define some general principles for a eutectoid reaction to be a feature of a binary system?

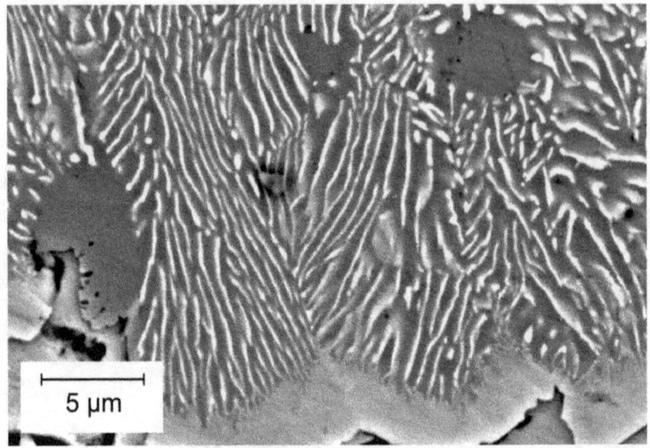

Figure 15.9 The FeO-Fe eutectoid structure produced by oxidising a pure iron sample at 900 °C for 10 min. and then isothermal transformation at 500 °C for 10^5 s. Image courtesy of Professor Guangming Cao, with further details in Wang et al. [27].

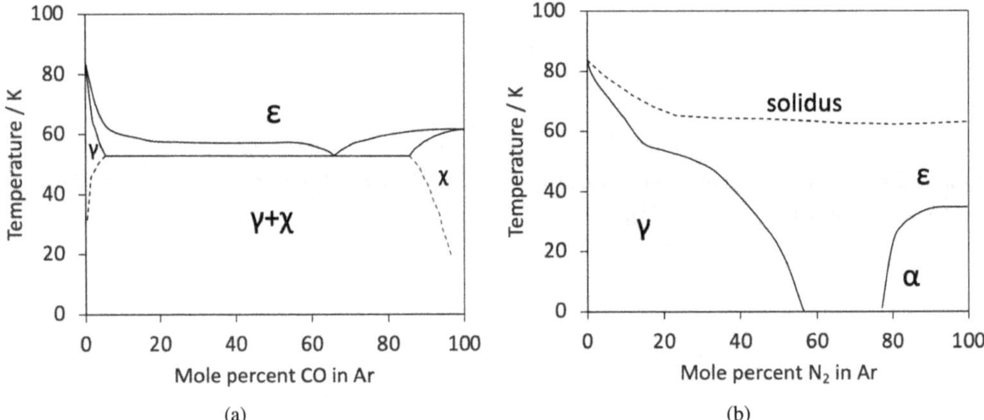

Figure 15.10 Measured phase diagrams. (a) Argon-carbon monoxide – see Equation 15.8 for the structures associated with the symbols. (b) Argon-nitrogen. The symbol ϵ represents a hexagonal close-packed structure ($P6_3/mmc$), α a cubic lattice ($Pa3$) and γ a cubic lattice ($Fm\overline{3}m$). The plots have been constructed using selected data from figures by Barrett and Meyer [28, 29].

Three solid phases must be capable of coexisting in equilibrium at a unique temperature T_E. This is not possible in the Ar-N_2 system because each $T - x$ domain contains only one phase. Transformation into a different crystal structure is induced when a phase boundary is crossed. This can happen without any change in phase-composition if a given alloy is heated or cooled, rather like the pressure-induced phase transformations in pure iron.

Using γ, α, θ in a generic sense, it must at $T > T_E$ be possible for the parent phase (γ) to exist in equilibrium with each of the product phases (α, θ) in isolation, to define $\gamma + \alpha$ and $\gamma + \theta$ two-phase fields, with $x^{\gamma\alpha} \geq x^{\alpha\gamma}$ and $x^{\gamma\theta} \leq x^{\theta\gamma}$. It follows that the γ must be able to accommodate a greater range of solute concentration than either of the product phases.

The minor phase in the eutectoid mixture must have a sufficient volume-fraction to easily maintain continuity at the common transformation front (*cf.* Fe-Au, Section 15.10).

15.10 IRON-GOLD

Iron and gold have a eutectoid at 2.3Au at% and 868 °C [30]:

$$\underbrace{\gamma}_{\text{Cubic Fm}\bar{3}\text{m}} \rightarrow \underbrace{\text{Au}}_{\text{Cubic } Fm\bar{3}m} + \underbrace{\alpha}_{\text{Cubic } Im\bar{3}m} \tag{15.9}$$

so the amount of gold in the eutectoid mixture is rather small at about 2%. This means that the formation of Au-lamellae and indeed, cooperative growth at a common transformation front becomes impossible. Instead, discontinuous, faceted fibres of gold form when the alloy is transformed in a temperature gradient, Figure 15.11. The transformation in a temperature gradient was intended to produce an aligned eutectoid, in analogy with previous work on directionally solidified eutectics. However, there are significant differences between the eutectoid and eutectic growth processes [30]. During eutectic solidification, S_I tends to be much coarser because diffusivity in the liquid ahead of the transformation front is rapid, so the total interfacial energy within the product mixture can be minimised by tolerating a larger interlamellar spacing.

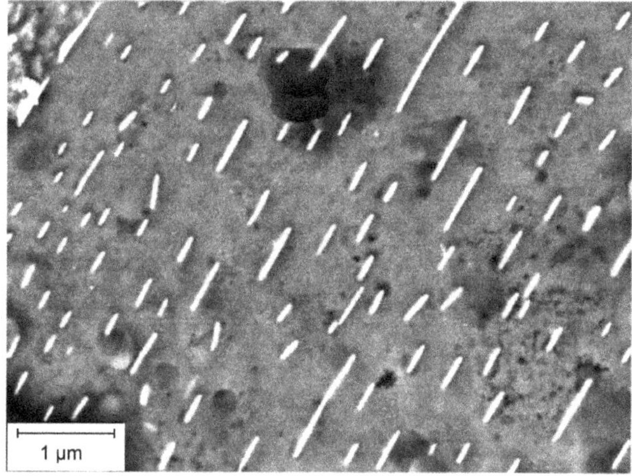

Figure 15.11 Fe-2.3Au at.% alloys with a final heat treatment of 950 °C for 3 h and the fully austenitic sample then slowly lowered through a temperature gradient. Image reproduced from Milenkovic et al. [31] with permission of Elsevier.

Clearly, one of the conditions for obtaining a lamellar structure is that the minor phase within the eutectoid must have a large enough volume fraction to form a plate-like structure rather than fibres, although crystallographic fit with the other phase also matters, p. 241. And since the gold fibres are facetted along their lengths with approximately square cross-sections, there may be an overriding role orientation-dependent interfacial energy in determining the shape.

15.11 EUTECTOID, APPARENTLY WITH THREE PRODUCT PHASES

In Fe-Mn-Al-C alloys that are rich in aluminium, it is common to observe the precipitation of κ-carbide with the approximate formula $(Fe,Mn)_3AlC$. It has been proposed that in some hypoeutectoid alloys, a proeutectoid reaction is followed on cooling by the 3-phase eutectoid [32]:

$$\underbrace{\gamma}_{\text{Cubic } Fm\overline{3}m} \rightarrow \underbrace{\alpha}_{\text{Cubic } Im\overline{3}m} + \underbrace{\kappa}_{\text{Cubic } Pm\overline{3}m} + \gamma' \quad \text{proeutectoid reaction, } T > T_{\text{E}} \tag{15.10}$$

$$\underbrace{\gamma'}_{\text{Cubic } Fm\overline{3}m} \rightarrow \underbrace{\alpha}_{\text{Cubic } Im\overline{3}m} + \underbrace{M_{23}C_6}_{\text{Cubic } Fm\overline{3}m} + \underbrace{\kappa}_{\text{Cubic } Pm\overline{3}m} \quad \text{eutectoid reaction } T < T_{\text{E}} \tag{15.11}$$

where γ' refers to the austenite of eutectoid composition, left after the proeutectoid reaction has occurred. It seems that the $\alpha + \kappa$ lamellae develop first and the $M_{23}C_6$ appears in the ferrite either in lamellar or particular form, so it may not be in contact with the transformation front. In that sense, it is difficult to regard it as an essential component in the cooperative growth of the $\alpha + \kappa$. The structure is illustrated in Figure 15.12 although it is emphasised that the shape of the $M_{23}C_6$ is not always lamellar, nor aligned to the $\alpha + \kappa$ lamellae [32].

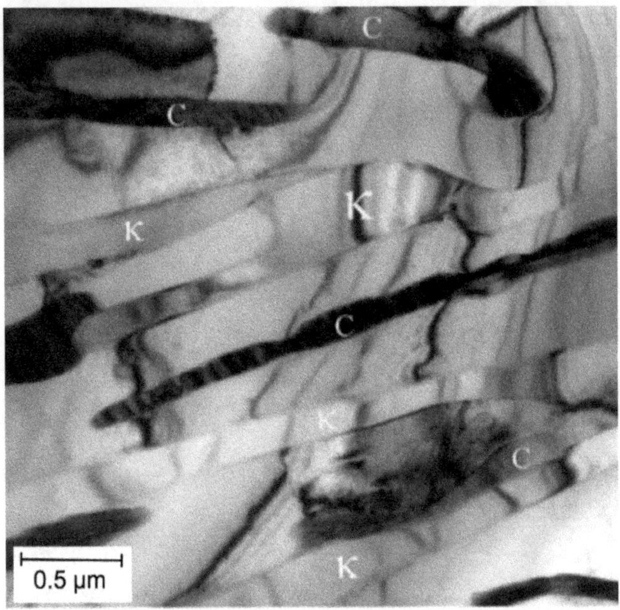

Figure 15.12 Eutectoid consisting of κ-carbide, $M_{23}C_6$ and ferrite following isothermal transformation from austenite, at 625 °C in Fe-13.5Mn-6.3Al-0.78C wt%. The 'C' refers to $M_{23}C_6$ and the unlabelled matrix is ferrite. Image reproduced from Cheng et al. [32] with permission of Springer Nature.

15.12 PEARLITIC PLESSITE

This occurs in iron meteorites that also have a substantial nickel concentration; in earth sciences terminology, austenite is referred to as *taenite* and ferrite as *kamacite*:

$$\underbrace{\text{parent taenite}}_{\text{Cubic } Fm\overline{3}m} \rightarrow \underbrace{\text{product taenite}}_{\text{Cubic } Fm\overline{3}m} + \underbrace{\text{kamacite}}_{\text{Cubic } Im\overline{3}m}. \tag{15.12}$$

The product taenite has a different chemical composition and crystallographic orientation than that of the parent and the taenite and kamacite grow together in approximately lamellar formations. The microstructure (Figure 15.13) often is described as 'pearlitic plessite', but because one of the product phases has the same crystal structure as the parent, it fits into the classification of discontinuous reactions [p.515, 33].

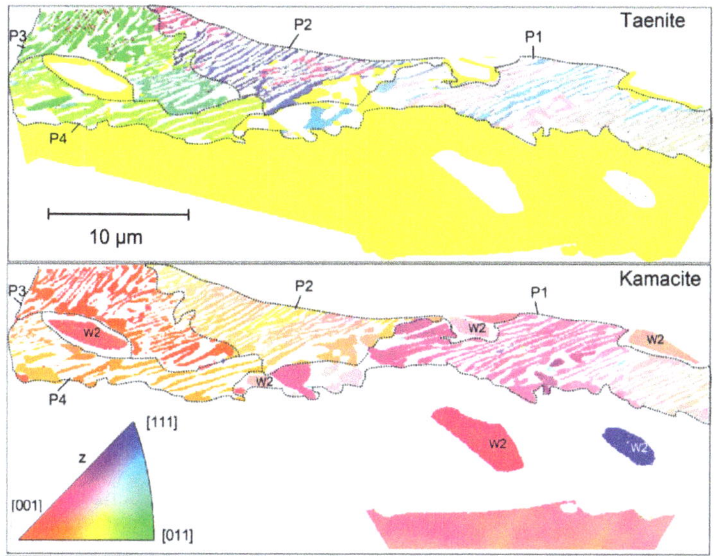

Figure 15.13 Orientation maps of pearlitic plessite in the Odessa meteorite. The top image shows the map for taenite but the lower yellow region is tetrataenite (FeNi, $P4/mmm$) which has interesting magnetic properties but requires an incredibly long time to form. The lower image shows the distribution of kamacite, but the regions labelled 'W' are kamacite that is not a part of the pearlitic plessite. Reproduced from Nichols et al. [34] under the CC BY 4.0 DEED license.

REFERENCES

1. K. Kuo: 'Metallography of delta ferrite part I – eutectoid decomposition of delta ferrite', *J. Iron and Steel Inst*, 1954, **15**, 433–441.
2. A. Nehrenberg, and P. Lillys: 'High temperature transformations in ferritic stainless steels containing 17 to 25% chromium', *Transactions of the American Society for Metals*, 1954, **46**, 1177–1213.
3. A. S. Chaus, and M. Sahul: 'On origin of delta eutectoid carbide in M2 high-speed steel and its behaviour at high temperature', *Materials Letters*, 2019, **256**, 126605.
4. T. Sakuma, M. Haga, and T. Nishizawa: 'Transmission electron microscopic study of a delta-eutectoid structure produced by carburization of Fe-Cr alloy', *Transactions of the Japan Institute of Metals*, 1975, **16**, 225–233.
5. T. Saito, T. Sakuma, and T. Nishizawa: 'Cellular growth of delta-pearlite formed by carburization of ternary iron alloys', *Transactions of the Japan Institute of Metals*, 1980, **21**, 483–494.
6. T. Saito, T. Sakuma, and T. Nishizawa: 'Morphology and growth kinetics of delta-pearlite formed by carburization of binary iron alloys', *Transactions of the Japan Institute of Metals*, 1980, **21**, 471–482.
7. J. Fridberg, and M. Hillert: 'On the eutectoid transformation of δ-ferrite in Fe-Mo-C alloys', *Acta Metallurgica*, 1977, **25**, 19–24.
8. K. Kuo: 'Metallography of delta-ferrite, part II, in 18-4-1-type high-speed steels', *Journal of the Iron and Steel Institute*, 1955, **181**, 128–134.
9. X. C. Xiong, A. Redjaïmia, and M. Gouné: 'Pearlite in hypoeutectoid iron-nitrogen binary alloys', *Journal of Materials Science*, 2009, **44**, 632–638.
10. H. Braune: 'Influence de l'azote sur le fer et l'acier (influence of nitrogen on iron and steel)', *Revue de Métallurgie*, 1905, **2**, 497–502.

11. W. Haidinger: 'VIII. Mineralogical account of the ores of manganese', *Earth and Environmental Science Transactions of The Royal Society of Edinburgh*, 1828, **11**, 119–142.
12. L. Kumar, R. Ramanujan, R. Tewari, P. Mukhopadhyay, and S. Banerjee: 'Active eutectoid decomposition in Zr-3 wt.% Fe', *Scripta materialia*, 1999, **40**, 723–728.
13. H. Donthula, B. Vishwanadh, T. Alam, T. Borkar, R. Contieri, R. Caram, R. Banerjee, R. Tewari, G. Dey, and S. Banerjee: 'Morphological evolution of transformation products and eutectoid transformation (s) in a hyper-eutectoid Ti-12 at% Cu alloy', *Acta Materialia*, 2019, **168**, 63–75.
14. S. Souza, C. Afonso, P. Ferrandini, A. Coelho, and R. Caram: 'Effect of cooling rate on Ti-Cu eutectoid alloy microstructure', *Materials Science and Engineering: C*, 2009, **29**, 1023–1028.
15. Y. Xu, J. Jiang, Z. Yang, Q. Zhao, Y. Chen, and Y. Zhao: 'The effect of copper content on the mechanical and tribological properties of hypo-, hyper-and eutectoid Ti-Cu alloys', *Materials*, 2020, **13**, 3411.
16. Y. Deng, G. Sheng, F. Wang, X. Yuan, and Q. An: 'Microstructure evolution and mechanical properties of transient liquid phase bonded Ti–6Al–4V joint with copper interlayer', *Materials & Design*, 2016, **92**, 1–7.
17. J. Ketelaar: 'Das elektrische leitvermögen des Ag_2HgJ_4: Ein beitrag zur frage nach dem leitungsmechanismus in den gutleitenden festen stoffen (the electrical conductivity of Ag_2HgJ_4: A contribution to the question of the conduction mechanism in highly conductive solid materials)', *Zeitschrift für Physikalische Chemie*, 1934, **26**, 327–334.
18. L. Suchow, and G. R. Pond: 'Electrical conductivity of Ag_2HgI_4, Cu_2HgI_4 and their eutectoid', *Journal of the American Chemical Society*, 1953, **75**, 5242–5244.
19. S. Hull, and D. Keen: 'Structural characterization of the superionic transition in Ag_2HgI_4, Cu_2HgI_4', *Journal of Physics: Condensed Matter*, 2000, **12**, 3751.
20. M. Friesel, B. Baranowski, and A. Lunden: 'Phase transitions of the system Ag_2HgI_4-Cu_2HgI_4 at normal and high pressure studied by differential scanning calorimetry', *Journal of physical chemistry (1952)*, 1990, **94**, 1113–1117.
21. A. Scherf, A. Kauffmann, S. Kauffmann-Weiss, T. Scherer, X. Li, F. Stein, and M. Heilmaier: 'Orientation relationship of eutectoid FeAl and $FeAl_2$', *Journal of Applied Crystallography*, 2016, **49**, 442–449.
22. G. Bastin, F. Van Loo, J. Vrolijk, and L. Wolff: 'Crystallography of aligned Fe-Al eutectoid', *Journal of Crystal growth*, 1978, **43**, 745–751.
23. J. Murray: 'The Al-Zn (aluminum-zinc) system', *Bulletin of Alloy Phase Diagrams*, 1983, **4**, 55–73.
24. W. A. Fischer, A. Hoffmann, and R. Shimada: 'Der wüstitzerfall unterhalb 570° in gegenwart von eisen', *Archiv für das Eisenhüttenwesen*, 1956, **27**, 521–529.
25. N. Otsuka, T. Doi, Y. Hidaka, Y. Higashida, Y. Masaki, N. Mizui, and M. Sato: 'In-situ measurements of isothermal wüstite transformation of thermally grown FeO scale formed on 0.048 mass% Fe by synchrotron radiation in air', *ISIJ international*, 2013, **53**, 286–293.
26. S. Hayashi, Y. Yamanouchi, K. Hayashi, Y. Hidaka, and M. Sato: 'Stress measurement in the iron oxide scale formed on pure Fe during isothermal transformation by in situ high-temperature X-ray diffraction', *Corrosion Science*, 2021, **187**, 109482.
27. H. Wang, G. Cao, W. Song, Z. Li, and Z. Liu: 'Lamellar spacing characteristics of eutectoids in oxide scale formed on iron after holding at 300–500°C', *High Temperature Corrosion of Materials*, 2023, **100**, 251–264.
28. C. Barrett, and L. Meyer: 'Phase diagram of argon-carbon monoxide', *The Journal of Chemical Physics*, 1965, **43**, 3502–3506.
29. C. S. Barrett, and L. Meyer: 'Argon-nitrogen phase diagram', *The Journal of Chemical Physics*, 1965, **42**, 107–112.
30. S. Milenkovic, A. Schneider, and A. W. Hassel: 'Gold nanostructures by directional solid-state decomposition', *Gold Bulletin*, 2006, **39**, 185–191.

31. S. Milenkovic, T. Nakayama, M. Rohwerder, and A. W. Hassel: 'Structural characterisation of gold nanowire arrays', *Journal of Crystal Growth*, 2008, **311**, 194–199.
32. W.-C. Cheng, Y.-S. Song, Y.-S. Lin, K.-F. Chen, and P. C. Pistorius: 'On the eutectoid reaction in a quaternary Fe-C-Mn-Al alloy: Austenite→ferrite+ kappa-carbide+ $M_{23}C_6$ carbide', *Metallurgical and Materials Transactions A*, 2014, **45**, 1199–1216.
33. J. W. Christian: Theory of Transformations in Metals and Alloys, Part I: 2 ed., Oxford, U. K.: Pergamon Press, 1975.
34. C. I. Nichols, R. Krakow, J. Herrero-Albillos, F. Kronast, G. Northwood-Smith, and R. J. Harrison: 'Microstructural and paleomagnetic insight into the cooling history of the IAB parent body', *Geochimica et Cosmochimica Acta*, 2018, **229**, 1–19.

Acronyms

bcc	Body-centred cubic
bct	Body-centred tetragonal
CTOD	Crack-tip opening displacement
CTSD	Crack-tip sliding displacement
EBSD	Electron back scattered diffraction
ECAP	Equal-channel angular processing
fcc	Face-centred cubic
hcp	Hexagonal close-packed
HV	Vickers hardness
HB	Brinell hardness
F	Face-centred
I	Body-centred
rpm	Revolutions per minute
TEM	Transmission electron microscopy
TTT	Time-temperature-transformation
UTS	Ultimate tensile strength

The use of braces, for example $x\{y\}$, implies a functional relationship, i.e., x is a function of y.

Methods of dealing mathematically with crystallography, interfacial structure and homogeneous deformations are described thoroughly in the book *Geometry of crystals, polycrystals and phase transformations*, so are not reproduced here.
The labelling of the orthogonal axes of the unit cell of cementite, is selected throughout to be consistent with the space group *Pnma* i.e., $a = 0.50837$ nm, $b = 0.67475$ nm and $c = 0.45165$ nm. Any crystallographic data referring to the alternative convention *Pbnm* $(a < b < c)$ have been translated appropriately.
A subscript on the symbol α_P is used when it is necessary to distinguish ferrite that is a part of pearlite.

Nomenclature

α	Ferrite
α'	Martensite
α_b	Bainite
α_P	Ferrite that is a part of pearlite
β	Fraction of plastic work dissipated as heat
β_e	2nd order nonlinear elastic coefficient in travelling wave equation
γ	Shear strain
γ	Austenite
γ_{poor}	Austenite that is rich in carbon, i.e. $x_\gamma < \overline{x}$
γ_{rich}	Austenite that is rich in carbon, i.e. $x_\gamma > \overline{x}$
γ_r	Retained austenite
Γ	Activity coefficient
δ_{ij}	Kronecker δ, equals zero when $i \neq j$ and one when $i = j$
δ	Delta-ferrite
ε_h	Heterogeneous elastic strain
ε_p	Plastic strain
ε_p^c	Critical value of plastic strain at the onset of adiabatic plastic instability
ε_{RA}	Fractional reduction of area in tensile test
ε_{TRIP}	Excess plastic strain due to phase transformation
ε_t^c	Total strain
ε_u	Uniform strain
ε_{vM}	von Mises equivalent strain
$\overline{\varepsilon}_p$	Average Plastic strain
ϵ	Hexagonal close-packed iron
ζ	Constraint factor in fracture mechanics
η	Parabolic thickening rate constant
θ	Cementite
λ_s	Wavelength of chemical segregation
λ_s	Wavelength
μ	Chemical potential
μ°	Chemical potential of pure substance
$\mu_\circ$	Permeability of free space
μ_B	Bohr magneton, $0.9 \times 10^{-23}\,\mathrm{A\,m^2}$
ν	Poisson's ratio
ρ	Density
ρ_e	Electrical resistivity
ρ_f	Density of forest dislocations
σ^f	Friction stress in tension

σ_a	Applied stress amplitude, in fatigue or tensile test
σ_e	Endurance or fatigue limit when cyclically loading smooth specimens
σ_f	Fracture stress
σ_{max}	Maximum stress amplitude in fatigue
σ_{op}	Crack opening stress
σ_{op}^{∞}	Crack opening stress for a long crack
σ_{se}	Standard error in regression analysis
σ_s	Surface energy per unit area
σ_{UTS}	Ultimate tensile strength
σ_{vM}	von Mises equivalent stress
σ_y	Yield strength or 0.2% proof strength
$\sigma_{\alpha\theta}$	α/θ interfacial energy per unit area
τ	Shear stress in general
τ^f	Friction stress in shear
τ_a	Applied shear stress
τ_y	Yield stress in shear
Φ	Kinetic strength of a heat treatment
ϕ	Diameter
a_C^{γ}	Activity of carbon in γ
a_X	Atomic percent of solute identified in the subscript
A	Area
A_A	Area fraction
Ac_3	Temperature at which a sample becomes fully austenitic during heating
Ae_1	Temperature separating the $\alpha+\gamma$ and α phase fields for a specific alloy
Ae_3	Equilibrium temperature above which only austenite is stable in a specific alloy
Ae_3	Temperature separating the $\alpha+\gamma$ and γ phase fields for a specific alloy
Ar_3	Temperature at which γ begins to transform to α during cooling
b	Magnitude of Burgers vector
b_1	Empirical constant. b_{31}
B	Magnetic flux density
$B_{saturation}$	Saturation magnetisation
$\overline{c}_i$	Average concentration of i in alloy, moles per unit volume
c	Molar concentration per unit volume
$c^{\gamma\alpha}$	Composition of γ that is in equilibrium with α
c_A	Amplitude of composition wave
C_g	Contiguity
C_P	Specific heat capacity at constant pressure
C_V	Number of grain corners per unit volume
$\overline{D}$	Interdiffusion coefficient defined relative to the laboratory frame of reference
D	Chemical or interdiffusion coefficient for a binary system
E	Young's modulus of elasticity
E_s	Shear modulus

f_A	Ratio of cross-sectional areas in the necked and uniform regions of tensile specimen
ΔG	Driving force, i.e., reduction in free energy following a reaction
$\Delta G^{\gamma\alpha}$	Gibbs free energy change for $\gamma \rightarrow \alpha$ transformation, $= G_\alpha - G_\gamma$
G^*	Activation free energy, e.g., for diffusion
I	Nucleation rate (particles per second). Any subscript indicates rate per unit area (A) or volume (V). Superscript identifies the phase nucleating.
H	Magnetic field strength
H_c	Magnetic coercivity
h_V	Number of holes per unit volume
j	Electrical current density
J_M	Polar moment of inertia
J_m	Magnetisation
$k_i^{\theta\alpha}$	Partition coefficient y_i^θ / y_i^α
k_{HP}	Hall-Petch coefficient
ΔK_Q	Apparent fracture toughness using a sample that does not satisfy dimensional requirements for K_{IC}.
ΔK_{th}	Threshold stress-intensity range in fatigue
K_{IC}	Fracture toughness in mode I loading
K_I	Stress intensity in mode I loading
L'_γ	A measure of austenite grain size that is not a mean lineal intercept
L_γ	A measure of austenite grain size that is not a mean lineal intercept
$\overline{L}$	Mean lineal intercept
$\overline{L}_{colony}$	Mean lineal intercept defining pearlite colony size
L	Length
L_V	Length of grain edges per unit volume
m	Exponent in equation relating stress to strain rate during superplasticity
M	Taylor factor for relating the strengths of single crystals to that of polycrystalline aggregate
M_F	Martensite-finish temperature
M_S	Martensite-start temperature
M_T	Torsional moment
n	Strain hardening coefficient
N	Number
N_e	Number of fatigue cycles below which failure is avoided
N_f	Number of fatigue cycles to failure
N_g	Number of grains
N_L	Cracks per unit length
$\overline{P}_L$	Mean number of points of intersection per unit length of test line in stereology
Q	Activation energy
r	Radius
r_A	Reduction of area measured in a uniaxial tensile test
R_σ	Ratio of maximum to minimum stress amplitude in fatigue

r	Radial coordinate
$S_{I,t}$	True interlamellar spacing
S_I	Interlamellar spacing
S_V	Interfacial area per unit volume
$S_{I,a}$	Apparent interlamellar spacing
t_γ	Austenitisation time
t_f	Transformation-finish time
t_s	Transformation-start time
T_γ	Austenitising temperature
T_C	Curie temperature
T_{DB}	Ductile-brittle transition temperature
T_E	Eutectoid temperature
T_m	Melting temperature
T_r	Room temperature
u_X	Ratio of atomic fractions of substitutional solute 'X' to (Fe+X)
V^*	Activation volume
v_P	Growth rate of pearlite
V_a	Volume per atom
V_V	Volume fraction
w_X	Weight percent of the solute identified in the subscript
$\overline{x}$	Average mole fraction
x	Mole fraction
x^θ	Mole fraction in phase θ
$x^{\theta\alpha}$	Mole fraction in θ that is in equilibrium with α
X	Substitutional solute
y_i	Concentration of substitutional solute *i* expressed as site fraction on a lattice
Y	Compliance function in fracture mechanics treatment of toughness
z	Position coordinate
$z_\perp$	Mean free distance for dislocation motion
z_ℓ	Half length of internal crack
$z_{\ell p}$	Half the sum of crack length and the plastic zones at the crack tips
z_c	Half the length of a crack as measured on the sample surface
z_{ex}	Magnetic exchange length
z_o	Length of tetrakaidecahedron edge
$\mathbb{Z}$	Integer

Indices

Author index

Subject index

For Product Safety Concerns and Information please contact our EU
representative GPSR@taylorandfrancis.com
Taylor & Francis Verlag GmbH, Kaufingerstraße 24, 80331 München, Germany

www.ingramcontent.com/pod-product-compliance
Lightning Source LLC
LaVergne TN
LVHW081313110826
845149LV00006B/1499

* 9 7 8 1 0 3 2 6 3 1 9 7 4 *